CAROLE DRAKE

THE BIOMECHANICS OF SPORTS TECHNIQUES

JAMES G. HAY
University of Iowa

THE BIOMECHANICS OF SPORTS TECHNIQUES

Second Edition

PRENTICE-HALL, INC., Englewood Cliffs, N.J. 07632

Library of Congress Cataloging in Publication Data

HAY, JAMES G 1936–
The biomechanics of sports techniques.

Includes bibliographical references and index.
1. Sports—Physiological aspects. 2. Human mechanics. I. Title.
RC1235.H38 1978 612'.76 77-15607
ISBN 0-13-077164-3

Printed in the United States of America

10 9 8 7 6 5

PRENTICE-HALL INTERNATIONAL, INC., *London*
PRENTICE-HALL OF AUSTRALIA PTY. LIMITED, *Sydney*
PRENTICE-HALL OF CANADA, LTD., *Toronto*
PRENTICE-HALL OF INDIA PRIVATE LIMITED, *New Delhi*
PRENTICE-HALL OF JAPAN, INC., *Tokyo*
PRENTICE-HALL OF SOUTHEAST ASIA PTE. LTD., *Singapore*
WHITEHALL BOOKS LIMITED, *Wellington, New Zealand*

To my parents,

IRENE and FRANK HAY

CONTENTS

TABLES

PREFACE TO THE SECOND EDITION

This second edition of *The Biomechanics of Sports Techniques* has been prepared for a number of reasons. First, extensive use of this text over the last four years has revealed ways in which some of the material might be presented with greater clarity. Second, a great deal more information is now available as the result of a marked upsurge in biomechanics research activity. Finally, there have been substantial changes in many of the sports techniques considered in the second half of this book.

Many changes have been made to clarify basic concepts. Some of these are major changes—e.g., the sections on distance, displacement, speed, velocity, and projectile motion have been all but entirely rewritten—and others are relatively minor. Corresponding efforts have been made to improve the quality and clarity of the illustrations.

Over the past few years there has been a great deal of research activity on problems associated with sports techniques. Much of this work has been concerned with highly practical problems—those repeatedly faced by teachers, coaches, and athletes. Where appropriate, the major findings from such studies have been woven into the discussion of the techniques concerned.

The techniques employed in sports sometimes change at an almost bewildering rate, so that those concerned have a difficult time keeping abreast of them. For example, in the few years since this text was first published, the grab start has become almost universally accepted as the fastest starting technique in swimming; the rotational technique has become accepted as a viable alternative to the long-dominant O'Brien technique in shot putting; the standing start, recently thought to be a similarly viable alternative to the traditional crouch start in sprinting, has been outlawed by a rule change; and the somersault long-jumping technique has arrived, been banned, and departed. Efforts have been made in this second edition to keep pace with such changes, and new techniques such as the grab start and the rotational style of shot putting have been discussed at some length.

Many of the changes in this second edition are the direct result of constructive comments made by readers of the previous edition. My sincere thanks are extended to those who have taken the time and trouble to acquaint me with their ideas; such comments will always be welcome.

Finally, it is my pleasant duty to extend thanks to those who have contributed significantly to the production of this second edition—Sandi Dillon, who prepared the manuscript; Barry Wilson, who produced much of the artwork; and Jesus Dapena, who contributed in a number of ways. Their contributions are truly appreciated.

JAMES G. HAY

PREFACE TO THE FIRST EDITION

The science of biomechanics is concerned with the forces that act on a human body and the effects these forces produce. Physical education teachers and coaches of athletic teams, whether they recognize it or not, are likewise concerned with forces and effects. Their ability to teach the basic techniques of a sport or physical activity depends very largely on their appreciation of both the effects they are trying to produce and the forces that cause them. It seems only logical therefore that physical educators, coaches, and athletes should turn to biomechanics to provide a sound, scientific basis for the analysis of the techniques used in sports. It is the purpose of this book to present such a scientific basis and to demonstrate how it can be used to advantage in the analysis of sports techniques.

Part I is concerned with formally defining biomechanics, with establishing the role biomechanics can play in analyzing sports techniques, and with determining the importance of a knowledge of biomechanics to the physical educator, the coach, and the athlete.

Part II is devoted to a discussion of those basic biomechanical concepts which are of greatest importance in the analysis of the techniques used in sports. Particular emphasis is placed on concepts associated with projectile motion, elastic impact, Newton's Laws and their angular analogues, mechanical energy, center of gravity, buoyancy and flotation, and fluid resistance.

The material is presented in such a way that the reader, in proceeding from the simple to the complex, needs only the information previously given to understand each new concept as it is introduced. For example, unlike many other texts in the field, the discussion on centers of gravity is delayed until the reader has been acquainted with the concepts of force, weight, moment, resultant moment, and equilibrium—concepts that must be clearly understood before a complete understanding of the center of gravity concept can be obtained.

Several important topics, not normally covered in a text of this kind, are considered. Among these are the concepts of impulse, mechanical energy, lift, and drag. The segmentation method of locating the center of gravity of an athlete is also covered in some detail.

Part III contains a detailed analysis of the techniques used in nine major sports—gymnastics, golf, skiing, swimming, and track and field among the individual sports and baseball, basketball, football, and softball among the team games.

The writer of a text of this kind is faced with deciding whether to write a short (and, necessarily, superficial) analysis of a large number of sports or whether to write a longer, more detailed analysis of just a few. The latter course has been chosen here in the belief that superficial treatments, by their very nature, rarely contain much of real value and in the hope that the more detailed approach may make the analysis of value to a wide range of readers.

The analysis of the techniques involved in each sport is divided into two parts. The first part, entitled Basic Considerations, is concerned with enumerating the basic factors involved in the performance of each technique and with showing how these factors interrelate with one another to produce the desired result. The second part contains a detailed discussion of the techniques themselves with particular emphasis on those areas where there are known to be disagreements among teachers and coaches. Where possible these disagreements are resolved, either by the simple application of the basic concepts of biomechanics or by reference to research findings on the subject. Where further research is needed before a disagreement can be resolved, this fact is drawn to the attention of the reader in the hope that this will not only acquaint him with the limits of present knowledge in the area, but perhaps also stimulate him to embark on an appropriate research project in quest of a solution.

Many people have contributed to the writing of this book in one way or another—some by the inspiration, guidance, and opportunities provided the author over the years; some by their direct contributions in terms of artwork, typing services, and assembling of material; and some by their patient acceptance that it all takes time and effort away from other important responsibilities. The author would like to extend his very sincere thanks to:

Frank Sharpley, whose unremitting search for biomechanical solutions to technique problems in track and field provided the author with his initial motivation to take up study in this area and whose enthusiasm, energy, and sheer innovative genius have been a repeated source of inspiration ever since.

Louis Alley, for his guidance during the author's years as a graduate student, and for his subsequent efforts to provide the author with every opportunity to pursue his work in this field.

Philip Smithells, who has done more to mold the author's philosophy of physical education and athletics than has any other man and who provided the author with time and facilities to work on this book.

Ross Hemera, the young artist from New Zealand who prepared most of the drawings.

David East and Roger Simmons, able and willing assistants, who did much of the "legwork" involved in assembling and preparing material.

Linda Godfrey, who prepared the graphs, and Ruth Bonar, Barbara Gorman, and Elnora Swartzendruber, who typed the manuscript.

And finally to:

Hilary, Linda, and Karen Hay, who patiently suffered the disruption of their family life that this book might be written.

JAMES G. HAY

THE BIOMECHANICS OF SPORTS TECHNIQUES

Part I

INTRODUCTION

Chapter 1

BIOMECHANICS IN PHYSICAL EDUCATION AND ATHLETICS

What Is Biomechanics?

In the course of the vast knowledge explosion currently engulfing us, new areas of knowledge are being developed and old ones reassessed. In this continuing process, the language of science (that area in which unmistakably the most dramatic advances are being made) is in a marked state of flux. As a new area of knowledge is developed and becomes recognized, it is given a name, and the scope of existing areas in close proximity to the new one comes under careful scrutiny. To compound the confusion that often results from this process there appears to be some competition for the honor of having bestowed a name on the new area. What results then is a multiplicity of new terms, usually with varying shades of meaning, all purporting to describe most suitably the subject matter within the new area.

The increasing development of a scientific approach to the analysis of human movements has been bedeviled by this very problem. At one time, the term *kinesiology* (literally, the science of movement) was used to describe that body of knowledge concerned with the structure and function of the musculo-skeletal system of the human body. Later the study of the mechanical principles applicable to human movement became widely accepted as an integral part of kinesiology. Later still the term was used much more literally to encompass aspects of all the sciences that impinged in any way on human movement. At this point it became clear that kinesiology had quite lost its usefulness to describe specifically that part of the science of movement concerned with either the musculo-skeletal system or the mechanical principles applicable to human movement. Several new terms were suggested as substitutes, and anthropomechanics, anthropokinetics, biodynamics, biokinetics, homokinetics, and kinanthropology all had their proponents. From all this there ultimately emerged one term that gained wider acceptance than any other—*biomechanics*.

The term biomechanics has been variously defined as follows:

> The mechanical bases of biological, esp. muscular activity [and] the study of the principles and relations involved.[1]
>
> The application of mechanical laws to living structures, specifically to the locomotor system of the human body.[2]

However, since this book is concerned principally with the application of biomechanics to a quite limited field, i.e., to the analysis of the techniques employed in sports, a further and more restricted definition is offered here:

> *Biomechanics is the science that examines the internal and external forces acting on a human body and the effects produced by these forces.*

(It should perhaps be noted that none of the definitions above is universally accepted and that already much broader areas than they suggest have been ascribed to biomechanics. Regrettably, it seems that the term may be taking the first steps along the same road to obsolescence that was followed earlier by kinesiology.)

What Is the Function of Biomechanics?

Physical education teachers and coaches are continually confronted with problems related to the techniques used in the various physical activities with which they are concerned.

Some years ago, the world's leading high jumper was a Russian named Valeriy Brumel. Brumel held the world and Olympic records for high jumping and was so superior to his contemporaries that he was considered virtually without equal. At that time, as there is today, there was a widespread tendency for coaches and athletes to adopt slavishly the methods of the current champion and to model themselves (or their charges) on him. Now many of the methods employed by Brumel were relatively new to high jumping and so he was copied with perhaps even more diligence than most. Some jumpers copied his lengthy, fast approach and his double-arm swing at takeoff. Others copied some of his training methods in which he jumped with weights attached to his body and jump-kicked to touch a basketball hoop with his leading foot. Still others attempted to copy his mental approach to competition and sat apart from the other jumpers with their backs to the bar. Almost everything that Brumel did was noted carefully and copied by jumpers around the world.

Some years before all this, and even more outstanding in his events than was Brumel in the high jump, a Czechoslovakian runner named Emil Zatopek completely dominated and revolutionized distance running. Like Brumel later, he too was copied. Some copied his surging race tactics. Others copied

[1] *Webster's Third New International Dictionary of the English Language* (Springfield, Mass.: G. & C. Merriam Co., 1961).

[2] *Dorland's Illustrated Medical Dictionary* (Philadelphia: W. B. Saunders Co., 1965).

(or attempted to copy!) his grueling interval training workouts, which included such incredible loads as 5 × 200 m, 60 × 400 m, and 5 × 200 m at racing speed, with fast 200-m jogs in between each repetition. Few, if any, copied his running action—and one needs but to have seen it (or even pictures of it) to understand why. No one, it seems, could think of any reason why bobbing the head, rolling the shoulders, flailing the arms, and screwing up the face could possibly help a person to run faster. It was therefore concluded, presumably, that Zatopek was successful despite these odd quirks in his running action and not because of them.

And herein lies one of the basic differences between Zatopek and Brumel—some of Zatopek's faults were obvious and his imitators carefully avoided them. Those who imitated Brumel, however, copied not only the good points but the faults as well—faults that, although less obvious, were probably more important in fixing the level of his performance than were the patently obvious ones of Zatopek.

All of this illustrates one of the problems that physical educators and coaches face. How can they determine which features of a champion's technique contribute to the high quality of his performance—and thus are possibly worth copying—and which are faults limiting that performance? The answer lies with the science of biomechanics, which provides the *only* sound, logical basis upon which to evaluate the techniques brought to our attention by champions.

Sometimes new techniques are brought to our attention by new rules. (In swimming, for example, the introduction of various new techniques of turning resulted from changes to the rules governing this part of a race.) The development of new equipment also produces changes in techniques. (The sponge-faced table tennis bat, the fiberglass vaulting pole, and the steel-edged ski are but a few examples of equipment changes that have had a very marked effect on sports techniques.)

How are physical educators and coaches to arrive at the best technique to use when new rules and new equipment invite or require changes in technique? Here again, biomechanics provides the basis upon which to make such decisions.

Having once established what technique should be used in a given instance, teachers and coaches are confronted with the task of detecting and correcting faults in an athlete's performance. The greatest difficulty here—even if it is seldom recognized as such—is that of locating the cause of the faults observed. For while it is not too difficult for an experienced eye to detect a gross fault in an athlete's technique (particularly when that technique is well known and widely used), the cause of that fault may be very difficult to locate. One of the reasons is that the cause is often far removed from the effect. (In jumping, tumbling, and diving, for example, effects observed in the air or on landing are almost always caused by faults in the technique of the takeoff or the run preceding the takeoff.) Now many teachers and coaches attempt to correct the effect that they have detected and give little or no thought to the underlying cause that has produced it. In general, such attempts are quite ineffectual—the athlete's performance being more likely to deteriorate as his mentor adds to his problems instead of helping him resolve them.

How can a teacher or coach improve his ability to locate the causes of the various faults he observes in an athlete's performance? The answer, again, is: via the science of biomechanics. For just as a knowledge of the scientific bases of motor learning equips a teacher or coach to make sound judgments concerning methods of instruction, length, frequency, and nature of practice, etc., and a knowledge of physiology equips him to make sound judgments concerning the amount and type of training to prescribe in a given case, a knowledge of biomechanics equips him to choose appropriate techniques and to detect the root causes of faults that may arise in their performance. In short, just as motor learning may be regarded as the science underlying the acquisition of skills, and physiology the science underlying training, biomechanics is the science underlying techniques. (*Note:* It is as well to recognize here that these sciences do not have ready-made answers to *all* the problems that confront physical education teachers and coaches. However, where they are unable to provide an answer immediately, they do offer the means whereby an answer can ultimately be obtained.)

How Important Is a Knowledge of Biomechanics?

Although all sorts of people are interested in sports techniques in one way or another, three groups readily distinguish themselves—physical educators, coaches, and athletes. Since each of these groups tends to view sports techniques in a somewhat different way, the importance of a knowledge of biomechanics to each of them will be considered in turn.

To the Physical Educator. Physical education and its aims have been variously described:

> Physical education is a way of education through physical activities which are selected and carried on with full regard to values in human growth, development, and behavior.[3]

> Physical education . . . has as its aim the development of physically, mentally, emotionally and socially fit citizens through the medium of physical activities . . . .[4]

[3] W. K. Streit and Simon A. McNeeley, "A Platform for Physical Education," *Journal of Health, Physical Education, Recreation*, XXI, March 1950, p. 136.

[4] Charles A. Bucher, *Foundations of Physical Education* (St. Louis: The C.V. Mosby Co., 1968), p. 21.

> The fundamental purpose of physical education is to promote through selected physical activity the establishment and maintenance of competencies, attitudes, ideals, drives and conditions . . . .[5]

The seemingly endless array of such statements have at least one thing in common: they all visualize realization of the aims of physical education "through the medium of physical activities." It should thus be clear that the success that any physical educator achieves must be conditional on his knowledge of this particular medium—the techniques, teaching and training methods involved, and the sciences upon which they are based. It can therefore be stated quite emphatically that a knowledge of biomechanics (and of motor learning and physiology) is absolutely essential to the physical educator who is not content to limit his effectiveness by making critical judgments based on guesswork. (*Note:* Since knowledge of a specific technique necessarily precedes any attempt to teach it to others or to train others to higher levels of achievement in its use, an argument might be made that a knowledge of biomechanics is of prime importance to a physical educator. However, because the final results of his work clearly depend not on his knowledge of any one scientific discipline but on the sum of his knowledge of several, this kind of argument serves little purpose.)

To the Coach. The importance of a knowledge of biomechanics to a coach depends to a certain extent on the sport that he is coaching. A cross-country coach, concerned primarily with cardiovascular and muscular endurance and only to a very limited extent with techniques, will clearly benefit less from a knowledge of biomechanics than will a coach of baseball, football, gymnastics, or swimming, in all of which techniques play a much larger role.

Another factor here is the level at which the individual is working. The physical educator generally works with beginners or near-beginners, and so he is concerned with the broad fundamentals of sports techniques and the broad biomechanical principles underlying them. The coach, on the other hand, works at increasingly more advanced levels and hence is concerned not only with broad fundamentals but also with precise details. As the level of performance increases, so does the coach's need for a more thorough knowledge of biomechanics. At the highest levels of sports in which techniques play a major role, improvement comes so often from careful attention to detail that no coach can afford to leave these details to chance or guesswork. For him, a knowledge of biomechanics might be regarded as essential.

At this point, there will possibly be some who are inclined to scoff at such a claim and to cite cases of highly successful coaches who have little or no apparent knowledge of biomechanics. Their success merely emphasizes that other things are also important—something that has never been denied here.

[5]Clyde Knapp and Patricia Hagman Leonard, *Teaching Physical Education in Secondary Schools* (New York: McGraw-Hill Book Co., 1968), p. 77.

What is worthy of thought, however, is the level to which the performances of their athletes might rise, if to all these coaches' attributes was added a knowledge of biomechanics!

To the Athlete. While the importance of a knowledge of biomechanics to physical educators and coaches is fairly generally agreed upon, there is no such agreement on its importance to athletes.

A number of studies have been conducted in an attempt to determine the value of a knowledge of biomechanics in the learning of a physical skill. However, since the subjects of almost all of these studies were complete beginners, they shed little light on the importance of a knowledge of biomechanics to the skilled athlete.

Some leading authorities have expressed opinions concerning this question:

> As the learner progresses or gets older and more generally experienced, verbal directions and an analysis of movement can help more in increasing the meaningfulness of the skill and in giving new insights into it.[6]

> The experimental literature does not cover the value of mechanical analysis for the advanced student, but empirical evidence seems to indicate somewhat greater value at the higher skill levels.[7]

Thus, although no conclusive evidence is available, it would appear that a knowledge of the biomechanical principles involved might well enhance the performance of an already skilled athlete.

[6]B. Knapp, *Skill in Sport: The Attainment of Proficiency* (London: Routledge and Kegan Paul, 1966), p. 28.

[7]John D. Lawther, *The Learning of Physical Skills* (Englewood Cliffs, N. J.: Prentice-Hall, Inc., 1968), p. 101.

Part II

BASIC CONCEPTS

Chapter 2

FORMS OF MOTION

In general all motion may be described as translation or rotation or some particular combination of these two.

Translation

Translation (or *linear motion*) takes place when a body[1] moves so that all parts of it travel exactly the same distance, in the same direction, in the same time. One way of deciding whether the motion of a particular body is translatory is to consider the motion of an arbitrary straight line drawn on the body. If during a given motion this line remains of the same length and is always parallel to the previous positions it occupied, the motion is one of translation. That this criterion is in effect the same as that given earlier (i.e., the same distance, same direction, same time criterion) can readily be seen in the example of translation shown in Fig. 1. Here the line that has been arbitrarily selected is one joining the right shoulder and right hip joints. (Any other line might equally well have been chosen.) The dotted lines—in this case straight and parallel—show the paths taken by the two joints mentioned.

In Fig. 2, the motion of the skydiver during free fall is also an example of translation, as successive positions of the line joining his hip and shoulder joints again indicate. This time, however, the path taken by each of these joints is a curved line rather than a straight one. (*Note:* The straight-line and curved-line types of linear motion are frequently referred to as *rectilinear translation* and *curvilinear translation*, respectively.)

Finally, in Fig. 3, although the first and last positions shown might suggest that translation has taken place, consideration of the intermediate positions clearly indicates that this is not the case.

[1] In biomechanics the term *body* is used to refer to both inanimate objects (such as items of sporting equipment) and animate objects (such as the human body or parts thereof). In this latter regard it is important to realize that while in some instances it is convenient to consider the human body in its entirety, in others it is helpful to consider it instead as a system comprised of a series of separate bodies (e.g., head, trunk, arm, forearm, etc.).

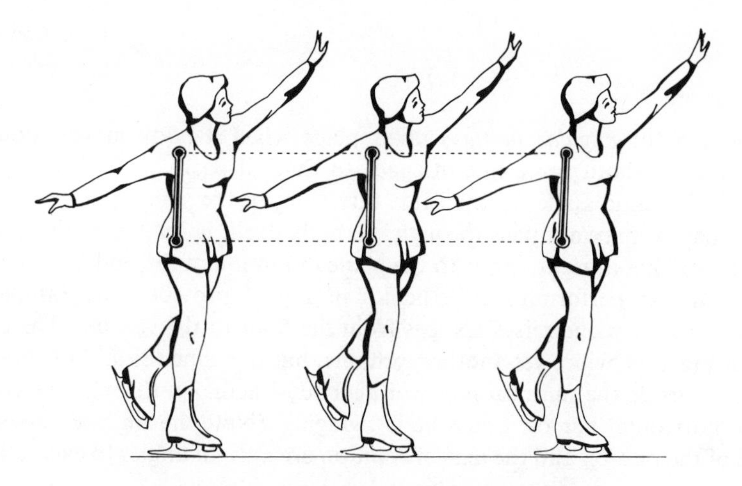

Fig. 1. Straight-line (or rectilinear) translation

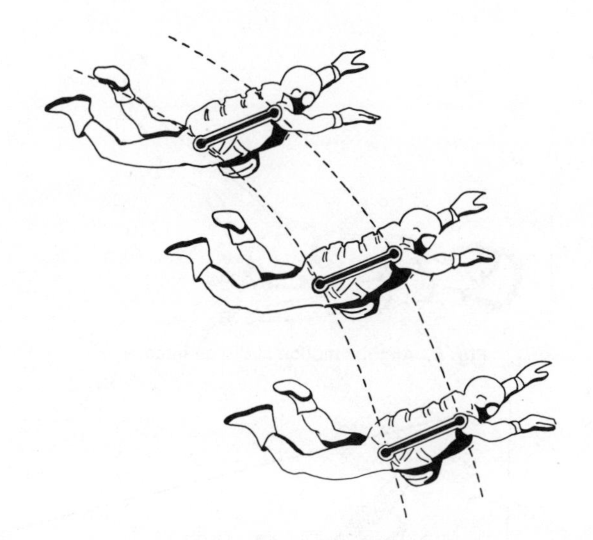

Fig. 2. Curved-line (or curvilinear) translation

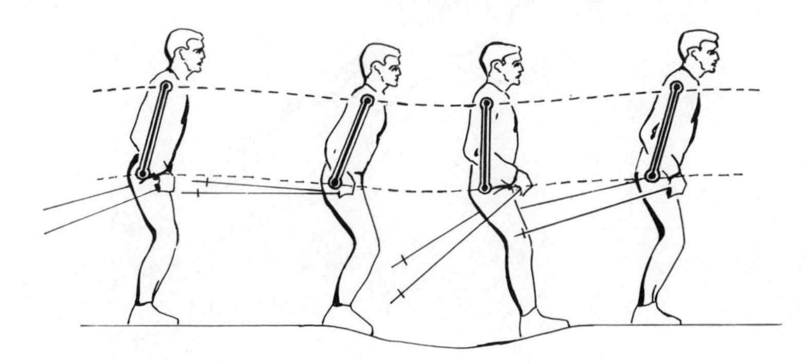

Fig. 3. Nonlinear motion

Rotation

Rotation (or *angular motion*) takes place when a body moves along a circular path about some line in space so that all parts of the body travel through the same angle, in the same direction, in the same time. This line, which may or may not pass through the body itself, is known as the *axis of rotation* and lies at right angles to the plane of motion of the body.

The athlete performing calisthenics in Fig. 4 provides an example of angular motion as he raises his legs from the floor to the vertical. The gymnast in Fig. 5 provides yet another example, but this time the axis of rotation lies just outside the physical limits of his body—actually, through the center of the horizontal bar on which he is swinging. (Note that in both cases the plane of the motion and the axis of rotation are at right angles to each other.)

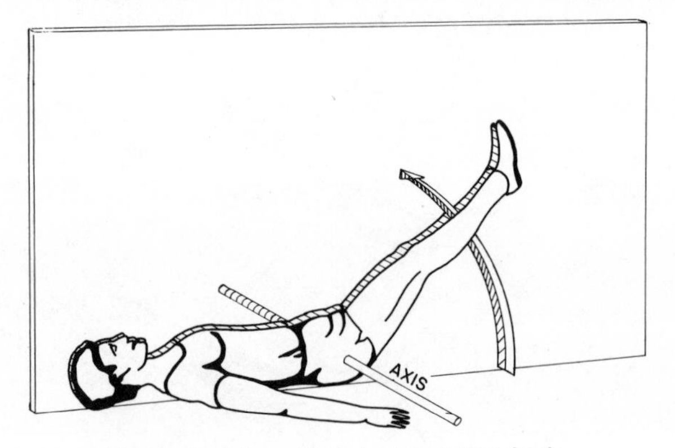

Fig. 4. Angular motion about an internal axis

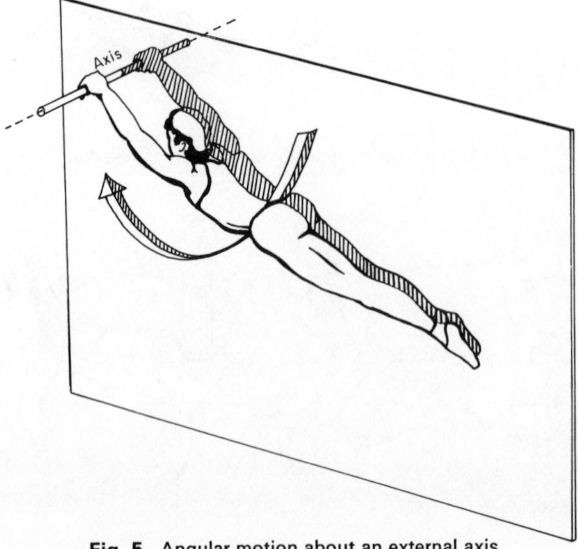

Fig. 5. Angular motion about an external axis

Fig. 6. General motion—translation and rotation combined

General Motion

While rotation is a good deal more common than translation as far as sports techniques are concerned, much more common than either is what will be called here *general motion*, which is some combination of the two. The racing cyclist, for example, translates his upper body as a direct result of the rotary movements of his legs (Fig. 6).

Together with this blending or combining of translation and rotation there is often a further blending—this time of several rotations. Consider the action of one of the cyclist's legs (Fig. 7). Here there are at least three simultaneous rotations taking place. First, there is the rotation of his thigh about an axis through his hip joint (which is itself translating). Then three is the rotation of his leg about his knee joint, and finally there is the rotation of his foot about his ankle joint. As can probably be imagined, a study of the combined motions of the elements within such a system can become quite complex.

Fig. 7. A complex general motion

Chapter 3

LINEAR KINEMATICS

Kinematics is that branch of biomechanics that is concerned with describing the motion of bodies. Thus kinematics deals with such things as how far a body moves, how fast it moves, and how consistently it moves. It is not concerned at all with what causes a body to move in the way it does. This latter aspect of motion is the preserve of kinetics—a complementary branch of biomechanics that will be considered later in this book.

Linear kinematics deals with the kinematics of translation or linear motion, while angular kinematics (see Chap. 4) deals with the kinematics of rotation or angular motion.[1]

Distance and Displacement

Distance and displacement are quantities commonly used to describe the extent of a body's motion. When a body moves from one location to another, the distance through which it moves is simply the length of the path that it follows. The displacement the body undergoes in the course of the same motion is found by measuring the length of a straight line joining its initial and final positions and noting the direction that this line takes. In lay terms, the displacement is a measure of the motion "as the crow flies."

The concepts of distance and displacement are perhaps best understood in terms of an example. Consider the maps of the two marathon courses shown in Fig. 8. In both instances, and in accord with long-established tradition, the competitors are required to run the rather odd distance of

[1]The concepts introduced in this chapter—and those introduced in the chapter on linear kinetics—apply equally to the translatory motion of a body that also experiences rotation as they do to the motion of one that experiences translation alone. In other words, they are applicable not only to the analysis of linear motion—which, after all, occurs only occasionally in sports—but also to the analysis of the linear aspects of general motion.

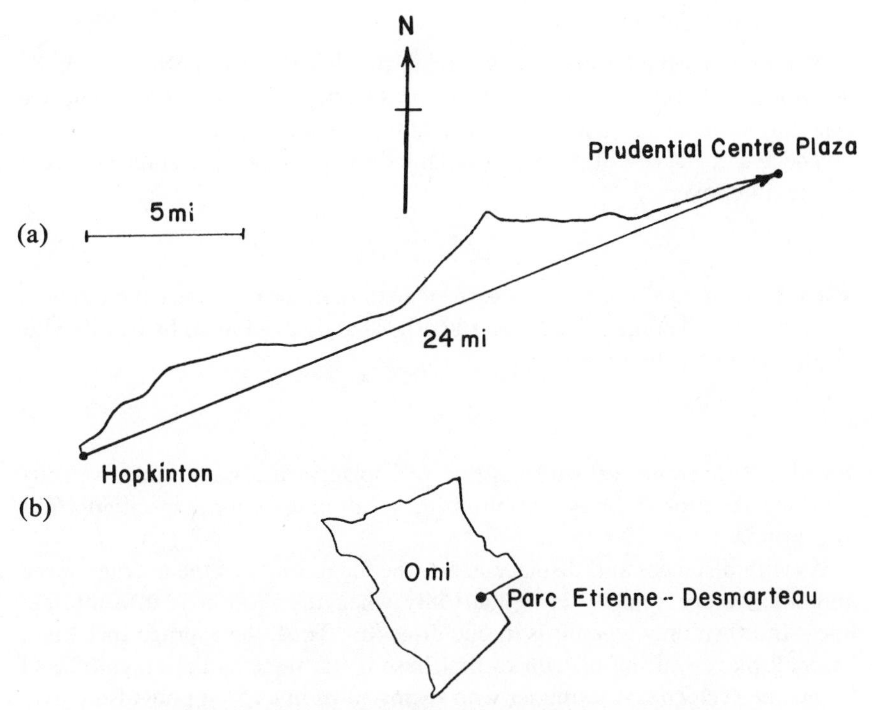

Fig. 8. The displacement a runner experiences in the process of completing a marathon depends on the nature of the course. In (a) the Boston marathon, his displacement is approximately 24 mi ENE. In a race over (b) the 1976 Olympic Games course in Montreal, his displacement is 0 mi.

26 mi 385 yd in order to complete the race. The displacements they undergo in the process depend on the nature of the course. In the first case shown (the course for the annual Boston marathon) those who finish undergo a displacement of 24 mi in an ENE direction in making their way from start to finish. In the second case (the course for the 1976 Olympic Games marathon) the starting and finishing lines coincide. Thus, although a competitor may spend 2–3 hr running more than 26 mi, the displacement he experiences in the course of the event is zero (ignoring any small differences in the positions at which he crosses the line at start and finish).

When a body moves in a straight line, as in a 100-yd dash, the distance over which it travels and the displacement it experiences have the same magnitude, namely 100 yd. The two quantities differ, however, in that the displacement must contain reference to the direction of the motion (as well as to its magnitude) whereas the distance is completely defined by its magnitude alone. Thus, a sprinter who runs 100 yd in a northerly direction covers a distance of 100 yd and undergoes a displacement of 100 yd, North.

Speed and Velocity

The rate at which a body moves from one location to another is usually described with reference to its speed or velocity—two quantities that are generally thought to be identical and that are frequently not.

The average speed of a body is obtained by dividing the distance covered by the time taken,

$$\bar{s} = \frac{l}{t} \tag{1}$$

where $\bar{s}$ = average speed, l = length of path (i.e., the distance covered), and t = time. The average velocity, on the other hand, is obtained by dividing the displacement by the time taken,

$$\bar{v} = \frac{d}{t} \tag{2}$$

where $\bar{v}$ = average velocity and d = displacement, and, since velocity is a measure of a body's motion in a given direction, specifying this direction.

As with distances and displacements, the magnitudes of the average speed and the average velocity are equal only when the motion is in a straight line—and then only when it is in one direction. Thus, the average speed of a baseball player during his run to first base is the same as the magnitude of his average velocity. A swimmer who swims 50 yd in a 25-yd pool also moves in an essentially straight line but, because he reverses direction at the midpoint in his swim, has an average speed and an average velocity that differ markedly from one another. If he completes the swim in, say, 30 sec, his average speed is

$$\begin{aligned}\bar{s} &= \frac{50 \text{ yd}}{30 \text{ sec}} \\ &= \frac{150 \text{ ft}}{30 \text{ sec}} \\ &= 5 \text{ fps}\end{aligned}$$

His average velocity, however, is

$$\begin{aligned}\bar{v} &= \frac{0 \text{ ft}}{30 \text{ sec}} \\ &= 0 \text{ fps}\end{aligned}$$

This somewhat surprising result comes about because the swimmer's displacement is zero. (*Note:* The small difference between his position on the block at the start and in the water at the finish has been ignored.) Thus, while increasingly greater average speeds are needed to break records in swimming, the average velocities generally remain unaltered at 0 fps.

At this point one might question the value of computing average velocities since they appear to convey very little information of interest. The merits of computing average speeds might also be questioned because, typically, the average speed over an extended distance is a very poor indicator of how that distance was covered. Consider, for example, the world records for 1500 m set by swimmers Steve Holland (Australia) in 1973 and Tom Shaw (U.S.A.) in 1975. The graph showing the average speeds of the two swimmers for the full distance of the event [Fig. 9(a)] conveys no more than was already known—that is, one man was faster overall than the other. However, when the average speeds for successive 300-m segments of the full distance are considered [Fig. 9(b)], the different ways in which the two races were swum begin to emerge. Specifically, it can be seen that Shaw swam the first and last 300 m markedly faster than Holland and that the two differed relatively little during the middle stages. When this process is continued one step further and the average speeds for every 100 m are considered [Fig. 9(c)], an even more detailed comparison becomes possible. This idea of obtaining an ever-clearer picture of what took place by progressively reducing the distance over which times are taken leads directly to the concept of *instantaneous speed.*

The instantaneous speed of a body is equal to its average speed over such a very short distance (starting from the position occupied by the body at the instant in question) that the speed will not have time to change. In similar fashion the instantaneous velocity of a body is defined as its average velocity over such a very short distance that the velocity will not have time to change. Further, as is the case with average speeds and velocities when a body is moving in a straight line (see earlier mention), the instantaneous speed of a body is equal to the magnitude of its instantaneous velocity.

Time and again in sports techniques it is this concept of an instantaneous speed or velocity (rather than an average value) that is critical. For instance, any time an athlete jumps or throws, it is the velocity of the body in question at the instant of takeoff or release that to a very large extent determines the final outcome.

[*Note:* (a) Physics and mechanics texts usually define instantaneous speed in terms of a "very short interval of time" rather than a "very short distance." The latter has been used here because it is believed to be a little easier to understand and because measurements of instantaneous speed taken in biomechanics frequently make use of the concept of a "very short distance." (b) When average and instantaneous quantities are represented here in algebraic form, average values are distinguished from instantaneous values by the addition of a line (or bar) above the appropriate letter. Thus, for example, s is used to represent the instantaneous speed and $\bar{s}$ (read "s bar") the average speed.]

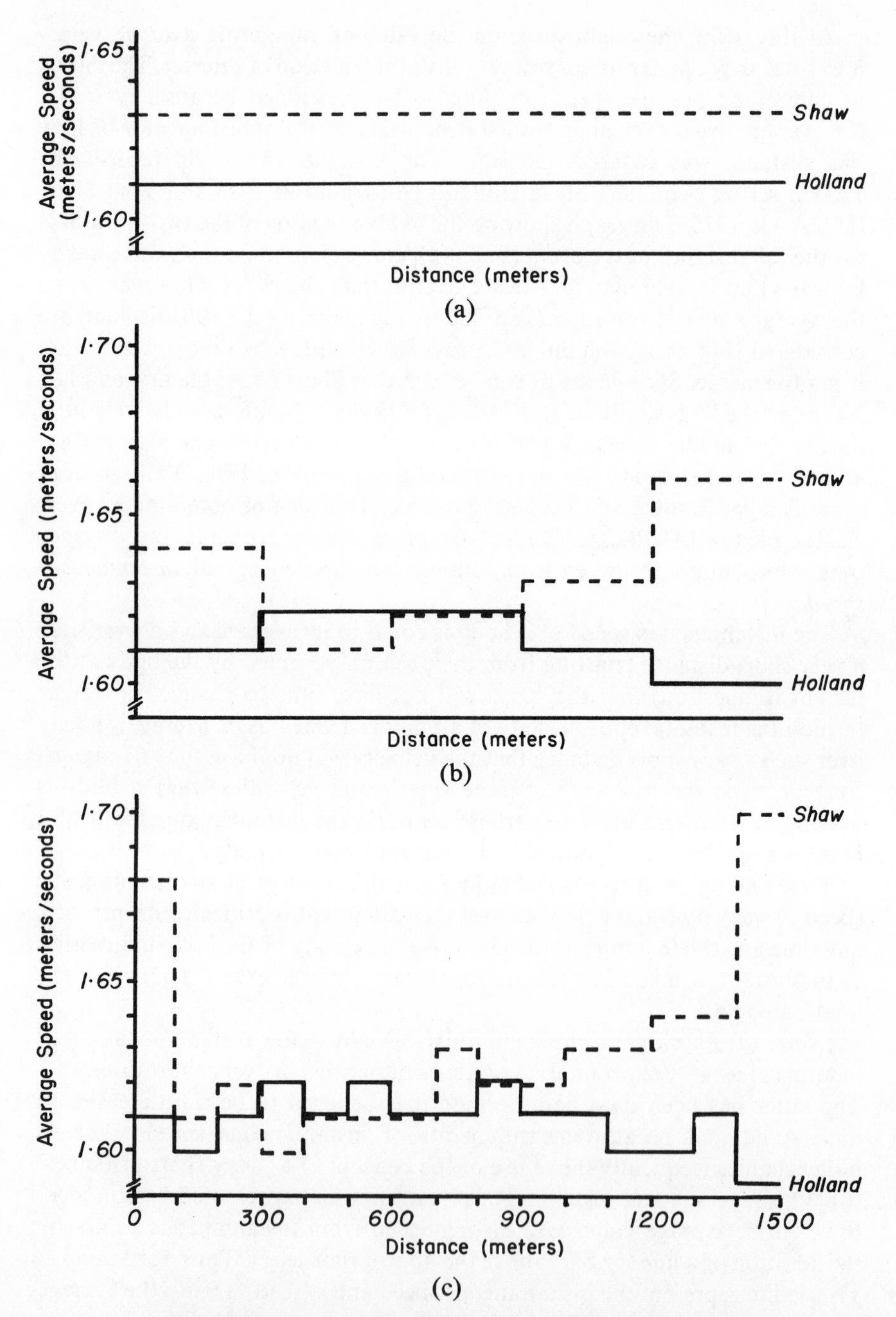

Fig. 9. The average speed becomes a progressively more useful indicator of performance as the distance (or time) over which it is computed is decreased.

Acceleration

In many sports activities the success enjoyed by an athlete is directly related to his ability to rapidly increase or decrease his velocity. The football lineman seeks to build up as much velocity as he can before he makes contact with his opponent. The runner stealing second builds up velocity as quickly as he can, in order to minimize the time available for the defense to react; and then reduces his velocity at the end of his run to avoid overrunning the bag. Basketball players must likewise be able not only to build up velocity quickly but also to "stop on a dime." All of these people, and many more in other sports, are concerned with *acceleration.*

Acceleration (which, like displacement and velocity, necessarily includes the element of direction as well as of size) is defined as the rate at which the velocity changes with respect to time. In algebraic form,

$$\bar{a} = \frac{v_f - v_i}{t} \tag{3}$$

where $\bar{a}$ = the average acceleration, v_f = the final velocity, v_i = the initial velocity, and t = the elapsed time.

A close look at Eq. 3 should show that it is possible to have positive, negative, and zero values for the acceleration. Anytime that v_f is greater than v_i, the numerator of the right-hand side of the equation will be positive and hence the acceleration itself will be positive. Conversely, when v_i is greater than v_f, the numerator and thus the acceleration are both negative. Finally, when v_f and v_i are equal, there has clearly been no change in the velocity and hence, by the very definition of acceleration, this latter must be zero.

All of this presents no problems when the motion takes place in a straight line and in one direction. Under such circumstances, it is customary to speak of negative acceleration as *deceleration* or *retardation* or just plain "slowing down." As an example, the base runner (Fig. 10) starts from a position of zero velocity and then undergoes positive acceleration until he is very close to second base. He then experiences negative acceleration as he goes into the

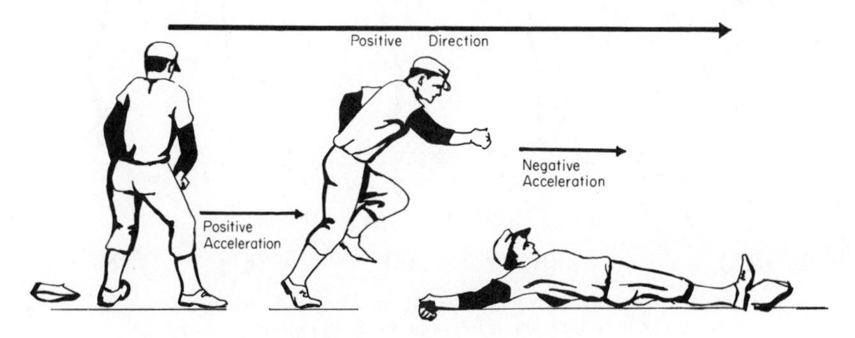

Fig. 10. Positive and negative acceleration in running between bases

slide that will ultimately bring him to rest in contact with the bag. (*Note:* At the end of his period of positive acceleration there will be an instant when he is no longer increasing his forward velocity nor has yet started to decrease it, i.e., a moment of zero acceleration.)

Problems may arise, however, when the acceleration of a body that moves first in one direction and then in the directly opposite direction is being considered. Consider the volleyball blocker in Fig. 11. If the upward direction is regarded as positive, the player's acceleration between positions (a) and (b) is positive because in position (b) she has a high velocity in the upward (positive) direction and in position (a) she has zero velocity:

$$\bar{a} = \frac{\text{large positive velocity} - 0}{t_1} = \text{some positive value}$$

Her acceleration in the interval between positions (b) and (c) is negative:

$$\bar{a} = \frac{0 - \text{large positive velocity}}{t_2} = \text{some negative value}$$

So far, all is consistent with what was said earlier about the baserunner—when the athlete "speeds up," she experiences positive acceleration; con-

Fig. 11. Blocking in volleyball. The blocker experiences positive acceleration each time she is in contact with the ground and negative acceleration while airborne.

versely, when she "slows down," she experiences negative acceleration. Now, however, the difficulty arises. Between positions (c) and (d) the athlete's acceleration is again negative. In position (c) she has zero velocity and in position (d) she has a high velocity in a downward (or negative) direction. Hence:

$$\bar{a} = \frac{\text{large negative velocity} - 0}{t_3} = \text{some negative value}$$

Between positions (d) and (e) the athlete's high negative velocity is reduced to zero, and in this process she once again experiences positive acceleration:

$$\bar{a} = \frac{0 - \text{large negative velocity}}{t_4} = \text{some positive value}$$

In other words, during these last two intervals the pattern established earlier has been completely disrupted—the athlete has experienced negative acceleration while "speeding up" and positive acceleration while "slowing down."

It is thus important to recognize that when directions are designated as positive and negative, these adjectives can no longer be used in conjunction with acceleration to indicate, respectively, "speeding up" and "slowing down."

Units in Linear Kinematics

With one exception, the units used in the measurement of the quantities described thus far generally provide few problems. In the United States, measurements of length are made in inches (in.), feet (ft), yards (yd), and miles (mi) and measurements of time in seconds (sec), minutes (min), and hours (hr). The units of measurement for those quantities that are some combination of these basic (or fundamental) quantities of length and time are derived in logical fashion. For example, since speed is determined by dividing a measurement of length by a measurement of time (average speed = distance/time), the appropriate unit for speed is logically one in which a unit of length is divided by a unit of time. Thus the foot per second (fps) and the mile per hour (mph) are units of speed.

With distance and displacement quite obviously measured in units of length, and speed and velocity both measured in the same units (fps, mph, etc.), the only quantity so far unaccounted for—and the only one likely to create any difficulty—is acceleration. A glance at the right-hand side of the equation for average acceleration (Eq. 3) reveals that the numerator is the difference between two velocities and thus will logically be measured in a unit appropriate to velocities. The denominator is a simple measurement of time and will likewise be measured in some appropriate unit. If the velocity unit used is the foot per second and the time unit is the second, the unit for acceleration will be the foot per second per second (fps/sec, ft/sec/sec, or fps^2).

An example may help to clarify this. Consider a swimmer whose forward velocity is 6 fps, 2 sec after the start of a race (Fig. 12). Since his velocity was 0 fps when the gun went off (he's an honest swimmer!), his change in forward velocity in the first 2 sec is obviously 6 fps. If this change is averaged out over the 2-sec time period, it can be seen that on the average he changed his forward velocity by 3 fps for each of the 2 sec. In other words, his average forward acceleration (i.e., his average change in forward velocity with respect to time) was 3 ft/sec/sec (or 3 fps²).

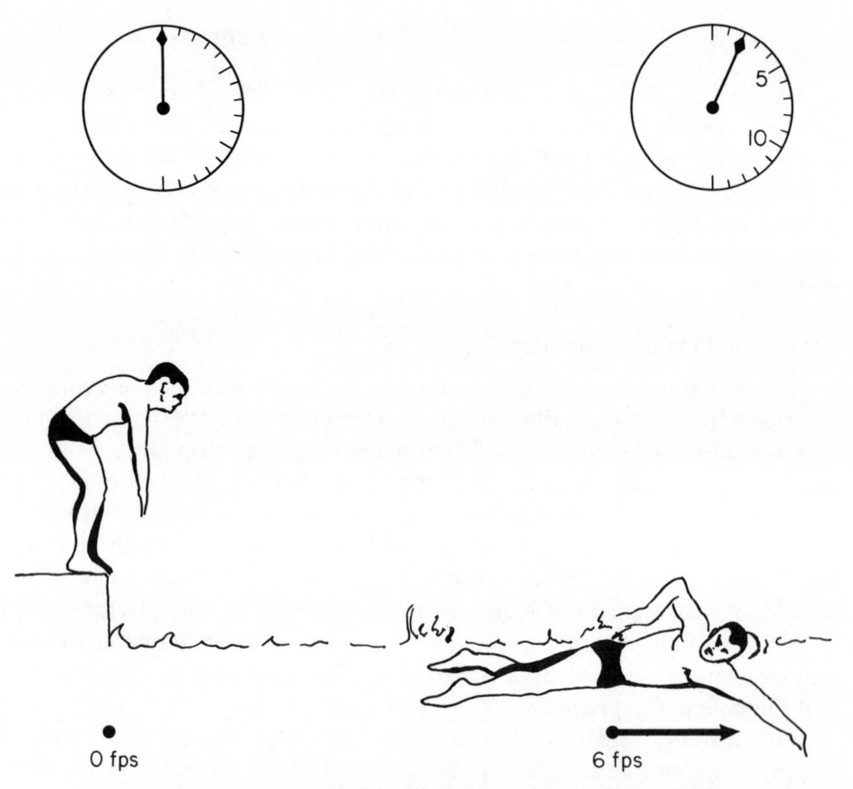

Fig. 12. Acceleration is the rate of change of velocity. This swimmer's forward velocity changed by 6 fps in 2 sec or, on the average, 3 fps each second. His average acceleration is thus 3 fps².

Acceleration Due to Gravity

The downward acceleration that a body experiences while in the air is due to the influence of the earth on all bodies near to its surface, an influence known as *gravity*. The acceleration due to gravity is essentially constant—it varies slightly from place to place on the earth's surface—and, because it is of especial significance in so many situations, is generally designated by a separate letter g. The magnitude of the acceleration due to gravity (and it is important to remember that g represents simply an acceleration and nothing else) is approximately 32.2fps^2.

Vectors and Scalars

Most of the kinematic (and kinetic) quantities considered in this text may be classified into two groups. Those, such as distance and speed, that can be completely described in terms of size (or magnitude) are known as *scalars*, while those, such as displacement, velocity, and acceleration, that require specification of both a magnitude and a direction are called *vectors*.[2]

Because arrows also have both magnitude (i.e., length) and direction, it is possible, and often very useful, to represent vectors by arrows. The velocities of the various bodies depicted in Fig. 13 are represented by the arrows shown. The length of each arrow represents the magnitude of the velocity to some chosen scale (in this case, 1 in. = 67 fps) and the direction in which the arrow is drawn indicates the direction in which the vector is acting.

Fig. 13. Velocities of release represented in vector form

[2]Not all quantities that can be described in terms of a magnitude and a direction are in fact vector quantities. Some, such as angular displacement (see pp. 50–52), have a magnitude and a direction but do not qualify as vectors because they do not also operate in accordance with the parallelogram of vectors (see the next section).

Resultant Vector

The place-kicker in Fig. 14 imparts a velocity to the football in the direction of the middle of the goalposts. Viewed from above, and all else being equal, the ball will then travel in a straight line in that direction [Fig. 14(a)]. If, however, there is a crosswind blowing, this too will impart a velocity to the football. At the instant shown in Fig. 14(b) the football has a velocity, represented by the vector OK, imparted to it by the kicker, and a second velocity imparted to it by the wind and represented by the vector OW. The net effect (or *resultant*) of these two velocity vectors can be found by completing the parallelogram of which OK and OW are adjacent sides and then constructing the diagonal through the point O. This diagonal represents in magnitude and direction the resultant velocity of the ball. In other words, as a result of the combined effects of the kicker and the wind, the ball moves in the direction indicated by the diagonal OR and with a speed represented by

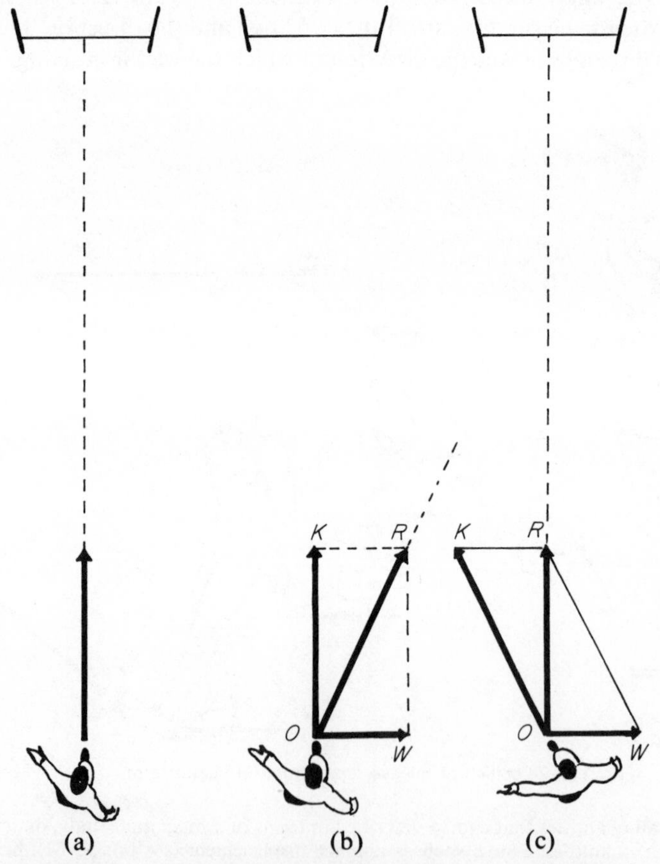

Fig. 14. The parallelogram construction to find the summed effect of two vectors

the length of that diagonal. Of course, a good kicker knows all this from experience and thus makes due allowance for the wind in directing his kick [Fig. 14(c)].

A parallelogram such as the one obtained in the example above is known as a *parallelogram of vectors*. Fortunately it is not necessary to go to the trouble of carefully measuring lines and angles and constructing such a parallelogram in order to determine the magnitude and direction of the resultant. Instead this can be found with relative ease by using an appropriate formula. When the angle between two vectors (A and B—Fig. 15) is a right angle, the magnitude of their resultant (R) can be found by using the theorem of Pythagoras:

$$R = \sqrt{A^2 + B^2} \tag{4}$$

and the direction in which this resultant acts can be found using simple trigonometry.[3] If the angle formed between vector B and the resultant is θ,

$$\tan \theta = \frac{A}{B}$$

and, therefore,

$$\theta = \arctan \left(\frac{A}{B}\right) \tag{5}$$

If these formulas are now applied to the example of Fig. 14(b), the magnitude of the resultant and the direction in which it acts are given by

$$OR = \sqrt{OK^2 + OW^2}$$

and

$$\theta = \angle ROW = \arctan \left(\frac{OK}{OW}\right)$$

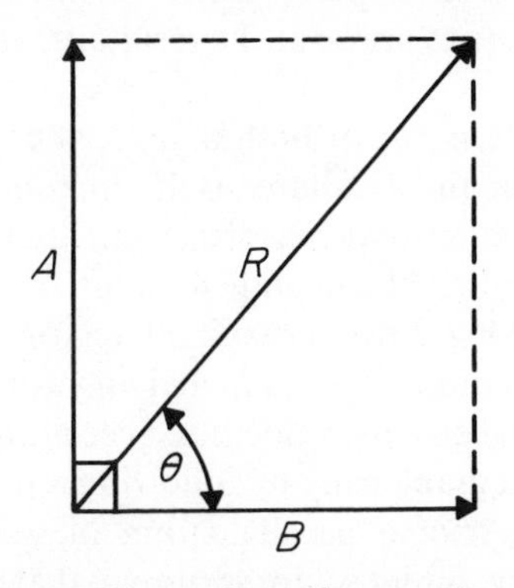

Fig. 15. The resultant of two vectors at right angles to each other

[3]Readers without a background in elementary trigonometry should refer to Appendix A.

If the angle (β) between the two vectors A and B is not a right angle, the process of arriving at the magnitude and direction of their resultant is a little more involved. The magnitude of the resultant, found using a trigonometrical identity known as the *cosine rule*, is given by

$$R = \sqrt{A^2 + B^2 + 2AB \cos \beta} \tag{6}$$

and the direction in which the resultant acts is given by

$$\theta = \arctan \left(\frac{A \sin \beta}{B + A \cos \beta} \right) \tag{7}$$

where θ is again the angle formed between vector B and the resultant R.

Vector Components

In football the object of the exercise is to advance the ball down the field and across the opponent's goal line. In the course of doing this, however, the ball is generally moved both forward and sideways as it is advanced and only rarely is it moved in a straight line directly downfield. Consider, for example, the favorite play of Fran Tarkenton shown in Fig. 16. In this play Tarkenton receives the ball from the center, drops back into the protective "pocket" formed by his teammates, and then throws a pass to his halfback who has swept around the right end of the line. If the play starts (as it always does) with the ball as P in the hands of the center and ends in this instance at Q, it can be seen that the effect of this play (i.e., the resultant displacement) can be represented by an arrow joining points P and Q. Looked at in a slightly different way, it can also be seen that the effect of the play has been to give the ball a certain displacement (represented by the arrow PR) toward the right sideline and a certain displacement (represented by the arrow PT) down the field. These two separate displacements, which can be added by means of a parallelogram of vectors to arrive at the resultant, are said to be *components* of the resultant.

In many sports activities one or both of a set of components are of principal concern rather than the resultant itself. In football, as can be readily appreciated, it is not the resultant displacement that determines the success of any given play but rather the downfield component of this displacement.

The process of breaking down a resultant vector into two components is most commonly used in biomechanics in arriving at the horizontal and vertical components of displacements, velocities, accelerations, etc. These components can be determined graphically by following a procedure that essentially reverses the construction of a parallelogram of vectors or by the use of elementary trigonometry. Suppose, for example, that it is desired to compare the horizontal and vertical velocities obtained by a champion high jumper with those obtained by a champion long jumper, at the instant of takeoff in their respective events. Suppose too that a velocity of 17 fps at an angle of 60° to the horizontal is taken as representative of championship performance

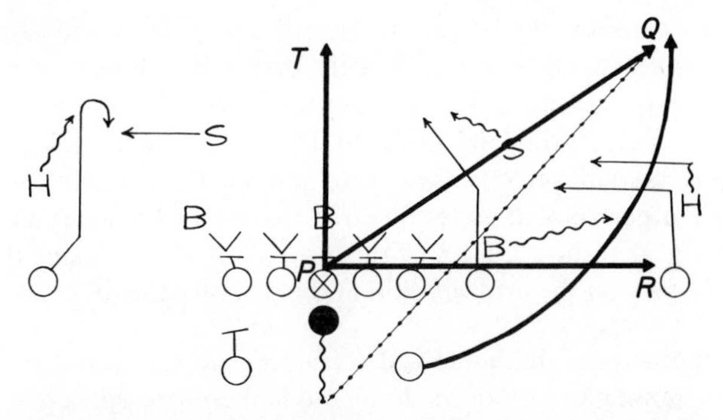

Fig. 16. Fran Tarkenton's favorite play—an example in which a component of a vector is more important than the vector itself. [*Adapted from Murray Olderman,* The Pro Quarterbook (*Englewood Cliffs, N.J.: Prentice-Hall, Inc., 1966*).]

in the high jump and that a velocity of 30 fps at an angle of 25° to the horizontal is representative of championship performance in the long jump.[4]

To arrive at the horizontal and vertical components of the high jumper's takeoff velocity using the graphical method, the following steps would be necessary (Fig. 17):

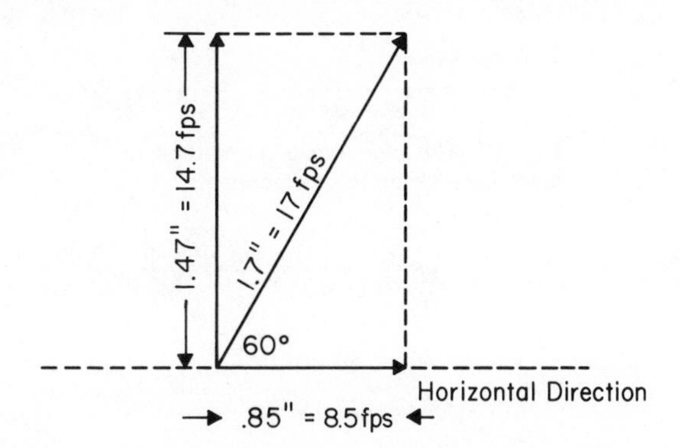

Fig. 17. The graphical method for resolving a vector into components

1. Choose a suitable linear scale (e.g., 1 in. = 10 fps).

2. Using the chosen scale, draw an arrow to represent the jumper's takeoff velocity.

3. Through each end of this arrow, draw straight lines to represent the horizontal and vertical directions. (*Note:* The acute angle between the horizontal lines and the line representing the takeoff velocity must be equal to the angle of takeoff, in this case 60°.)

[4]Assuming appropriate values for the other factors that have a bearing on the outcome, the above would result in a high jump of 7 ft 5 in. and a long jump of 27 ft 4 in.

4. Measure carefully the lengths of two adjacent sides of the parallelogram thus formed and, using the linear scale originally chosen, convert these measurements to velocities. The velocity determined from the side of the parallelogram representing the horizontal direction is the horizontal component of the takeoff velocity (or, more simply, the horizontal velocity at takeoff). The other velocity determined is the vertical velocity at takeoff. It should be noted that the accuracy of the results obtained in using this method hinges very largely on the drafting skill of the person using it.

The trigonometrical method for determining the component velocities is less time-consuming and more accurate and is therefore the better of the two methods. To find the horizontal and vertical components of the long jumper's takeoff velocity using this method, a rough sketch of the parallelogram constructed in using the graphical method is first drawn and the relevant information is attached (Fig. 18).

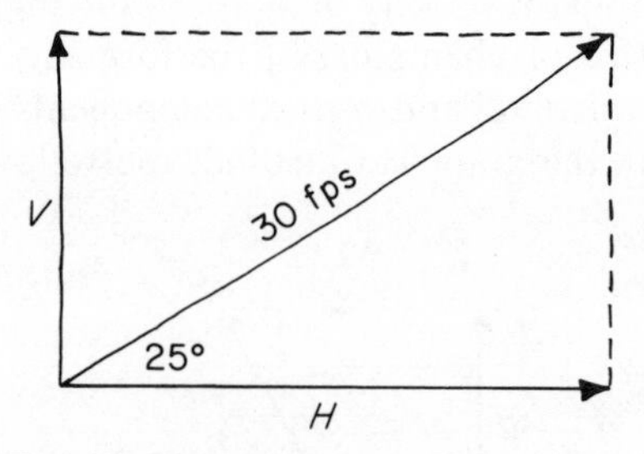

Fig. 18. The trigonometrical method for resolving a vector into components

Then, since

$$\frac{H}{30} = \cos 25^\circ$$

$$H = 30 \cos 25^\circ$$

$$= 27.19 \text{ fps}$$

and, similarly,

$$V = 30 \sin 25^\circ$$

$$= 12.68 \text{ fps}$$

It is perhaps of interest to note that the ratio of horizontal velocity to vertical velocity for one event is essentially the reverse of that for the other. This fact has some important implications for the teaching and coaching of these events.

Uniformly Accelerated Motion

When a body experiences the same acceleration (in both magnitude and direction) throughout some interval of time, its acceleration is said to be *constant* (or *uniform*). Under such circumstances the average acceleration of the body is exactly the same as its acceleration at any instant during the period involved. This rather obvious fact permits three important relationships—known as equations of uniformly accelerated motion—to be obtained. These equations, in which, *for a given direction*, v_i = the initial velocity (the instantaneous velocity at the start of the motion under consideration), v_f = the final velocity (the instantaneous velocity at the end of the motion), d = the displacement the body has undergone, t = the time involved, and a = the acceleration, are as follows:

$$v_f = v_i + at \tag{8}$$

$$d = v_i t + \tfrac{1}{2}at^2 \tag{9}$$

$$v_f^2 = v_i^2 + 2ad \tag{10}$$

While in some activities (such as wrestling, fencing, bowling, and weight lifting) in-the-air motion is of little, if any, concern, in others (such as diving, tumbling, ski-jumping, and all the field events) such motion is the very essence of the activity. It is in the analysis of these in-the-air activities (where the uniform acceleration is that due to gravity) that the equations of uniformly accelerated motion have their most important applications. However, before considering some of the factors of importance in such cases, it may be well to consider some simple examples and to develop a procedure for attacking problems of this kind.

The motion-picture camera is probably the most widely used tool in the analysis of sports movements. It is used extensively by coaches to record the action in games, meets, and workouts. It is also used by research workers who are concerned with obtaining precise measurements from the resulting films. So that the speed at which bodies are moving can be accurately determined from such films, it is often necessary to find out how many frames of film are being exposed each second. One way of doing this is to film a heavy weight (e.g., a shot) being dropped to the ground from some known height. The time taken for the shot to fall this distance can be determined using the following general procedure:

1. Write the information that is required:

The time taken: $t = ?$

2. Write the information that is known:

Initial velocity: $v_i = 0$ fps

Acceleration: $a = 32.2$ fps^2

Displacement: $d = 7$ ft, say

3. Determine which of the three equations of uniformly accelerated motion contains all the variables recorded in steps 1 and 2 above:

$$d = v_i t + \tfrac{1}{2}at^2$$

4. Substitute the appropriate values in this equation, and determine the information required:

$$7 = (0 \times t) + (\tfrac{1}{2} \times 32.2 \times t^2)$$

$$7 = 16.1t^2$$

$$t = \sqrt{\frac{7}{16.1}}$$

$$= 0.66 \text{ sec}$$

A count of the number of frames of the film taken between the instant of release of the shot and the moment of it hitting the ground is then made. (The frame showing the shot at the instant of release is not included in arriving at this total, but all subsequent frames up to and including the one showing the shot at the instant it touches the ground are included.) This figure (say, 42) can then be used together with the computed time to determine the number of frames per second at which the camera is filming and the average time interval between frames:

$$\text{Camera speed} = \frac{42}{0.66}$$

$$= 63.63 \text{ frames per second}$$

$$\text{Average interval between frames} = \frac{0.66}{42}$$

$$= 0.016 \text{ sec}$$

Another application of these equations is found in the simple ruler-dropping test used to measure a person's response time (Fig. 19). In this test a ruler (or rod) is held by the tester between the outstretched index finger and thumb of the subject's dominant hand, so that the top of the subject's hand is level with the bottom of the ruler. The subject is instructed to catch the ruler just as soon as he can after it has been released by the tester. The distance between the bottom of the ruler and the top of the subject's hand when the ruler has been caught (i.e., the distance the ruler fell) is used to determine the subject's response time.

So that a direct reading of the response time can be obtained, an appropriate scale is marked on the ruler. If it is desired to have the scale graded in $\frac{1}{10}$-sec intervals, the distance of the first gradation from the bottom of the ruler can be determined as follows:

Displacement: $d = ?$

Time: $t = 0.1$ sec

Acceleration: $a = 32.2$ fps^2

Initial velocity: $v_i = 0$ fps

Using Eq. 9,

$$d = v_i t + \tfrac{1}{2}at^2$$

$$d = (0 \times 0.1) + (\tfrac{1}{2} \times 32.2 \times 0.1 \times 0.1)$$

$$= 0.16 \text{ ft}$$

$$= 1.92 \text{ in.}$$

The location of the other gradations on the scale can be determined in similar fashion.

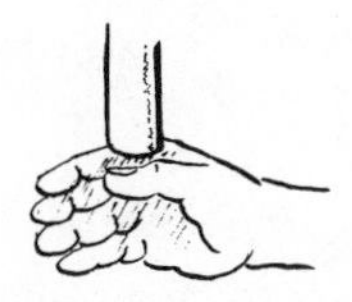

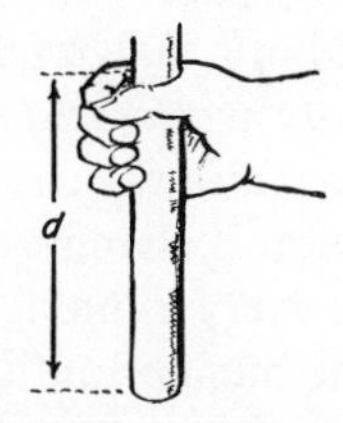

Fig. 19. Handgrip response time—an example of the application of the equations of uniformly acelerated motion

While the equations discussed in this section are important in the analysis of many sports techniques (and this will become especially evident in the next section), considerable care must be exercised in using them. Consider a track coach who is anxious to increase the horizontal velocity with which his long jumper leaves the takeoff board. Seeking some fairly precise measure of this velocity, he decides to time the athlete's approach run and use this time in computing his horizontal velocity at takeoff:

Final velocity: $v_f = ?$

Displacement (i.e., his run-up length): $d = 150$ ft, say

Initial velocity: $v_i = 0$ fps

Time: $t = 6$ sec, say

Using Eq. 9, he first finds the acceleration since he must have this before he can determine the athlete's horizontal velocity:

$$d = v_i + \tfrac{1}{2}at^2$$

$$150 = (0 \times 6) + (\tfrac{1}{2} \times a \times 6^2)$$

$$150 = 18a$$

$$a = \frac{150}{18} \text{ fps}^2$$

Then, using Eq. 8, he computes the final horizontal velocity:

$$v_f = v_i + at$$

$$= 0 + \frac{150}{18} \times 6$$

$$= 50 \text{ fps}$$

Now on the surface this appears to be a singularly useful technique. It would enable the coach to abandon such subjective comments as "Yes, that looked a bit faster" and "I don't think you were quite so fast that time" in favor of quite objective observations like "That was 3 fps faster than your last one." Unfortunately it has one limitation that renders the whole method utterly useless: *the equations of uniformly accelerated motion (as their name suggests) apply only in the case of a body whose velocity is changing at a constant rate.* Since a long jumper's acceleration is changing throughout his approach run (in fact, throughout each and every running stride), it is clear that these equations cannot be used in the manner outlined above. (Incidentally, such misuse would almost inevitably result in a much higher value for the horizontal velocity than was actually obtained. While the figures used in the example above are reasonable enough, the resulting horizontal velocity is well in excess of that ever achieved by any sprinter!)

Projectiles[5]

Many sports involve the projection of a body into the air. In shot-putting, baseball, soccer, and tennis the body projected into the air (the *projectile*) is an inanimate one. In diving, gymnastics, and the jumping events (and occasionally in the other sports already mentioned) the projectile is the performer himself. In all of these sports and in many others the quality of performance depends very largely on the performer's ability to control and/or predict the outcome of the projectile motion involved. The tennis player executing a delicate drop shot must stroke the ball in such a way as to ensure that it

[5]For the sake of simplicity, the effects of air resistance will be ignored throughout the ensuing discussion of projectile motion. These effects will be considered in Chap. 7, "Fluid Mechanics."

travels high enough and far enough to clear the net and yet not so far as to make it easy for his opponent to return. In short, he must exert a very fine degree of control over the motion of the ball. A kick-return man on a football team has no control whatever over the flight of the ball but he must successfully predict its motion and position himself appropriately to receive it. To these people, and all others concerned with sports involving projectile motions, an understanding of the factors that govern the behavior of projectiles is of critical importance.

For purposes of analysis, the horizontal and vertical motions of a projectile may be considered separately. For example, when a stationary soccer ball is kicked upfield from a goal kick, the velocity imparted to the ball generally acts in a direction at some angle θ to the horizontal [Fig. 20(a)]. If this velocity is resolved into its horizontal and vertical components, the effect of the kick can be studied by considering each of these components in turn.

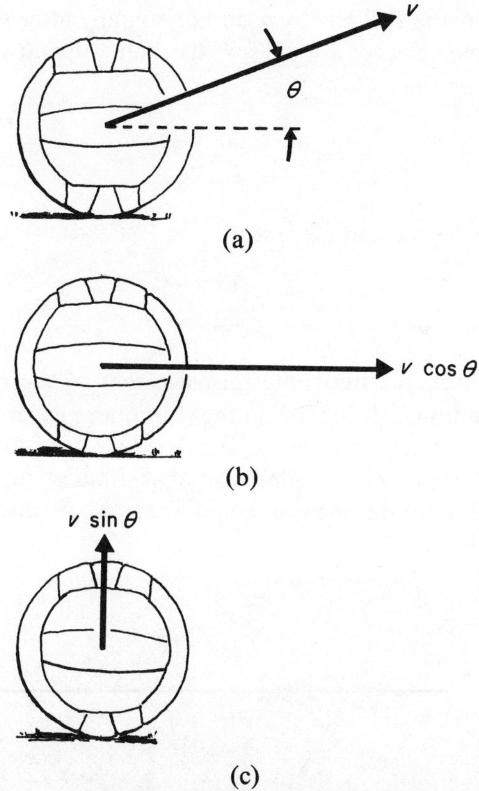

Fig. 20. Resultant, horizontal, and vertical velocities of a soccer ball at the moment of "release" in a goal-kick

Horizontal Motion. The horizontal velocity of the soccer ball in Fig. 18 at the instant it is released or projected into the air is $v \cos \theta$ fps. Further, since there is nothing which tends to change the rate at which it is moving horizontally (ignoring air resistance), it maintains this same horizontal velocity throughout its flight. This means, of course, that its average horizontal velocity for the period during which it is a projectile is also $v \cos \theta$ fps.

Now, a simple rearrangement of Eq. 2 reveals that the distance a body will travel in a given direction is simply a product of the average velocity at which it is traveling in that direction and the length of time involved:

$$d = \bar{v}t$$

or, in the present case,

$$d_H = v \cos \theta \times t \tag{11}$$

where d_H = the horizontal displacement. This last equation can be used to determine how far the ball has traveled horizontally after given intervals of time. For example, if $v \cos \theta = 50$ fps, the ball will have been displaced 5 ft horizontally by the end of $\frac{1}{10}$ sec:

$$\begin{aligned} d_H &= 50 \times \tfrac{1}{10} \\ &= 5 \text{ ft} \end{aligned}$$

10 ft horizontally by the end of $\frac{2}{10}$ sec:

$$\begin{aligned} d_H &= 50 \times \tfrac{2}{10} \\ &= 10 \text{ft} \end{aligned}$$

and so on. In short, the horizontal displacement of the ball increases by 5 ft for every additional $\frac{1}{10}$ sec it is in flight. In more general terms, the ball travels equal horizontal distances in equal intervals of time. The concept that a projectile, for which the effects of air resistance can be ignored, will travel equal horizontal distances in equal intervals of time is illustrated in another context in Fig. 21.

Fig. 21. When the effects of air resistance are negligible, a projectile travels equal horizontal distances in equal intervals of time.

The total horizontal distance the ball travels while it is in the air, a distance known as the *range* of the projectile, can also be found using Eq. 11. If T = the time of flight of the projectile and $v \cos \theta$ = the horizontal velocity, as before, the range of the projectile, R, is given by

$$R = v \cos \theta \times T \tag{12}$$

From this equation it is apparent that if a soccer player making a goal-kick (or any other performer who projects a body into the air) wishes to alter the range of the projectile, he can do so only by altering the horizontal velocity of release and/or the time of flight.

Since the time of flight is obviously of great importance in sports that require controlling or predicting the range of a projectile, those involved with such sports must have an understanding of the factors that determine the time of flight. Such understanding is also needed in sports such as diving and trampolining, where, although the range is only a minor consideration, the time of flight is a critical factor in determining the quality of performance.

Vertical Motion. The horizontal velocity of the soccer ball in Fig. 18 acts parallel to the ground and thus has no tendency to lift the ball into the air. The lifting of the ball is due entirely to the vertical velocity imparted to it prior to release. The time of flight of the ball is also largely determined by the vertical velocity at release.

Logically, the time of flight is equal to the time it takes the projectile to reach the peak of its flight (designated here as t_{up}) plus the time it takes to return from this peak to the point of landing (t_{down}):

$$T = t_{up} + t_{down}$$

The time to reach peak height, t_{up}, can be found using the procedure outlined on pp. 29–30.

Time: $t = t_{up} = ?$

Initial velocity: $v_i = v \sin \theta$
(i.e., the vertical velocity at release)

Final velocity: $v_f = 0$
(i.e., the vertical velocity at peak)

Acceleration: $a = -g$

Using Eq. 8,

$$v_f = v_i + at$$

$$0 = v \sin \theta - g t_{up}$$

$$t_{up} = \frac{v \sin \theta}{g} \tag{13}$$

The time of descent, t_{down}, can be expressed as a function of the vertical displacement, d_{down}, that the projectile undergoes during the descent:

Time: $t = t_{\text{down}} = ?$

Initial velocity: $v_i = 0$
(i.e., the vertical velocity at peak)

Acceleration: $a = -g$

Displacement: $d = -d_{\text{down}}$

Using Eq. 9,

$$d = v_i t + \tfrac{1}{2}at^2$$

$$-d_{\text{down}} = -\tfrac{1}{2}gt_{\text{down}}^2$$

$$t_{\text{down}} = \sqrt{\frac{2d_{\text{down}}}{g}} \tag{14}$$

Now, the vertical displacement, d_{down}, depends on the height at which the projectile lands. If it lands above the level from which it was released (as, for example, when the ball lands on the hoop during a free throw in basketball), d_{down} is less than d_{up}, the vertical displacement the projectile experiences in rising to its peak height. If it lands at the level from which it was released (as is frequently the case when a goal kick is taken in soccer) d_{down} is equal to d_{up}. Finally, if it lands below the level at which it was released (as in shot putting) d_{down} is greater than d_{up}.

When the projectile lands at the level from which it was released (the simplest of the three possibilities referred to above), d_{down} can readily be found by determining d_{up}:

Displacement: $d = d_{\text{up}} = ?$

Initial velocity: $v_i = v \sin \theta$

Final velocity: $v_f = 0$

Acceleration: $a = -g$

Using Eq. 10,

$$v_f^2 = v_i^2 + 2ad$$

$$0 = (v \sin \theta)^2 - 2gd_{\text{up}}$$

$$d_{\text{up}} = \frac{(v \sin \theta)^2}{2g}$$

Then, since $d_{\text{down}} = d_{\text{up}}$, substitution of this expression into Eq. 14 yields a new expression for t_{down}:

$$t_{\text{down}} = \sqrt{\frac{2}{g} \cdot \frac{(v \sin \theta)^2}{2g}}$$

$$= \frac{v \sin \theta}{g} \tag{15}$$

Comparison of Eqs. 13 and 15 quickly reveals that, when release and landing are at the same level, it takes exactly the same time for a projectile to reach the peak of its flight as it does for it to return to its original level. This finding has important implications in a number of sports. For example, a trampolinist who is attempting to do a double back somersault should have half the stunt (i.e., the first somersault) completed by the time he reaches the peak of his flight. Similarly, an outfielder racing to position himself to catch a high fly ball should be at least halfway to the required position by the time the ball reaches its peak height. (*Note:* In this latter example, any difference between the height of release and the height of "landing" can safely be ignored, since the distances involved are huge by comparison.)

With t_{up} and t_{down} known, the time of flight can easily be computed:

$$T = t_{up} + t_{down}$$

$$= \frac{v \sin \theta}{g} + \frac{v \sin \theta}{g}$$

$$= \frac{2v \sin \theta}{g} \qquad (16)$$

This result also has important implications. Since g is constant for any given location (see pp. 22–23), the time of flight depends solely on $v \sin \theta$, the vertical velocity at release. Thus, any performer who wishes to alter the time of flight of a projectile can do so only by altering this parameter. The punter who wishes to increase the time of flight of a football so that his teammates can get down the field to "cover the punt" must somehow increase its vertical velocity at the instant it leaves his foot; the gymnast who needs more time in the air so that he can complete a tumbling stunt currently beyond him must increase the vertical velocity with which he leaves the ground, and the basketball player who wishes to decrease the time the ball is in the air when he makes a long pass must correspondingly decrease the vertical velocity with which the ball leaves his hands.

If a projectile is released at a height above or below the level at which it lands, d_{down} is given by

$$d_{down} = d_{up} + h$$

$$= \frac{(v \sin \theta)^2}{2g} + h \qquad (17)$$

where h = the height at which the projectile was released minus the height at which it landed—or, in other words, the height of release relative to the height of landing. From this definition, it can be seen that h is negative when the projectile lands above its point of release and positive when it lands below that point. (*Note:* For the sake of simplicity, h will hereinafter be referred to as the height of release. It is essential, however, that its complete definition—the height of release *relative to the height of landing*—be remembered throughout.)

Substituting Eq. 17 into Eq. 14 and combining the result with Eq. 13 yields the following somewhat lengthy expressions for the time of descent and the time of flight when the points of release and landing are not at the same level:[6]

$$t_{\text{down}} = \sqrt{\frac{(v \sin \theta)^2 + 2gh}{g}} \tag{18}$$

$$T = \frac{v \sin \theta + \sqrt{(v \sin \theta)^2 + 2gh}}{g} \tag{19}$$

This latter result indicates that where release and landing are not at the same level, the time of flight can be modified by altering the vertical velocity of release and/or the height of release. Thus a shot-putter who wants to increase the time of flight of the shot and thereby enable it to carry further (see Eq. 12) has two means at his disposal—he can increase the vertical velocity of the shot at the instant of release or he can increase the height from which he releases it. The relationship between height of release and time of flight is also well known to tumblers and divers, who frequently increase their heights of release during the initial stages of learning a new stunt or dive. Tumblers do this by using elevated takeoffs (the top of a vaulting box, a springboard, or a mini-tramp), while divers achieve the same effect by "taking the dive up" to a higher board than they ultimately plan to use.

Range of a Projectile. The range of a projectile was shown earlier (p. 35) to be equal to the product of the horizontal velocity at release and the time of flight:

$$R = v \cos \theta \times T$$

Substitution of the time of flight for a projectile that lands at the level from which it was released (Eq. 16) yields an expanded expression for the range in such cases:

$$R = v \cos \theta \times \frac{2v \sin \theta}{g}$$

$$= \frac{v^2 \, 2 \sin \theta \cos \theta}{g}$$

This expression can be simplified by using the trigonometrical identity

$$\sin 2\theta = 2 \sin \theta \cos \theta$$

and obtaining

$$R = \frac{v^2 \sin 2\theta}{g} \tag{20}$$

[6]It may be of interest to note here that when release and landing are at the same height; i.e., $h = 0$, Eq. 19 reduces to Eq. 16, thus confirming what has already been demonstrated in such cases.

Thus, with g constant, the horizontal range depends on v and θ, the magnitude and direction of the initial velocity, respectively—or, to put it in even simpler terms, on the speed and angle of release. It is clear too that, in general, the greater the speed of release, the greater will be the horizontal range, and this, of course, is consistent with everyday experience; e.g., the harder a ball is thrown or kicked, the farther it will go.

A quick look at a table of sine values (Appendix A) shows that they vary from 0.0000 (when the angle is 0°) to 1.0000 (when the angle is 90°). The maximum value that can be obtained if the sine of an angle is taken is therefore 1. Thus, for any given speed of release, the maximum value of the horizontal range will be obtained when

$$\sin 2\theta = 1$$

This occurs when

$$2\theta = 90^\circ$$

i.e., when

$$\theta = 45^\circ$$

Hence, if all else is equal, the optimum angle at which to project a body in order to obtain the maximum horizontal range possible is 45°.

This applies, however, only to situations in which release and landing are at the same level. Where the level from which the body is projected is above the level at which it lands, the horizontal range is obtained by use of Eq. 12 and 19:

$$R = v \cos\theta \times \left(\frac{v \sin\theta + \sqrt{(v \sin\theta)^2 + 2gh}}{g}\right)$$

which reduces to

$$R = \frac{v^2 \sin\theta \cos\theta + v \cos\theta \sqrt{(v \sin\theta)^2 + 2gh}}{g} \qquad (21)$$

Close inspection of Eq. 21 shows that in this situation there are just three factors that have any influence on the horizontal range of a projectile—the speed, angle, and height of release. (g is once again taken to be a constant.) Of these three it can readily be seen that the greater the speed and the height of release, the greater will be the resulting range. (Assuming θ is constant and less than 90°; if $\theta = 90°$, R will be zero irrespective of the speed or the height of release. The body will go straight up and straight down again!)

Regrettably the optimum angle of projection cannot be found as readily as in the case in which release and landing are at the same level. In the present case the optimum angle depends on both the speed and the height of release. Table 1 shows how the optimum angle of release varies with these two factors in shot putting. From this table it can be seen that:

1. The optimum angle of release is always less than 45°.

2. For any given height of release, the greater the speed of release, the more closely the optimum angle approaches 45°.

3. For any given speed of release, the greater the height of release, the less is the optimum angle.[7]

4. Equal increases in either height of release or speed of release do not yield consistently equal changes in the optimum angle or the resulting distance.

(*Note:* These conclusions hold true in general and not just for those values included in Table 1.)

*Table 1. Variation of Optimum Angle with Height and Speed of Release**

Height of Release	*Speed of Release* (fps)					
	28	32	36	40	44	48
8 ft 0 in.	38°	39°	40°	41°	42°	42°
	(31.34 ft)	(38.99 ft)	(47.58 ft)	(57.13 ft)	(67.65 ft)	(79.15 ft)
7 ft 6 in.	38°	39°	40°	41°	42°	42°
	(30.95 ft)	(38.58 ft)	(47.15 ft)	(56.69 ft)	(67.21 ft)	(78.69 ft)
7 ft 0 in.	39°	40°	41°	41°	42°	42°
	(30.55 ft)	(38.16 ft)	(46.73 ft)	(56.25 ft)	(66.76 ft)	(78.23 ft)
6 ft 6 in.	39°	40°	41°	42°	42°	43°
	(30.16 ft)	(37.75 ft)	(46.29 ft)	(55.81 ft)	(66.31 ft)	(77.78 ft)

*The distances obtained by the indicated combinations of speed of release, height of release, and optimum angle are shown in parentheses. These distances do not include the extra distance (approximately 1 ft) that the shot is in advance of the inside edge of the stop board at the instant of release.

Having determined that three factors influence the horizontal range of a projectile, it is now of some importance to consider just how much influence each of these factors has. The field events coach should know, for example, whether it is best for his shot-putter to concentrate his main efforts toward obtaining (1) a greater speed, (2) a more nearly optimum angle, or (3) a greater height, at the instant of release. The graph in Fig. 22 compares the results obtained when a 63-ft shot-putter increases each of these factors by a comparable amount.

The curve labeled $(\theta + 5\%\,\theta)$ shows the effects of increasing the angle of release by 5% while keeping the other two factors (speed and height of release) constant. If the angle of release is 0° and this is increased by 5%,

[7]The rounding of the optimum angle "correct to whole degrees" masks this trend in the case of the 44-fps release speed.

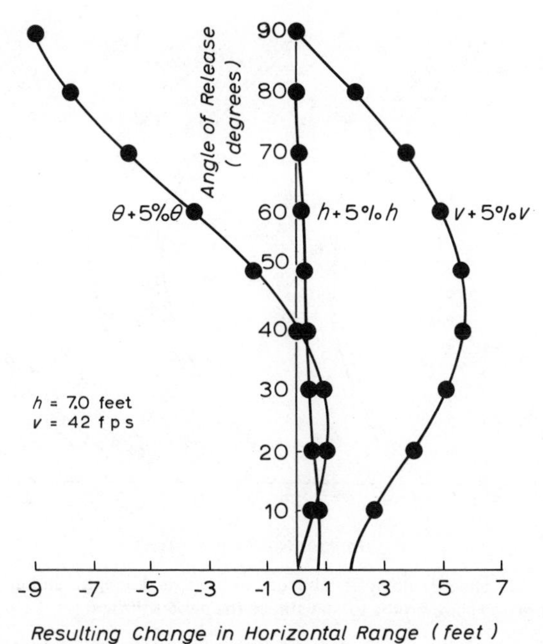

Fig. 22. The effects produced by 5% changes in the speed. height, and angle at which a shot is released

there is clearly no change in either the angle of release $[0° + (\frac{5}{100} \times 0°) = 0°]$ or the horizontal range. The curve therefore passes through the origin, 0. If the angle of release is 20° and this is increased by 5% (to 21°), the horizontal range is increased by approximately 1 ft. For release angles of greater than 40°, a 5% increase actually reduces the horizontal range.

The curve labeled $(h + 5\% h)$ shows the effects of increasing the height of release by 5% while keeping the other two factors constant. If $v = 42$ fps and $\theta = 0°$, this 5% increase yields an increase of approximately 0.65 ft in the horizontal range. As θ becomes larger, the 5% increase in h yields progressively smaller increases in the horizontal range.

Finally the curve labeled $(v + 5\% v)$ shows the effects of increasing the speed of release by 5% while keeping the height and angle of release constant.

These curves (which, incidentally, are typical of those that would be obtained for any normal combination of release speed and height—see Fig. 23)[8] show very clearly that the increase in release speed is much more effective in terms of increasing the horizontal range than comparable increases in either the angle or height of release. Thus, in general, shot-putters are best advised to concentrate their attention more on developing faster release speeds than on either increasing the height of release or optimizing the angle of release.

[8]Figure 23 shows the effect of comparable changes in the speed, angle, and height of release on the horizontal range, for three different speeds of release. The pattern shown in Fig. 18 is clearly evident in all three instances.

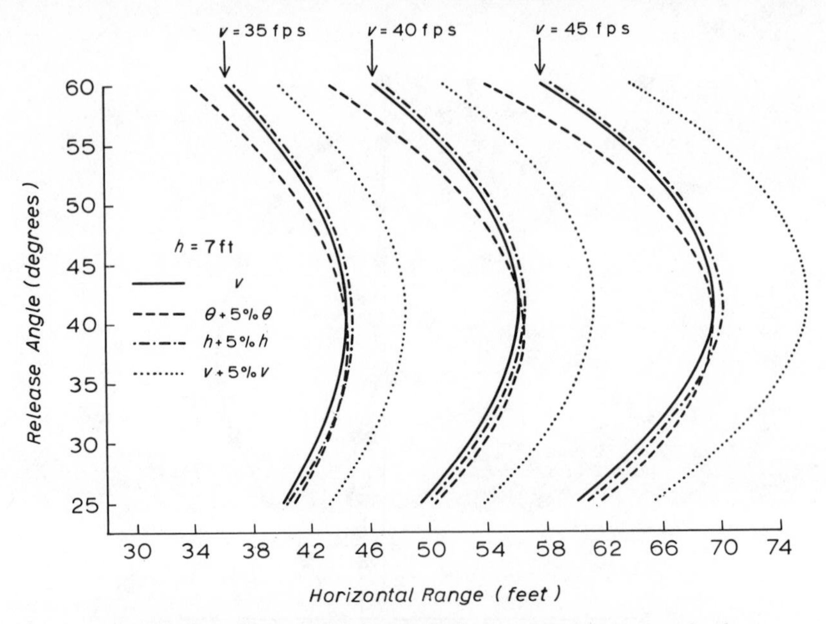

Fig. 23. The effects of 5% changes in the speed, height, and angle of release of a shot, on the total distance the shot is thrown

Trajectory. The path that the projectile follows in its passage through the air (the *trajectory* of the projectile) is also of some interest. Consider again the soccer ball that is kicked upfield from a goal-kick, and let it be supposed that a speed of 64 fps is imparted to the ball at an angle of 30° to the horizontal. The horizontal and vertical displacements at the end of each successive $\frac{1}{10}$ sec of the flight can readily be obtained using Eq. 11 and 9, respectively. Values for these displacements are shown in Table 2.

If these displacements are represented graphically, the form of the graph is a smooth, symmetrical curve (Fig. 24). If, too, the same process is followed for a series of release speeds and angles, the resulting curves are all of this same general form (Fig. 25). These curves are members of a special class of curves known in geometry as *parabolas.* And, since each curve is no more than a map or representation of the path followed by the body concerned, it can be seen that the flight path of a projectile is parabolic in form.

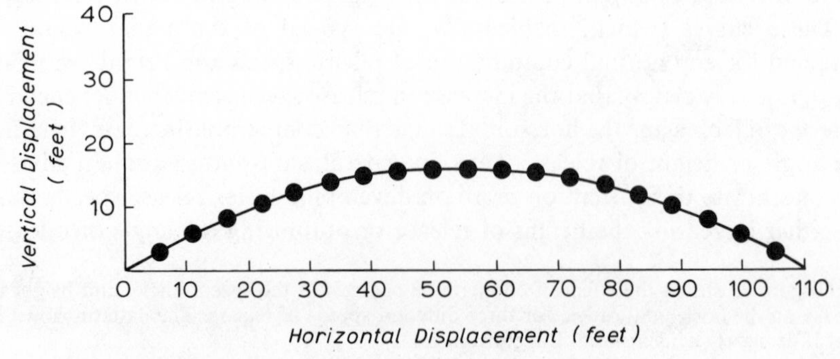

Fig. 24. Trajectory of a soccer ball following a goal-kick

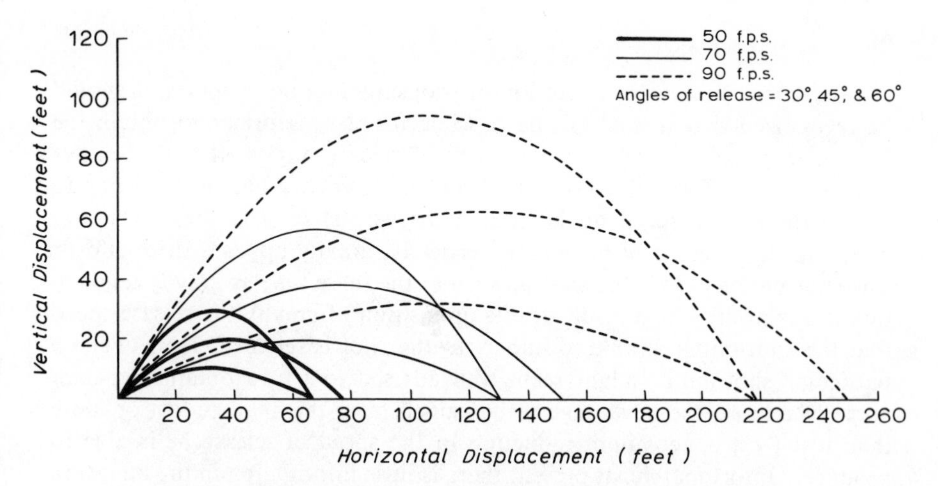

Fig. 25. Parabolic paths followed by projectiles released at a variety of speeds and angles

*Table 2. Horizontal and Vertical Displacements at End of Each 0.1 sec of Flight**

Time from Instant of Release (sec)	*Horizontal Displacement* (ft)	*Vertical Displacement* (ft)
0.0	0.00	0.00
0.1	5.54	3.04
0.2	11.08	5.76
0.3	16.63	8.15
0.4	22.17	10.22
0.5	27.71	11.98
0.6	33.25	13.40
0.7	38.80	14.51
0.8	44.34	15.30
0.9	49.88	15.76
1.0	55.42	15.90
1.1	60.97	15.72
1.2	66.51	15.22
1.3	72.05	14.39
1.4	77.59	13.24
1.5	83.14	11.78
1.6	88.68	9.98
1.7	94.22	7.87
1.8	99.76	5.44
1.9	105.31	2.68
1.99	110.29	0.00

*Soccer ball kicked from ground level with a speed of 64 fps at an angle of 30° to the horizontal.

Limitation. In any consideration of projectile motion in sports, it should be recognized that it may not be possible for the performer to obtain the same release speeds over a wide range of release angles. If a shot-putter directs all his efforts in a horizontal direction, gravity has no tendency to reduce the release speed of the shot. However, as soon as the shot-putter begins to direct some of his efforts vertically, gravity opposes these efforts. The shot-putter pushes the shot upward at the same time as gravity tends to pull it downward. As a result of this opposition of gravity the release speed that the shot-putter is able to impart to the shot is reduced. Thus, while a particular shot-putter might seem well advised to try to obtain a greater angle of release, the advantage to be gained from this may be lost or more than lost by a concomitant reduction in the speed of release he is able to produce. Unfortunately, at present there is little information in the literature on this aspect of projectile motion in sports and until relevant experimental evidence becomes available there is little that can be done beyond being aware of the problem.

Chapter 4

ANGULAR KINEMATICS

The basic concepts involved in the description of angular motion (i.e., in angular kinematics) are very closely related to those encountered in the description of linear motion (i.e., in linear kinematics).

Angular Distance and Angular Displacement

When a rotating body moves from one position to another, the *angular distance* through which it moves is equal to the angle between its initial and final positions, *measured following the path taken by the body*. Consider, for example, the multiple-exposure photograph of a gymnast performing a forward giant circle on the horizontal bar in Fig. 26. The first exposure shows the gymnast with his body in a momentary handstand position atop the bar. The sixteenth and final exposure shows him with about 10° yet to travel in order to complete the full counterclockwise circle. The angular distance through which his body rotated between exposures one and sixteen is therefore equal to 350°—i.e., to the angle between his body positions in these first and last exposures measured following the counterclockwise path of the motion.

The *angular displacement* that a rotating body experiences is equal in magnitude to the smaller of the two angles between the body's initial and final positions. In the example of Fig. 26, the angle between the body's initial and final positions is 350° when measured counterclockwise and 10° when measured clockwise. The magnitude of the angular displacement is, therefore, 10°.

The direction of an angular displacement can be stated as clockwise or counterclockwise or, more concisely, as positive or negative. In the latter case, counterclockwise is generally designated as the positive direction. The angular displacement of the gymnast in Fig. 26 can thus be stated as 10° clockwise or as −10°. (*Note:* The stated direction is the one that would need

Fig. 26. The angular distance through which the gymnast moves and the angular displacement he experiences are equal in magnitude only under certain conditions.

to be followed in moving the body from its initial position, through 10°, to its final position. It is not necessarily the direction in which the body actually moved.)

The angular distance through which a body moves and the angular displacement it experiences are equal in magnitude if it rotates through an angle equal to or less than 180° and if the rotation is in one direction. That these two quantities can be equal in magnitude only if the motion is in one direction can readily be appreciated if one considers the actions of a child playing on a swing. Here, although the angle through which the body swings is almost always less than 180°, it is only up to the point at which the direction of motion is first reversed that the angular distance and angular displacement involved are equal in magnitude.

Angular Speed and Angular Velocity

The average angular speed of a body is obtained by dividing the angular distance through which the body moves by the time taken:

$$\bar{\sigma} = \frac{\phi}{t} \tag{22}$$

where $\bar{\sigma}$ = average angular speed and ϕ = angular distance.

The average angular velocity is obtained in similar fashion by dividing the angular displacement by the time taken:

$$\bar{\omega} = \frac{\theta}{t} \tag{23}$$

where $\bar{\omega}$ = average angular velocity and θ = angular displacement, and specifying the direction.

As with angular distances and angular displacements, and for obvious reasons, the magnitudes of the average speed and angular velocity are equal if the body rotates through an angle equal to or less than 180° and if the rotation is in one direction.

Both average and instantaneous values may be determined for angular speed, angular velocity, and angular acceleration (see next section) and, as is the case with their linear counterparts (speed, velocity, and acceleration), it is the instantaneous values that generally yield the most meaningful information.

Angular Acceleration

The angular acceleration, $\bar{\alpha}$, is the rate at which the angular velocity of a body changes with respect to time. In algebraic form,

$$\bar{\alpha} = \frac{\omega_f - \omega_i}{t} \tag{24}$$

where ω_i = the initial angular velocity, and ω_f = the final angular velocity. Taking the gymnast in Fig. 26 as an example, if his angular velocity is 250°/sec at the instant he passes through the horizontal on his downward swing and 400°/sec as he passes below the bar 0.3 sec later, his average angular acceleration is

$$\bar{\alpha} = \frac{400°/\text{sec} - 250°/\text{sec}}{0.3\ \text{sec}}$$

$$= 500°/\text{sec}^2$$

Linear and Angular Kinematics

This chapter began with a reference to the close relationship between the quantities used in the description of linear motion and those used in the description of angular motion. Now that the latter have been defined (thus permitting an appropriate comparison to be made), an examination of Table 3 on page 48 will readily reveal the truth of this statement.

Units in Angular Kinematics

Perhaps surprisingly, no less than three units are commonly used in the measurement of angular distance. Only two of these are used with any fre-

Table 3. Quantities Used in Linear and Angular Kinematics

Linear Kinematics	*Angular Kinematics*
Distance	Angular distance
Displacement	Angular displacement
Speed $= \frac{\text{distance}}{\text{time}}$	Angular speed $= \frac{\text{angular distance}}{\text{time}}$
Velocity $= \frac{\text{displacement}}{\text{time}}$	Angular velocity $= \frac{\text{angular displacement}}{\text{time}}$
Acceleration $= \frac{\text{final velocity} - \text{initial velocity}}{\text{time}}$	Angular acceleration $= \frac{\text{final angular velocity} - \text{initial angular velocity}}{\text{time}}$

quency in sports, but because the derivation and use of at least one important relationship hinges on a knowledge and understanding of the third unit, all three will be considered here.

In diving, it is customary to speak of "full-twisting one-and-a-half's," "inward two-and-a-half's," and other dives described in similar terms. Although direct reference is very rarely made to the unit of angular distance used in such descriptions, this is universally understood to be one *revolution* (rev). Thus in a "double-twisting one-and-a-half" the diver executes two complete turns or revolutions about the long axis of his body (the double twist) and one and one-half revolutions about a horizontal axis parallel to the end of the diving board (the one-and-a-half somersault).

A much smaller unit, the *degree* ($= \frac{1}{360}$ rev), is widely used in everyday life and in certain situations in sports.[1] The loft of a golf club, for example, is normally specified in terms of the number of degrees that the clubface is set back from the vertical.

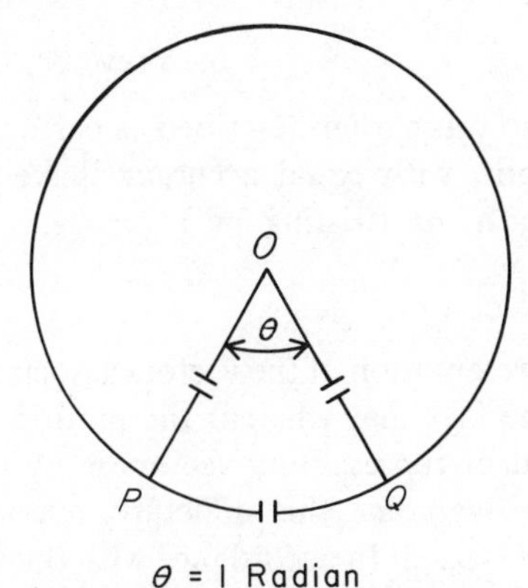

θ = 1 Radian

Fig. 27. A radian is the angle subtended at the center of a circle by an arc equal in length to the radius.

The third unit, the *radian* (rad), is rarely used in connection with sports techniques but is widely used in engineering and other fields. The radian is probably best defined with reference to a figure. Consider the circle in Fig. 27. If the arc PQ is equal in length to the radius of the circle center O, then the angle POQ—formed by the radii joining the ends of the arc to the center of the circle—is equal to 1 rad.

Unfortunately, the radian can be a somewhat mystifying unit unless some basis for comparison with other units is available. A rough approximation of the number of degrees in a radian can be obtained if one considers that

[1] While it is rarely, if ever, used in a practical teaching or coaching situation, an even smaller unit, the *minute* ($\frac{1}{60}$°) is sometimes used for research purposes when a highly precise measurement is required.

since $OP = OQ =$ the arc PQ, the figure OPQ closely approximates an equilateral triangle. And, because each of the angles within an equilateral triangle is equal to 60°, it can be surmised that 1 rad must be approximately equal to 60°—actually slightly less than 60° because PQ is a curved line rather than a straight one and this causes a reduction in the size of the angle POQ.

A much more precise figure for the number of degrees in a radian can be obtained by considering how many times an arc of a length equal to the radius can be divided into the circumference of a circle:

$$\frac{\text{Circumference}}{\text{Radius}} = \frac{2\pi r}{r}$$

$$= 2\pi$$

From this it can be seen that 2π rad must be equal to 360°, or 1 rev. By simple division it can be determined that

$$1 \text{ rad} = 57.3°$$

$$= 0.16 \text{ rev}$$

And so, the diver who was earlier described as performing a "double-twisting one-and-a-half" might, with equal accuracy, have been said to execute a "720 twisting 540" or a "4π twisting 3π"!

Angular-Motion Vectors

The graphical representation of the vectors associated with angular motion is complicated by the fact that whereas the motion of the body is circular, the standard method of representing vectors is by means of a straight line (i.e., an arrow). To overcome this difficulty, a convention known as the *right-hand thumb rule* is used. In accordance with this rule, an angular-motion vector is represented by an arrow drawn so that if the curled fingers of a person's right hand point in the direction of rotation, the direction of the arrow coincides with the direction indicated by the extended thumb (Fig. 28). The magnitude of the vector is represented by the length of the arrow in the usual way.

Any angular-motion vector (such as angular velocity or angular acceleration) can be represented in this way and can be either added to a corresponding vector to obtain a resultant or can be resolved into components, in precisely the same manner as outlined in the previous chapter (Fig. 29).

(*Note:* Although angular displacement has both a magnitude and a direction and can therefore be represented by an arrow, it is not a vector, because angular displacements cannot be summed using the parallelogram of vectors. This can readily be demonstrated by taking a book and turning it through 180° about one edge and then through another 180° about a second edge [Fig. 30(a)]. If these two angular displacements are appropriately represented by arrows and a parallelogram is constructed in the usual manner, the diagonal of that parallelogram [Fig. 30(b)] is in no way representative of the

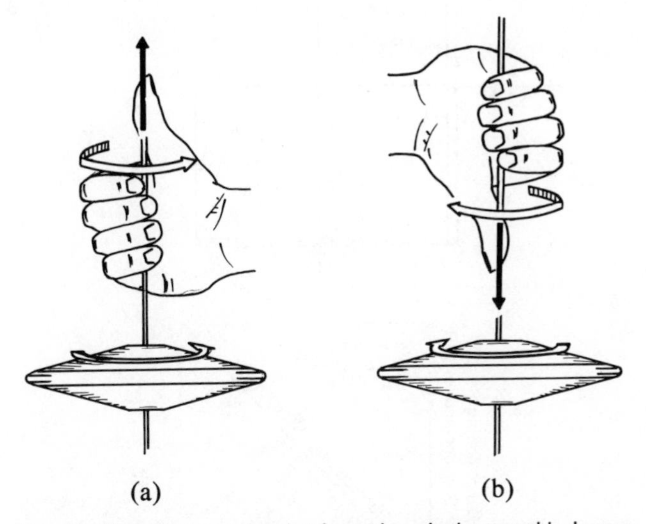

Fig. 28. The right-hand thumb rule used to obtain a graphical representation of the angular velocity of a discus. [*Note:* The discus is shown rotating as it would if thrown away from the observer by (a) a left-handed thrower and (b) a right-handed thrower.]

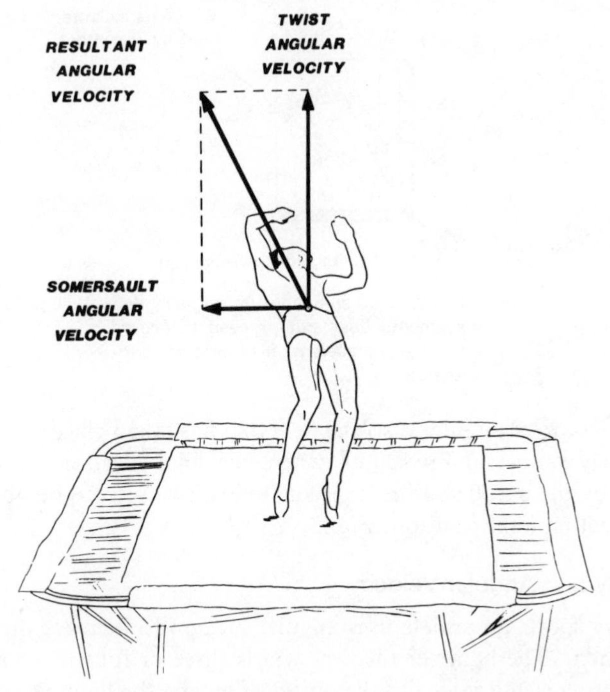

Fig. 29. The parallelogram of vectors is used here to determine the resultant angular velocity of a gymnast about to perform a double-twisting back somersault on the trampoline.

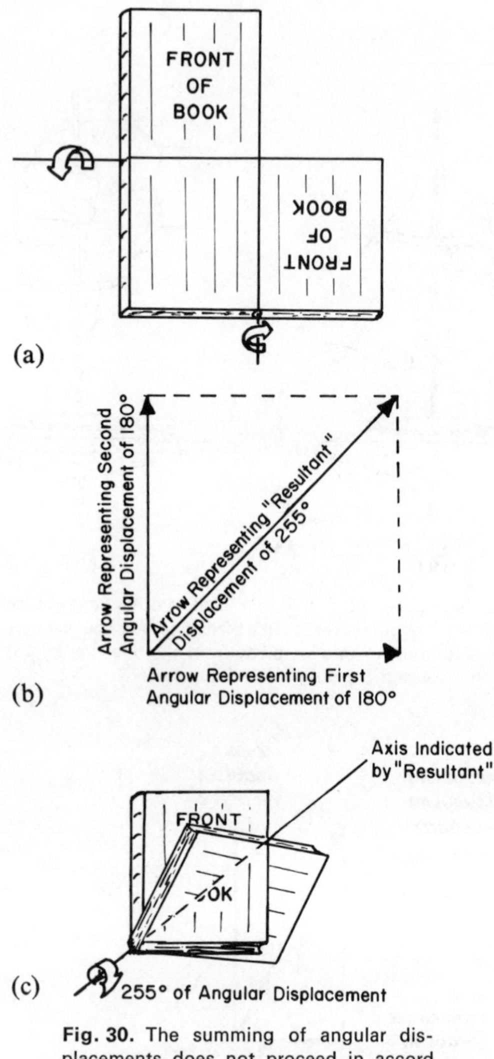

Fig. 30. The summing of angular displacements does not proceed in accord with the parallelogram-of-vectors construction.

"resultant" or total angular displacement that has taken place. This is perhaps most clearly shown in Fig. 30(c), where the final position of the book as indicated by the parallelogram construction can be seen to be quite different from the actual final position, Fig. 30(a)].)

Velocity and Angular Velocity

In many sports an athlete uses angular motion to increase the velocity of an implement. The hammer thrower whirls three or four times in the circle at an ever-increasing rate with the object of having the hammer moving at as great a velocity as possible at the moment he releases it. The golfer does the same kind of thing when he swings his driver in an arc from the limit of his backswing around to the point of contact with the ball. A softball pitcher also seeks to have his fast ball moving at the maximum possible velocity as he releases it near the low point of the angular motion of his arm. In almost all projectile activities in sports the performer relies on angular motion(s) preceding the release to obtain a specified (in the above examples, a maximum) velocity of projection. It is therefore important that the relationship between velocity and angular velocity be clearly understood.

If the golfer of Fig. 31 moves the clubhead from P to Q in some time t, the average speed $\bar{s}$ of the clubhead is given by

Fig. 31. The velocity of the clubhead is equal to the product of its angular velocity and the radius of rotation.

$$\bar{s} = \frac{\text{distance}}{\text{time}} = \frac{\text{arc } PQ}{t}$$

The average angular speed of the club during the same time is given by

$$\bar{\sigma} = \frac{\text{angular distance}}{\text{time}} = \frac{(\text{arc } PQ/r)}{t} = \frac{\text{arc } PQ}{rt}$$

(*Note:* The angular distance is measured in radians; the number of radians involved is found by dividing the length of the arc PQ by the radius r.) If the first of these equations is now substituted into the second, a relationship between the average speed and the average angular speed is obtained:

$$\bar{\sigma} = \frac{\bar{s}}{r}$$

or, rearranging,

$$\bar{s} = \bar{\sigma} r$$

If the distance (or time) over which the average speed and the average angular speed are computed is so small that these latter quantities have no opportunity to change, a similar equation relating the instantaneous speed and angular speed can be derived:

$$s = \sigma r \tag{25}$$

Finally, since the instantaneous speed and the instantaneous velocity of a body are equal in magnitude (p. 17) and the instantaneous angular speed and instantaneous angular velocity are similarly equal in magnitude, Eq. 25 can be rewritten as follows:

$$v_T = \omega r \tag{26}$$

where v_T = the velocity of the clubhead tangential to its path. As already suggested, this relationship between velocity and angular velocity is of considerable importance in many sports activities. In hammer throwing, for example, it summarizes the athlete's whole credo—to obtain the maximum velocity of the hammer (i.e., maximum v_T) by sweeping the hammer through the widest possible radius (maximum r), while turning as fast as can be controlled (maximum ω).

This equation also shows that for a constant angular velocity, the longer the radius, the greater the velocity—a fact that, again, has not been lost on the hammer throwers. Felton,[2] for example, has computed that "a six-inch increase in the hammer's effective radius can produce more than 30–40 feet in distance thrown, provided turning speed and release angle remain constant."

Velocity and Angular Acceleration

The angular acceleration of a body can be considered in terms of components acting along and at right angles to the curved path followed by the body—the *tangential component* and the *radial component*, respectively. Consider the example of the bowler in Fig. 32. During the delivery phase the ball moves vertically downward at one point near the start of the delivery, and then just before release it moves horizontally forward. In between these two points it moves in a series of directions between downward and forward. Now, a change in the direction of motion of a body requires that it be accelerated, just as surely as does a change in its rate of motion. In the case of the bowling ball, the change in direction comes about because the restraining effect of the bowler's arm will not allow the ball to travel along the same line for any two consecutive instants. This restraining effect causes the ball to be accelerated toward the center of the circle along which it is moving (i.e., toward an axis through the shoulder joint). This acceleration is called the *radial acceleration* (a_R), and its magnitude is found using the equation

$$a_R = \frac{v_T^2}{r} \tag{27}$$

[2]Sam Felton, "The Hammer Throw," in *International Track and Field Coaching Encyclopedia*, ed. Fred Wilt and Tom Ecker (West Nyack, N.Y.: Parker Publishing Co., Inc., 1970), p. 345.

where v_T = the velocity of the ball tangential to its path and r = the length of the radius. Thus, if the tangential velocity of the ball at A (Fig. 32) is 21 fps and the distance from the center of the ball to the shoulder joint (i.e., the length of the radius) is 2.5 ft,

$$a_R = \frac{(21 \text{ fps})^2}{2.5 \text{ ft}}$$

$$= 176.4 \text{ fps}^2$$

In the normal bowling action both the magnitude and the direction of the ball's motion change continuously as the ball is swung downward and forward to the point of release. The rate at which the velocity of the ball changes as it moves along its curved path is the *tangential acceleration* and is given by

$$\bar{a}_T = \frac{v_{Tf} - v_{Ti}}{t} \tag{28}$$

where $\bar{a}_T$ = the average tangential acceleration, v_{Ti} = the initial tangential velocity, v_{Tf} = the final tangential velocity, and t = the time during which this change in velocity occurs.

Thus, if the ball is moving with a tangential velocity of 21 fps at A and with a velocity of 22 fps at B, 0.02 sec later

$$\bar{a}_T = \frac{(22 - 21) \text{ fps}}{0.02 \text{ sec}}$$

$$= 50 \text{ fps}^2$$

(*Note:* The tangential acceleration of the ball during its passage from A to B is accompanied by a substantial increase in the radial acceleration—assuming that the length of the radius remains constant, the radial acceleration at B is $22^2/2.5 = 193.6 \text{ fps}^2$.)

Fig. 32. Angular acceleration during the delivery in bowling

Chapter 5

LINEAR KINETICS

Inertia

When a body is lying at rest, it is reluctant to do anything other than to remain at rest. A heavy barbell lying on the floor of a weight-training room shows this reluctance by the resistance it provides when attempts are made to move it. A body in motion is similarly reluctant to change what it is doing, as anyone who has thrust out a bare hand to stop a hard-driven baseball or cricket ball can testify. This characteristic of a body (its reluctance to change whatever it is doing) is known as its *inertia*.

Mass

The quantity of matter of which a body is composed is called its *mass* and is a direct measure of the inertia that the body possesses. Thus, a lightly laden (or less massive) barbell is easier to lift than a heavily laden (or more massive) one. Similarly, it is easier to alter the motion of a running back who has a relatively small mass than it is to effect the same alteration in the motion of a lineman who has a larger mass.

Force

A body's state of being "at rest" or "in motion" can be changed by the action of some other body. The pushing or pulling effect that this other body has and that causes the change is termed a *force*. Thus, a body at rest can be made to move when some other body exerts a force on it. Similarly, a body in motion can be slowed, speeded, or have the direction of its motion altered if another body exerts a force on it.

However, while the introduction of a force can produce or alter motion, not all forces are sufficiently large to have this effect. Consider again the heavy barbell lying on the floor of the weight-training room. Two weight lifters take turns at attempting to lift it. The less strong of the two struggles hard but ultimately fails to get the barbell off the floor. By his efforts, however, he does *tend* to move the barbell, for he gets it closer to that point where it would move upward than it would have been if he'd left it alone. The stronger weight lifter succeeds in lifting the barbell from the floor, thereby changing it from a state in which it is lying at rest to one in which it is in motion. Thus, although force was exerted on the barbell in an upward direction in both cases, the results obtained were somewhat different. These two different results provide a basis for a more formal and complete definition of force than that already given—force is that which alters or tends to alter a body's state of rest or of uniform motion in a straight line.

Internal and External Forces

When a snooker player hits the cue ball into the tightly packed triangle of reds at the start of a game, the cue ball exerts a force on each of the balls with which it makes contact. In turn, these balls exert forces on those they contact. If it is said that the 16 bodies involved (the 15 red balls and the cue ball) comprise a system, it is normal to refer to the forces that they exert on one another as *internal forces*—that is, forces that are internal to the system. When the scattering balls make contact with the padded cushions surrounding the table and exert forces on them, these are termed *external forces*, because the bodies involved (the cushions and the snooker balls) are not all within the system. If one of the reds or the cue ball hit, say, the pink or the blue ball, the forces exerted would similarly be termed external forces.

It should be fairly obvious that whether forces are regarded as internal or external depends entirely on how the system is defined at the outset. If all 22 of the snooker balls are regarded as bodies within the system, the forces that they exert on one another are rightly classified as internal forces. If the scope of the system is further enlarged to include the whole snooker table as well, then the forces between the cushions and the snooker balls are also internal forces.

It can thus be seen that the classification of forces into internal and external is purely a matter of convenience. In biomechanics it is generally regarded as convenient to consider the constituent parts of the human body as "the system" and any force exerted by one part on another as an internal force. For example, whenever the contraction of a muscle causes forces to be exerted on the bones to which it is attached or on the cartilage within a joint or on the ligaments surrounding a joint, these forces are regarded as internal forces. Conversely, the forces due to air resistance, gravity, and contact with the ground or some other body are regarded as external forces.

Newton's First Law of Motion

The ancient Greeks believed that a body moved when there was a force acting on it and that it ceased moving if the force was removed. This belief was rejected by the great Italian scientist Galileo (1564–1642) and subsequently replaced by what has since become known as Newton's first law of motion, first formulated by Sir Isaac Newton (1642–1727). This law may be expressed as follows:

> *Every body continues in its state of rest or motion in a straight line unless it is compelled to change that state by external forces exerted upon it.*

It is perhaps of interest to note that this law, which so succinctly summarizes many of the concepts outlined in the preceding sections, has not been proved directly. Since it is impossible to produce here on earth a situation in which there are no forces acting on a body, it has obviously not been possible to arrive at those conditions necessary to scientifically test the theory. However, this is an academic rather than a practical limitation, for the law has never been shown to be other than consistent with experience.

Newton's Law of Gravitation

In considering sports techniques it is common to think in terms of forces resulting from the contact between one body and another—the tennis racket makes contact with the ball and exerts a force upon it, a basketball strikes the backboard and exerts a force against it, and the wrestler takes a grip on his opponent and exerts forces on him.

In addition to forces that are the direct result of contact between two bodies, there are other forces that exist whether or not the bodies are in contact. These are forces that tend to make bodies gravitate toward each other. The nature of these forces was first described by Newton who, so the well-known story has it, was hit on the head by an apple falling from a tree and after reflecting on this incident formulated what is now known as *Newton's law of gravitation:*

> *Any two particles of matter attract one another with a force directly proportional to the product of their masses and inversely proportional to the square of the distance between them.*

Expressed algebraically, the law reduces to

$$F \propto \frac{m_1 m_2}{l^2} \tag{29}$$

where $F =$ the force acting on each particle, m_1 and $m_2 =$ their respective masses, and $l =$ the distance between them.

In sports the total of all the attractive forces that the particles of one body exert on the particles of any other body is generally so small that its effect is imperceptible. Thus, although each of the balls on a billiards or snooker table exerts on each of the others a force "directly proportional to the product of their masses and inversely proportional to the square of the distance between them," these forces are so small that they can be disregarded.

The one body whose effect on others cannot be disregarded, and the one that makes a consideration of Newton's law of gravitation of some significance in the analysis of sports techniques, is the earth. The attraction that the earth has for all other bodies is known as *gravity* and, as indicated by Eq. 29, varies directly with the mass of the body involved and inversely with its distance from the earth (strictly speaking, with its distance from the center of the earth). Thus, if all else is equal, a massive heavyweight weight lifter experiences a much greater attractive (or gravitational) force than does a much less massive jockey. On the other hand, a skydiver leaping out of an aircraft at 10,000 ft is subjected to a lesser gravitational attraction than if he merely leapt from a 10-ft high diving board. The difference in the forces in the latter case is fairly small, because even though values of d are squared in arriving at the magnitudes of the forces, the addition of 10,000 ft to the radius of the earth—a radius of approximately 3959 mi (or 20,903,520 ft)—makes very little difference.

The effects that differences in d have on sports performances are nonetheless the subject of press comment before most Olympic Games. This interest on the part of the press stems from the fact that because the earth has the appearance of having been pushed in or flattened at the poles, some parts of the earth's surface are farther from the earth's center than others. Competitors in a shot-put competition at the equator, for instance, are approximately 13 mi further from the earth's center than they would be if they competed at either of the poles. This means that the gravitational force that pulls the shot down to the ground is slightly less and thus more favorable at the equator than at the poles.

Momentum

Every body in motion—from a track sprinter to a long-distance swimmer to a bowling ball rolling down a lane—has a certain mass and a certain velocity, and the product of these two is known as the *momentum*, or quantity of motion, that the body possesses.

The momentum of a body is generally of little importance in sports unless that body becomes involved in a collision with another body. Then, the result of the collision hinges very largely on how much momentum each of the bodies had just before the collision took place. The greater the momentum of a body, the more pronounced the effect that it produces on other bodies in its path. If, for example, two bowlers use identical techniques and each releases the ball at precisely the same velocity, the bowler who is using the ball that has the greater mass (and therefore the greater momentum) is more likely to score well than his counterpart with the less massive ball. This comes about because the ball with the greater momentum has the tendency to cause the pins to fly about more dramatically, knocking the other pins down and contributing to a better score than does the ball with less momentum.

A difference in momentum may also result from a difference in the velocity at which a body moves. In softball or baseball, for example, the batter controls the momentum the bat has at the instant of contact with the ball, by controlling its velocity. If he wants to hit a home run, he tries to have the bat moving at a very high velocity as it strikes the ball. Conversely, if he wants to bunt, he tries to have the bat moving at a very low velocity.

Newton's Second Law of Motion

A little experimentation with a putter and a golf ball quickly reveals that if the ball is lying on a flat, level green and is struck by the putter, it will move off in the direction in which it has been struck—or, to be more precise, in the direction of the force that has been applied to it. One does not have to be especially perceptive to note also that the harder the ball is hit (i.e., the greater the force that is applied to it), the faster it will move off across the green. Equivalent observations might be made in many similar situations—e.g., in passing in soccer or basketball, and in hitting in squash, hockey, or volleyball.

What may not be quite so obvious, but is true nonetheless, is that when the body to which a force is applied is already moving, the same two things occur—it moves in the direction in which the force acts, and its change in speed in that direction is related to the size of the force. The truth of this statement is often not obvious because while the body does move in the direction in which the force acts, it may also retain some motion in another direction. The defensive basketball player in Fig. 33 provides an example. Imagine that a pass is thrown in such a way that he can just get his outstretched hand to it. If he exerts a force on the ball in the direction OA, he will cause the ball to move in that direction. The ball will, however, retain some motion in the direction in which it was traveling originally—the direction OB. If the arrows OX and OY represent the velocity vectors due to the force exerted by the defensive man and to the original motion of the ball, respectively, the ball will be deflected in the direction of their resultant OR.

Newton summarized these various effects in precise scientific fashion when he formulated his *second law of motion*. This law may be stated as follows:

> *The rate of change of momentum of a body (or the acceleration for a body of constant mass) is proportional to the force causing it and takes place in the direction in which the force acts.*

Expressed in algebraic form,

$$F \propto \frac{m_1 v_f - m_2 v_i}{t}$$

or, for a body of constant mass,

$$F \propto m \frac{v_f - v_i}{t}$$

$$\propto ma$$

Now in mathematics a statement of this kind can be changed from an "is proportional to" to an "is equal to" statement by multiplying one side by a constant. Thus, if k is the symbol used for the constant,

$$F = kma \tag{30}$$

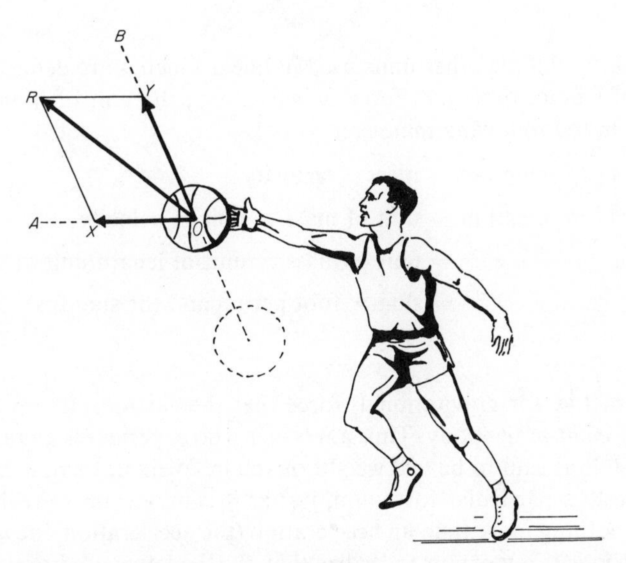

Fig. 33. A deflected pass in basketball—the applied force accelerates the ball in the direction in which the force acts.

Units in Linear Kinetics

Mass. The unit of mass is the *slug*, a term that derives from the words *sluggish* or *sluggishness* and thus implies that characteristic resistance to change to which reference has already been made.

Force. It is normal practice to define units of force in terms of the accelerations they produce. The unit of force in the American, or gravitational, system of units—the system used in this text—is the *pound* (lb) and is defined as the force that will produce an acceleration of 1 fps^2 in a body of 1 slug mass.

It might also be noted that this statement can be rearranged to provide a definition of the unit of mass; viz., a mass of 1 slug is the mass that, when acted on by a force of 1 lb, will have an acceleration of 1 fps^2.

It is pertinent now to reconsider the statement of Newton's second law as it appears in Eq. 30. If the values discussed in the preceding paragraphs are substituted into this equation, one arrives at

$$F = kma$$

$$1 \text{ lb} = k \times 1 \text{ slug} \times 1 \text{ fps}^2$$

i.e.,

$$1 = k \times 1 \times 1$$

from which it can be seen that the value of k is 1 and that Eq. 30 can be reduced to the well-known form

$$F = ma \tag{31}$$

Other units. All the other units used in linear kinetics are defined in terms of those of length, time, and force or mass; e.g., the unit of momentum is arrived at in the following manner:

$$\begin{aligned} \text{Momentum} &= \text{mass} \times \text{velocity} \\ \text{Unit of momentum} &= \text{unit of mass} \times \text{unit of velocity} \\ &= \text{unit of mass} \times \text{unit of length/unit of time} \\ &= \text{slug} \times \text{foot per second (or slug-fps)} \end{aligned}$$

Weight

The attractive (or gravitational) force that the earth exerts on a body is called the *weight* of the body. Thus a wrestler who experiences a gravitational force of 160 lb is said to have a weight of 160 lb. Weight, then, is merely the name given to a particular force and, just as it is important to realize that g stands for nothing more than an acceleration (the acceleration due to gravity, p. 22), so too is it important to realize that W (the letter used to designate a body's weight) represents nothing more than a particular force.

Newton's law of gravitation (pp. 58–60) indicates that the force of attraction the earth exerts on a body, the force just now defined as the weight of the body, varies slightly depending on its location. For example, a wrestler who weighs 225 lb in Nairobi, Kenya, would weigh 1 lb more than this in Helsinki, Finland. The mass of a body, a quantity often misunderstood and confused with its weight, differs in this respect. Whereas the weight of a body changes according to where it is located, its mass remains constant irrespective of location. When it is recalled that the mass of a body was described (p. 56) as the quantity of matter of which it is composed, the truth of the last statement is perhaps easier to accept, for it would seem logical to expect that the amount of matter in a body would not be altered simply because the body had been shifted from one place to another.

Although mass and weight differ in this way, there is a clear-cut relationship between the two quantities. Consider a trampolinist at the peak of his flight. The earth exerts a downward force W on him, and as a result he is accelerated back toward the bed with an acceleration g. The relationship between his weight, his mass, and the acceleration due to gravity is evident from Newton's second law. Substituting in Eq. 31,

$$F = ma$$

$$W = mg \tag{32}$$

or, if this is rearranged,

$$m = \frac{W}{g}$$

Thus, a trampolinist who weighs 161 lb can be seen to have a mass of 5 slugs:

$$m = \frac{161}{32.2}$$

$$= 5 \text{ slugs}$$

or one of mass 4 slugs to weigh 128.8 lb:

$$W = 4 \times 32.2$$

$$= 128.8 \text{ lb}$$

Newton's Third Law

When an athlete runs, he pushes down and back against the ground, thereby exerting a force against it in that direction. The athlete himself goes up and forward as a result of this driving action. It can easily be reasoned from Newton's first law that this up-and-forward motion could only result from the athlete having had a force exerted on him in that direction [Fig. 34(a)]. A weight lifter performing a bench press applies force to the barbell to lift it. The barbell in turn "pushes down" on the hands of the weight lifter [Fig. 34(b)]. A basketball player dribbling the ball exerts force on it to push it down toward the floor. As he does so, the ball resists his action slightly and exerts some force against his hand. This he senses as a slight increase in the pressure against his fingers. When the ball strikes the floor, it exerts another force, this time against the floor. After first contacting the floor, the ball slows down and changes the direction in which it is moving. Again, from Newton's first law, it is clear that this could happen only if some body had exerted a force on the ball [Fig. 34(c)].

An endless number of similar examples could be cited, where one body exerts a force against another and receives in return a force exerted in the opposite direction. This characteristic action of two bodies exerting forces on each other forms the basis of Newton's third law of motion, which is generally stated in the form:

To every action there is an equal and opposite reaction.

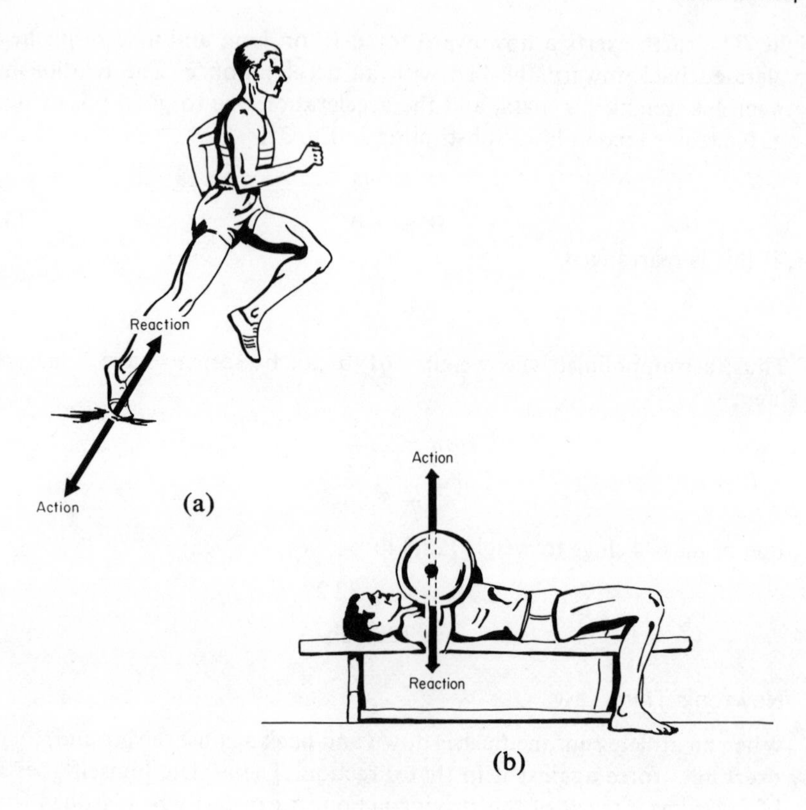

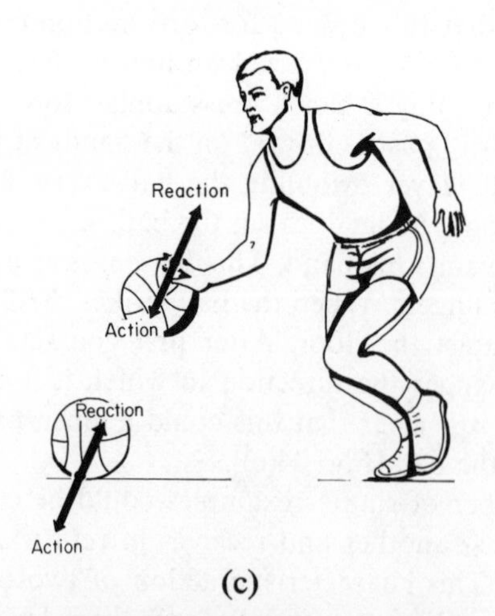

Fig. 34. Examples of Newton's third law (a) in running, (b) in performing a bench press, and (c) in dribbling a basketball

However, because the terms "action" and "reaction" have no precise meanings in biomechanics, and their use can therefore lead to unnecessary confusion, the law above is more meaningfully expressed as:

For every force that is exerted by one body on another there is an equal and opposite force exerted by the second body on the first.

Irrespective of which of the statements above is used, it is customary to call one of the two forces involved the "action" and the other the "reaction" although there are no universally accepted rules regarding which is which.

Newton's third law goes beyond the observations made concerning the examples depicted in Fig. 34, for not only does it refer to opposing forces but it also indicates that these are equal in magnitude. The truth of this latter condition is sometimes difficult for people to accept. The main reason appears to be that the effects that two bodies produce on each other are often quite different. The runner bounds forward and nothing much seems to happen to the earth—the other body involved; the barbell is pressed to full arm's length and the body appears unaffected, and so on. These seeming contradictions can all be explained in terms of Newton's second law, which indicates that for a constant force the acceleration that a body experiences is inversely proportional to its mass:

$$F = ma$$

$$\therefore \quad a = \frac{F}{m}$$

and

$$a \propto \frac{1}{m}$$

In other words, the larger the mass, the less the acceleration. Now if both the earth and the runner have a force of the same size exerted against them, it is reasonable to expect that the effect would be more obvious in the case of the runner than in the case of the earth. The earth in fact would seem to be unaffected—but only because its colossal mass, compared with that of the runner, would not make apparent the acceleration it experienced.

Exactly the same thing is true in the other cases mentioned. In the case of the weight lifter his body position is such that, at least with regard to vertical forces, he is firmly anchored or fixed to the earth. This means, in effect, that the two bodies interacting with each other are the barbell and the lifter-plus-earth, and again the effect on the first is more apparent than that on the second. This process of adding the mass of a large body to the mass of a smaller one is very important in many sports and probably none more so than rifle shooting. When a rifle is fired, equal and opposite forces are exerted against the bullet and the rifle. Because the mass of the bullet is small, it acquires a high velocity as a result of the force exerted against it.

The rifle, on the other hand, acquires a lesser velocity in keeping with its greater mass. This lesser velocity is still sufficient to send the rifle traveling backward fast enough to deliver a painful blow to the shoulder of the marksman foolish, or inexperienced, enough to hold it incorrectly. To avoid having this happen, the good marksman effectively increases the mass of his rifle by holding it firmly into his shoulder, thus making the two bodies involved the bullet and the rifle-plus-man. If the marksman shoots from a prone-lying position, the mass is again increased to be rifle-plus-man-plus-earth, and the acceleration that this "body" experiences is small indeed.

Difficulties often arise in understanding how this third law of Newton's is applied in specific situations. Consider, for example, the athlete in Fig. 35(a). This man is doing some heavy resistance training aimed at strengthening his legs. The question frequently asked in such situations is, "If the rope pulls on the man with a force exactly equal and directly opposite to that with which the man pulls on the rope, how can he ever move forward?" The flaw in the thinking that leads to this kind of question is that due consideration has not been taken of all the forces that act on each body. The athlete has four external forces exerted upon him:

1. the pulling force P along the line of the rope,
2. his body weight W,
3. a ground-reaction force, R [cf. Fig. 34(a)], and
4. an air-resistance force, A.

These forces are shown in Fig. 35(b). What happens to the athlete—whether he moves forward or struggles without avail—depends only on the resultant of these forces. If the horizontal components of A and P combine to be equal in magnitude to the horizontal component of R, the athlete is unable to move forward. (His weight, of course, has no horizontal component and thus has no direct influence on motion in a horizontal direction.) If, on the other hand, the combined horizontal components of A and P are greater than the horizontal component of P, the athlete moves forward. The motion of the tray that the athlete is striving to pull along can be predicted in a similar fashion.

[*Note:* In deciding how a body will move under specific conditions, it is usually helpful to start by drawing what is known as a *free-body diagram.* This is a diagram like Fig. 35(b), in which the body of interest is depicted completely removed or free from its environment and in which *all* the external forces acting on the body are represented in appropriate vector form. The components of the external forces that act in a given direction are then examined carefully in order to determine how the body will move in that direction or, if the exact magnitude of the forces is not known, to determine what conditions are necessary to get the body moving as required.)

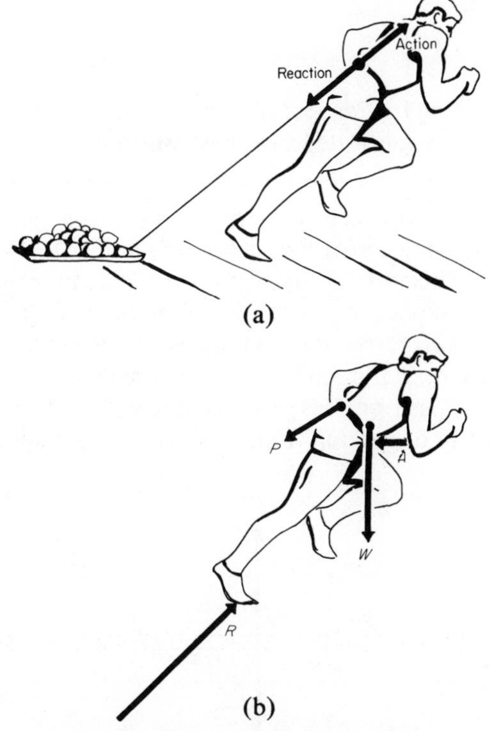

Fig. 35. The action and the reaction act on different bodies and thus do not forestall motion by canceling each other out.

Statements such as "in long jumping, a forceful stamp of the takeoff foot accentuates the upward motion" and "in the high jump, stamp the takeoff foot hard, so that the push up will be as forceful as possible (Newton's third law)" appear from time to time in the literature and also result from a misunderstanding of Newton's third law. The error stems from the fact that the ground reaction to a foot stamp (and, in truth, the reaction to any force that is ever applied) occurs at the same instant that the force is applied. Therefore, any such strong ground reaction would occur far too early in the sequence of movements at takeoff to be anything less than an embarrassment to the jumper. (From a physiological standpoint, the wisdom of stamping the foot and jarring the leg in this way would also seem open to question.)

Friction

The force of attraction that the earth exerts on a body is called the body's weight. A number of other forces are similarly given special names. One of these is the force that arises whenever one body moves or tends to move across the surface of another. This force, which always opposes the motion or impending motion, is called *friction.*

Because there are some quite distinct differences in the nature of the friction that arises under varying circumstances, two types of friction will be considered here: sliding friction and rolling friction. A third type, the friction that is present in fully lubricated bearings, will not be considered because of its complexity and its very limited application to sports techniques.

Sliding Friction. Friction acts only when a body is in motion or has some tendency to start moving across the surface of another body. A barbell disc lying on the floor is acted upon by two forces: one, its weight *W*, and the other *R*, an upward supporting force exerted by the floor [Fig. 36(a)]. Under the action of these two forces the disc has no tendency to slide across the floor and thus there is no friction acting to oppose this tendency. If a weight lifter gives the disc a push with his foot, the disc will tend to slide. Only then will friction (F) act in opposition to this tendency [Fig. 36(b)].

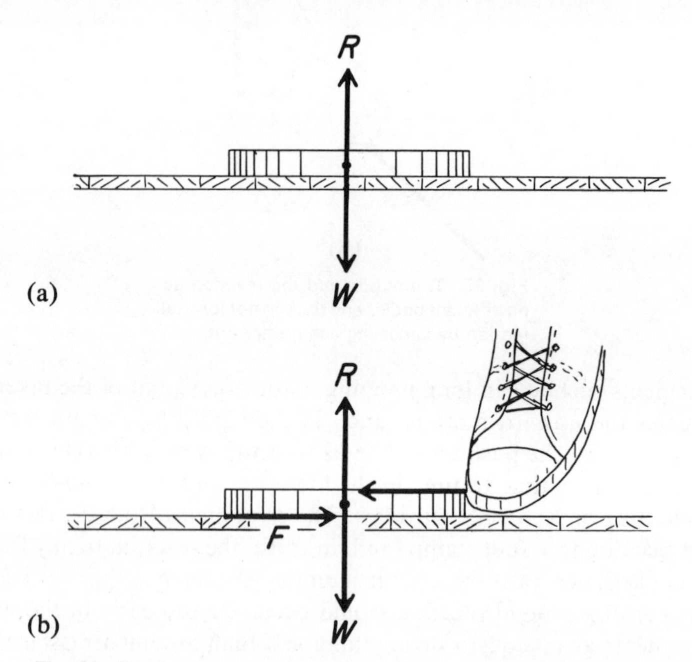

Fig. 36. Friction acts to oppose forces that cause, or tend to cause, the sliding of one body over another.

Another important characteristic of friction is that, until sliding actually commences, the magnitude of the friction is exactly equal to that of the force tending to cause the body to slide. In other words, up to that point friction effectively cancels out the force tending to cause the body to slide, and no sliding takes place. Once the friction has reached its upper limit in magnitude (*limiting friction*), sliding is about to commence. A lineman pushing against a blocking sled (Fig. 37) can be used to illustrate this concept. Suppose the magnitude of the limiting friction is 280 lb. If the lineman exerts a horizontal

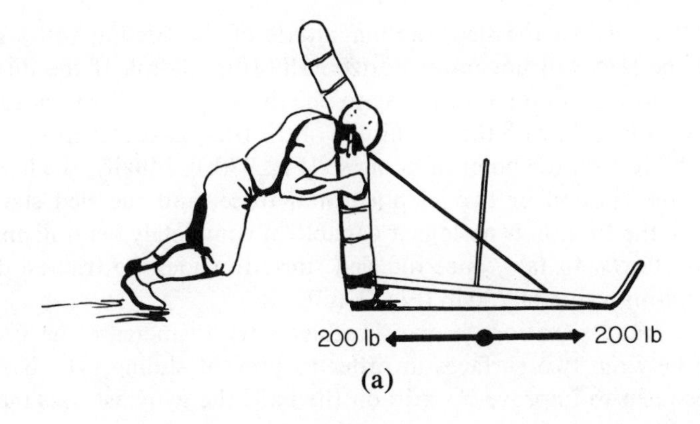

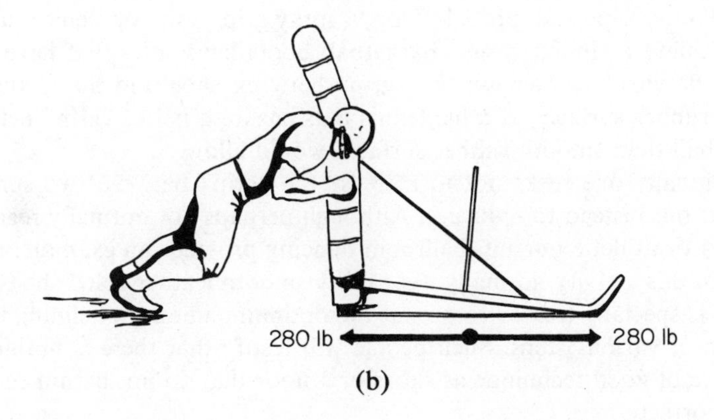

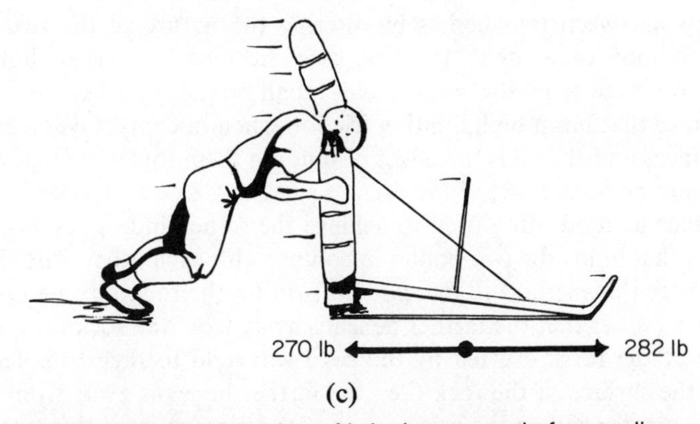

Fig. 37. Until sliding is imminent, friction increases as the force tending to cause the body to slide increases. Once the latter force exceeds the maximum friction (limiting friction), the body begins to slide. At this point, the opposing friction decreases in magnitude.

force of 200 lb against the sled, the magnitude of the friction will also be 200 lb and the sled will not move horizontally [Fig. 37(a)]. If the lineman increases the horizontal force he exerts to 280 lb, the sled still will not move horizontally, although, with the magnitude of the friction stretched as it were "to the limit," it is on the point of doing so [Fig. 37(b)]. Finally, the lineman musters another pound or two of horizontal force, and the sled starts to slide, because the friction is no logner capable of completely neutralizing the effect of his efforts. In fact, once the sled starts to slide, the friction drops below its limiting value of 280 lb [Fig. 37(c)].

There are many situations in which athletes try to increase the friction (or "grip") between two surfaces in order to prevent sliding. The baseball pitcher uses resin to improve his grip on the ball; the gymnast uses magnesium chalk to improve his grip on the apparatus; and the pole-vaulter uses sticky adhesive tape and, probably too, a spray grip, resin, or Venice turpentine to achieve a similar result. Basketball boots have soles that have been specially designed to increase the "grip" between shoe and floor, and the pimpled rubber surfaces of table tennis bats ensure a better "grip" between bat and ball than smooth rubber surfaces would allow.

Occasionally one seeks not to increase the "grip" between two surfaces or objects but instead to reduce it. Although perhaps not normally regarded as a sport or athletic pursuit, ballroom dancing provides an example. Competitors in this activity normally wear fairly smooth leather-soled shoes, and the floor is specially treated to permit the optimum amount of sliding in the execution of various steps. Such people will testify that there is nothing as destructive of good technique as shoes or a floor that do not permit the feet to slide correctly.

In all the examples above, the performer attempts to increase or decrease the "grip" between two bodies by altering the nature of the two surfaces that bear upon each other. In some cases he modifies one or both of the surfaces involved (e.g., the soles of basketball boots); in others he interposes a substance that has a high limiting friction when in contact with each of the two surfaces and thus has the effect of binding them together (e.g., resin and magnesium powder.)

Another method often used to achieve the same kind of result is to alter the force that holds the two bodies in contact with each other. The climber in Fig. 38 uses this method to increase his grip on the rock face he is climbing down. He knows that the farther he leans away from the rock face, the more the line of the force exerted by the rope will tend to thrust his feet firmly against the surface of the rock (i.e., the farther he leans away from the rock face, the greater will be the component of the force exerted by the rope that will be acting at right angles to the face). Thus, the rope helps to hold the two surfaces (soles of boots and rock) together and thereby to reduce the tendency for the boots to slip. (Incidentally, as the climber leans progressively farther away from the rock face, the magnitude of the force exerted by the rope also increases. The force holding his feet against the surface of the rock is thus enhanced in two ways—by changes in both the magnitude and the direction of the force exerted via the rope.)

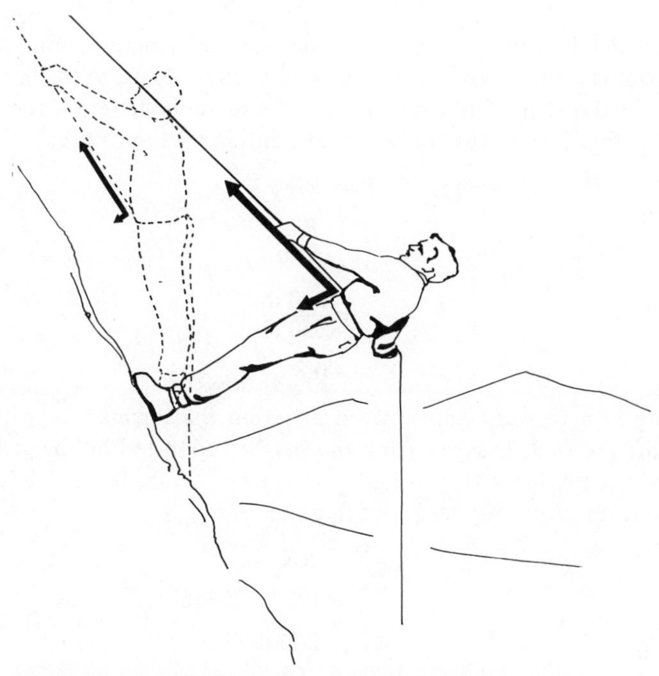

Fig. 38. The rock climber increases the limiting friction between his feet and the rock by leaning well away from the rock face.

The friction that the lineman has to overcome in order to start the blocking sled moving (Fig. 37) can be increased or decreased in similar fashion. He can reduce the force holding the two surfaces (sled and ground) together by driving forward and upward instead of directly forward. This, he soon learns, reduces the friction he is fighting to overcome. His coach, however, has a simple answer to this—he can increase the friction by standing on the sled and forcing the sliding surfaces more firmly together.

From these examples it is apparent that there are two ways in which the friction between bodies can be modified: (1) by altering the nature of the bearing surfaces and (2) by changing the forces that hold these surfaces together. These "findings" are summarized by two very similar statements.

The first, sometimes referred to as the *first law of friction*, states that

> *For two dry surfaces, the limiting friction is equal to the normal reaction*[1] *multiplied by a constant, the value of this constant depending only on the nature of the surfaces.*

That is,

$$F = \mu R \tag{33}$$

where $F =$ the limiting friction, $R =$ the normal reaction, and $\mu =$ the constant known as the *coefficient of limiting friction*. Thus, in the case of the lineman and the sled, if the weight of the sled (and therefore the normal reaction) is, say, 350 lb, the coefficient of limiting friction is 0.8:

$$\begin{aligned} F &= \mu R \\ \mu &= \frac{F}{R} \\ &= \frac{280 \text{ lb}}{350 \text{ lb}} \\ &= 0.8 \end{aligned}$$

By pushing in a forward and upward direction the lineman can, in effect, partially lift the sled, thus reducing the weight supported by the ground. If in doing this he reduces the normal reaction by 100 lb, he also reduces the limiting friction—but only by 0.8 of that amount:

$$\begin{aligned} F &= \mu R \\ &= 0.8 \times 250 \text{ lb} \\ &= 200 \text{ lb} \\ \text{Reduction in } F &= (280 - 200) \text{ lb} \\ &= 80 \text{ lb} \end{aligned}$$

(*Note:* Use of this example presupposes that it is reasonable to regard the base of the sled and the ground as "two dry surfaces.")

One important feature of Eq. 33 is that it shows the limiting friction is independent of the area of contact between the two bodies. Thus, if all else were equal, a man walking down a steep slope would have the same tendency to slip regardless of whether his shoe size was 5 or 15!

The second statement is practically identical to the first and concerns the

[1]When a body lies or moves on the surface of another, its weight (or a component of it) acts on this second body in a direction at right angles to the surface of contact. The second body exerts an equal force on the first body, in the opposite direction (the *normal* direction). This force is called the *normal reaction* (Fig. 39). External force may be used to add to or subtract from the weight component. If this is done, the normal reaction is similarly affected.

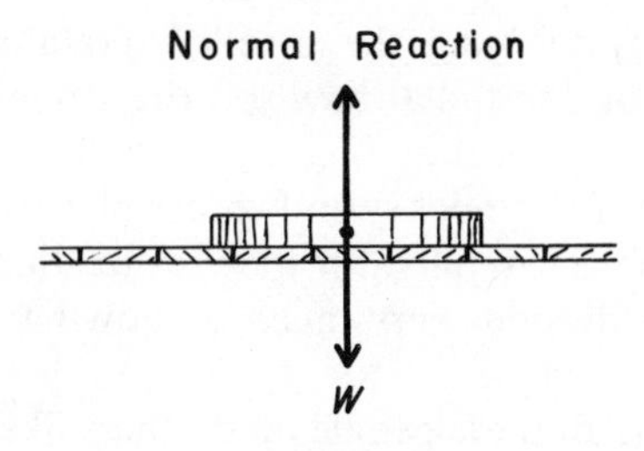

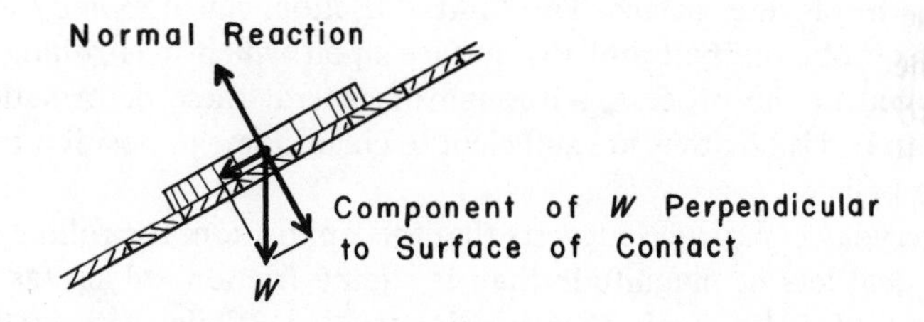

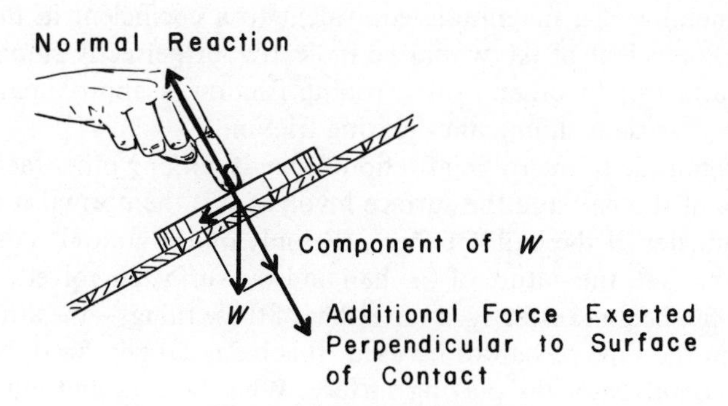

Fig. 39. The normal reaction

magnitude of the friction when a body is actually sliding. This is given by the equation

$$F_s = \mu_s R \tag{34}$$

where F_s = the sliding friction, R = the normal reaction, as before, and μ_s = the *coefficient of sliding friction.* In any given case, the value of this latter coefficient is less than the value for the coefficient of limiting friction. This is in accord with the everyday experience that it is easier to keep a body sliding than it is to start it sliding in the first place.

Rolling Friction. The experienced golfer carefully studies the approach, the path that his ball will follow, before making his putt. He looks closely at the length of the grass and the way it is lying (the so-called "grain"). He looks, too, at whether the grass is wet or dry and at how soft the ground is, for he knows that all of these things have a bearing on how easily his ball will roll across the green toward the hole. Field hockey and soccer players also study the condition of the playing surface carefully, for they realize that how well the ball rolls across this surface depends very much on how hard, smooth, and dry it is.

Whether they realize it or not, all these people (and others like them who are concerned in some way with a ball rolling across a surface), are actually considering the friction that will oppose the motion of the ball when it rolls across the playing surface. This kind of friction, called *rolling friction*, occurs because both the ball and the surface upon which it is rolling are slightly deformed in the process. Although in general these deformations are too small to be visible, they are sufficient to create some opposition to the motion of the ball.

Everyday experience suggests that this opposition, the rolling friction, is a good deal less in magnitude than is sliding friction, which (as has already been noted) is less again than limiting friction. While coefficients of limiting and sliding friction are normally within a range from 0.1 to 1.0, rolling friction is generally of a magnitude equivalent to a coefficient in the order of 0.001. (A coefficient of 0.0 would be indicative of perfectly smooth or frictionless surfaces.) In other words, rolling friction is approximately 100 to 1000 times less than sliding and limiting friction.

The magnitude of the rolling friction depends, among other factors, on (1) the nature of the ball and the surface involved, (2) the normal reaction, and (3) the diameter of the ball. Of these, the only one commonly considered in sports is the first, the nature of the ball and the surface involved. In general, very little can be done directly about either of these things—the athlete having little say in the type of ball to be used (this being largely fixed by the rules) or in the condition of the playing surface. What he can, and must, do if he wishes to achieve the best of which he is capable, is to check the prevailing conditions and adjust his game accordingly. If the playing surface is heavily grassed and is soft and wet, the rolling friction will be relatively high and the player will have to exert more force than usual to offset the effect of this high rolling friction; e.g., if he is a golfer and is putting, he should endeavor to stroke the ball more firmly, under such conditions. If, on the other hand, the surface is hard and fast, rolling friction will be relatively low and the forces required less than usual. An alternative to modifying the forces exerted is to change completely the technique that is used; e.g., a field hockey player, involved in a match on a soft ground, may want to use aerial (or "flick") passes in preference to drives that might be stopped prematurely in the "holding" conditions.

Impulse

The shot-putter in Fig. 40 is performing a standing throw from a specially designed platform that measures the forces he exerts against it. These forces are detected by sensitive strain gauges built into the platform and are then recorded on a long sheet of graph paper that is moving at a fixed speed past the recording pens. Figure 40 shows a record of the horizontal forces that are exerted in the line of the throw—positive forces are those acting in the direction of the throw and negative forces are those acting in the opposite direction. From this record it can be clearly seen that the magnitude of the force is changing continually during the throw and that the direction of the force also changes. This continually changing horizontal force may be thought of as having a certain magnitude for a very small period of time, then a slightly different magnitude for the next small period of time, and so on. If one of these small time periods is considered, the product of the force (F) and the small time during which the force acts (t) is defined as the *impulse* of the force:

$$\text{Impulse} = F \times t \tag{35}$$

This product of a constant force and the time during which it acts is also equal to the narrow rectangular area under the force-time curve for that time interval. The total impulse is the sum of all the infinite number of such smaller

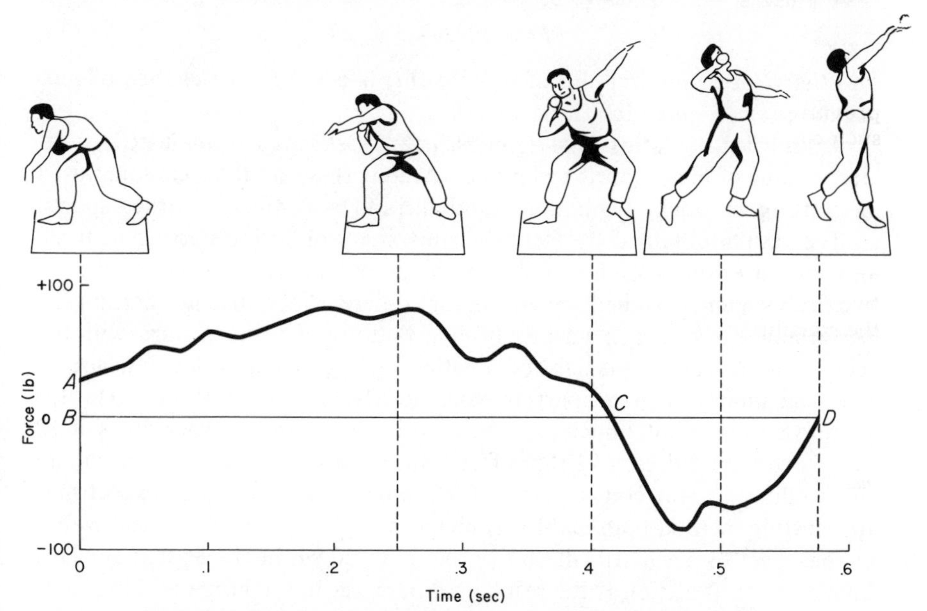

Fig. 40. The horizontal impulse exerted against the ground during the execution of a standing shot put

impulses and can be shown mathematically to be equal to the total area under the force-time curve. Thus in Fig. 40 the area bounded by the force-time curve and the lines *CB* and *BA* represents the total impulse (65.33 lb-sec) in the positive direction, and that bounded by the curve and the line *CD* represents the total impulse (24 lb-sec) in the negative direction. The algebraic sum of these two values is the total impulse:

$$\text{Total impulse} = (65.33 - 24) \text{ lb-sec}$$
$$= 41.33 \text{ lb-sec}$$

A useful relationship involving impulse can be obtained by rearranging the algebraic statement of Newton's second law:

$$F = ma$$

Since a, the acceleration, has previously been equated with $(v_f - v_i)/t$ (Eq. 3), this expression can be substituted for a in the previous equation:

$$F = \frac{m(v_f - v_i)}{t}$$

or

$$F = \frac{mv_f - mv_i}{t}$$

Rearranging this expression yields

$$Ft = mv_f - mv_i \tag{36}$$

In other words, the impulse of a force (Ft) is equal to the change of momentum ($mv_f - mv_i$) that it produces.

A knowledge of this *impulse-momentum relationship* is basic to an understanding of many sports techniques. Among these are the techniques used in starting in track, swimming, football, and a number of other sports.

The results obtained by Henry[2] in his study of sprint starting in track provide an example of how the impulse-momentum relationship applies in such cases. Henry studied the effects that different foot spacings had on the performance of a crouch start and found that use of a bunch start (with the feet 11 in. apart) got his subjects off the starting blocks faster than use of either medium (feet 16 in. apart) or elongated (feet 21 in. apart) starts. He also found that use of the bunch start resulted in significantly slower times at 10 and 50 yd than did use of either of the other two starts. At first glance these two findings appear contradictory, for it would seem logical to expect that the starting method that enabled runners to clear their blocks fastest would enable them to get to 10 yd, and probably to 50 yd, in the least time. This apparent contradiction is the result of differences in the horizontal impulses exerted against the blocks. When the subjects used the bunch start, the

[2]Franklin M. Henry, "Force-Time Characteristics of the Sprint Start," *Research Quarterly*, XXIII, October 1952, pp. 301–18.

horizontal impulses they exerted against the blocks were limited by the relatively short time in which they were in contact with the blocks and thus in a position to exert horizontal forces against them. This in turn limited the horizontal velocities of the subjects as they left the blocks. (*Note:* Because the initial horizontal momentum is zero and the mass of the subject is constant in any given case, the horizontal velocity on leaving the blocks is directly proportional to the horizontal impulse exerted against the blocks—cf. Eq. 36.) An explanation for Henry's seemingly odd results is now apparent—although the subjects cleared their blocks soonest when using the bunch start, the slight time advantage they had was soon offset because their horizontal velocities as they left the blocks were relatively small.

Conservation of Momentum

When a bowling ball strikes a pin, the force exerted by the ball on the pin is exactly equal and directly opposite to that exerted by the pin on the ball (Newton's third law). The time during which these forces act is also exactly the same—either the two bodies are in contact or they're not, and each will exert force on the other only when they are in contact. Since the impulse is the product of the force and the time, the impulse each body receives is exactly equal in magnitude and opposite in direction to that which the other body experiences. Furthermore, according to the impulse-momentum relationship, the respective changes in momentum of the two bodies must also be equal and opposite. Thus, because the momentum lost by the ball is equal and opposite to that gained by the pin, the total momentum of the system (ball-plus-pin) is unaltered by the impact. These ideas are summarized in the *principle of conservation of momentum*:

> *In any system of bodies that exert forces on each other, the total momentum in any direction remains constant unless some external force acts on the system in that direction.*

In the example of the bowling ball and pin, as in most other examples in sports, the total momentum is only approximately constant, because the external forces of friction and air resistance are acting. However, since the magnitude of the external forces is so small, it would be reasonable to expect the approximation to be a fairly close one in this instance.

Impact

There is a large group of sports in which one body collides (or impacts) with another and in which the success of a participant depends very largely on his ability to predict the outcome of such impacts. In squash, handball, and paddleball, players are continually called upon to predict where the ball will go following an impact with a wall, the floor, and even the ceiling (in the latter two games) and to position themselves ready for their next shot in accordance with this prediction. If they misjudge the outcome of the impact, they are very likely to find themselves in a position from which it is difficult, perhaps impossible, to make a suitable return shot. Assuming a player does position himself correctly, his next task is to play the ball in such a manner as to obtain the best results. To do this, he must know how the ball will react to the various ways in which he might hit it and then choose the way that is most appropriate to the situation.

Tennis and table tennis players have very similar problems to those of their squash, handball, and paddleball counterparts, for, in addition to predicting the outcome of an impact between the ball and some part of the playing court, each of these people also faces the problem of what will happen to the ball once it strikes a racket or bat. In a somewhat different way, golfers, hockey players, and batters in softball or baseball also face this kind of problem, as do athletes kicking in football, heading in soccer, or passing and spiking in volleyball. In view of the prominent role of the impact situation in so many sports, it is important to consider the factors that influence the outcome when two bodies collide.

Elasticity. When a ball hits a fixed surface, both the ball and the surface are slightly compressed. Then, since most bodies tend to return to their original shape after they've been slightly deformed, the ball rebounds from the surface as both bodies strive to restore themselves to their former shape. The same sequence of compression and restitution takes place when two moving bodies (e.g., a bat and a ball) impact with each other. The property of a body that causes it to endeavor to regain its original shape once it has been deformed is called its *elasticity*, a property possessed by most of the bodies involved in impacts in sport.

Coefficient of Restitution. The tendency of a body to return to its normal shape once it has been deformed (i.e., its elasticity) differs from one body to another. Some return very quickly to their original shape, while others do so much less quickly. Because there is no way of directly calculating the elasticity of a body, it is necessary to rely on the results of experiments to help predict the outcome of any given impact.

Sir Isaac Newton investigated the properties of elastic bodies and the results of impacts between them and formulated the following empirical law (Newton's law of impact):

> *If two bodies move toward each other along the same straight line, the difference between their velocities immediately after impact bears a constant relationship to the difference between their velocities at the moment of impact.*[3]

[3]If the two bodies are not traveling along the same straight line before impact, their component velocities along a line perpendicular to the surface of contact obey this law.

In algebraic terms,

$$v_1 - v_2 = -e\,(u_1 - u_2)$$

or

$$\frac{v_1 - v_2}{u_1 - u_2} = -e \qquad (37)$$

where v_1 and v_2 = the velocities immediately after impact of bodies 1 and 2, respectively; u_1 and u_2 = their respective velocities immediately before impact; and e = a constant known as the *coefficient of restitution.*

Because this law indicates that how two bodies move after impact depends on how they were moving before impact and on a coefficient e, it is important to consider those factors on which the value of e depends. Probably the easiest way of doing this is to examine what happens in a very simple impact situation when various conditions are modified. Consider a ball that is dropped onto a fixed surface (e.g., the floor). If the ball is called body no. 1 and the floor is body no. 2, the velocities of body no. 2 before and after the impact are both zero, for all practical purposes; i.e., $u_2 = v_2 = 0$. Equation 37 then reduces to

$$\frac{v_1}{u_1} = -e \qquad (38)$$

Because velocities are somewhat more difficult to measure than are distances, it is desirable to convert this form of the law into yet another form. This is done by considering the ball's motion before and after impact with the floor and by using the appropriate equation of uniformly accelerated motion (Eq. 10) to arrive at expressions for v_1 and u_1. From Fig. 41 it can be seen that

$$u_1 = \sqrt{2gh_d} \qquad (39)$$

and it can similarly be shown that

$$v_1 = \sqrt{2gh_b} \qquad (40)$$

where h_d and h_b are, respectively, the height from which the ball was dropped and the height to which it subsequently bounced. If now these expressions are substituted in Eq. 38, this latter becomes

$$e = \sqrt{\frac{h_b}{h_d}} \qquad (41)$$

[*Note:* Because the velocity of the ball after impact is in the negative direction (Fig. 38), the negative value of the square root is taken in Eq. 40.]

The law expressed in this form suggests immediately a way in which the factors that influence the value of e can be examined. If a ball is dropped from a known height and the height to which it bounces is noted, the value of e for that situation can easily be computed. If slight changes are made in the conditions (e.g., using a different ball or a different landing surface), the effects that these changes have on the value of e can be noted. The results of

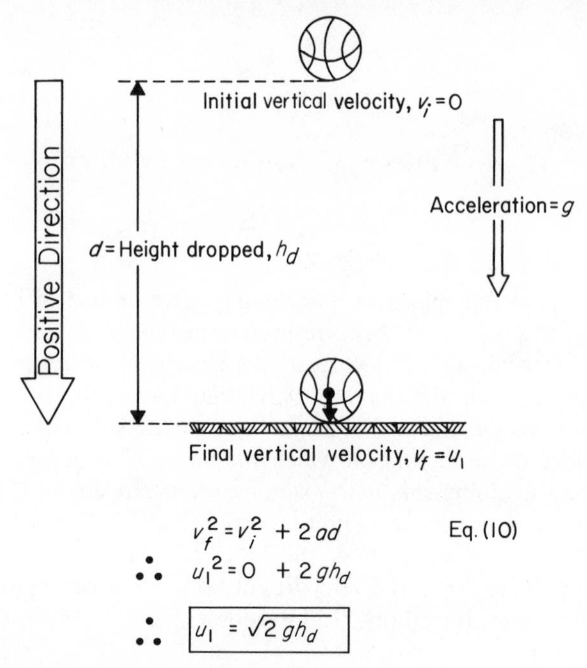

Fig. 41. Deriving an expression for the velocity at impact when a ball is dropped from a known height

two such experiments are shown in Tables 4 and 5, and from these results it is quite evident that the nature of the two impacting bodies determines to a very large extent what value *e* takes. Also, it is clear that it would be incorrect to refer to "the coefficient of restitution of a body," for this coefficient depends not just on one of the impacting bodies but on both of them.

Table 4. The Coefficient of Restitution for Balls Dropped from a Height of 72 in. onto a Hardwood Floor

Type of Ball	*Height Bounced* (in.)	*Coefficient of Restitution**
"Super ball"	56.75	0.89
Basketball	41.75	0.76
Soccer	41.50	0.76
Volleyball	39.75	0.74
Tennis—well-worn	36.00	0.71
—new	32.00	0.67
Lacrosse	27.50	0.62
Field hockey	18.25	0.50
Softball	7.25	0.32
Cricket	7.00	0.31

*Values for the coefficient of restitution vary from 0.0 when the impact is said to be inelastic, because the bodies do not separate after the impact, to a theoretical and never-attained limit of 1.0.

Table 5. The Coefficient of Restitution for a Volleyball Dropped from a Height of 72 in. onto Various Surfaces

Type of Surface	*Height Bounced* (in.)	*Coefficient of Restitution*
"Proturf"	41.25	0.76
Wood	40.66	0.75
"Uniturf"	40.41	0.75
Steel plating	40.00	0.74
Concrete	39.50	0.74
Tumbling mat (1 in. thick)	32.66	0.67
Gravel	26.33	0.60
Grass	13.50	0.43
Gymnastic landing mat (8 in. thick)	13.00	0.42

Some years ago in major league baseball there was a brief flare-up between the Chicago White Sox and the Detroit Tigers over charges leveled by the Tigers that their opponents had been artificially cooling the balls used in a five-game series played in Chicago. The White Sox countercharged that the balls used the previous weekend in a four-game series in Detroit had been heated and dried out and that this had led to a 53-run scoring spree quite out of keeping with the total of 17 runs scored in Chicago. Whether the temperature of the balls used in these matches was deliberately or accidentally changed may never be revealed. What is known, though, is that a ball will become more "lively" than usual if it is heated and less so if it is cooled. Baila,[4] a schoolboy who was especially interested in the White Sox-Tigers affair, has demonstrated these effects using the same simple ball-dropping experiment already described here (Table 6). The effect of changes in temperature is also well known to squash players, for whom the pregame warm-up serves to prepare the ball for the game (by getting it warm) as much as it does the players.

Another factor that influences the value of e is the velocity at which the two bodies concerned meet each other. The extent to which this velocity before impact affects the value of e has been well demonstrated by Plagenhoef,[5] who reported values obtained when a number of balls were dropped from a height of 100 in. onto a "firm, wood floor" and when the same balls were "kicked or thrown to obtain velocities between 50 and 60 mph." In each case the value for e was less when the velocity before impact was between

[4]Dorel L. Baila, "Project: Fast Ball—Hot or Cold?" *Science World*, September 16, 1966, pp. 10–11.

[5]Stanley Plagenhoef, *Patterns of Human Motion: A Cinematographic Analysis* (Englewood Cliffs, N.J.: Prentice-Hall, Inc., 1971), pp. 82–83.

*Table 6. The Effect of Temperature Changes on the Coefficient of Restitution**

Type of Ball	Coefficients of Restitution (The height of the bounce when each ball was dropped from a height of 72 in. is shown in parentheses.) Cooled (1 hr in freezer)	Normal	Heated (15 min at 225°)
Baseball	0.50 (18 in.)	0.53 (20 in.)	0.55 (22 in.)
Solid rubber ball	0.57 (23 in.)	0.73 (38 in.)	0.80 (46 in.)
Golf ball	0.67 (32 in.)	0.80 (46 in.)	0.84 (51 in.)
"Super ball"	0.91 (59 in.)	0.91 (60 in.)	0.95 (65 in.)

*Adapted from data in Dorel L. Baila, "Project: Fast Ball—Hot or Cold?," pp. 10–11.

50 and 60 mph than when the ball was dropped from 100 in. (equivalent to a velocity before impact of approximately $\frac{1}{3}$ mph). The differences in the value of e ranged from 0.02 (0.6 to 0.58) for a golf ball to 0.3 (0.8 to 0.5) for a handball.

Direct and Oblique Impact. There are a few examples in sports where two bodies collide with each other directly. That is to say, they are either both moving along the same straight line immediately prior to impact or one of them is at rest and the other is moving along a line at right angles to the surface where contact occurs. In Fig. 42 are shown a number of cases of this so-called *direct impact.*

Far more common in sport than such *direct impacts* are those in which the two bodies do not collide directly or "head-on." This type of collision is called an *oblique impact.* A bounce pass in basketball is an example of an oblique impact, because before contacting the floor the ball travels at some angle to it other than a right angle. In short, the ball approaches the floor from an oblique angle. The same thing is true of a lay-up shot, because the ball is thrown or placed against the backboard at an oblique angle. Most of the shots played in racket games (tennis, squash, etc.) also involve oblique impacts, for only rarely is the racket moved to meet the ball along the same line the ball is traveling. This is because players of these games generally try to hit the ball away from their opponents and this means striking it obliquely so that it will not merely return in the same direction from which it came.

In order to analyze oblique impacts in some detail, it is convenient to consider them under two headings:

Fig. 42. Examples of direct impact: (a) a soccer goalkeeper punching away a lobbed shot at goal, (b) a basketball official testing the bounce of the ball, and (c) the cue striking a ball in billiards or snooker.

1. oblique impact with a fixed surface (e.g., a bounce pass in basketball),
2. oblique impact with a moving body (e.g., as in tennis when the racket makes an oblique impact with the ball).

Oblique Impact with a Fixed Surface. When a ball (or, for that matter, any other body) strikes the floor (or any other fixed surface), it exerts forces on it. These forces can be resolved into components that act along the surface and components that act at right angles to the surface. If, however, both the ball and the surface were smooth (i.e., their coefficient of limiting friction was the impossible 0.0), the ball would not be able to exert force along the surface of the floor because it would not be able to "grip" the floor and "push" it in that direction. The floor, similarly, would not be able to exert any force against the ball in the opposite direction. There would be, in fact, a total absence of friction between the two bodies. In this kind of theoretical situation, the only forces that the ball could exert on the floor (and, of course, that the floor could exert on the ball in reaction) would be those that acted at right angles to the surface. In order to understand what happens when a body makes an oblique impact with a fixed surface (and, in the next section, with a moving body), it is convenient to consider first the theoretical case where there can be no frictional forces acting and thus where only forces at right angles to the surface are involved. In practice, and especially where a ball has had spin imparted to it, the effect of friction (i.e., of the forces acting along the surface) can be quite pronounced indeed. These effects, which can be simply added to those produced by the forces acting at right angles to the surface, will be referred to later.

Figure 43 shows a squash ball bouncing on the floor in the course of a rally. The velocity it has at the instant it makes contact with the floor of the court is indicated by the vector u and its horizontal and vertical components by u_H and u_V, respectively. Now, because the ball and the floor are imagined to be perfectly smooth and there are thus no horizontal forces acting on either body, there are no forces that can alter the horizontal motion of the ball. The horizontal velocity must therefore be the same after impact as it was before, i.e., $v_H = u_H$. The same is not true, however, of the vertical velocity, for both its direction and magnitude are changed markedly as a result of the impact. In the first place, the force that the floor exerts on the ball causes the ball to completely reverse the direction of its vertical motion. Then, in addition, the elasticity of each of the bodies concerned (ball and floor) modifies the magnitude of the vertical velocity in accordance with Eq. 38:

$$\frac{v_V}{u_V} = -e$$

or

$$v_V = -eu_V$$

where v_V equals the vertical velocity of the ball immediately after impact.

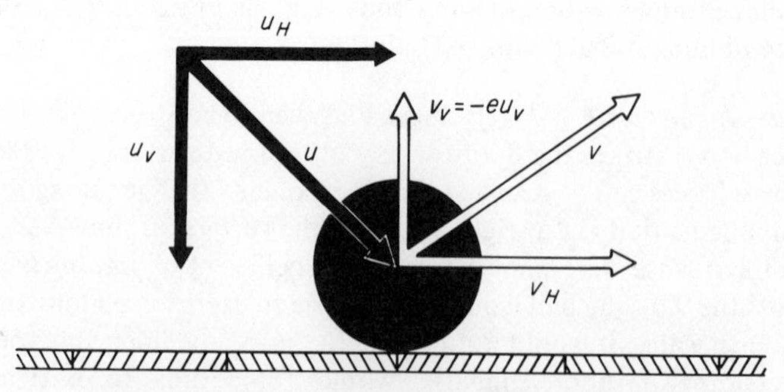

Fig. 43. Velocity changes during an oblique impact

The important thing about this relationship is that since e is always less than one, the right-hand side always has a value less than u_V. (*Note:* The negative sign on the right-hand side merely indicates that the direction of the motion has been reversed.) In other words, v_V, the vertical velocity after impact, is always less in magnitude than u_V, the vertical velocity before impact. How much less one is than the other depends obviously on the value of e. In the case of squash balls the value of e, found by dropping a ball onto a wooden floor from a predetermined height, is of the order of 0.6. Thus the vertical velocity after impact is approximately six-tenths of the vertical velocity before impact.

By combining the horizontal and vertical velocities, the resultant velocity after impact can be determined and its direction compared with that of the resultant velocity before impact. For this latter purpose it is customary to define these two directions in terms of the angle each makes with a line perpendicular to the surface at the point of contact (i.e., the common normal). The angle that the direction of the velocity before impact makes with this perpendicular is called the *angle of incidence*, and the angle that the direction of the velocity after impact makes with it is called the *angle of reflection* (Fig. 44).

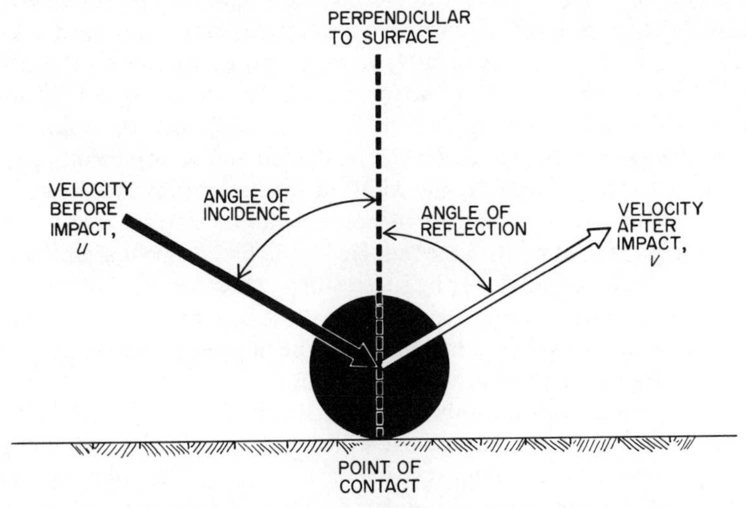

Fig. 44. The angles of incidence and reflection in an oblique impact

Because the horizontal velocities before and after impact are equal in magnitude, and the vertical velocity after impact is less in magnitude than that before impact, the angle of reflection is greater than the angle of incidence. It is important to note that this last statement applies only to oblique impact on a fixed surface, when the effects of friction are negligible or not considered. When frictional effects are taken into account, the relationship between the angles of incidence and reflection is modified in a manner dependent entirely on the situation. Under such circumstances no blanket statement can be made concerning how these two angles compare. Certainly the often-made statement that "the angle of incidence is equal to the angle of reflection" is utterly wrong except

1. in that very rare situation where the effect of friction on the horizontal velocity exactly balances that of elasticity on the vertical velocity, and
2. when there is a direct impact between two bodies and there is no friction involved.

In this latter case, which has very limited practical significance indeed, the angle of incidence (0°) exactly equals the angle of reflection (0°).

Possibly one of the best examples of how a knowledge of the relationship between angles of incidence and reflection influences success can be seen in the game of snooker. For instance, consider the situation represented in Fig. 45. Here the player must hit the white cue ball *W* so that the first ball it hits is the red ball *R*. Directly in the path between these two is a ball *C* of some other color. In order to successfully execute the shot, the player clearly has to "bank" the cue ball off the cushion so that it passes around the ball *C* and hits the red. To do this successfully, he must aim at a point on the cushion such that the angle of incidence so formed will lead to an angle of reflection consistent with the cue ball hitting the red. If he misjudges the point of contact, the cue ball will pass to either side of the red and he will incur a penalty. Fortunately for him, however, the width of the ball means that he can have the cue ball hit any of a number of points within a narrow range and still make subsequent contact with the red. He has, in fact, some margin for error. But all this is quite negative, for by successfully executing the shot and having the cue ball hit the red, the player merely avoids being penalized. To make a more positive contribution to his total score he must endeavor to put the red into one of the six pockets around the table. If he decides to try to put the ball into the center pocket on the left-hand side in Fig. 45, the cue ball must contact the red in such a way that it will then follow the appropriate path. This means that the cue ball must strike the red at a precise point (or, at best, within a very narrow range of points), and this in turn reduces drastically the range of acceptable points of contact of the cue ball with the cushion. Clearly, then, the success of the snooker player hinges first on his ability to select the correct point of aim from his knowledge of the angles involved and then, of course, on his ability to execute the shot in a manner consistent with his intention.

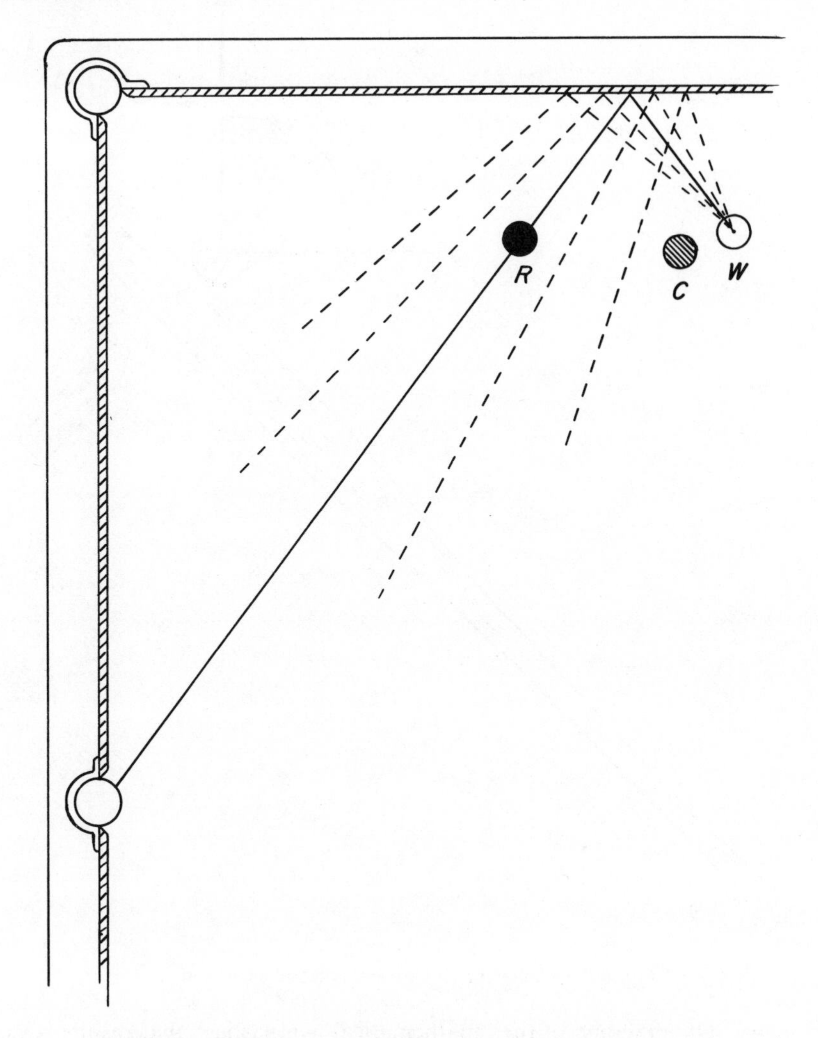

Fig. 45. A player's success in snooker depends largely upon his ability to select an appropriate angle of incidence.

Oblique Impact with Moving Bodies. Undoubtedly the most involved cases of oblique impact between two bodies are those in which both bodies are moving or are free to move. All the cases of hitting, kicking, and heading come within this category.

In order to determine what happens in these cases it is necessary to use the equation that defines the coefficient of restitution (Eq. 37), another that expresses the conservation-of-momentum principle, and some basic trigo-

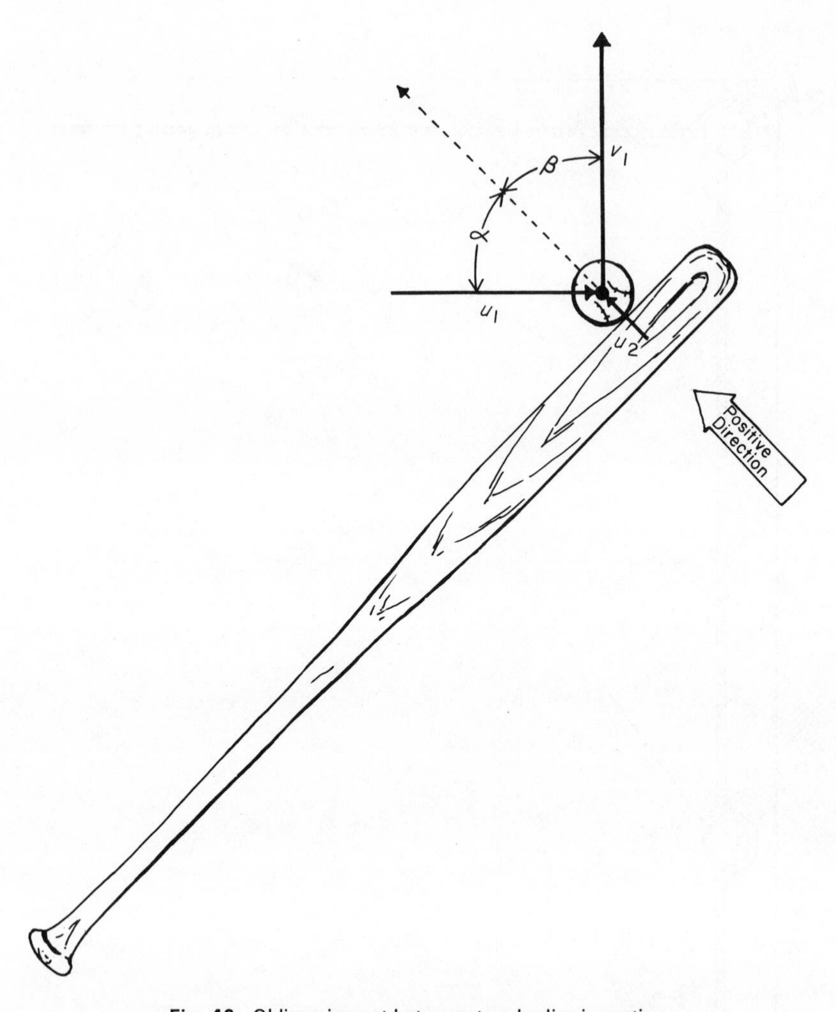

Fig. 46. Oblique impact between two bodies in motion

nometry. The outcome of the "mathematical gymnastics" that results is two rather involved equations for the speed and direction of a body after such impact. In Fig. 46 a baseball bat is shown at the moment it makes an oblique impact with a ball. The velocity vectors u_1 and u_2 represent the velocity of the ball and the bat, respectively. If the acute angle between these two vectors is α, the respective masses of the two bodies are m_1 and m_2, and their mutual coefficient of restitution is e, the speed and direction of the ball after impact are given by the following lengthy expressions:

$$v_1 = \sqrt{\left[\frac{m_2u_2(1+e) + u_1 \cos\alpha(em_2 - m_1)}{m_1 + m_2}\right]^2 + (u_1 \sin\alpha)^2} \tag{42}$$

$$\beta = \arctan\left[\frac{(u_1 \sin\alpha)(m_1 + m_2)}{m_2u_2(1+e) + u_1 \cos\alpha\,(em_2 - m_1)}\right] \tag{43}$$

[*Note:* For the sake of simplicity it has been assumed (1) that there is no friction involved when the two bodies impact and (2) that the velocity of the part of the bat that makes contact with the ball, i.e., u_2, acts in the direction of the common normal.]

Equations 42 and 43 are especially helpful if one is interested in carefully analyzing what happens in a particular case. Suppose, for example, one wants to analyze what happened in the case of the baseball bat striking a ball. First a set of values is chosen for each of the six variables involved. Then five of the variables are kept constant while the value of the sixth is changed. The speed and direction of the ball after impact are calculated for each of a series of values of this sixth variable. This, in effect, amounts to saying, "If all else is equal, what happens to the speed and direction of the ball after impact, as such-and-such a value increases or decreases?" This process is repeated until each of the six variables has been varied while the other five remained constant. A summary of the sort of results that can be obtained from such an analysis is presented in Table 7. The initial values chosen in this example were

$$m_1 = 0.010 \text{ slug}$$
$$m_2 = 0.058 \text{ slug}$$
$$u_1 = 120 \text{ fps}$$
$$u_2 = 50 \text{ fps}$$
$$e = 0.5$$
$$\alpha = 30°$$

The results in Table 7 (p. 90) show in unmistakable terms what happens under these circumstances. For instance, they show that if all else is equal, the speed of the ball after impact can be increased by

1. increasing the mass of the bat,
2. decreasing the mass of the ball,
3. increasing the initial velocity of the bat,
4. increasing the initial velocity of the ball,
5. increasing the angle of incidence,or
6. increasing the value of the coefficient of restitution.

These results have clear-cut implications for hitters who wish to improve their ability to hit the ball "out of the park." They indicate that to achieve this objective a hitter should use a more massive bat, pick a ball that is moving at a high velocity, and swing the bat harder so that it has a greater velocity at the moment of impact. (*Note:* It is assumed that the mass of the ball and the coefficient of restitution are both effectively beyond the control of the batter and thus of little practical significance. In addition, the idea of

Table 7. Speed and Angle of Reflection of a Baseball Following Oblique Impact with a Bat

Quantity Varied			*Speed of Ball after Impact* (fps)	*Angle of Reflection* (degrees)
Mass of bat (slugs)	0.039	(20)	100.63	36.60
(weight in ounces	0.049	(25)	106.91	34.14
in parentheses)	0.058	(30)	111.50	32.56
	0.068	(35)	115.00	31.45
	0.078	(40)	117.74	30.64
Mass of ball (slugs)	0.006	(3)	121.78	29.52
(weight in ounces	0.008	(4)	116.45	31.01
in parentheses)	0.010	(5)	111.50	32.56
	0.012	(6)	106.91	34.14
	0.014	(7)	102.65	35.77
Velocity of bat (fps)	30		90.89	41.31
	40		100.90	36.49
	50		111.50	32.56
	60		122.53	29.32
	70		133.89	26.62
Velocity of ball (fps)	60		84.63	20.76
	90		97.56	27.47
	120		111.50	32.56
	150		126.13	36.49
	180		141.22	39.59
Angle of	0		98.57	0.00
incidence (degrees)	10		100.24	12.00
	20		104.87	23.04
	30		111.50	32.56
	40		118.95	40.42
	50		126.11	46.80
	60		132.03	51.92
Coefficient of	0.3		90.38	41.59
restitution	0.4		100.63	36.60
	0.5		111.50	32.56
	0.6		122.83	29.24
	0.7		134.49	26.50

increasing the speed of the ball after impact by delaying the swing and thus increasing the angle of incidence would appear to have relatively little to recommend it. For it seems very likely indeed that if the bat were brought into contact with the ball early in the swing, it would only be at the expense of a considerable reduction in the velocity of the bat. And this, of course, would tend to negate any advantage to be derived from increasing the angle of incidence. Furthermore, even in the unlikely event that the velocity were not reduced, a concomitant increase in angle of reflection would also increase the possibility that the ball would end its flight in foul territory.)

Spin and Friction. The previous discussion of direct and oblique impact has assumed that no friction has been involved. In point of fact, however, friction plays an important part whenever two bodies collide with each other. Consider a table tennis ball bouncing on a table. When it strikes the table, it has a tendency to keep moving at the same speed and in the same direction as it was immediately before impact (Newton's first law). Because of this tendency the ball exerts a force on the table in that direction. In reaction the table exerts an equal and opposite force on the ball (Newton's third law). The effects produced by the vertical component of that reaction force were considered in a previous section. The horizontal, or frictional, component opposes the horizontal motion of the ball and causes a reduction in its horizontal velocity. In addition to this (and for reasons that will be elaborated on in the next chapter) the friction tends to impart some spin to the ball.

If the ball is spinning at the moment it makes contact with the table, the magnitude of the friction is modified according to the rate and direction of that spin. As a direct consequence, the speed and direction of the ball after impact are also modified. The table tennis player in Fig. 47 is driving the ball back into his opponent's court. To do this he swings his bat in a forward and upward direction across the line of flight of the ball so that when his bat makes contact with the ball, it not only sends it back toward the other end of the table but also imparts a large amount of spin to it. This spin, which results from the back of the ball (i.e., that part in contact with the bat) being lifted upward and forward during the stroke, is known as *topspin*. Viewed from the same position as taken by the artist who drew Fig. 47, a ball with topspin appears to be rotating in a clockwise direction. (It is sometimes not realized that the direction in which a body appears to be rotating depends on the position of the observer. Had the artist taken up his position on the other side of the table, the same topspin drive would have appeared to lead to a counterclockwise rotation of the ball.) When the ball subsequently strikes the table, it has both linear and angular motions—a linear motion along its direction of travel and an angular motion about an axis through the center of the ball.

Fig. 47. A forehand topspin drive in table tennis. The application of topspin to the ball ensures that it will come off the table fast and at a low angle and thus be difficult to return.

Consider the part of the ball that makes contact with the surface of the table. As a result of its linear motion this part of the ball, like all its other parts, has a forward velocity in the horizontal direction. In addition, however, it has a horizontal velocity in the backward direction, due to its angular motion. (With a ball that has had topspin imparted to it, the part of the ball that is lowermost at any instant is traveling directly backward, relative to the center of the ball. At the same time each of the other parts is momentarily traveling in some other direction—Fig. 48.) Clearly, then, it is the resultant of these two velocities that indicates what horizontal velocity that part of the ball has as it strikes the table. This in turn determines in which direction the friction the table exerts on the ball will act and, to a large extent, how the ball will move after impact. If the backward horizontal velocity due to the ball's spin is greater than the forward horizontal velocity due to its linear motion, the friction acts in a forward direction. Under these conditions, friction acts not to decrease the horizontal velocity but rather to increase it. The ball, in fact, comes off the table with a greater horizontal velocity than it possessed when it hit the table. Because the horizontal velocity has been increased (and the vertical motion unaltered) by the spin, the angle of reflection is also greater than it would have been had no spin been imparted to the ball. These effects are well known to table tennis players, who deliberately apply heavy topspin with their offensive strokes so that the ball will come off the table fast and at a low angle and thus be difficult to return.

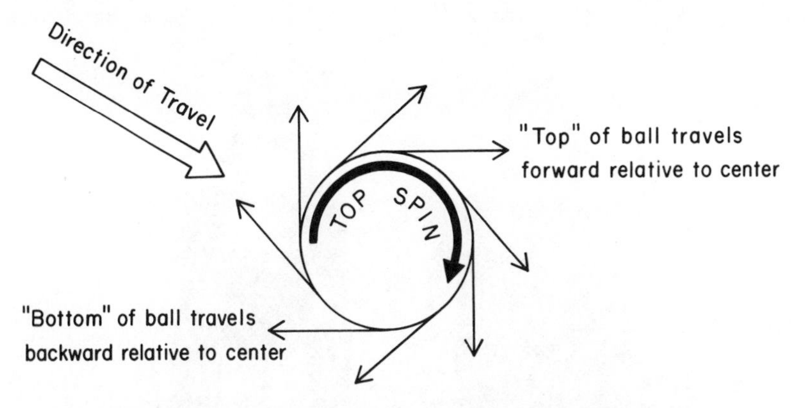

Fig. 48. Points on the surface of a spinning ball move with different velocities.

The analysis above can be similarly applied when the spin of the ball is in other directions. A defensive chop in table tennis is executed by bringing the bat downward and forward across the path of the ball. This imparts a *backspin* (i.e., rotation in exactly the reverse direction to topspin) to the ball, which on landing tends to make the ball slow down (as increased friction opposes its horizontal motion) and to decrease the angle of reflection. Because a ball hit like this travels forward comparatively slowly after impact with the table, this type of backspin stroke is a very effective one from a defensive standpoint, for the ball cannot now be hit back with as great a speed as if it had been traveling faster. (See similar example of hitting in baseball, pp. 87–91). In some sports (notably tennis, table tennis, and cricket) a *sidespin* is imparted to the ball and this, for the very same reasons outlined above, causes the ball to change its motion in a lateral direction as a result of impact with the playing surface. In such cases, the ball is said to "kick" or "break" to one side or the other.

Pressure

When the girl in Fig. 49 stands erect, her body weight of 135 lb is supported on the soles of her feet, an area of some 27 sq in. If she lies down, her body weight is, of course, still 135 lb, but the area over which this weight is supported is increased to 180 sq in. If the force exerted on a body (the ground-reaction force equal to the girl's weight, in this example) is divided by the area involved, the average *pressure* or force per unit area is obtained. Thus, when the girl is standing,

$$\text{Pressure} = \frac{\text{force}}{\text{area}} \tag{44}$$

$$= \frac{135\ \text{lb}}{27\ \text{sq in.}}$$

$$= 5\ \text{lb/sq in.}$$

(usually abbreviated to psi)

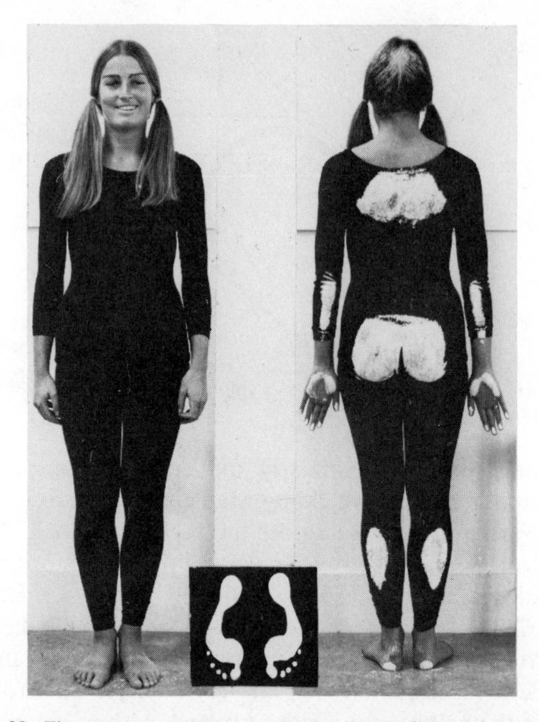

Fig. 49. The average pressure on a supporting surface varies with the area of the surface involved. When the girl stands erect, the soles of her feet (see inset) support her weight and the average pressure is relatively large because the area involved is small. When she lies on her back, the area of the supporting surface (shown in white in the right-hand photograph) is markedly increased and the average pressure is correspondingly reduced.

and when she is lying,

$$\text{Pressure} = \frac{135\ \text{lb}}{180\ \text{sq in.}}$$
$$= 0.75\ \text{psi}$$

This concept of pressure is of particular importance with respect to safety measures in sports, and failure on the part of an athlete to minimize the pressure that any one part of his body must withstand can lead to very serious injuries indeed. In parachuting, performers frequently find it necessary to spread the forces that their bodies must withstand over a large area by initiating a rolling action as soon as they contact the ground. Failure to execute this technique correctly has been known to result in fractures of one or both legs. High jumpers and pole-vaulters similarly spread the forces involved when they land flat on their backs on the soft landing pads now available. In the days when dirt and sand pits were used for these events, the forces exerted on the jumper at landing were so great that if the weaker parts of the body had been called upon to withstand their share, serious injury would certainly have resulted. Landing on the back was thus rarely seen. Instead, athletes in these events generally tried to land on their feet and to cushion the shock as best they could by dropping onto other parts of the body as soon as they landed. Exponents of judo are another group for whom the concept of spreading force over a large area is of fundamental importance. This is readily apparent in the considerable emphasis placed on use of the correct techniques of falling in the early stages of learning the sport and, in fact, at all levels at which the sport is practiced. The same basic concept is used in most of those sports where athletes wear special protective clothing. Such clothing serves to spread the force over a large area in cases where the athlete cannot do this effectively by moving his body. In baseball, for instance, a batter hit on the head by the ball has no way of effectively spreading the force by moving his body. He can wear a helmet, though, which when it is hit transmits the force over a large area via the suspension system inside it. The same thing is the basis for the use of protective helmets in football and motorcycling; for the wearing of gloves in baseball, cricket, and boxing; and for other similar items of equipment in football (various joint pads), baseball (face masks, chest protectors, etc.), and other sports.

Work

Many words used in everyday language take on much more precise meanings when used in science. One such word is *work*. In everyday usage, work is anything in which physical or mental effort is used to achieve some goal. In biomechanics, the term *work* has a much more limited meaning:

> *Whenever a force acts on a body, the work done by the force is equal to the product of its magnitude and the distance that the body moves in the direction of the force, while the force is being applied to it.*

Expressed in algebraic form,

$$W = Fd \tag{45}$$

where W = the work done by the force, F = the magnitude of the force, and d = the appropriate distance.

Consider a weight lifter performing a two-hand snatch, a lift in which a barbell is raised overhead in one uninterrupted movement. If he exerts a constant upward force of 400 lb against the barbell, while lifting it the first 2 ft from the floor, the work done by the upward force during this portion of the lift is

$$\begin{aligned} W &= 400 \times 2 \\ &= 800 \text{ ft-lb} \end{aligned}$$

(*Note:* The unit used is the foot-pound and not, as might logically have been expected, the pound-foot. This is because the pound-foot unit is reserved for use with another quantity that will be discussed in Chap. 6.)

Now, if the force acts in the same direction as that in which the body moves, the work done by the force is regarded as *positive work*. If the force acts in the opposite direction to that in which the body moves, *negative work* is said to have been done by the force. Thus in the example of the weight lifter the work done by the upward force is positive and the work done by gravity (the weight of the barbell, say, 300 lb) is negative; i.e.,

$$\begin{aligned} W &= -(300 \times 2) \\ &= -600 \text{ ft-lb} \end{aligned}$$

The total work is the sum of these two (i.e., 200 ft-lb) and is equal to the work done by the resultant of the two forces:

$$\begin{aligned} W &= (400 - 300) \times 2 \\ &= 200 \text{ ft-lb} \end{aligned}$$

When a man does a pull-up on a horizontal bar, the muscles of his arms and shoulders provide a vertical force that lifts his body. Because the direction in which the man's body moves is the same as that in which the force exerted by the muscles acts, these muscles can be said to do positive work. When the man then lowers his body, these same muscles exert an upward force to help control the descent. (In the absence of this force, the body would plummet down and the man would feel a very abrupt jerk when his arms became fully straight.) During the descent these muscles are doing negative work.

Power

In arriving at the amount of work done by a force, no account is taken of the length of time that is involved in getting the work done. Thus, if a weight lifter does, say, 1100 ft-lb of work in raising a barbell overhead, the amount of work is in no way dependent on how long he took to perform the feat. Whether the lift took 0.5 sec, 1 sec, or even 2 sec, the amount of work done is still 1100 ft-lb. What does change though is the *power*, or the rate at which the work is performed, for to arrive at this value the work done is divided by the time taken:

$$P = \frac{W}{t} \tag{46}$$

where $P =$ the power developed, $W =$ the work done, and $t =$ the time taken. Thus, although the work done in the example cited above is constant throughout, the power developed changes as follows:

Time (sec)	*Work* (ft-lb)	*Power* (ft-lb/sec)
0.5	1100	2200
1.0	1100	1100
2.0	1100	550

Because the numbers often become rather large when the foot-pound per second is used as the unit of power, it is quite common to use a larger unit, the horsepower (hp), when measuring the rate at which work is done:

$$1 \text{ hp} = 550 \text{ ft-lb/sec}$$

Thus the weight lifter develops 4 hp if he completes the lift in 0.5 sec and 1 hp if he takes 2.0 sec to do so.

Energy

The term *energy* is another that is widely used outside technical discussions. This usage, though, is not as frequently at variance with the technical definition of the word as is the case with work. Energy formally defined is "the capacity to do work" and thus, when a person says he has no energy or, alternatively, that he is bursting with energy, he could reasonably be interpreted as meaning he had either no capacity for work or a very great capacity to do work.

Because there are forms of energy other than those of concern in biomechanics, it is customary to refer to the three types of energy that are of interest in this field as types of *mechanical energy*. These three types are those in which bodies have energy by virtue of their motion, their position, and their state of deformation.

Kinetic Energy. The energy that a body has because it is moving is known as its *kinetic energy*. The amount of kinetic energy that a body possesses is given by the formula

$$\text{K.E.} = \tfrac{1}{2}mv^2 \tag{47}$$

where K.E. = the kinetic energy, m = the mass of the body, and v = the velocity at which it is moving. Thus, a skier who has a mass of 5 slugs (equivalent to a weight of 161 lb) and who is traveling at a velocity of 80 fps has a kinetic energy of 16,000 slugs-(fps)^2:

$$\begin{aligned}\text{K.E.} &= \tfrac{1}{2} \times 5 \times 80 \times 80 \\ &= 16{,}000 \text{ slugs-(fps)}^2\end{aligned}$$

The unit of kinetic energy given here derives logically from considering the units used to measure the various quantities on the right-hand side of Eq. 47. It is, however, a rather clumsily worded unit and for this reason is rarely, if ever, used. Instead, because energy is "the capacity to do work" and this latter quantity is measured in foot-pound units, it is convenient to use the same unit in measuring all three kinds of mechanical energy. This change in the unit used is greatly helped by the fact that

$$1 \text{ slug-(fps)}^2 = 1 \text{ ft-lb}$$

Therefore, whenever the kinetic energy of a body is being computed, and the mass is expressed in slugs and the velocity in feet per second, the answer can be taken to be in foot-pound units.

Potential Energy. A trampolinist at the peak of his flight has the capacity to do work because of his position relative to the surface of the earth. When he falls back toward the trampoline bed, his weight does work "equal to the product of its magnitude and the distance that the body moves in the direction of the force." The energy due to the position that a body occupies relative to the earth's surface is called *potential energy* and can be determined by multiplying the weight of the body (i.e., the force) by its height above the surface (i.e., the distance):

$$\text{P.E.} = Wh \tag{48}$$

where P.E. = the potential energy, W = the weight of the body, and h = its height above the ground.

If the trampolinist is performing on a pit (or ground level) trampoline and is, say, 150 lb in weight and 10 ft above the bed at the peak of his flight, his potential energy at this time is

$$\begin{aligned} \text{P.E.} &= 150 \times 10 \\ &= 1500 \text{ ft-lb} \end{aligned}$$

Unless he also has some horizontal velocity at this instant (and, in general, only poor trampolinists do!), his velocity and kinetic energy are both zero. As the trampolinist falls back toward the bed, the potential energy he possessed at the peak of his flight is gradually reduced. After he has fallen 1 ft, it is 1350 ft-lb (i.e., 150 × 9 ft-lb); after 2 ft, it is 1200 ft-lb (150 × 8 ft-lb); and so on until finally he reaches the bed of the trampoline with no potential energy at all. During this process, however, his body is being accelerated toward the bed by gravity and, as it gets progressively faster, it acquires more and more kinetic energy. If the appropriate equations are used (Eq. 10 and 47), it can be computed that after he has fallen 1 ft, his kinetic energy is 150 ft-lb; after 2 ft, it is 300 ft-lb; and so on until at the moment he reaches the bed it is no less than 1500 ft-lb. Addition of the kinetic energy and potential energy at each of these heights in turn reveals that this sum is constant and equal to 1500 ft-lb:

At 10 ft:	K.E. = 0 ft-lb	P.E. = 1500 ft-lb	Sum = 1500 ft-lb
At 9 ft:	K.E. = 150 ft-lb	P.E. = 1350 ft-lb	Sum = 1500 ft-lb
At 8 ft:	K.E. = 300 ft-lb	P.E. = 1200 ft-lb	Sum = 1500 ft-lb
.	.	.	.
.	.	.	.
.	.	.	.
At 1 ft:	K.E. = 1350 ft-lb	P.E. = 150 ft-lb	Sum = 1500 ft-lb
At 0 ft:	K.E. = 1500 ft-lb	P.E. = 0 ft-lb	Sum = 1500 ft-lb

If the trampolinist's ascent, from takeoff to peak of flight, were similarly analyzed, exactly the reverse process could be observed. At takeoff his kinetic energy is 1500 ft-lb and his potential energy is zero and, as he rises, this kinetic energy is gradually transformed into potential energy until at the peak of his flight he has 1500 ft-lb of potential energy and no kinetic energy. Again, as in the case of the descent, the sum of the kinetic and potential energies throughout the ascent is a constant 1500 ft-lb.

These characteristic changes in kinetic and potential energies during the performance of a stunt on the trampoline have been well demonstrated in a study by Baker[6] (Fig. 50, p. 100). The following points should be noted:

1. In the airborne phases the kinetic energy decreases and the potential energy increases during the ascent, and the reverse process occurs during the descent.

2. The sum of the kinetic and potential energies has some constant value during each of the airborne phases.

3. The gymnast retains some potential energy on landing due to the fact that he is performing on a trampoline bed that is some distance above ground level. Differences in the value of this potential energy as he makes each of his three landings depend on his body position.

A little thought on the matter suggests that the changes in kinetic and potential energies experienced by a trampolinist must also be experienced by any body that becomes airborne, and confirmation of this is contained in the *law of conservation of mechanical energy*, which, for present purposes, may be stated as

> *When gravity is the only external force acting on it, the mechanical energy of a body is constant.*

Thus, when a body is airborne (and the effects of air resistance are small enough to be ignored), the sum of the body's kinetic and potential energies is constant.

[6]James Baker, "Cinematographic Analysis of a Rebound Tumbling Event" (Unpublished term paper, University of Otago, New Zealand, 1969).

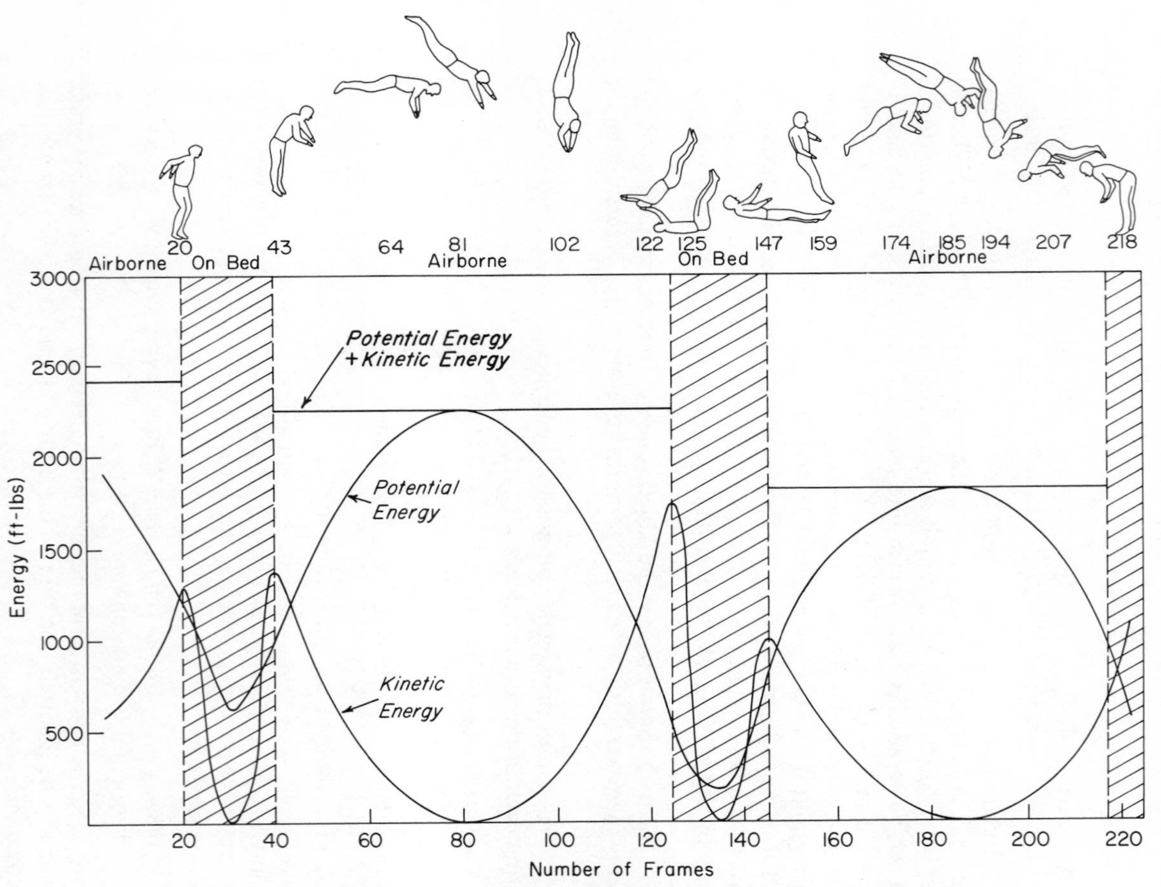

Fig. 50. Mechanical energy changes during the execution of a trampoline stunt

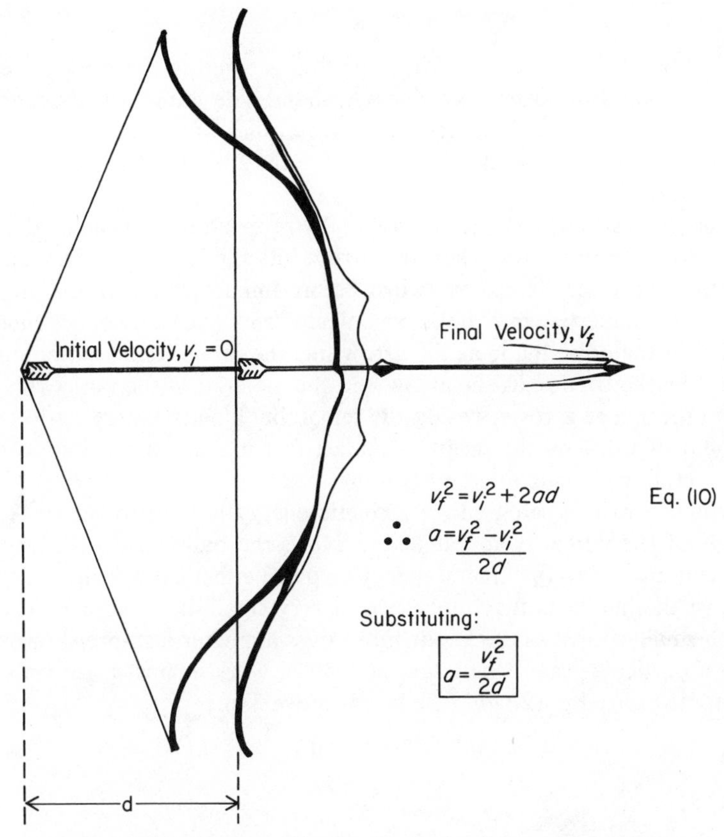

Fig. 51. Deriving an expression for the acceleration experienced by an arrow as it leaves the bow

Work-Energy Relationship. There is an important relationship among work, kinetic energy, and potential energy that has several useful applications in the analysis of sports techniques. Consider an arrow about to be fired horizontally from a bow. Immediately before the bowstring is released, the arrow is at rest and therefore possesses no kinetic energy. Then, when the archer releases the bowstring, work is done on the arrow until it loses contact with the string, by which time it has come to possess a certain amount of kinetic energy. Looking at this process in some detail (Fig. 51) it can be seen that the average acceleration of the arrow is $v_f^2/2d$. If now this value is substituted into the equation for Newton's second law (Eq. 31),

$$F = \frac{mv_f^2}{2d}$$

where F equals the average horizontal force exerted on the arrow, and this is rearranged,

$$Fd = \tfrac{1}{2}mv_f^2 \tag{49}$$

the left-hand side of the resulting equation is equal to the work done on the arrow by the bowstring, and the right-hand side is the kinetic energy that the arrow possesses as it leaves the bow. In other words, as it leaves the bow, the arrow possesses kinetic energy equal to the amount of work that has been done on it.

The process outlined here is essentially reversed during the final stages of the arrow's motion, for when the arrow hits the target, it gives up all the kinetic energy that it possesses just before impact. A small amount of this energy is converted into nonmechanical forms such as sound energy (reflected in the noise made as the arrow hits the target) and heat energy (generated by the friction as the arrow and the material of the target rub against each other). The arrow gives up the remaining kinetic energy to do an equal amount of work on the target. This last fact has important implications in such activities as catching and landing.

When a ball is being caught, kinetic energy is used to do work on the hands of the person who catches it. Now, the ball can do the amount of work equivalent to the kinetic energy involved either by exerting a large force against the hands as they move over a very small distance or by exerting a much smaller force as the hands move over a greater distance. For example, if in a particular case a ball does 30 ft-lb of work against a person's hands, the force exerted is 120 lb if the hands move 3 in.:

$$W = Fd$$

$$F = \frac{W}{d}$$

$$= \frac{30}{0.25}$$

$$= 120 \text{ lb}$$

and only 15 lb if the hands move 2 ft:

$$F = \frac{30}{2}$$

$$= 15 \text{ lb}$$

It is for this reason that skilled catchers allow their hands to move (or "give") with the ball as they catch it, for they know that otherwise the force that the ball exerts on the hands may be sufficient to cause them to fumble the catch and perhaps even to injure their hands.

The sport of boxing provides another example in the fighter with a so-called "glass jaw," a man who is especially vulnerable to a punch to this part of his anatomy. This condition probably stems from a lack of mobility in the boxer's neck, which does not allow his head to travel or "ride" with a punch so that the force of the blow can be reduced to a tolerable level. If such a lack of mobility does not allow him to do this effectively, the force of each of the blows he receives is correspondingly increased, and under such circumstances it is little wonder that he soon earns a reputation for being easy to knock out.

To date this discussion has been confined to changes in the kinetic energy of a body as work is done on or by that body. However, changes in potential energy must also be taken into account, if the body experiences changes in its elevation as well as changes in its velocity.

Consider again the bow and arrow of Fig. 51, and imagine the arrow directed vertically upward rather than horizontally. If the speed at which the arrow leaves the bowstring is v_f, the average acceleration of the arrow during the period that the bowstring is applying force to it is $v_f^2/2d$, as before. The average force exerted on the arrow during the same period is equal to $F - W$, where F is the average upward vertical force exerted via the bowstring and W is the weight of the arrow. If these values are now substituted into Eq. 31,

$$F - W = \frac{mv_f^2}{2d}$$

and rearranged,

$$(F - W)d = \tfrac{1}{2}mv_f^2$$

$$Fd = \tfrac{1}{2}mv_f^2 + Wd \tag{50}$$

it can be seen that the work done is equal to the sum of the changes in kinetic and potential energy that result. (This is the so-called *work-energy relationship.*) Thus, for example, if a 161-lb basketball player raised his body 1 ft and acquired a velocity of 12 fps by the time he left the ground in a center jump, the work done could be computed as follows:

$$\begin{aligned} \text{Work} &= (\tfrac{1}{2} \times 5 \times 12^2) + (161 \times 1) \\ &= 360 + 161 \\ &= 521 \text{ ft-lb} \end{aligned}$$

Similarly, the work done by a weight lifter in raising a 250-lb barbell a distance of 6 ft to a held position overhead (i.e., $v_f = 0$) can be computed using Eq. 50:

$$\begin{aligned} \text{Work done on barbell} &= \left(\frac{1}{2} \times \frac{250}{32.2} + 0\right) + (250 \times 6) \\ &= 0 + 1500 \\ &= 1500 \text{ ft-lb} \end{aligned}$$

If the athlete himself weighs 200 lb and raises his body 2 ft in the course of getting the barbell aloft, the total work done is

$$\begin{aligned} \text{Work done on barbell} + \text{work done on lifter} &= 1500 + 400 \\ &= 1900 \text{ ft-lb} \end{aligned}$$

Strain Energy. Whenever a body has the capacity to do work because it has been deformed and has a tendency to return to its original shape, it is said to possess *strain energy*. A fully drawn bow has the capacity to do work by virtue of the deformation it has undergone. This capacity is clearly apparent when the bowstring is released and the bow does work on the arrow. At this point the strain energy (or at least a large part of it) possessed by the bow is used to give kinetic energy to the arrow. When a trampoline bed is depressed by a gymnast landing on it, it possesses a certain amount of strain energy as a result of the deformation that it undergoes. Then, as it returns to its normal state, it does work on the gymnast, causing him to gain both kinetic and potential energy. A vaulter using a fiberglass pole puts a deep bend in it as a result of the work he does in the early stages of the vault. The strain energy that the pole then possesses enables it to do work on the athlete as it straightens later in the vault. In a well-executed vault this work done by the pole is clearly evident in the marked upward surge that the vaulter experiences as the pole recoils.

Chapter 6

ANGULAR KINETICS

The terms or basic concepts involved in an explanation of the causes of angular motion, i.e., in angular kinetics, are very closely related to those encountered in linear kinetics.

Eccentric Force

Imagine a landing mat lying on the floor of a gymnasium and suppose that a gymnast gives it a sharp kick in order to alter its position. What happens to the mat as a result of the force the gymnast exerts upon it depends on the line of action of this force—i.e., on the straight line along which the force is directed. If the line of action passes through the center of the mat [Fig. 52(a)], the mat is translated some distance in the direction in which the force acts. If it doesn't pass through the center of the mat, the mat is translated and rotated simultaneously, with the direction of the rotation depending on which side of the center the line of action passed [Fig. 52(b) and (c)]. (*Note:* Unless a body is fixed at some point, any rotation imparted to it occurs about an axis passing through the center of the body—strictly speaking, through the center of gravity of the body, a point whose precise location will be considered in some detail in a later section, p. 120.)

A force whose line of action does not pass through the center of gravity of the body on which it acts is called an *eccentric force.*

Couple

To continue with the movement of gymnastic equipment, suppose that two gymnasts are trying to turn a side horse at right angles to its present position. To achieve their objective they arrange themselves as shown in Fig. 53 and exert equal and opposite parallel forces (F_1 and F_2) against the apparatus. Considering each of these as a separate eccentric force, it can be

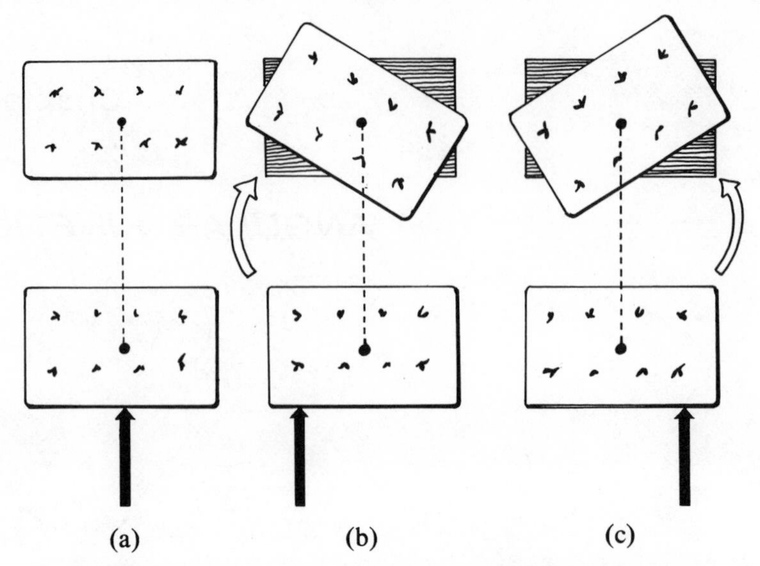

Fig. 52. Eccentric forces cause translation and rotation while those that act through the center of a body cause only translation.

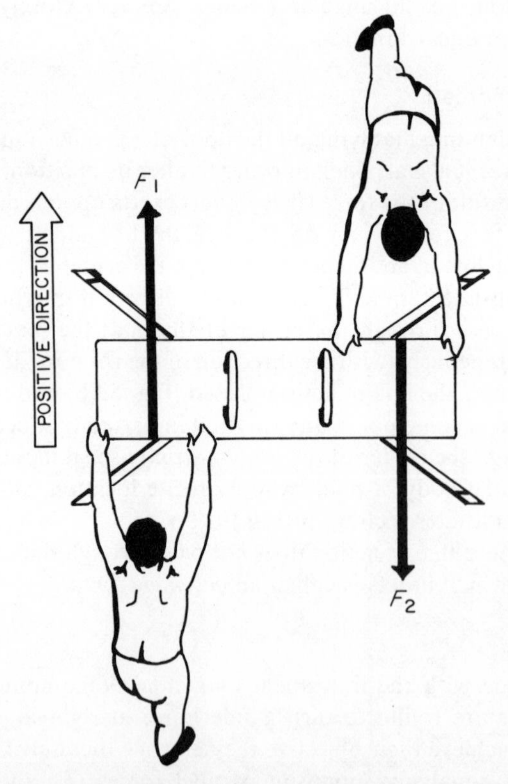

Fig. 53. Equal, oppositely directed, and parallel forces constitute a couple. (*Note:* F_1 and F_2 are the resultant horizontal forces exerted by the respective gymnasts on the horse.)

seen that F_1 tends to translate the horse in a positive direction and to rotate it clockwise, while F_2 tends to translate the horse in a negative direction and rotate it clockwise. Because the two forces are equal in magnitude, their tendencies to translate the horse in opposite directions are also equal and thus effectively cancel each other out. The remaining tendency of each force, to rotate the apparatus clockwise, is in no way hindered (and, in fact, is enhanced) by the presence of the other, and the horse is simply rotated on the spot. An arrangement like this of two equal and opposite parallel forces is termed a *couple*.

To summarize: it has been shown in these last two sections that

1. a force directed through the center of a body tends to cause translation,
2. a couple tends to cause rotation, and
3. an eccentric force tends to cause simultaneous translation and rotation.

These are the basic causes of the three forms of motion commonly observed (see Chap. 2).

Moment

When a couple is exerted on a body, it tends to produce angular motion. The extent of this tendency depends on two factors related to the nature of the couple. The first is the magnitude of the forces involved—the greater the forces, the greater their tendency to produce angular motion. In the case of the two gymnasts, the harder they push, the more likely are they to cause the horse to rotate. The second is the distance between the lines of action of the two forces constituting the couple—the greater the distance, the greater the tendency to produce angular motion. If the two gymnasts push against the horse along the lines of the two pommels, their tendency to cause the horse to rotate is less than if they exert the same forces nearer the ends of the apparatus. The product of these two factors is a measure of the turning effect that a couple possesses and is called the *moment* of the couple, or the *torque*. Thus:

$$M = F \times \text{M.A.} \tag{51}$$

where $M =$ the moment of the couple, $F =$ the magnitude of one of the forces involved, and M.A. $=$ the shortest, or perpendicular, distance between the lines of action of the two forces (the so-called *moment arm*).

The words moment and torque are also used to describe the turning effect produced when a force is exerted on a body that pivots about some fixed point. Consider the case of an oarsman who, at the completion of the pull phase, lifts the blade of the oar out of the water preparatory to moving it back into position to begin the next stroke [Fig. 54(a)]. To get the oar into this position, he pushes forward and laterally on the oar handle. The resultant force that he exerts on the oar handle [F in Fig. 54(a)] tends to rotate the oar counterclockwise and to translate it in the direction in which the force acts. Since the

Fig. 54. The moment of a couple producing rotation about a fixed point is sometimes referred to as the moment of the applied force (*Photographs courtesy of Donald C. MacKay*).

oar simply rotates about an axis through the swivel (the fixture that supports the oar partway along its length), it must be concluded that when the oarsman exerts force on the oar handle, the swivel exerts an equal, opposite, and parallel force [F' in Fig. 54(a)] on the shaft of the oar. This latter force cancels out the translatory tendency of the force applied by the oarsman and, together with that force, makes up a couple that produces the observed rotation of the oar. In cases such as this, it is customary to ignore the force exerted at the pivot and, when dealing with the magnitude and direction of the turning effect, to speak simply of the *moment of the applied force*. For example, the moment of the force exerted by the oarsman in Fig. 54 is $F \times d$, where d is the distance from the pivot to the line of action of F. (*Note:* This custom or convention in no way alters the magnitude of the moment—whether it is considered as the moment of F about the axis through A or the moment of a couple FF', its magnitude is still $F \times d$.)

In the case illustrated in Fig. 54(a) the oarsman exerted force on the oar handle in a direction at right angles to the midline of the oar. If the force he exerted was 25 lb and the distance from its line of action to the swivel (i.e., the moment arm) was 3 ft, the moment of the force was 75 lb-ft.

If the oarsman were to apply a force of the same magnitude in a direction other than at right angles to the oar (say, at 65°), the resulting moment could be computed in either of two ways.

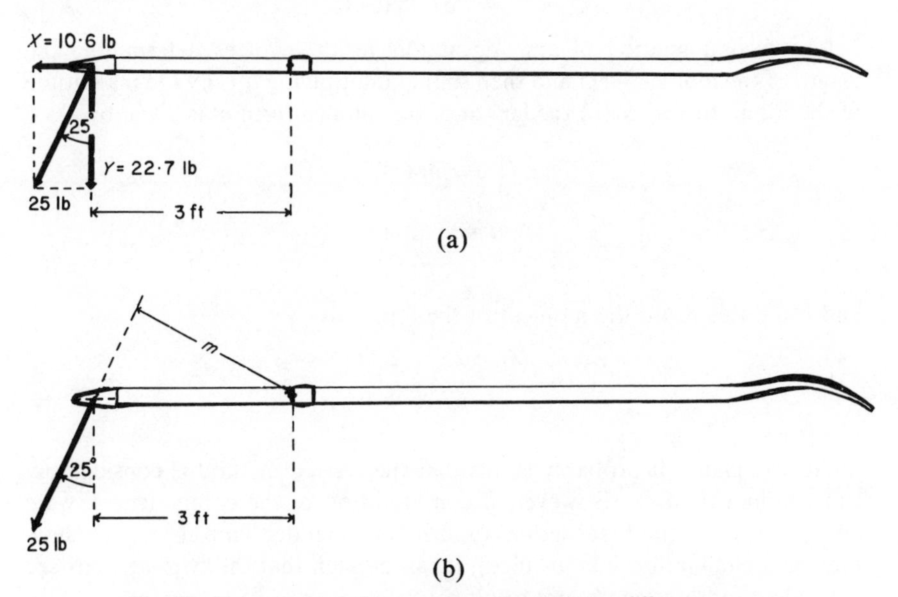

Fig. 55. The moment of a force may be computed in two different ways.

In the first of these the force is resolved into two components, one acting along the midline of the oar and the other at right angles to this midline. The moment of each of these components can then be considered separately. For an oarsman exerting a force of 25 lb at an angle of 65° [Fig. 55(a)], the component X acting along the midline is given by

$$\frac{X}{25} = \cos 65°$$

$$\therefore \quad X = 25 \times \cos 65°$$

$$= 25 \times 0.423$$

$$= 10.575 \text{ lb}$$

and the component Y acting at right angles is given by

$$\frac{Y}{25} = \sin 65^\circ$$

$$\therefore \quad Y = 25 \times \sin 65^\circ$$

$$= 25 \times 0.906$$

$$= 22.65 \text{ lb}$$

Because its line of action passes through the axis at O, the component X has a zero moment arm and thus no tendency to cause the oar to rotate. The only rotational effects must therefore be those due to component Y:

$$M_Y = 22.7 \times 3$$

$$= 67.95 \text{ lb-ft}$$

The second method of arriving at this result involves determining the length of the moment arm and then simply multiplying this by the magnitude of the force. In Fig. 55(b) the length of the moment arm m is given by

$$\frac{m}{3} = \sin 65^\circ$$

$$\therefore \quad m = 3 \times \sin 65^\circ$$

$$= 2.718 \text{ ft}$$

and using this result the moment of the force is

$$M = 25 \times 2.72$$

$$= 67.95 \text{ lb-ft}$$

as before.

At first glance it probably seems that the second method is considerably shorter than the first. However, if consideration of the component X were eliminated from the first method (and it was included here merely to show that such elimination was justified), it can be seen that the two methods are equivalent in both length and result.

From a practical standpoint, if an oarsman wants to modify the moment of the force that he applies to the oar, he has two basic alternatives—he can alter the magnitude of the force or he can adjust the length of its moment arm. With respect to this second alternative he has at least three ways in which he can make this adjustment:

1. he can alter the point at which the force is applied by moving his hand along the oar handle,
2. he can adjust the position of the button on the oar and thereby alter the distance from the end of the oar handle to the swivel, and
3. he can modify the direction of the force that he exerts.

There are numerous other examples in sports where one part of a body is "fixed" and an applied force causes the body to rotate. In most of these examples performers can alter either the magnitude of the force or the length of its moment arm in order to modify the turning effects that their actions produce. In some cases, though, there is only one alternative. The diver in Fig. 56 is a case in point. At the instant shown he is acted on by a moment about an axis through his toes equal to the product of his body weight and the moment arm x and, because his weight is constant, the only way in which he can modify this moment is to alter the length of the moment arm.

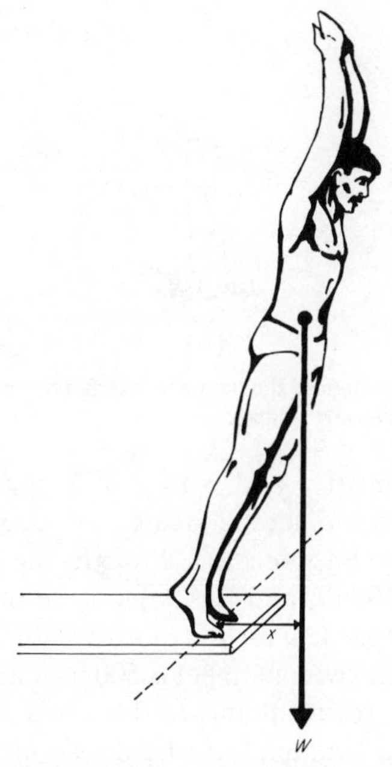

Fig. 56. The diver is acted upon by a moment equal to the product of his weight and the horizontal distance from his line of gravity to the transverse axis about which he rotates.

Resultant Moment

When a body is acted on by a number of forces each of which has a tendency to cause it to rotate about some point, the net effect of these various tendencies is obtained by summing the moments about the point in question. Consider the seesaw in Fig. 57(a). The boy on the left creates a moment equal to his weight multiplied by the distance x and tends to rotate the seesaw in a counterclockwise direction. Similarly, the moment attributable to the boy on the right is equal to the product of his weight and the distance y and acts

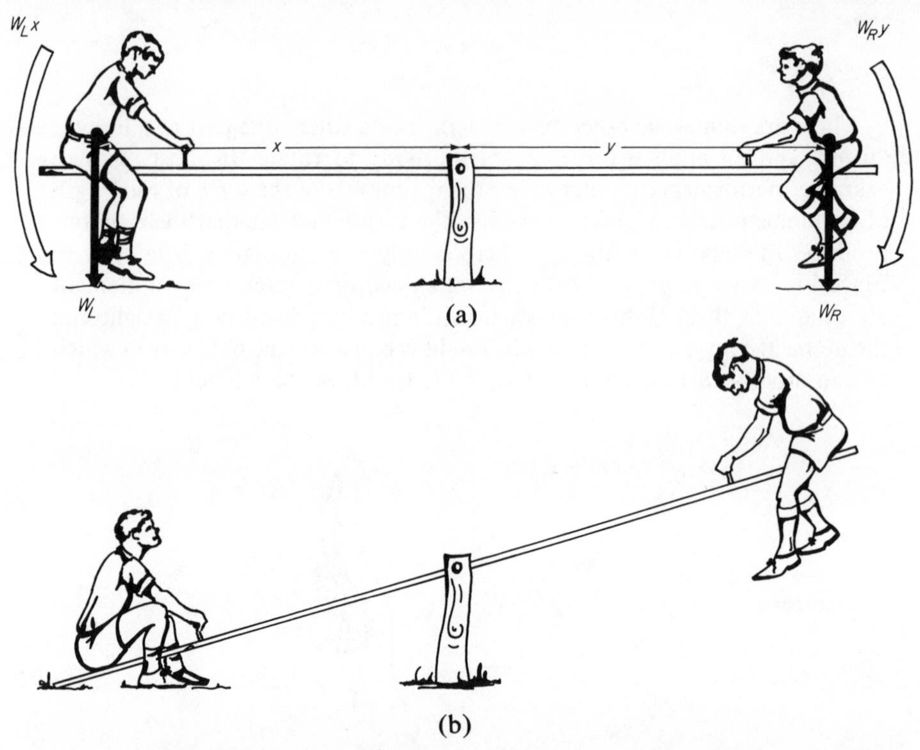

Fig. 57. The resultant of the moments exerted by each boy determines which way the seesaw rotates.

in a clockwise direction. If x and y are equal, and they are approximately so in most cases, which moment is greater will depend only on the relative weights of the two boys. Suppose $x = y = 5$ ft, the boy on the left-hand end of the seesaw weighs 100 lb, and the boy on the other end weighs 80 lb. If counterclockwise is regarded as the positive direction and clockwise as negative, the counterclockwise moment is 500 lb-ft and the clockwise moment is -400 lb-ft. The sum (or resultant) of these two moments is 100 lb-ft and, being positive, acts in a counterclockwise direction. The outcome, therefore, is that the left-hand end of the seesaw will move down and the right-hand end will move up.

When the seesaw reaches the position shown in Fig. 57(b), it tends to remain there. Now this could not happen if there existed a resultant moment acting on the seesaw. The conclusion one must come to, therefore, is that the ground exerts a force on the seesaw and that this force has a moment exactly equal to the difference between the moments of the weight forces. Under these circumstances, and in the absence of any additional forces being introduced, the seesaw would remain in the same position indefinitely. This would be very dull sport indeed, and so as the seesaw approaches the position depicted, the boy on the left-hand end places his feet on the ground and drives hard with his legs. The ground reaction that this evokes is generally sufficient to produce a resultant moment that reverses the direction of the rotation and sends the right-hand end of the seesaw down to the ground.

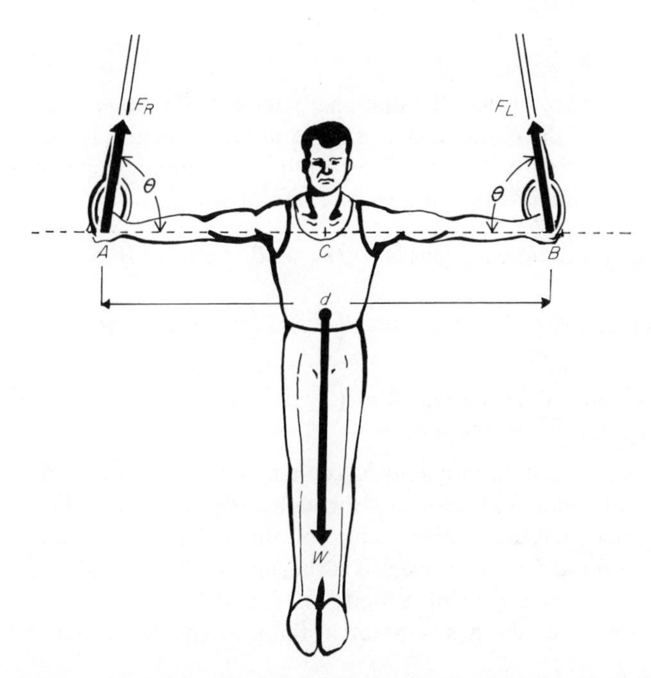

Fig. 58. A gymnast performing a cross on the rings is in equilibrium under the influence of the various forces that act upon him.

Equilibrium

When a body is at rest, i.e., it is neither translating nor rotating, it is said to be in a state of equilibrium. The gymnast performing a cross on the still rings (Fig. 58) provides a suitable example to illustrate some of the characteristics of a body in equilibrium. The body of the gymnast is acted upon by three external forces—F_R and F_L (the forces exerted on his hands by the rings and acting at some angle θ to the horizontal) and W (his weight, or the force of gravity).

From the information contained in Fig. 58, the resultant force in any direction can be computed, e.g.,

$$\text{Resultant vertical force} = F_R \sin\theta + F_L \sin\theta - W$$

$$\text{Resultant horizontal force} = F_R \cos\theta - F_L \cos\theta$$

Suppose that the resultant vertical force works out to be some positive value. Then, from Newton's second law (p. 60), it is apparent that the gymnast's body is being accelerated upward due to the resultant force acting in that direction. Alternatively, if the resultant force were some negative value, the acceleration would be in the reverse direction. Both of these alternatives are unacceptable, however, because they are at odds with the facts—the gymnast's body is not being accelerated at all. Therefore, if the resultant vertical force can be neither positive nor negative, the only possible conclusion is that it must be zero. The same reasoning leads to the conclusion that the resultant horizontal force (and, for that matter, the resultant force in any direction) must also be zero. Hence, the first characteristic of a body in equilibrium—the resultant of the components of force in any direction is zero.

The information of Fig. 58 also permits the resultant moment about any point in the plane of the figure to be computed; e.g.,

$$\text{Resultant moment about } A = (F_L \sin\theta \times d) - \left(W \times \frac{d}{2}\right)$$

$$\text{Resultant moment about } B = \left(W \times \frac{d}{2}\right) - (F_R \sin\theta \times d)$$

$$\text{Resultant moment about } C = \left(F_L \sin\theta \times \frac{d}{2}\right) - \left(F_R \sin\theta \times \frac{d}{2}\right)$$

(midpoint of the line AB)

If any of these resultant moments has either a positive or a negative value, the gymnast will rotate about an axis through the point in question. Since it is already known that the gymnast is not rotating at all, these resultant moments (like the resultant forces considered earlier) must all be equal to zero. Hence, the second characteristic of a body in equilibrium—the resultant of the moments of force, about any point in the same plane as that in which the forces act, is zero.

When a body is in equilibrium, the characteristics cited above can often be used to work out the magnitude of any unknown forces. Suppose, for example, that the gymnast in Fig. 58 weighs 150 lb, that $\theta = 75°$, and that $d = 6$ ft. Then, by taking the moments about the point A,

$$(F_L \sin\theta \times d) - \left(W \times \frac{d}{2}\right) = 0$$

$$(F_L \sin 75 \times 6) - (150 \times 3) = 0$$

$$F_L = \frac{450}{6 \times 0.966}$$

$$= 77.64 \text{ lb}$$

The magnitude of F_R, which by now should be intuitively obvious, could be confirmed by taking moments about B (or C) or by using one of the resultant force equations; e.g.,

$$F_R \cos\theta - F_L \cos\theta = 0$$

$$F_R = \frac{F_L \cos\theta}{\cos\theta}$$

$$= 77.64 \text{ lb}$$

The inward and upward components of the forces exerted by the rings on the gymnast's hands are, respectively, $F \cos\theta = 20.11$ lb and $F \sin\theta = 75$ lb, the latter showing that the gymnast's weight is supported equally by each of the rings—a fact suggested by the symmetry of the gymnast's body position.

The conditions of equilibrium expressed here are often used in problems concerned with the amount of force necessary to upset a body's equilibrium, e.g., in the construction of hurdles. The international rules relating to the construction of hurdles for track events state that a hurdle "shall be of such a design that a force of at least 3.6 kilogrammes (8 lb) and not more than 4 kilogrammes (8 lb 13 oz) applied to the centre of the top edge of the crossbar is required to overturn it."[1] To prevent hurdles from overturning when forces of less than 8 lb are applied to the crossbar, counterweights are placed on the supporting legs of the hurdle. But how should these weights be arranged? Consider the 3-ft 6-in. hurdle in Fig. 59. When a force of 8 lb is applied to the top of the hurdle, the counterclockwise moment about the axis *AB* is $8 \times 3.5 = 28$ lb-ft. Therefore, because the hurdle should not overturn when a force of less than 8 lb is applied to it, the resultant clockwise moment must be 28 lb-ft. If each of the supporting legs weighs 2 lb and is 2 ft in length (maximum allowable is 2 ft $3\frac{1}{2}$ in.), the clockwise moment due to them is 4 lb-ft. It is now pertinent to consider how heavy the counterweights must be and where they should be placed. Since the maximum moment arm for each weight is 2 ft, the length of the supporting leg, and the clockwise moment each must produce is at least $[(28 - 4)/2 =]$ 12 lb-ft, the very lightest the weights can be is just less than $12/2 = 6$ lb. If the counterweights are heavier

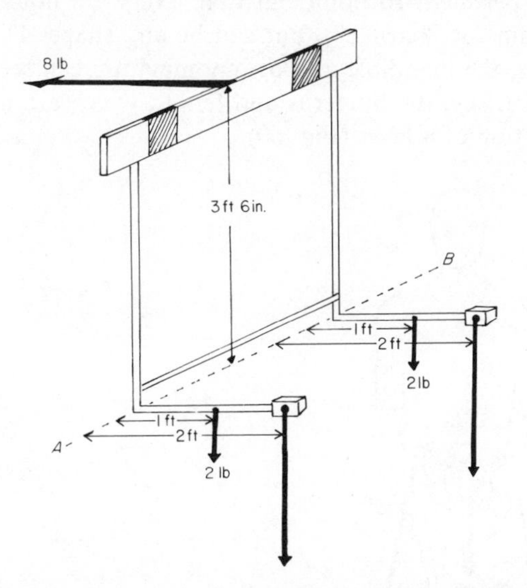

Fig. 59. To conform with the rules a hurdle must be so constructed that the resultant moment will cause it to overturn when a given force is applied to the rail.

[1]Since the rule does not state in which direction this 8 lb-8 lb 13 oz force is to be considered to act, it is quite meaningless if taken literally. For the remainder of this discussion, however, it will be assumed that the intent is that the force should be considered to act in a forward horizontal direction. It is also assumed that the word "overturn" is used to mean the movement of the vertical parts of the hurdle beyond the vertical rather than the complete upsetting of the hurdle so that it cannot return to its original upright orientation.

than this, the moment arm will have to be reduced in order to conform to the rules. Similarly, if the height of the hurdle is lowered to 3 ft, the counterweights must be brought closer to the axis, for now the "overturning moment" is reduced to 24 lb-ft and the opposing moment must be likewise reduced.

Levers

According to dictionary definitions, a lever is a "bar or some other rigid structure, hinged at one point, and to which forces are applied at two other points." The hinge or pivot point of a lever is known as the *fulcrum*, one of the forces that acts on the lever as the *weight* (or *resistance*) that opposes movement and the other as the *force* that causes or tends to cause the lever to move.

Oars, vaulting poles, baseball bats, and ski poles are examples of levers used in sports. However, by far the most important levers in any analysis of human movements are those within the body itself—the bones. For whatever influence the body has on external levers such as those just mentioned, ultimately this influence results from the action of these internal levers. At this stage it is pertinent to point out that levers are not necessarily long, straight, and thin (or "barlike"), but can be any shape. Thus, despite their irregular shapes, the mandible and os innominatum can serve as levers just as readily as can, say, the humerus and femur, which are more akin to the popular conception of a lever (Fig. 60).

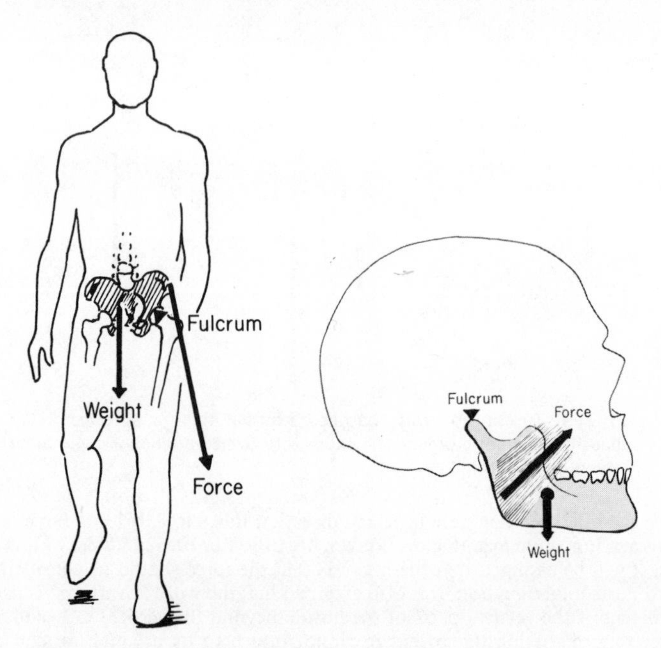

Fig. 60. Levers are not necessarily long, thin, and barlike.

The functions of levers are basically twofold:

1. A lever can serve to increase the effect produced when a force is exerted on a body. Unless some mechanical aid is used, it is generally necessary to exert a force at least as large as the weight of a body if one is to hold or lift it. With the assistance of a lever, however, a body can be lifted or held in position by exerting forces that may be considerably less than the weight of the body. For example, the footballer in Fig. 61 maintains his head in the position shown by using his skull as a lever, with the base of the skull as the fulcrum about which the lever tends to turn. The weight of his head plus the helmet he wears (say, 20 lb) is one of the two forces applied to the lever and the other is that applied by the trapezius muscle to the back of the skull. If the forces and fulcrum are disposed as indicated, the force F that the trapezius must exert in order to keep the lever in equilibrium is given by

$$F \times 2\tfrac{1}{2} = 20 \times 1$$

$$F = \frac{20 \times 1}{2\frac{1}{2}}$$

$$= 8 \text{ lb}$$

Thus, because the skull acts as a lever, the muscular force that must be exerted in order to maintain a 20-lb body in equilibrium is no more than 8 lb. In other words, the lever has served to markedly increase the effect that a relatively small force can produce.

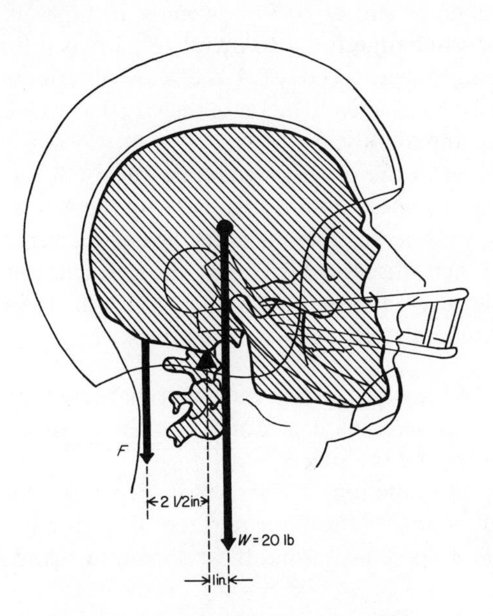

Fig. 61. A lever may be used to increase the effect that a small force is capable of producing.

2. A lever can serve to increase the distance through which a body can be moved in a given time or, expressed another way, to increase the speed with which a body can be moved. Figure 62 shows the legs of a footballer in the process of punting. In this example the lever under consideration is that formed by the part of his kicking leg below the knee. This lower leg lever rotates about a fulcrum at the knee joint as a result of the force exerted by the muscles that extend the knee and against the resistance of the force exerted by the ball on the foot.[2] If the lever moves through an angle of 30° during the time the ball is in contact with the foot, the point A to which the muscular force is applied moves through an arc of 1.05 in. (i.e., assuming the dimensions shown in Fig. 62).

$$\text{Arc } AA' = \text{circumference} \times \frac{30}{360}$$

$$= \pi \times \text{diameter} \times \frac{30}{360}$$

$$= 3.14 \times \overset{1}{\cancel{4}} \times \frac{\overset{1}{\cancel{30}}}{\underset{\underset{3}{\cancel{12}}}{\cancel{360}}}$$

$$= 1.05 \text{ in.}$$

During this same time the point B at which the ball makes contact with the foot moves through an arc of 10.5 in., exactly 10 times as great. Because it takes exactly the same time for A to travel 1.05 in. as it does for B to travel 10.5 in., the average linear speeds of A and B are also in the ratio 1 : 10. Thus a lever can be used to effectively increase the speed at which a muscular force is capable of moving a body.

(*Note:* The lever in the previous example consisted not of one bone but rather of a series of bones "bound together" by the action of ligaments and muscles. It was, in a sense, a momentary lever—a lever "assembled" for a specific task and then "disassembled" once the task had been completed. This kind of composite lever, although perhaps not described as such, appears with great regularity in analyses of sports techniques.)

For the sake of brevity, the two purposes that levers may serve and that have been described here will henceforth be referred to as (1) increasing the force and (2) increasing the speed.

Which of the two functions a lever serves depends only on the distances from the lines of action of the force and the resistance to the fulcrum—the so-called force and resistance arms. If, as shown in the example of Fig. 61,

[2]For the sake of simplifying the example, the additional resistance provided by the weight of the lower leg is disregarded here.

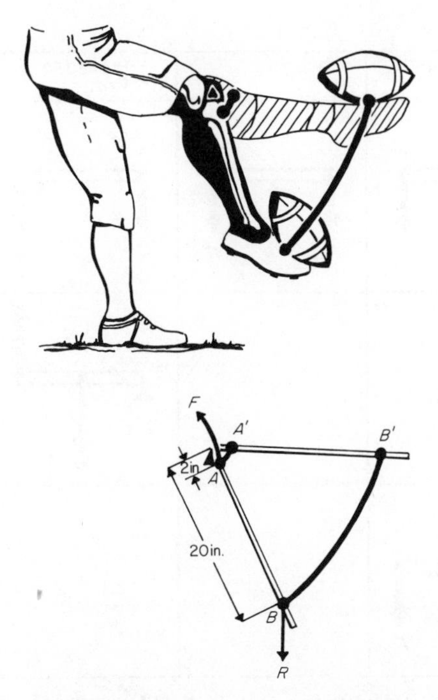

Fig. 62. An example of the use of a lever to increase the speed at which a body part is moved

the force arm is longer than the resistance arm, the function of the lever is to increase the force. On the other hand, if the force arm is shorter than the resistance arm (Fig. 62), the lever serves to increase the speed. Finally, if both arms are of equal length, no advantage is gained by using the lever—it serves to increase neither the force nor the speed. (*Note:* Because most of its bony levers have a force arm that is shorter than the resistance arm, the human body is generally considered to be much better equipped to make fast movements than to make forceful ones.)

Levers have been classified into three orders (or classes) according to the relative location of the points at which the force, fulcrum, and resistance act. A lever in which the fulcrum lies between the points at which the force and the resistance are applied is termed a *first-class lever*; that in which the fulcrum is at one end and the resistance applied closer to it than is the force is termed a *second-class lever*; and that in which the fulcrum is again at one end but the relative positions of the force and resistance are reversed is termed a *third-class lever*. However, since the significant thing about a lever is not its geometry but the function it serves (Fig. 63), this arbitrary, nonfunctional classification of lever types would seem to add nothing to warrant its further consideration here.

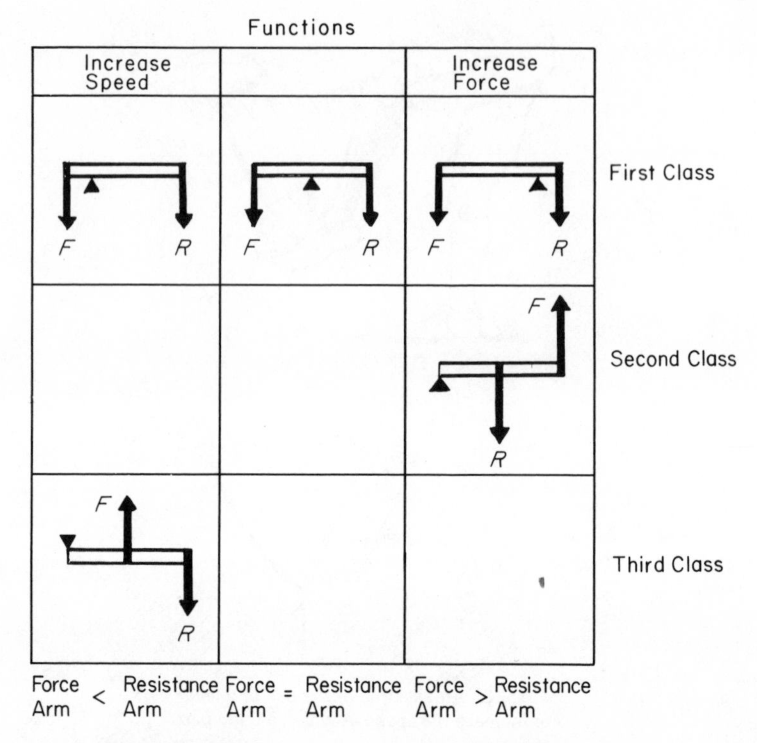

Fig. 63. Levers—functional and geometric classifications

Center of Gravity

When a body is acted upon by gravity, every particle of which it is composed experiences an attraction toward the earth (Newton's law of gravitation, p. 58). The resultant of all these attractive forces is the weight of the body, and the direction of the resultant is parallel to that of the lines of action of the individual forces. This much, at least, would seem to be fairly obvious. What is not generally so obvious is the answer to the question, "Through what points does the line of action of this resultant force act?" This section considers the answer to that question.

Consider a practice relay baton, consisting of a 1-ft length of broomstick. The location of the line of action of its weight W can be found by attempting to balance the baton on a sharp edge, say, the edge of a high-jump crossbar. Apart from the force applied at the edge (or fulcrum), the only force acting on the baton in this situation is its weight. Therefore, if the baton rotates in a clockwise direction upon being released, it must be assumed that the line of action of the weight passes to the right of the fulcrum, for only in this way could the weight have a clockwise moment of force [Fig. 64(a)]. Similarly, if the baton rotates in a counterclockwise direction, the line of action of the weight must pass to the left of the fulcrum [Fig. 64(b)]. Finally, if the baton maintains its position once it has been released, the moment of the weight about the fulcrum must be zero [Fig. 64(c)]. Since this could occur only if the length of the moment arm was zero, the line of action of the weight has thus been established. It is a vertical line (sometimes referred to as the *line of gravity* or *gravity line*) passing through the baton and through the point on the fulcrum upon which the baton is balanced.

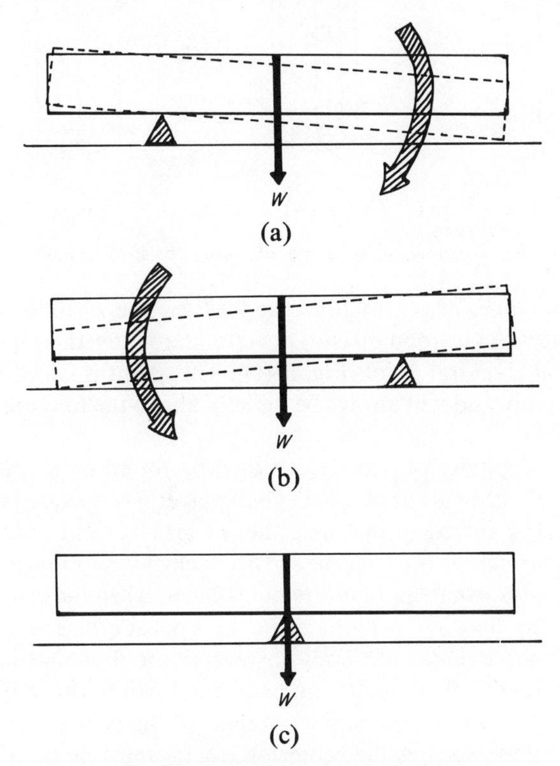

Fig. 64. Determining the gravity line

When the baton is turned so that it stands on one end, the line of gravity runs down its length rather than across it as before. Suppose that the balancing procedure is followed with the baton in this position. When the baton is balanced on the fulcrum, there is not just one vertical line directly above the fulcrum but instead a whole plane [Fig. 65(a)]. Which of the vertical lines in this plane is the line of gravity can be established by rotating the baton and then rebalancing it. In this way a second vertical plane containing the line of gravity is found [Fig. 65(b)]. Since the line of gravity must lie in the second plane as well as in the first one, it must be the line formed where the two planes intersect (the only line that does lie in both planes). And so, once again, the position of the line of gravity has been established.

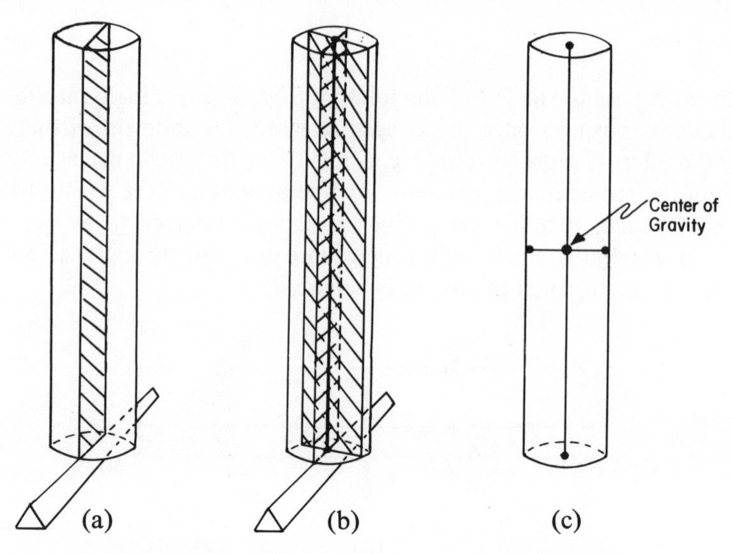

Fig. 65. Determining the center of gravity (balance method)

When the two lines of gravity found so far (i.e., one when the baton lay on its side and one when it stood on end) are considered together, it is found that they intersect at a central point [Fig. 65(c)]. This point is called the *center of gravity* of the body and will always be directly above the fulcrum if the baton is balanced.

The center of gravity of a body can also be found by suspending it. In Fig. 66 the cardboard cutout of a volleyball player is suspended from a point *A*. If the cutout is moved so that its center of gravity lies to the right of the vertical line, through *A* (i.e., the line *AB*), a clockwise moment exists that causes the cutout to rotate in that direction. Then, when the center of gravity swings across the line *AB*, a moment in the opposite direction comes into effect and this soon causes the body to reverse the direction in which it is swinging. Eventually, after oscillating back and forth in this way, the cutout comes to rest. When this occurs, the moment of the weight about the pivot at *A* must be zero, and thus the center of gravity must lie on a line directly below this point. The direction of the line *AB* is marked on the cutout, which is then suspended from a second point (not on the line that has just been drawn) and made to oscillate anew. When the suspended cutout again comes to rest, a second vertical line is drawn through the point of suspension. The point at which the two drawn lines intersect is the center of gravity of the cutout (i.e., assuming that the cutout is thin and that the third dimension can therefore be disregarded).

Two important facts concerning centers of gravity can be demonstrated using these methods.

The first of these is that the center of gravity need not lie within the physical limits of a body. If each of the bodies in Fig. 67 (the Roman ring, basketball, football helmet, and boomerang) were systematically balanced or suspended and its center of gravity located, it would be found in every case

Fig. 66. Determining the center of gravity (suspension method)

that this point lay not within the material of the body but somewhere in the space within, or around, it.

Occasionally the center of gravity is referred to as the "point-of-balance" of a body and it is either stated or implied that this is the point on which it is possible for it to be balanced or supported. However, the fact that the center of gravity can lie outside the body (i.e., beyond its physical limits) renders this interpretation quite meaningless, as attempts to balance a ring by providing support at its center of gravity readily show.

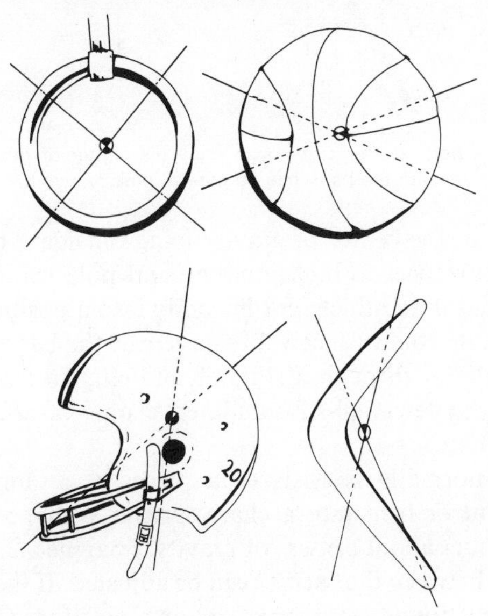

Fig. 67. The center of gravity of a body need not lie within the physical substance of the body.

Fig. 68. The Fosbury flop technique of high jumping affords the *possibility* of an athlete passing over the bar while his center of gravity passes through it, or beneath it.

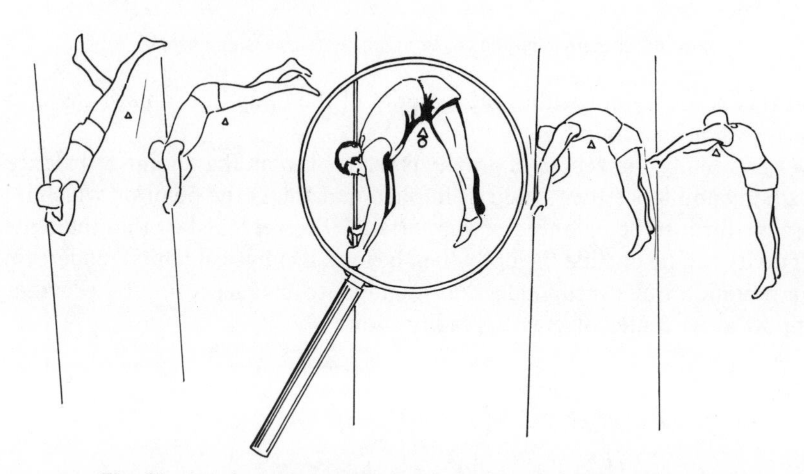

Fig. 69. A pole vaulter *might* also have his center of gravity pass through or beneath the bar while he himself passes over it.

The possibility of the center of gravity lying outside a body has aroused some interest with respect to high jumping and pole vaulting, where it has been contended that if an athlete got his body into a position similar to that of the boomerang, he could conceivably pass over the bar while his center of gravity passed under or through it (Fig. 68 and 69). To date, however, there appear to have been few well-substantiated claims of athletes having succeeded in doing this.

A boomerang normally consists of a piece, or a number of laminated pieces, of wood cut or bent into a characteristic V shape. To illustrate the second important fact about centers of gravity, imagine a special boomerang in which the angle between the "arms" can be adjusted. If the center of gravity of this boomerang is found for a variety of arm positions (Fig. 70), it can be seen that its location changes as the positions of the arms are altered. Initially

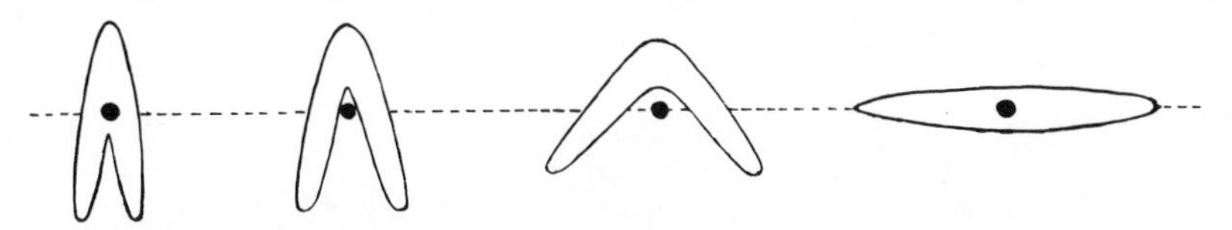

Fig. 70. The center of gravity of a body moves as the parts of the body move.

the center of gravity lies within the substance of the body, then, after moving outside it as the angle between the arms is increased, it returns to lie within the body as the angle approaches 180°.

This same phenomenon can be demonstrated using an athlete as a subject. The athlete in Fig. 71(a) is lying on a plank balanced on the edge of a piece of crossbar. With both arms extended above his head his center of gravity lies approximately 47 in. from the soles of his feet. (The alternate black and white stripes in Fig. 71—and, too, in Fig. 73—are each 3 in. in width.) If he lowers one arm and puts it by his side, and the system is once again balanced, it is found that his center of gravity has moved toward his feet. [In the case of the subject pictured in Fig. 71(b), the center of gravity was approximately 46 in. from the soles of his feet, in this second position.] Thus, as before, the location of the center of gravity of a body whose parts are free to move depends on the orientation of those parts. As the parts move, so too does the body's center of gravity.

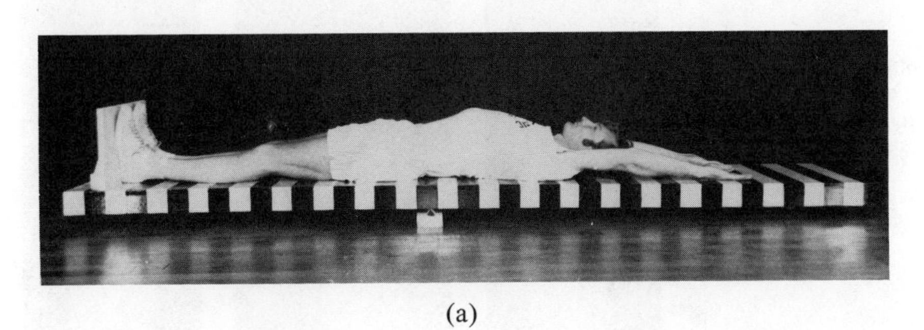

(a)

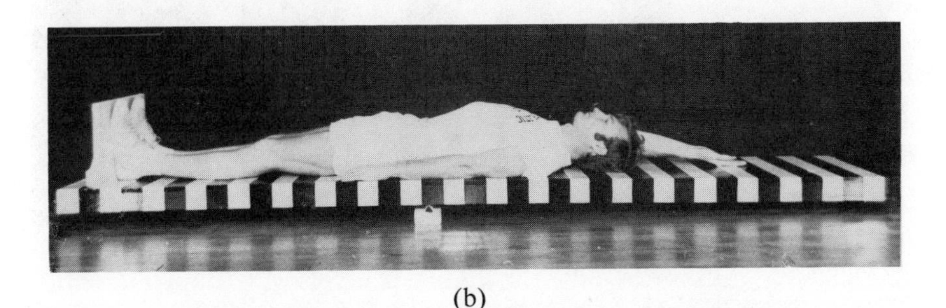

(b)

Fig. 71. From a position in which both arms are extended overhead, the lowering of one arm to a position by the side causes the athlete's center of gravity to move 1 in. closer to his feet. The distance to which the athlete can reach beyond his center of gravity is increased by the same amount.

Basketball players in a jump-ball situation, volleyball players leaping to spike, and line-out forwards in rugby all make good use of this fact. Take, for example, the basketball player. When he jumps for the ball, he swings both arms forward and upward to assist him in gaining height. Then, once he has left the floor, he allows one arm to drop down by his side while he strives to get maximum reach with the other. Apart from any other consideration, this dropping of one arm definitely acts to his advantage. Consider the athlete in Fig. 71. With both arms extended overhead he can reach to a distance of 49 in. beyond his center of gravity. (His fingertips are 96 in. and his center of gravity 47 in. above the soles of his feet.) With one arm lowered, this reach-beyond-center-of-gravity distance is increased by 1 in. (His fingertips still reach to 96 in. but his center of gravity is now only 46 in. above the soles of his feet.) Now in an actual jump-ball situation, the maximum height to which an athlete's center of gravity rises is determined by the velocity and height of his takeoff; he is simply a projectile and thus subject to the laws governing projectile motion (see pp. 32–44). How much beyond this maximum height he is capable of reaching depends on how he orients his body (Fig. 72).

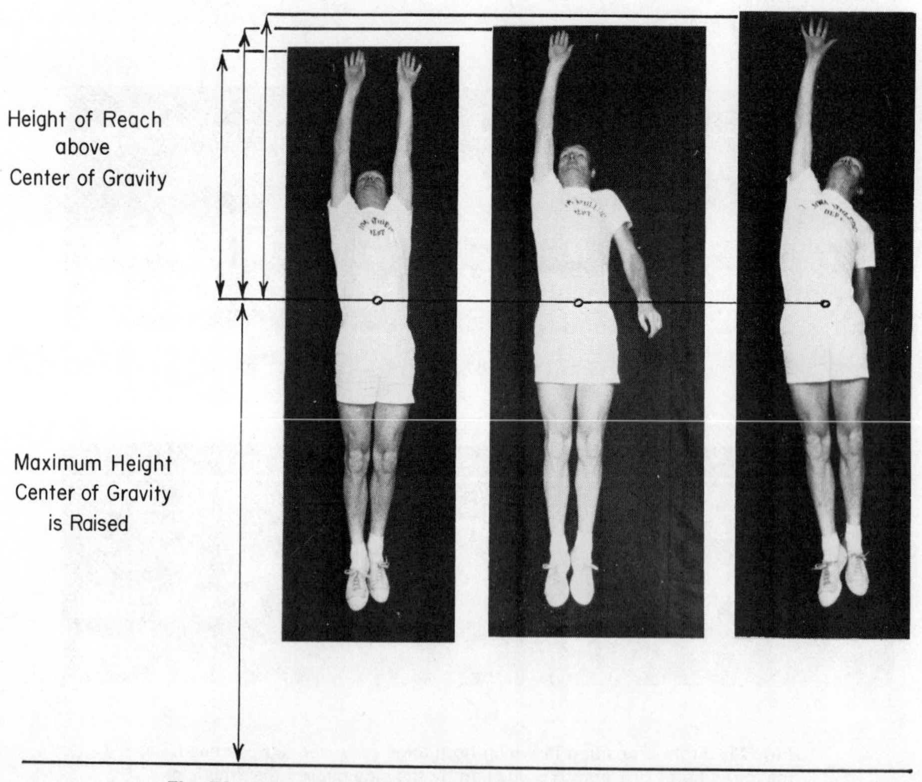

Fig. 72. Adjusting the body position to obtain the highest reach possible from a given jump

The location of the center of gravity of a human body, at rest and in motion, has been the subject of much interest over the years and is of considerable significance in many analyses of sports techniques. Many methods have been devised to estimate the position of the center of gravity of a human body. [*Note:* The word "estimate" is used advisedly because the location of a human body's center of gravity changes with each beat of the heart, with each breath the person takes, with the ingestion of food, and with changes in the disposition of the various body fluids. As a result, the accuracy with which its position can be determined at any one instant is very much open to question. Duggar, the author of an excellent review of the literature in this area, has stated that "the most exacting measurements will not ensure pin-pointing the location of the CG of a particular man to within $\frac{1}{3}$ cm"[3] ($\frac{1}{8}$ in.). Indeed, variations of the order of $\frac{1}{2}$ to $\frac{2}{3}$ cm have been reported as resulting from forced inspiration alone.[4]]

Reaction-Board Methods. Possibly the simplest method used to find the center of gravity of a living human subject is that which involves the use of a so-called *reaction board.* This board is supported horizontally on two "knife-edges," one of which rests on a block of wood while the other rests on the platform of a set of scales. After (R_1) the initial reading on the scales has been noted, the subject lies in a supine position on top of the board with the soles of the feet pressed against a vertical fitting placed so that the surface in contact with the feet lies directly above one of the knife-edges [Fig. 73(a)]. The new reading on the scales (R_2) is then noted.

Now, because the board was in equilibrium before the subject lay upon it, and because the board-plus-subject is in equilibrium when the subject does lie down, it must be concluded that whatever new forces are introduced by the subject are themselves in equilibrium. These new forces are (1) the reactions to the increases in weight borne by the platform of the scales and by the wooden block and (2) the weight of the subject. Because these forces are in equilibrium, the sum of their moments about any point in the plane in which they act is equal to zero. This latter fact is used to gain information concerning the location of the body's center of gravity. Consider the moments of these newly introduced forces about the point A in Fig. 73. The moment of the force $R_2 - R_1$ applied by the platform of the scales to the knife-edge resting upon it is $72(R_2 - R_1)$ lb-in. and acts in a counterclockwise direction. The moment of the weight force is Wx lb-in., where x in. is the unknown horizontal distance of the center of gravity from the point A. This moment acts in a clockwise direction. Because the body is in equilibrium, these moments must

[3]Benjamin C. Duggar, "The Center of Gravity of the Human Body," *Human Factors,* IV, June 1962, p. 131.

[4]A. Mosso, "Application de la balance a l'étude de la circulation du sang chez l'homme," *Archives Italiennes de Biologie,* V, 1884, pp. 130–43.

be equal in magnitude; i.e.,

$$Wx = 72(R_2 - R_1)$$

$$\therefore \quad x = \frac{72(R_2 - R_1)}{W} \text{ in.} \tag{52}$$

As an example, consider the data obtained for the subject in Fig. 73(a):

Weight: $W = 101.25$ lb

Initial scale reading: $R_1 = 19.5$ lb

Final scale reading: $R_2 = 68$ lb

(a)

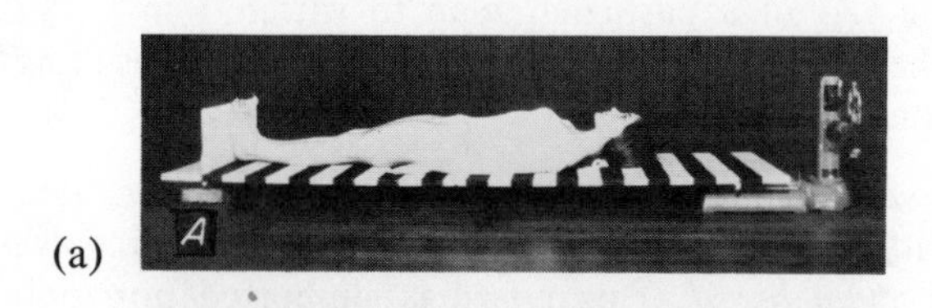

(b)

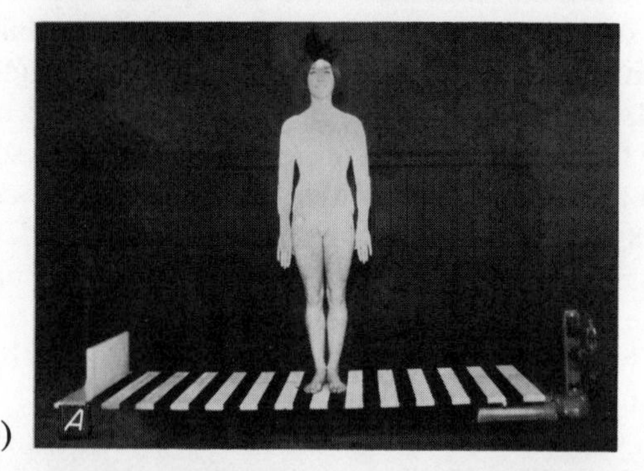

(c)

Fig. 73. Using a reaction board to determine the location of the center of gravity in one dimension

The horizontal distance of the subject's center of gravity from A is then computed, using these data:

$$x = \frac{72(R_2 - R_1)}{W}$$

$$= \frac{72(68 - 19.5)}{101.25}$$

$$= 34.5 \text{ in.}$$

It will be readily appreciated that this procedure determines the position of the center of gravity only in terms of distance from the soles of the feet. Its position laterally (i.e., relative to the left- or right-hand side of the body) or frontally (i.e., relative to the front or back of the body) can be obtained, using the same basic method, by having the subject assume an erect standing position on the reaction board [Fig. 73(b) and (c)]. [*Note:* While it would seem reasonable to assume that changing from the position shown in Fig. 73(b) to that of Fig. 73(c) would not alter the location of the center of gravity relative to the body, it would hardly seem reasonable to expect the same result when the body position was changed from supine to erect standing. Mainly because of shifts in the disposition of the body viscera and fluids, the distance of the center of gravity from the soles of the feet when a subject was standing would probably be slightly less than when she adopted a lying position.]

Another and similar method of locating the center of gravity of an individual involves the use of a large reaction board supported on scales at two or more points. In the example depicted in Fig. 74, the board has three metal spikes protruding from its undersurface. The spikes at the points A and B each rest upon the platform of a set of scales, while the remaining spike (at C) rests on a solid block. After the initial readings R_{A1} and R_{B1} are noted, the subject adopts the required position on top of the board. The scale readings R_{A2} and R_{B2} are then taken and the location of the center of gravity is determined by taking moments about the lines AC and BC. The distance of the center of gravity from each of these lines (x and y, respectively) is given by the equations:

$$x = \frac{(R_{B2} - R_{B1})h}{W} \tag{53}$$

and

$$y = \frac{(R_{A2} - R_{A1})h}{W} \tag{54}$$

where h equals the altitude of the equilateral triangle ABC and W equals the weight of the subject.

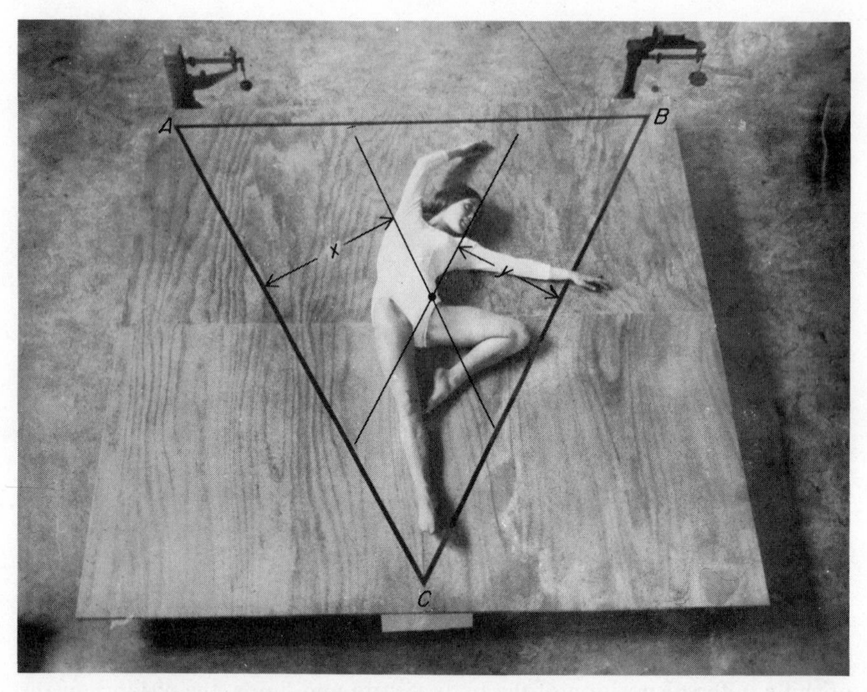

Fig. 74. Using a large reaction board to locate the center of gravity in two dimensions

This method, or some minor modification of it, has been used in a number of studies of sports techniques.[5,6,7] The procedure generally followed in such instances is:

1. The subject is filmed performing the required technique.
2. The film is viewed and the frames to be used in the analysis are selected.
3. The reaction board is used as a screen and a life-size image of the subject, as he appears in the frame to be analyzed, is projected onto the board.
4. A chalk line is drawn on the board around the projected image of the subject.
5. After the initial scale readings have been taken, the subject assumes the position represented by the chalk outline.
6. The position of the subject's center of gravity is then obtained using the method already described.
7. Steps 3 to 6 are repeated as required.

[5]William H. Groves, "Mechanical Analysis of Diving," *Research Quarterly*, XXI, May 1950, pp. 132–44.

[6]Peter C. McIntosh and H. W. B. Hayley, "An Investigation into the Running Long Jump," *Journal of Physical Education*, XLIV, November 1952, pp. 105–8.

[7]A. H. Payne and F. Blader, "A Preliminary Investigation into the Mechanics of the Sprint Start," *Bulletin of Physical Education*, VIII, April 1970, pp. 21–30.

There are a number of practical limitations inherent in this method. Chief among these are the need to have the subject available throughout and the considerable time involved in locating the center of gravity for each position.

Mannikin Methods. In an attempt to overcome these limitations, a number of workers[8,9,10] have devised mannikins that can be posed in a variety of positions and whose center of gravity can quickly be determined using the methods of suspension or balance. Unfortunately, the use of a mannikin introduces new problems at the same time as it overcomes old ones. Among these is the problem that arises if the weight of the mannikin is not distributed in the same way as the weight of the human body that it represents. For example, Page[11] has reported that when a simple cardboard mannikin, posed in an erect standing position, has both arms moved to a full extended position overhead, its center of gravity is raised the equivalent of some 5 in. On the other hand, studies by Swearingen[12] and by Santschi and his co-workers[13] have yield mean values of 2.25 in. and 2.4 in., respectively, for the shift in the center of gravity when human subjects perform the same action.

While not a particularly easy undertaking, it is possible to overcome this problem by weighting the various parts of the mannikin so that its weight distribution closely approximates that of a human body.[14] Even when this is done, however, the usefulness of a mannikin is restricted by the fact that it can assume only a limited number of positions. As a result, other methods are needed to locate the center of gravity for body positions that are not part of the mannikin's "repertoire."

Segmentation Method. Probably the most versatile method for finding the center of gravity of a human body is the *segmentation method,* which hinges upon a simple, yet important, mathematical relationship. Consider the barbell in Fig. 75(a). Each of the discs weighs 40 lb, the bar itself weighs 20 lb, and the lines of action of these weight forces are as shown. If a line AB is drawn in a plane perpendicular to that of the weight forces and at a horizontal

[8]W. M. Abalakov, cited by D. D. Donskoi in *Biomechanik der Körperübungen* (Berlin: Sportverlag, 1961), pp. 174–75.

[9]W. T. Dempster, "Free Body Diagrams as an Approach to the Mechanics of Human Posture and Motion," in *Biomechanical Studies of the Musculo-Skeletal System,* ed. F. Gaynor Evans (Springfield, Ill.: Charles C Thomas, Publisher, 1961), pp. 105–6.

[10]R. L. Page, "The Movement of the Centre of Gravity of the Human Body," *The Leaflet,* March 1968, pp. 18–19.

[11]R. L. Page, *Ibid.,* pp. 18–19.

[12]J. J. Swearingen, *Determination of Centers of Gravity of Man,* Civil Aeromedical Research Institute, Oklahoma City, Okla., August 1962.

[13]W. R. Santschi, J. DuBois, and C. Omoto, *Moments of Inertia and Centers of Gravity of the Living Human Body,* AMRL Technical Documentary Report 63–36, Wright-Patterson Air Force Base, Ohio, May 1963.

[14]W. T. Dempster, "Free Body Diagrams as an Approach to the Mechanics of Human Posture and Motion," pp. 105–6.

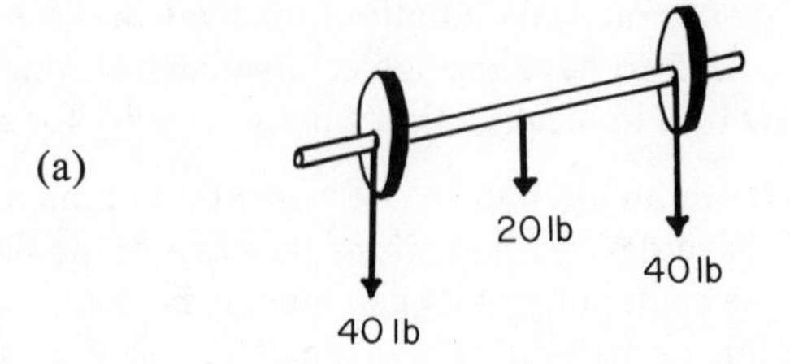

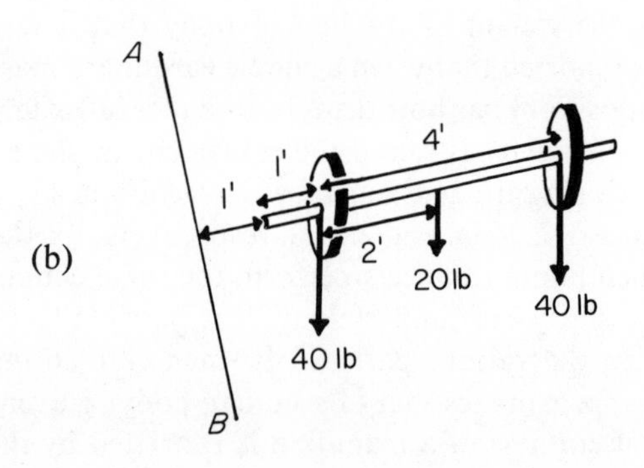

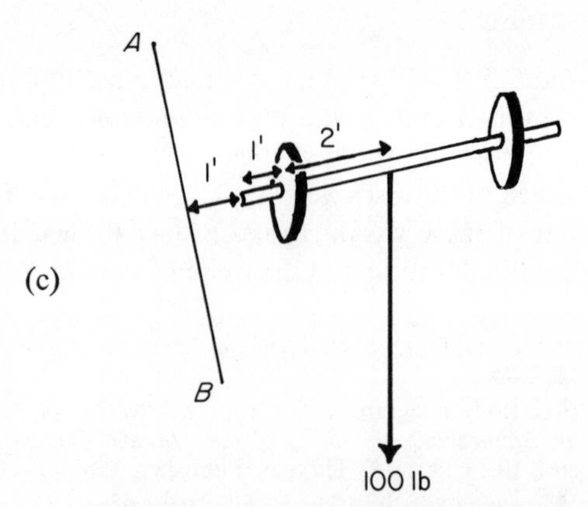

Fig. 75. A simple example to demonstrate that the sum of the moments of a number of forces about a given axis is equal to the moment of their resultant about the same axis. ["The sum of the moments is equal to the moment of the sum (resultant)."]

distance of, say, 1 ft from the left-hand end of the bar,[15] the moments of the weights about this line are

$$40 \times 2 = 80 \text{ lb-ft}$$

$$20 \times 4 = 80 \text{ lb-ft}$$

$$40 \times 6 = 240 \text{ lb-ft}$$

$$\text{Sum of these moments} = 400 \text{ lb-ft}$$

The resultant weight of the discs and the bar is 100 lb and acts through the center of gravity of the loaded barbell at the midpoint of the bar [Fig. 75(c)]. The moment of this resultant about AB is

$$100 \times 4 = 400 \text{ lb-ft}$$

which is exactly the same as the sum of the moments of the weights considered separately.

This computational process could be repeated any number of times for any combination of disc and bar weights and for any position of the line AB (in a plane perpendicular to that of the weight forces) and exactly the same relationship would be found in each case; i.e., the sum of the moments of the separate weight forces would be equal to the moment of their resultant. This in itself would not seem to be of any great importance were it not that it provides a means of locating the center of gravity of a body. Suppose, for example, that it were desired to find the center of gravity of a barbell loaded as shown in Fig. 76. Suppose, too, that a line CD (equivalent to AB) were drawn at a distance of 2 ft from the left-hand end of the bar. The moment of the resultant weight (100 lb) about this line is $100x$ lb-ft, where x ft equals the horizontal distance between the center of gravity of the barbell and CD. The sum of the moments of the separate forces about CD is 560 lb-ft, i.e.,

$$\begin{aligned} \text{Sum of moments} &= (25 \times 3) + (20 \times 5) + (55 \times 7) \\ &= 75 + 100 + 385 \\ &= 560 \text{ lb-ft} \end{aligned}$$

The relationship between the moment of the resultant and the sum of the moments of the separate weights can now be used to determine x, the horizontal distance of the center of gravity from CD:

$$100x = 560$$

$$x = 5.6 \text{ ft}$$

[15]In this example, the line AB has been drawn at right angles to the line of the bar, in order to simplify the subsequent arithmetic [Fig. 75(b)]. This, however, is purely a matter of convenience—there is no restriction on the orientation of the line except that it must lie "in a plane perpendicular to that of the weight forces."

(a)

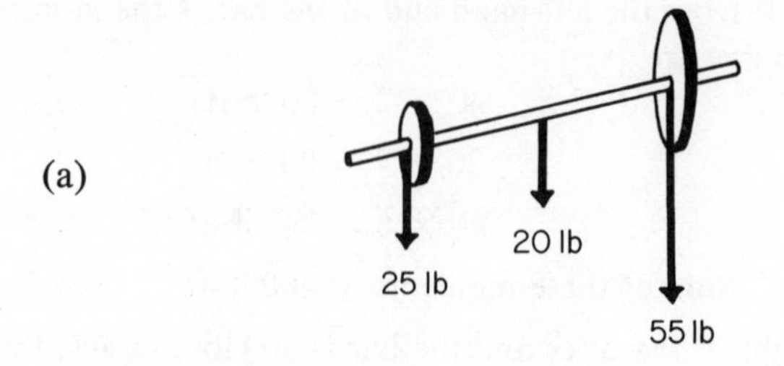

(b)

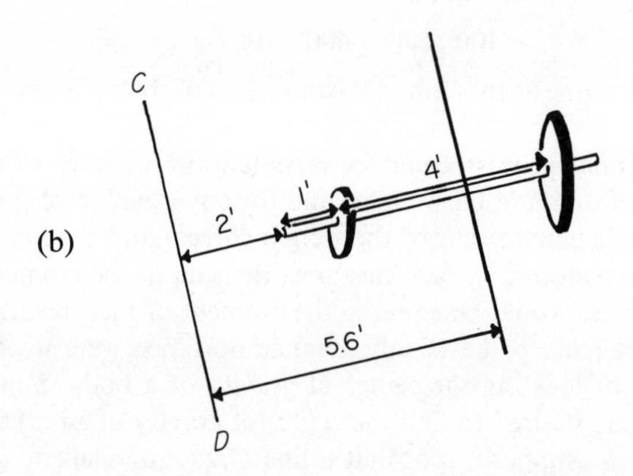

(c)

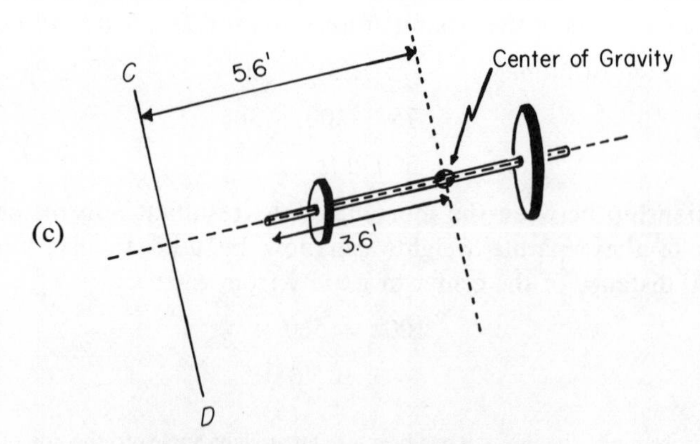

Fig. 76. Using the "... sum of the moments is equal to the moment of the sum" to locate the center of gravity of a barbell

A brief look at the arrangement of discs and bar should be all that is necessary to establish that the center of gravity of the loaded barbell must also be on the line running lengthwise through the center of the bar. Combining these two results reveals that the center of gravity of the loaded barbell lies in the center of the bar at a distance of 5.6 ft from *CD* or, more meaningfully, 3.6 ft from the left-hand end of the bar.

The method used here to find the distance of the center of gravity of a body from some arbitrarily chosen line can be used to find the center of gravity of an athlete in action. This can be done from a photograph, provided information is available concerning (1) the weights of the various parts or segments of the athlete's body (his arm, forearm, hand, thigh, etc.) and (2) the locations of the centers of gravity of these segments.

A number of studies have been conducted in which at least one of the objectives was to obtain average values for these quantities. Some investigators[16,17,18,19,20] have frozen cadavers, cut them into segments, weighed the segments, and balanced or suspended them to find their centers of gravity. Others[20,21] have had living subjects immerse selected segments and by noting either the volume of water displaced or the change in total body weight have computed the weight of these segments. Still others[22,23,24] have attempted to arrive at appropriate values by purely mathematical means.

Of all the data available on the weights of body segments and the locations of the centers of gravity of the segments, those obtained by Clauser et al.[18] (Tables 8 and 9) seem to be the most appropriate for use in analyses of sports techniques.

[16]W. Braune and O. Fischer, "The Center of Gravity of the Human Body as Related to the Equipment of the German Infantry." *Treatises of the Mathematical-Physical Class of the Royal Academy of Sciences of Saxony*, No. 7, Leipzig, 1889 (U. S. Army Air Forces, Air Material Command Translation No. 379, Wright Field, Dayton, Ohio).

[17]Wilfred T. Dempster, *Space Requirements of the Seated Operator*, WADC Technical Report 55–159, Wright-Patterson Air Force Base, Ohio, 1955.

[18]Charles E. Clauser, John T. McConville, and J. W. Young, *Weight, Volume and Center of Mass of Segments of the Human Body*, AMRL Technical Report 69–70, Wright-Patterson Air Force Base, Ohio, 1969.

[19]R. F. Chandler et al., *Investigation of Inertial Properties of the Human Body*. AMRL Technical Report 74–137, Wright-Patterson Air Force Base, Ohio, 1975.

[20]Henry G. Cleaveland, "The Determination of the Center of Gravity of Segments of the Human Body" (Ed. D. dissertation, University of California, Los Angeles, 1955).

[21]Wilfred T. Dempster, *Space Requirements of the Seated Operator*.

[22]Ernest P. Hanavan, "A Mathematical Model of the Human Body" (M.S. thesis, Air University, U.S.A.F., 1964).

[23]Philip V. Kulwicki, Edward J. Schlei, and Paul L. Vergamini, *Weightless Man: Self-Rotation Techniques*, AMRL Technical Documentary Report 62–129, Wright-Patterson Air Force Base, Ohio, October 1962.

[24]Charles E. Whitsett, *Some Dynamic Response Characteristics of Weightless Man*, AMRL Technical Documentary Report 63–18, Wright-Patterson Air Force Base, Ohio, April 1963.

*Table 8. Weights of Body Segments Relative to Total Body Weight**

Segment	*Relative Weight*
Head	0.073
Trunk	0.507
Upper arm	0.026
Forearm	0.016
Hand	0.007
Thigh	0.103
Calf	0.043
Foot	0.015

*Adapted from data presented in Clauser et al., *Weight, Volume and Center of Mass of Segments of the Human Body*, p. 59.

*Table 9. Location of Centers of Gravity of Body Segments**

Segment	*Center-of-Gravity Location Expressed as Percentage of Total Distance between Reference Points*
Head	46.4% to vertex; 53.6% to chin-neck intersect
Trunk	38.0% to suprasternal notch; 62.0% to hip axis
Upper arm	51.3% to shoulder axis; 48.7% to elbow axis
Forearm	39.0% to elbow axis; 61.0% to wrist axis
Hand	82.0% to wrist axis; 18.0% to knuckle III
Thigh	37.2% to hip axis; 62.8% to knee axis
Calf	37.1% to knee axis; 62.9% to ankle axis
Foot	44.9% to heel; 55.1% to tip of longest toe

*Adapted from data presented in Clauser et al., *Weight Volume and Center of Mass of Segments of the Human Body*, pp. 46–55.

As an example in the use of the segmentation method, suppose one wanted to locate the center of gravity of the athlete in Fig. 77. The steps to be taken are as follows:

1. Mark on the photograph the position of those reference points (Table 9) associated with each segment. The positions of reference points obscured by other body parts should be estimated carefully.

2. Construct a stick-figure representation of the subject by ruling straight lines between appropriate reference points. The head-and-trunk line is obtained by joining the midpoint of the line between the right and left hip joints to the middle of the trunk at the level of the suprasternal notch.

3. Measure the length of each segment line, and divide these various

Fig. 77. The segmentation method for locating the center of gravity

lengths in the appropriate ratio as indicated in Table 9. Mark the points of division (i.e., the centers of gravity of the segments) on their respective lines.

4. Rule two arbitrary axes (OY and OX), one to the left and one below the stick figure.

5. Prepare a form such as that shown in Table 10, and in Column 1 enter the weights of the segments.

6. For each segment, measure the perpendicular distance from the center of gravity to the line OY, and enter this distance in the appropriate place on the form (Table 10, Column 2).

7. To find the moments about OY, multiply the weight of each segment by the distance of its center of gravity from the line, and enter these values on the form (Table 10, Column 3).

8. Find the sum of the moments about OY by adding the contents of Column 3 on the form.

9. Add the contents of Column 1. If the procedure outlined here has been correctly followed, all parts of the body will have been taken into account and this total will be equal to 1 (i.e., the sum of the weights of all the body parts, expressed in terms of the total body weight).

Then, since the moment of the resultant weight about OY is equal to 1 multiplied by some unknown distance x, and since this is equal to the sum of the moments of the segments considered separately, $x =$ the sum found in step 8. (In the present example, $x = 6.478$ cm.)

10. Rule a line $O'Y'$ parallel to OY and at a distance x from it. The center of gravity of the subject lies on this line.

11. Repeat steps 5 to 10, taking moments about OX instead of OY. The center of gravity of the subject lies on the line $O'X'$ drawn parallel to OX and at the computed distance from it. And, finally, because the center of gravity lies on both $O'Y'$ and $O'X'$ and these two lines have only one point in common (the point where they intersect), it is here that the center of gravity is situated.

Table 10. Form for Computation of Center-of-Gravity Coordinates

Segment	*Column 1* *Segment Weight*	*Column 2* *Distance to* OY (cm)	*Column 3* *Moments about* OY	*Column 4* *Distance to* OX (cm)	*Column 5* *Moments about* OX
Head	0.073	6.8	0.496	6.9	0.504
Trunk	0.507	6.8	3.448	4.7	2.383
Right upper arm	0.026	7.5	0.195	6.3	0.164
Right forearm	0.016	7.8	0.125	7.1	0.114
Right hand	0.007	8.2	0.057	8.2	0.057
Left upper arm	0.026	6.2	0.161	6.4	0.166
Left forearm	0.016	5.2	0.083	7.3	0.117
Left hand	0.007	4.2	0.029	8.3	0.058
Right thigh	0.103	5.3	0.546	3.2	0.330
Right calf	0.043	3.1	0.133	3.4	0.146
Right foot	0.015	1.2	0.018	3.8	0.057
Left thigh	0.103	6.9	0.711	2.2	0.227
Left calf	0.043	7.8	0.335	1.2	0.052
Left foot	0.015	9.4	0.141	2.2	0.033
	1.000	Sum of moments =	6.478	Sum of moments =	4.408

While the segmentation method outlined here is both accurate and versatile, the large amount of measuring and computational work involved makes it very time-consuming. However, people conducting detailed analyses of sports skills are now able to enlist the aid of computers to overcome this problem.

Stability

The punching bag in Fig. 78 is in a state of equilibrium, under the action of two forces—its weight W and an equal and opposite force R exerted by the supporting rope. When a boxer punches the bag, it undergoes a slight displacement from its original position and its state of equilibrium is disrupted.

In this process the lines of action of the weight and rope forces are separated and the two forces act as a couple that tends to return the bag to its original position. A body, like the punching bag, that tends to return to its equilibrium position after being displaced is said to be in *stable equilibrium.*

Fig. 78. Stable equilibrium

The gymnast performing a balance on the beam (Fig. 79) is also in equilibrium under the action of a pair of forces. In her case, however, the effect of a small displacement from this equilibrium position produces quite different results, for the couple consisting of the forces R and W' tends to rotate her farther away from her original position rather than back toward it. A body like this, which tends to move away from its equilibrium position once it has been displaced, is said to be in *unstable equilibrium.*

Like the bodies in the two previous examples, the tennis ball of Fig. 80 is in a state of equilibrium. However, since it hes no tendency either to return to its original position or to move still further away from it, its response to a slight displacement is quite different from that of the other two bodies. This comes about because no matter in what position the ball is placed (i.e., on a level surface), the weight and reaction forces are exactly equal and opposite. Such a body is in a state of *neutral equilibrium.*

Fig. 79. Unstable equilibrium

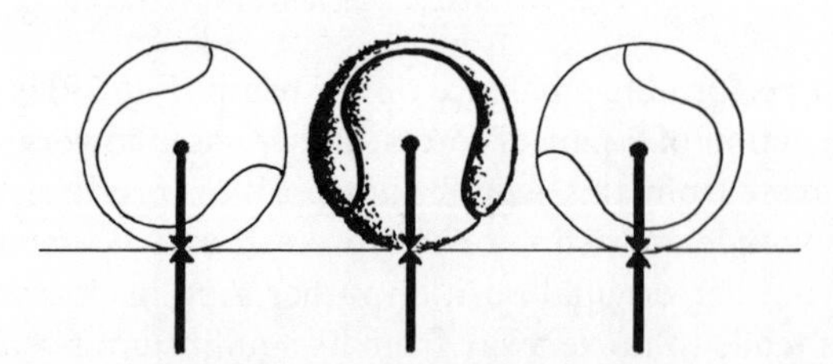

Fig. 80. Neutral equilibrium

The stability of a body in equilibrium (which is, of course, what the terms above refer to) depends on a number of factors. These are:

1. the position of the line of gravity relative to the limits of the base,
2. the weight of the body, and
3. the height of the center of gravity relative to the base.

The role of the first factor, the position of the gravity line relative to the limits of the base, may be considered in terms of the stance adopted by the wrestler in Fig. 81(a). Suppose this man's opponent exerted a horizontal force of 30 lb on him at a height of 3 ft from the ground. The moment of this force about an axis through *A* (the outer limit of the base) is 90 lb-ft. Opposing this moment, which tends to turn the wrestler counterclockwise about *A*, is the moment of his own weight, 75 lb-ft. Since this is less than the "overturning" moment, the wrestler tends to move away from his equilibrium position by rotating about an axis through *A*. Now, if the wrestler anticipates his opponent's move and shifts his center of gravity over his left foot to counter it, the moment of his weight about the axis through *A* approaches 150 lb-ft. And this, of course, is sufficient to counter the tendency of the 30-lb force (applied by his opponent) to rotate him about the axis through *A*. By shifting his weight in this manner, however, the wrestler makes himself vulnerable to attack from his right. For example, if his center of gravity is over his left foot and 2 in. from *B*, the limit of the base in that direction, the moment of his weight that resists the overturning tendency of a force applied to his right

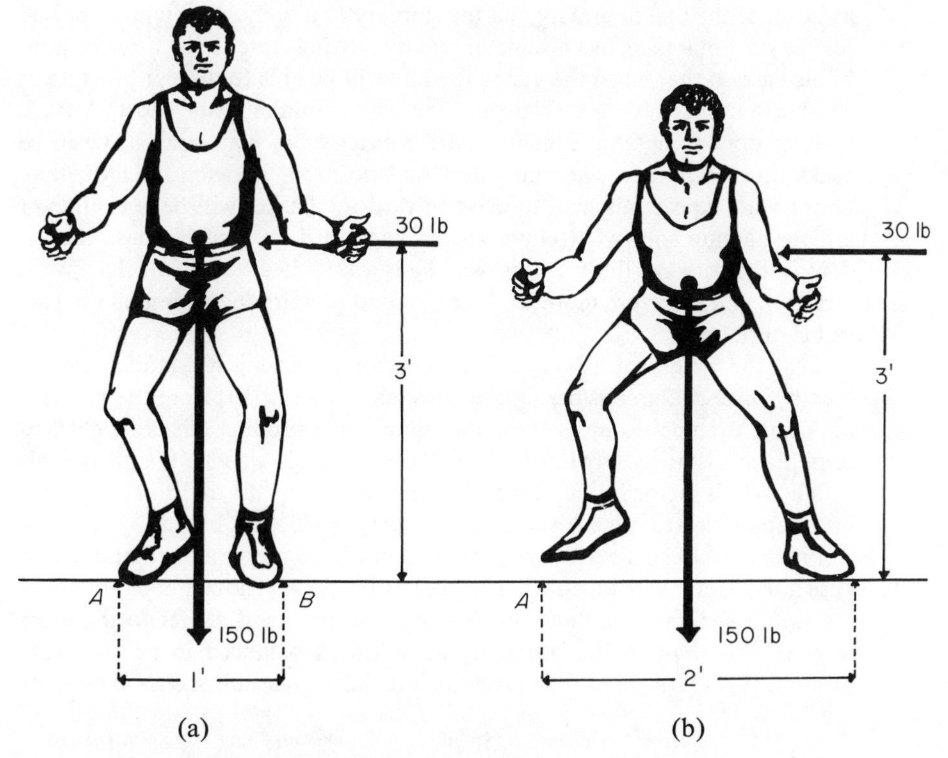

Fig. 81. The stability of equilibrium is influenced by the distance of the line of gravity from the limits of the base.

side is only 25 lb-ft. Thus a force of a little more than 5 lb applied at about shoulder height, i.e., about 5 ft from the ground, will be all that is necessary to set him rotating about an axis through *B*.

To increase his stability in one direction without simultaneously decreasing it in the opposite direction, the good wrestler increases the lateral distance between his feet [Fig. 81(b)]. Now, with the line of gravity equidistant from the lateral limits of the base, the moment resisting any tendency for him to be overturned is 150 lb-ft regardless of whether the force applied by his opponent is from his left or his right.

Because the area of the base (the area enclosed by a line drawn around the outermost limits of those points upon which the body is supported) is inevitably increased in this process, it is frequently stated that the area of the base is a factor influencing the stability of a body. This is only true to the extent that an increase in the area of the base increases the distance of the gravity line from the limits of the base. (*Note:* Readers interested in pursuing this matter further are referred to a study by Londeree[25] in which this and other matters related to the stability of equilibrium are considered at some length.)

Athletes in a variety of sports make use of the relationship between the location of the line of gravity and the stability of a body. A swimmer, poised for the start of a race, has his line of gravity passing close to the forward limit of his base so that when the gun is fired he will be able to disrupt his state of equilibrium and initiate forward motion with a minimum of effort. A track sprinter does something similar, and for exactly the same reason, when he rocks upward and forward into the "set" position. A defensive basketball player who expects his man to drive toward the basket will have his line of gravity passing somewhat closer to the backward limit of his base than it does to the forward limit. In this way he is a little less stable in a backward direction and therefore more ready to respond to a driving action on the part of his opponent.

That the weight of a body is also a factor governing its stability can be seen by once again considering the wrestler in Fig. 81(b). Here the moment opposing attempts to upset his equilibrium is the product of his weight and some fixed distance. Obviously then, the greater his weight, the greater his stability. This factor is one of the obvious reasons for the division of athletes into weight classes in such sports as wrestling, judo, and boxing.

Finally, imagine a skittle that could be balanced on either end and whose base dimensions were identical in both cases (Fig. 82). The center of gravity of the skittle lies on its midline (or axis of symmetry) and nearer to the more massive end than to the other. If the skittle is balanced in its "normal" position [Fig. 82(a)] and then given an angular displacement θ as shown, its

[25]Ben R. Londeree, "Principles of Stability: A Re-examination," *Research Quarterly*, XL, May 1969, pp. 419–22.

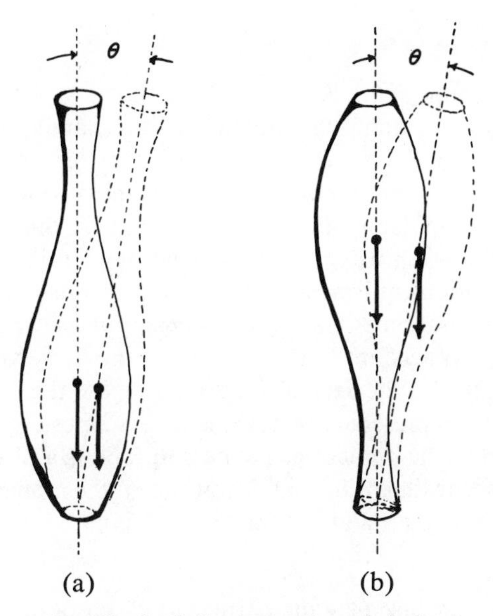

Fig. 82. If all else is equal—as it is in the case of the skittles depicted here—the higher the center of gravity of a body, the less its stability.

line of gravity still passes through the base and the skittle tends to return to its original equilibrium position. However, if the inverted skittle [Fig. 82(b)] is given the same angular displacement, its line of gravity passes outside the limits of the base and the skittle falls. The reason for the skittle being less stable in an inverted position than in a normal one can be deduced by considering the factors that have been shown to influence the stability of a body—the location of the line of gravity relative to the base limits and the weight of the body. Since these factors are identical, in both cases, the difference in stability must be due to some other factor. This factor, the only obvious difference between the two positions, is the height of the center of gravity above the base. If all else is equal, the lower the center of gravity, the more stable the equilibrium.

Moment of Inertia

The resistance of a body to changes in its motion is known as its inertia. In the case of linear motion, the inertia of a body is measured by its mass—the more massive a body, the greater its inertia and the more difficult it is to change its linear motion. With angular motion a similar state of affairs exists except that it is not only the mass of a body that determines its resistance to changes in its motion but also how this mass is distributed relative to the axis about which the body is rotating or tends to rotate. If the mass is concentrated close to the axis, it is much easier to alter the angular motion of a body than if the same mass is farther from the axis. When a child uses a full-sized baseball bat and "chokes up" on the handle, he is instinctively acknowledging and making use of this fact. For, by his actions, he moves the axis of rotation of the bat closer to the major mass of the bat and thereby makes the swing (angular motion) easier to control. Small children do the same kind of thing when using adult-sized eating utensils.

The angular motion equivalent of mass as a measure of a body's resistance to a change in its motion is termed the *moment of inertia* and, as has been indicated, takes into account both the mass of the body and how this mass is distributed relative to the axis of rotation. Suppose the axis of rotation for the bat in Fig. 83 is represented by the line XY. Suppose too that the particle of matter located at the point A has a mass m_1 and is a distance r_1 from the axis XY. Then, according to formal definitions of the moment of inertia, this particle's moment of inertia about the axis XY is

$$I_1 = m_1 r_1^2$$

Similarly, another particle (say the particle at B) has a moment of inertia

$$I_2 = m_2 r_2^2$$

and yet another (that at C), even though on the opposite side of the axis, has a moment of inertia

$$I_3 = m_3 r_3^2$$

If the moments of inertia of all the particles comprising the bat are summed, the result is the moment of inertia of the bat about XY:

$$\begin{aligned} I &= I_1 + I_2 + I_3 + \dots \\ &= m_1 r_1^2 + m_2 r_2^2 + m_3 r_3^2 \dots \\ &= \sum mr^2 \end{aligned} \tag{55}$$

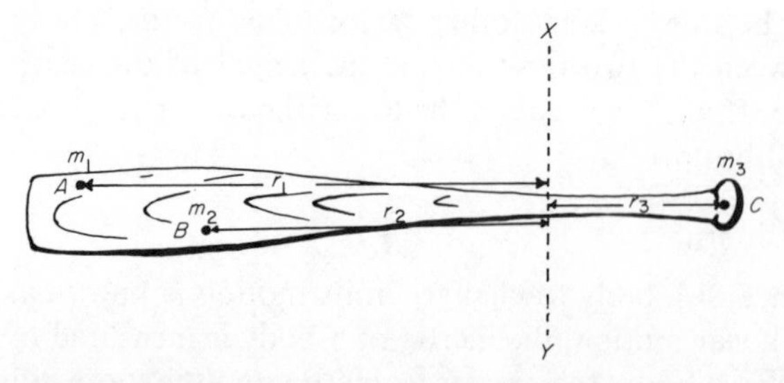

Fig. 83. Defining the moment of inertia of a baseball bat

The moment of inertia of a body may be determined in a variety of ways. If the body is regular in shape (e.g., square, rectangular, circular, or spherical) or is made up of parts that are themselves regular, its moment of inertia can be determined mathematically using Eq. 55. If, like many of the bodies involved in sports, it is irregular in shape, its moment of inertia is probably best found using an experimental method.

Experimental methods have been used to obtain values for the moments of inertia of segments of the human body (Table 11), and these values can be

*Table 11. Moments of Inertia of Selected Body Segments about Transverse Axes through Their Centers of Gravity**

Segment	*Moment of Inertia* ($slug\text{-}ft^2$)
Head	0.0183
Trunk	0.9300
Upper arm	0.0157
Forearm	0.0056
Hand	0.0004
Thigh	0.0776
Calf	0.0372
Foot	0.0028

*Adapted from Whitsett, *Some Dynamic Response Characteristics of Weightless Man*, p. 11.

used in determining the moment of inertia of a whole body. The procedure, somewhat akin to that of finding a body's center of gravity by the segmentation method, involves the use of a relationship known as the *parallel-axes theorem*. This theorem (which enables the moment of inertia of a body about any axis to be computed if its moment of inertia about a parallel axis through its center of gravity is known) is probably stated most simply in algebraic form:

$$I_A = I_{CG} + md^2 \tag{56}$$

where I_A = the moment of inertia of the body about an axis through the point A, I_{CG} = its moment of inertia about a parallel axis through its center of gravity, m = the mass of the body, and d = the distance between the parallel axes.

An example may help to clarify the meaning of this theorem. Suppose one wanted to compute the moment of inertia of a sprinter's leg as it rotated about an axis through his hip joint in the recovery phase of a running stride (Fig. 84). The moment of inertia of his thigh about a transverse axis through its center of gravity and parallel to the hip axis is given in Table 11 as 0.0776 slug-ft². The mass of his thigh is given in Table 8 as 0.103 times his total body mass. If this latter is, say, 5 slugs, the mass of his thigh is 0.515 slug. The distance between the thigh center of gravity axis and the hip axis is, say, 1 ft.

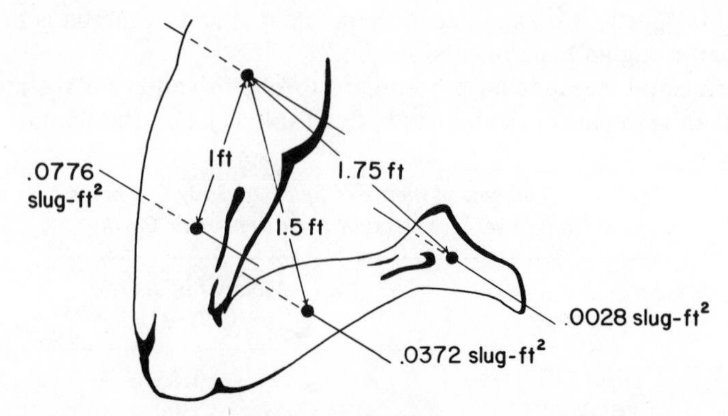

Fig. 84. The moment of inertia of a sprinter's leg can readily be determined using the parallel-axes theorem.

Then, according to the parallel-axes theorem, the moment of inertia of his thigh about the hip axis is

$$\begin{aligned} I'_{\text{hip}} &= I_{CG} + md^2 \\ &= (0.0776) + (0.515)(1)^2 \\ &= 0.5926 \text{ slug-ft}^2 \end{aligned}$$

Similar computations yield (a) the moment of inertia of the calf about the hip axis:

$$\begin{aligned} I''_{\text{hip}} &= I_{CG} + md^2 \\ &= (0.0372) + (0.043)(1.5)^2 \\ &= 0.1340 \text{ slug-ft}^2 \end{aligned}$$

and (b) the moment of inertia of the foot about the hip axis:

$$\begin{aligned} I'''_{\text{hip}} &= I_{CG} + md^2 \\ &= 0.0028 + (0.075)(1.75)^2 \\ &= 0.2325 \text{ slug-ft}^2 \end{aligned}$$

The moment of inertia of the whole lower limb about the hip axis is then found by simply adding the values for the three segments:

$$\begin{aligned} I_{\text{hip}} &= I'_{\text{hip}} + I''_{\text{hip}} + I'''_{\text{hip}} \\ &= 0.5926 + 0.1340 + 0.2325 \\ &= 0.9591 \text{ slug-ft}^2 \end{aligned}$$

Obviously this procedure could be extended to include all the segments of a body. In this way the moment of inertia of the whole body about an appropriate axis could be found. (Figure 85 shows the moments of inertia of a human body in some common diving and gymnastic positions.)

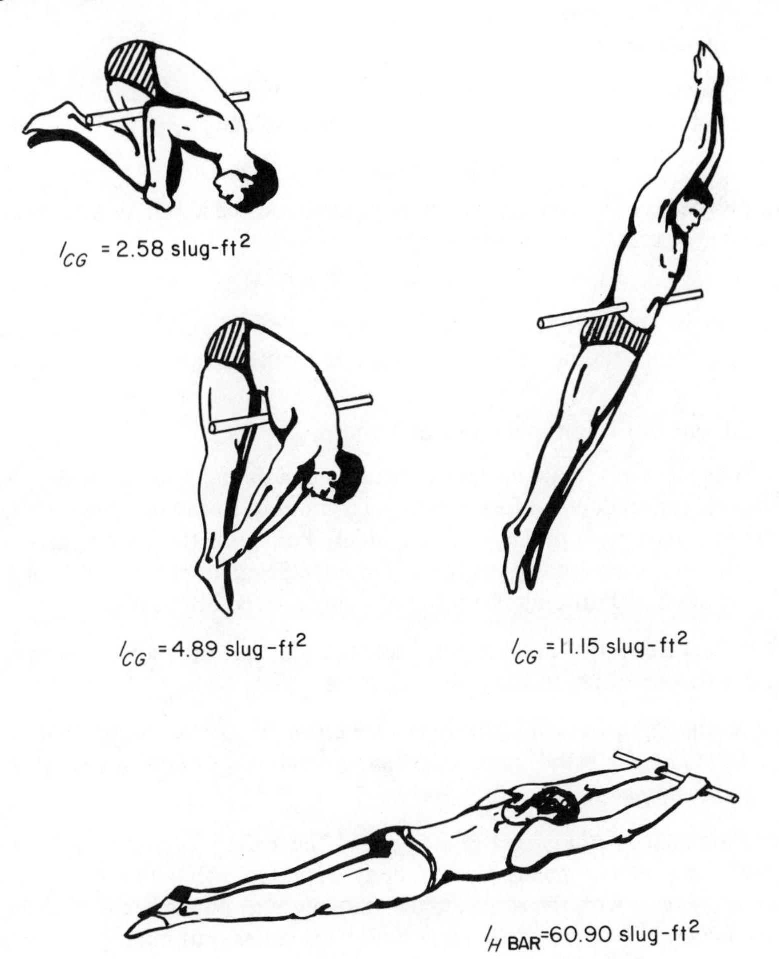

Fig. 85. Moments of inertia in some typical diving and gymnastic positions

Principal Axes

The rotation of a body is commonly described with reference to three axes (known as *principal axes*), which pass through its center of gravity. These axes are perpendicular to each other and are so located that the body's moment of inertia about one of them is as large as it could possibly be and about another is as small as it could possibly be—that is, as large or small as could possibly be for any axis that passed through the center of gravity of the body.

For a human body in an erect standing position the principal axes are closely approximated by lines drawn through the center of gravity and passing from the top of the head to the feet (the *longitudinal axis*), from left to right (the *transverse axis*), and from front to rear (the *frontal or anteroposterior axis*.)

Angular Momentum

The angular momentum possessed by a rotating body is equal to the product of its moment of inertia and its angular velocity:

$$\text{Angular momentum} = I\omega \qquad (57)$$

Thus, if the diver in Fig. 85(a) is rotating at 10 rad/sec about an axis through his center of gravity, his angular momentum is

$$2.58 \times 10 = 25.8 \text{ slug-ft}^2\text{/sec}$$

(Note again the parallel between corresponding quantities in linear and angular motion: momentum = mv; angular momentum = $I\omega$.)

Analogues of Newton's Laws of Motion

Just as most of the quantities of linear kinematics and linear kinetics have equivalent (or analogous) forms in angular motion, so too do Newton's laws of motion. Although perhaps not so widely known as the laws applying to linear motion, these analogous forms are nonetheless of considerable importance in obtaining an understanding of many sports techniques.

The First Law. For present purposes, the angular analogue of Newton's first law can be stated in the following form:

> *A rotating body will continue to turn about its axis of rotation with constant angular momentum, unless an external couple or eccentric force is exerted upon it.*

This statement, perhaps better known as the *principle of conservation of angular momentum*, means that a body spinning will continue spinning indefinitely (and with the same angular momentum) unless some other body exerts a couple or an eccentric force on it that causes it to modify its angular motion.

This fact is of particular significance to divers, gymnasts, jumpers, and other athletes who become airborne during the course of their events. Consider the example of a diver performing a tucked backward one-and-one-half somersault (Fig. 86). As he leaves the board, his extended body has certain amounts of linear and angular momentum—linear momentum to project him high into the air and to give him time to complete his mid-air actions, and angular momentum to rotate him through the required one-and-one-half somersaults. Shortly after takeoff the diver moves into a tucked position, thereby moving his body mass much closer to the axis of rotation and decreasing his moment of inertia. Because his angular momentum can be altered only by an external couple or eccentric force (principle of conservation of angular momentum), the internal forces involved in this tucking process have no effect upon it—it is the same when he is in the tuck as it was when he

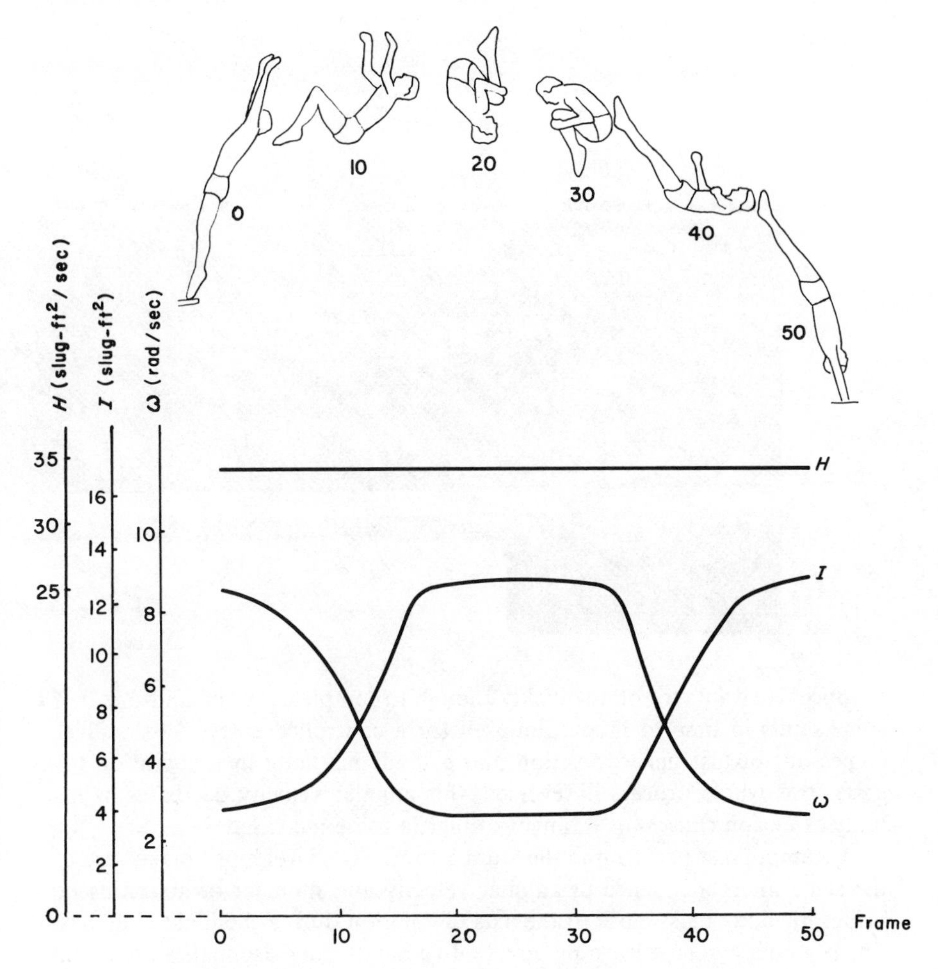

Fig. 86. A tucked backward one-and-one-half dive illustrates well the interplay between angular momentum, moment of inertia, and angular velocity.

left the board in an extended position. A quick glance at the equation for angular momentum, viz.,

$$\text{Angular momentum} = I\omega$$

shows that the only way the diver can decrease his moment of inertia and yet still retain the same angular momentum is to increase his angular velocity by a corresponding amount. And this is exactly what happens. At takeoff he has a relatively large moment of inertia and a relatively small angular velocity. When he tucks, he decreases his moment of inertia and increases his angular velocity—and, of course, it is this latter effect that he is trying to

produce. He wants to rotate quickly enough to complete his one-and-one-half somersaults in time to prepare himself for a controlled entry. As the diver comes out of his tucked position and extends his body in preparation for entry, this whole process is reversed—his angular velocity decreases as his body extension causes his moment of inertia to be increased.

A trampolinist performing the stunt known as "swivel hips" makes use of the same interdependence of angular velocity and moment of inertia as he brings his body mass close to the axis of rotation during the flight (Fig. 87). The beginner gymnast learning how to do a handspring also makes use of the same principle when he tucks in preparation for landing (Fig. 88). Instinctively he knows that he has insufficient angular momentum to enable him to complete the stunt in the approved manner, so he tucks and increases his angular velocity so that he can at least avoid crashing on his back.

Fig. 88. Good form in the handspring, contrasted with the poor form used by a beginner to "salvage" the stunt

Fig. 87. Swivel hips, another example of the conservation-of-angular-momentum principle

The Second Law. The angular analogue of Newton's second law can be stated as

> *The rate of change of angular momentum of a body is proportional to the torque causing it and has the same direction as the torque.*

Or it can be expressed algebraically in the form

$$T \propto \frac{I_2\omega_2 - I_1\omega_1}{t}$$

where T = the applied torque, I_1 and I_2 = the initial and final moments of inertia, and ω_1 and ω_2 = the initial and final angular velocities. If $I_1 = I_2$, this relationship reduces to

$$T \propto I\frac{(\omega_2 - \omega_1)}{t}$$

$$\propto I\alpha$$

where α = the angular acceleration. With the methods explained on pp. 61–62, this equation can be converted to the form

$$T = I\alpha \tag{58}$$

The diver in Fig. 56 provides a useful example of the application of Eq. 58. At the instant depicted, his body is being angularly accelerated about an axis through his feet. The torque causing this angular acceleration is equal to the product of the diver's weight and the horizontal distance x. If the diver is executing a dive requiring a large amount of rotation (say, a forward two-and-a-half), he will want this torque to be relatively large so that his body can leave the board with sufficient angular velocity to enable him to complete the dive. On the other hand, if he is executing a dive that involves very little rotation (say, a plain forward dive), he will have little need for angular acceleration during the takeoff and will want to keep the applied torque relatively small. Because his weight is constant, and the only other factor that influences the magnitude of the torque is the distance x, the only way in which he can control this torque is by making alterations in the magnitude of x. And this is exactly what he does. By varying the amount of forward body lean that he has at takeoff, he controls the magnitude of x and hence too the magnitude of the torque applied at this time.

The Third Law. The angular analogue of Newton's third law can be stated as

> *For every torque that is exerted by one body on another there is an equal and opposite torque exerted by the second on the first.*

Sports offer countless examples to illustrate this angular equivalent of Newton's third law. Probably the most common is that in which an athlete applies a torque to one part of his body by contracting a muscle (or group of muscles), thereby causing that part to rotate. The equal and opposite reaction to this applied torque causes some other part of his body to rotate, or tend to rotate, in the opposite direction.

When a long jumper swings his legs forward ready for landing, a torque equal and opposite to that exerted on his legs is applied to the remainder of his body. The net effect is that as the jumper swings his legs forward and upward, in say a clockwise direction, the remainder of his body moves forward and downward in a counterclockwise direction (Fig. 89).

It should be noted here that since angular accelerations obtained when equal torques are applied to two different bodies depend on their respective moments of inertia (Eq. 58), the effect on one body is rarely equal to that produced on the other—cf. the corresponding linear case, pp. 65–66. This is evident in a basketball jump shot, for example. When a player propels the ball toward the basket by extending his elbow and flexing his wrist, the remainder of his body is acted upon by torques equal in magnitude and opposite in direction to those causing these movements. Because the moment of inertia of the rest of the body is much greater than that of either the forearm or hand, the effects of these equal torques appear to be quite different. Whereas the forearm and hand sweep quickly through a relatively large angle, the remainder of the body exhibits only a slight tendency to rotate in the opposite direction.

Fig. 89. Angular action and reaction in a long-jump landing

The angular "action-reaction" effects are by no means limited to a forward-backward (or sagittal) plane. In an across-the-body (or frontal) plane these effects are very clearly seen in the instinctive responses of a gymnast who senses she is about to topple off a balance beam (Fig. 90). As soon as she feels she is starting to overbalance, she rotates her arms (and perhaps her nonsupporting leg) in the direction in which she is falling. The effect of these actions is to cause the rest of the body to rotate in the opposite direction. If this maneuver is successfully carried out, her tendency to overbalance is arrested.

The tennis player in Fig. 91 exerts torques to produce the clockwise backhand stroke depicted. These torques are accompanied by equal and opposite torques, which tend to turn the rest of his body in a counterclockwise direction. However, because his feet are in firm contact with the ground, this contrary tendency is transmitted to the ground. Now, the moment of inertia of the tennis player about the axis in question is relatively small and the angular acceleration he experiences is thus clearly apparent—one can readily see how he is affected by the torques produced by his muscles. In contrast, the moment of inertia of the earth is enormous, and the angular acceleration it experiences as a result of the torques transmitted via the player's feet is quite imperceptible. Players are often exhorted, therefore, to keep both feet in contact with the ground so that the reactions that accompany the strokes they make can be "absorbed" in this way.

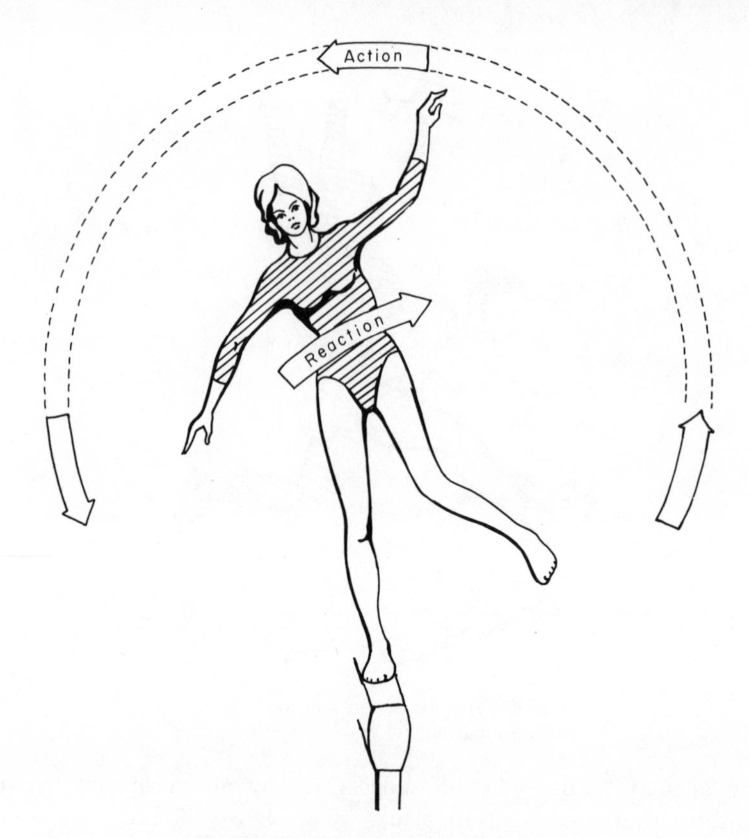

Fig. 90. The gymnast's instinctive actions create an angular reaction tending to restore her balance.

Fig. 91. The reaction to the tennis player's backhand drive is "absorbed" by the ground.

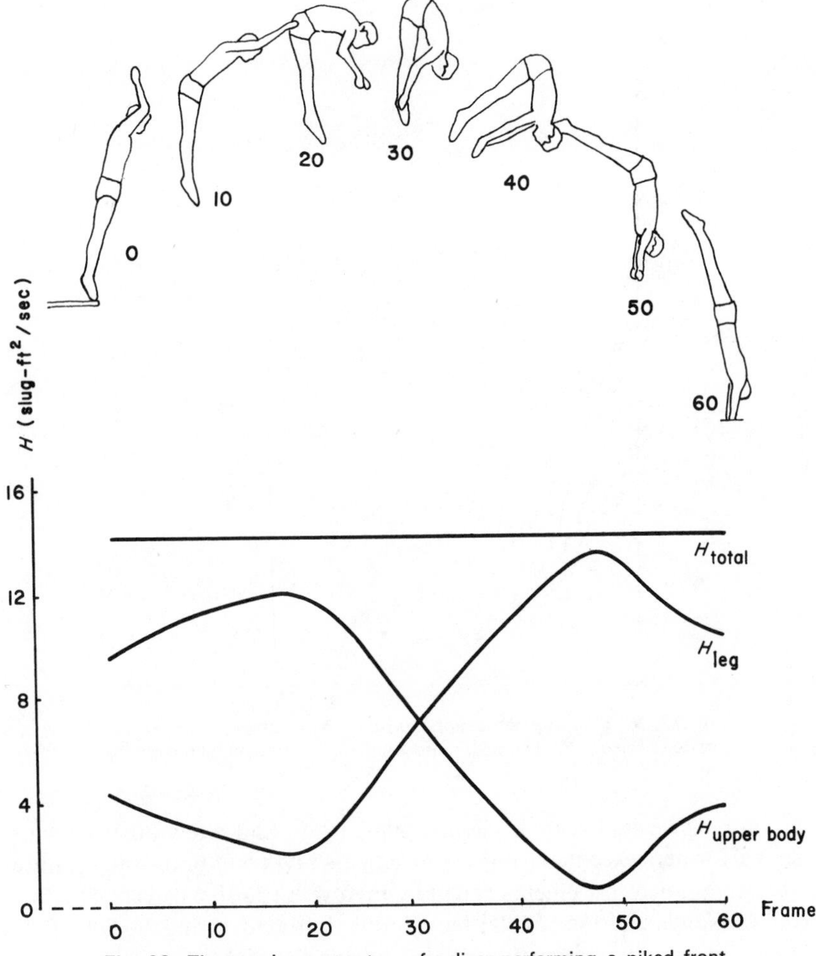

Fig. 92. The angular momentum of a diver performing a piked front dive is first localized in his upper body and then in his legs.

Transfer of Momentum

When a body is airborne and the angular momentum of one part of the body is decreased, some part (or all) of the rest of the body must experience an increase in angular momentum if the total angular momentum is to be conserved (or held constant). For example, consider the case of a diver performing a piked front dive (Fig. 92). As he goes into his piked position, the angular momentum of his legs is reduced to zero or near-zero—they appear to be stationary in the air—and the angular momentum of his arms and trunk is markedly increased. Then, as he assumes his position for entry, this process is reversed. The angular momentum of the arms and trunk is reduced to zero, or near-zero—and they appear to remain stationary—while the legs swing upward and into line with the rest of the body. This process whereby momentum is redistributed within a body is commonly referred to as a *transfer of momentum*.

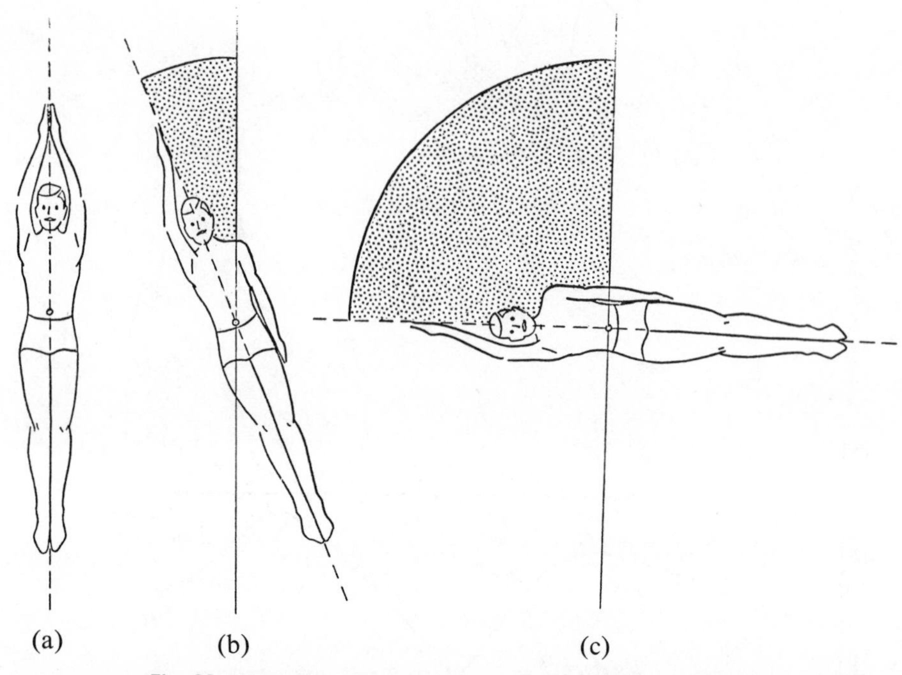

Fig. 93. With appropriate actions to rotate him sideways, a diver can "trade" somersaulting angular momentum for twisting angular momentum.

A similar process occurs in diving, gymnastics, and a few other activities where performers execute airborne movements involving both somersaulting and twisting. Such movements generally involve a transfer (or "trading") of part of the somersaulting angular momentum, initiated during the takeoff, for the twisting angular momentum needed during the flight.

Consider the case of a diver who leaves the board to execute a front dive [Fig. 93(a)] and then, while airborne, vigorously swings one arm sideways to bring it down alongside his body. The reaction to this motion is a contrary rotation of the rest of his body about his frontal axis—a rotation that tilts his body in the opposite direction [Fig. 93(b)]. Although virtually impossible in practice, suppose that the diver abducts his arm with such vigor that in reaction his body is brought into a horizontal position [Fig. 93(c)].

Now when the diver left the board he possessed a certain amount of angular momentum about a horizontal axis perpendicular to the line of the board (an axis sometimes referred to as the *axis of momentum*) and, because the transverse axis of his body coincided with this horizontal axis, he experienced a somersaulting rotation. When his body was subsequently rotated into a horizontal position, his longitudinal axis was brought into line with the axis of momentum. Thus, in keeping with the conservation-of-angular-momentum principle, his body acquired the same amount of angular momentum about its longitudinal axis as it had previously possessed about its transverse axis. In other words, by readjusting the position of his body relative to the axis of momentum, the diver was able to "trade" all his somersaulting momentum for twisting momentum.

In practice, the actions used to make the body tilt—generally both arms moved vigorously in the same angular direction—are insufficient to produce more than a very limited amount of side-somersaulting rotation. As a result, only part of the diver's somersaulting angular momentum is "traded" for twisting angular momentum, and the dive is executed with both somersaulting and twisting proceeding simultaneously. Finally, because the trading of somersaulting for twisting angular momentum requires the diver to move his body out of alignment with the vertical plane of his flight path, his body is not correctly positioned for entry once he has completed the prescribed number of twists and somersaults. To correct for this the good diver reverses the direction of his earlier arm action, thereby imparting sufficient side-somersaulting rotation to his body to bring it back into its original alignment for entry.

The concept of transferring momentum is most frequently used in explaining what takes place in situations other than those in which a body is airborne. For example, when a diver performs a backward dive, he swings his arms upward and backward prior to leaving the board. Then, as they near the limit of their range of motion in this direction, they begin to slow down. The angular momentum that the arms lose at this time (or at least a large portion of it) is "matched" by a corresponding increase in the angular momentum of the rest of the body. Now, although the angular momentum of a body is not necessarily conserved in situations like this (since the body is subject to external torques that tend to alter its angular momentum), the effect is qualitatively very similar to cases in which the angular momentum is conserved—one part of the body loses (or "gives up") angular momentum at the same time as another part experiences a gain in angular momentum.

Initiating Rotation in the Air

While the angular analogues of Newton's laws indicate quite clearly that a body cannot acquire angular momentum unless acted upon by an external torque, the possibility of initiating rotation in the air has been of considerable interest for some time. Sparking much of this interest has been the performance of cats, rabbits, guinea pigs, and other animals, which have the ability to right themselves when falling upside down. While initially some doubts were expressed that these animals initiated the turns in the air rather than at the time of takeoff, it is now widely accepted that this is the case.

How are these turns initiated in apparent defiance of Newton's laws? The answer most frequently put forward lies in the relationship between the moments of inertia of the body parts that interact when an angular action is initiated in the air. Consider the cat being dropped in Fig. 94. As he begins to fall, he bends (or pikes) in the middle [Fig. 94(a)], brings his front legs in

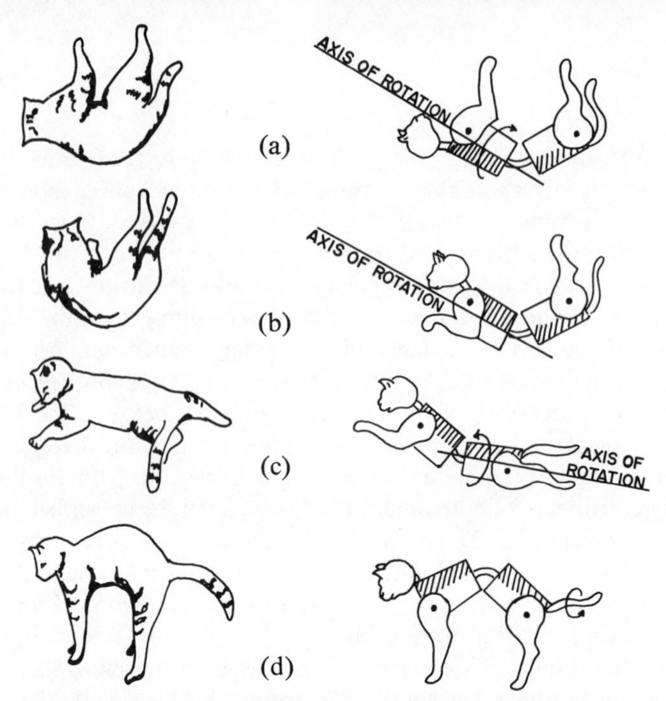

Fig. 94. A falling cat initiates rotation in the air in the absence of an external torque and in apparent defiance of the law of conservation of angular momentum. (*The figures from which these diagrams were redrawn first appeared in* New Scientist, *the weekly international review of science and technology, 128 Long Acre, London WC2, and appear with the publisher's permission.*)

close to his head, and rotates his upper body through 180° [Fig. 94(b)]. In reaction to this rotation, his lower trunk, hind legs, and tail, all of which are some considerable distance from the axis of rotation, rotate in the opposite direction. However, because the moment of inertia of these body parts is much greater than that of his upper body, the angular distance through which they move is correspondingly small ("about 5°" according to McDonald[26]). To complete the required 180° turn, the cat then brings his hind legs and tail into line with his lower trunk and rotates these body parts about an axis running longitudinally through his hindquarters [Fig. 94(c)]. The reaction is again very small, this time due to the disposition of the upper body relative to the axis. Finally, to make any minor adjustments necessary, the cat rotates his tail in a direction opposite to that in which it is desired to move the body. (Since Manx cats and cats completely without tails can right themselves if held upside down and dropped, these final movements of the tail are clearly not essential ingredients of the righting maneuver.)

It should be carefully noted that throughout this whole sequence of movements the angular momentum of one body part has always been "matched" with an equal and opposite angular momentum of some other part and that, as a consequence, the total angular momentum has been quite

[26]Donald McDonald, "How Does a Cat Fall on Its Feet? "*The New Scientist*, VII, June 30, 1960, p. 1647.

unaffected. In short, the cat has *not* defied Newton's laws but has merely appeared to do so. (*Note:* Research by Smith and Kane[27] and Kane and Scher[28] has suggested some alternative explanations of the process by which the cat rights itself when dropped in an inverted position.)

With all the interest in this question of how a falling cat rights itself, some interesting side results have been reported:[29]

1. If dropped upside down, a cat can turn over within its own standing height.
2. The cat's eyes and the mechanisms of its inner ear both play a part in sensing the need to initiate a turn. Of these, the eyes seem to be the more important—a blindfolded cat dropped from as low as 3 ft lands clumsily, while a cat without an intact inner ear mechanism can still right itself efficiently. However, a cat deprived of both sensory organs made no attempt to right itself when dropped upside down.
3. A blindfolded cat, rotated in a special apparatus in order to "confuse" the organs of its inner ear, was reported to have rotated over and landed on its back when it accidentally slipped feet first out of the apparatus!

The possibility of athletes' using similar techniques has been the source of some interest, and various writers have discussed the matter relative to diving,[30,31,32,33] trampolining,[34] and track and field.[35,36] The question has also interested those engaged in research related to man's ability to maneuver in a weightless state.[37,38,39]

[27]Preston G. Smith and Thomas R. Kane, *The Reorientation of a Human Being in Free Fall*, Technical Report No. 171, Division of Engineering Mechanics, Stanford University, May 1967.

[28]T. R. Kane and M. P. Scher, "A Dynamical Explanation of the Falling Cat Phenomenon," *International Journal of Solids and Structures*, V, July 1969.

[29]Donald McDonald, "How Does a Cat Fall on Its Feet?" pp. 1647–48.

[30]Geoffrey H. G. Dyson, *The Mechanics of Athletics* (London: University of London Press Ltd., 1973), pp. 105–9.

[31]George Eaves, *Diving, The Mechanics of Springboard and Firmboard Techniques* (London: Kaye & Ward, Ltd., 1969).

[32]F. R. Lanoue, "Mechanics of Fancy Diving" (M. Ed. thesis, Springfield College, 1936).

[33]Donald McDonald, "How Does a Man Twist in the Air?" *New Scientist*, X, June 1, 1961, pp. 501–3.

[34]Dennis E. Horne, *Trampolining: A Complete Handbook* (London: Faber & Faber, Ltd., 1968), p. 111.

[35]H. A. L. Chapman, "Rotation—Its Problems and Effects" in *International Track and Field Digest*, ed. by Don Canham and Phil Diamond (Ann Arbor, Mich., "Champions on Film," 1957), p. 243.

[36]Dyson, *The Mechanics of Athletics*, pp. 111–13.

[37]T. R. Kane and M. P. Scher, "Human Self-Rotation by Means of Limb Movements," *Journal of Biomechanics*, III, January 1970, pp. 39–49.

[38]Kulwicki, Schlei, and Vergamini, *Weightless Man: Self-Rotation Techniques.*

[39]Smith and Kane, *The Reorientation of a Human Being in Free Fall.*

From all this research and speculation several conclusions have been reached:

1. A man *can* initiate rotation while he is in the air. A simple demonstration of this can be given by a trained gymnast who does a series of consecutive, pike-to-front-drop movements on a trampoline and executes a half twist to land in a flat back position (instead of in a front drop) when called upon to do so. A shouted command to the gymnast shortly after he leaves the bed—and at which time he has zero angular momentum because that is what is required to perform a series of consecutive front drops—enables him to convincingly demonstrate the point (Fig. 95).

2. A man's ability to initiate rotation in the air is a function of how much training or practice he has had—in a plain jump from a 1-m board, a trained diver can initiate a twist in the air and turn through as much as 450°, while an untrained man can rarely exceed 90°.

Fig. 95. A trampolinist can initiate a rotation in the air, as in this front-drop pike half-twist flat-back sequence.

3. The basic mechanism involved appears to be similar to that used by the cat, although there are variations from movement to movement and from one individual to another in performing the same movement.

4. Starting positions in which the body is arched or piked facilitate the initiating of rotation. In addition, it appears that while a man can, with relative ease, initiate a twisting rotation while in the air, initiating rotations about either of the other two principal axes does not appear to be quite so easy.

The extent to which divers, trampolinists, and others involved in "aerial activities" actually use such techniques is anything but easy to determine. It is clear though that a considerable number of the movements in such activities are executed in a manner consistent with the initiation of catlike rotations in the air.

Centripetal and Centrifugal Force

When a tennis player executes a forehand drive, muscles of his trunk exert forces on his arm to cause it and the racket (which serves as an extension of the arm) to swing through an arc. Suppose that the vector F (Fig. 96) represents the forces exerted by the trunk on the arm and R and T represent the components of F along and at right angles to the arm. If the axis of rotation is a vertical one through the shoulder joint, T causes an angular acceleration about this axis that increases the tangential velocity of the racket. R, on the other hand, is the force responsible for the radial acceleration (see p. 54) that changes the direction in which the racket is moving. Because this latter force acts toward the center of rotation, the axis, it is known as the *centripetal* (or "center-seeking") *force*.

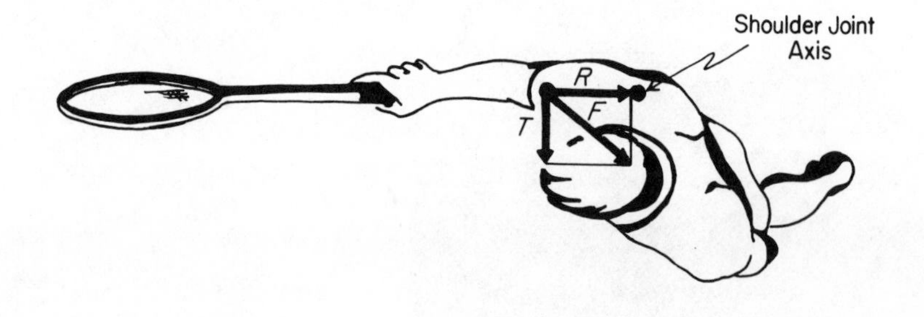

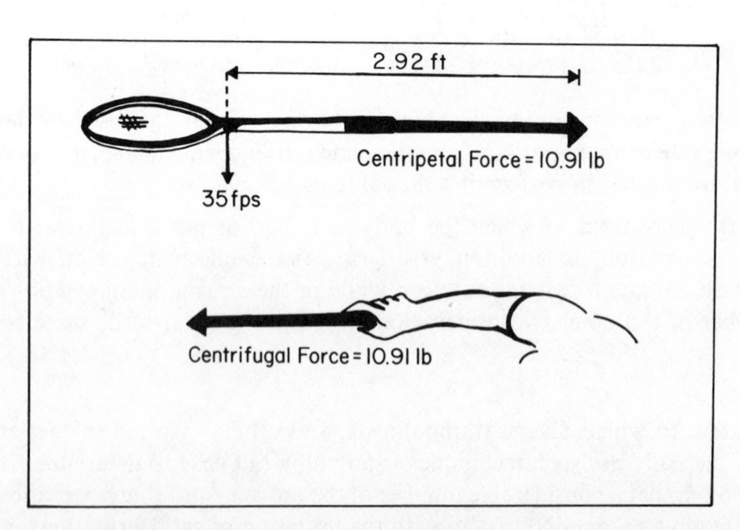

Fig. 96. Centripetal and centrifugal forces in a forehand shot in tennis

The magnitude of this force is obtained by combining the equations for Newton's second law ($F = ma$) and for radial acceleration ($a = v_T^2/r$). Thus:

$$\text{Centripetal force} = m\frac{v_T^2}{r} \tag{59}$$

Because $v_T = \omega r$ (Eq. 26), it is sometimes useful to substitute $\omega^2 r^2$ for v_T^2 in Eq. 59. The right-hand side of this equation is then expressed in terms of the angular rather than the linear velocity:

$$\text{Centripetal force} = mr\omega^2 \tag{60}$$

Suppose now that the racket has a mass of 0.026 slug and that its center of gravity is 2.92 ft from the axis and is moving with a speed of 35 fps. Then, according to Eq. 59, the centripetal force applied to the racket must be

$$\frac{0.026 \times (35)^2}{2.92} = 10.91 \text{ lb}$$

Further, since the man's hand is the only other body in contact with the racket, this 10.91-lb force must be applied to the racket by his hand. Now whenever one body exerts a force on another "there is an equal and opposite force exerted by the second body on the first" (Newton's third law). Thus, if the hand exerts a force of 10.91 lb on the racket, the racket in turn exerts a force of 10.91 lb in the opposite direction on the hand. This force, which always acts away from the center of rotation, is termed the *centrifugal* (or "center-fleeing") *force.*

The concept of a centrifugal force is frequently the source of confusion. The two principal reasons appear to be

1. A failure to recognize that centrifugal and centripetal forces do not both act on the same body. Apart from the fact that Newton's third law indicates that the "action" acts on one body and the "reaction" on another, it should be obvious that the resultant of two equal and opposite forces acting on a body is zero and that application of a zero force would not change the direction in which a body is moving.

2. A failure to realize that "actions" and "reactions" occur simultaneously. Thus, if a tennis player released his grip on his racket partway through a stroke, both the centripetal force exerted by his hand on the racket and the centrifugal force exerted by the racket on his hand would cease to exist at the same time. Under such circumstances, and in accord with Newton's first law, the racket would tend to continue traveling in the same direction as it was at the moment of release (i.e., tangent to the point on the arc at which it was located at that instant).

[*Note:* Contrary to what is often supposed in such cases, the racket does not have a tendency to travel radially outward under the action of a centrifugal force—(a) because the centrifugal force acts on the player's hand and not on the racket and (b) because, even if it did act on the racket, it would cease to exist at the moment the corresponding centripetal force was removed.]

Centripetal and centrifugal forces are exerted whenever a body moves on a curved path. In sports, however, there are times when these forces seem more important than others because athletes must make conscious adjustments in technique to allow for their existence. Probably the most striking examples are seen when track sprinters and cyclists negotiate a bend in the track. Here the only body that can exert the required centripetal force on them is the ground, the only body with which they are in contact. If the ground exerts an inward horizontal force against the foot of the runner (or against the wheels of the bike), this eccentric force will have the required effect of changing the direction of his motion. However, it will also have the undesirable effect of rotating him outward. (Remember, an eccentric force causes both translation and rotation—see p. 105). To combat this rotary effect the athlete leans inward so that the vertical component of the ground reaction will also act eccentrically and provide a moment in the opposite direction to that produced by the eccentric centripetal force. When the speed of athletes exceeds a certain limit (as it often does in cycling and motor sports) or the radius of the track is very small (as at many indoor track meets), the ground is no longer able to provide the necessary amount of centripetal force. This means that, unless some additional provision is made, the athletes will have to slow down going into the bends or risk failing to safely negotiate them. It is to avoid these problems that banked tracks are built for cycling velodromes and indoor track meets. In this way the component of the athlete's weight acting down the slope can contribute to the centripetal force required, and the need for an inward directed force exerted by the ground is at least reduced, and perhaps eliminated entirely.

Chapter 7

FLUID MECHANICS

All motion in sports is influenced by the fluid environment in which it takes place. A basketball player driving down the court is slowed a little by the need to push his way through the surrounding air, his particular fluid environment. A skin diver's motion is affected even more by the fluid environment (the water) in which he operates. A swimmer competes in two different fluid environments (the water and the air) simultaneously and his motion is influenced by the effects that each of these has on those parts of his body that move through it.

In a great many instances the effects produced by the fluid environment are so small that they can reasonably be disregarded in all but the most precise analysis. Garfoot,[1] for example, has computed that a 16-lb shot, otherwise destined to be thrown approximately 56 ft, would have this distance reduced by about 0.4 ft as a result of air resistance. As this constitutes less than 1 % of the total distance, the common practice of ignoring the effects of air resistance in shot-putting is clearly justified.

In other cases, the effects of fluid resistance may be very pronounced indeed. Among these are those (like swimming, rowing, canoeing, yachting, and water skiing) concerned with motion through water and those concerned with high-velocity motion through the air (either as projectiles in such sports as skydiving, ski jumping, archery, baseball, and golf or as nonprojectiles in skiing, ice skating, and motor racing).

Flotation

The ability of a body to float (i.e., to maintain a stationary position at the surface of the water) is of some importance in most aquatic sports. In swim-

[1]Brian P. Garfoot, "Analysis of the Trajectory of the Shot," *Track Technique*, No. 32, June 1968, p. 1006.

ming, for instance, a person's floating ability can influence his success at both beginner and championship levels. Logically, a person who is able to float with ease is likely to learn how to swim more readily than one who floats only with difficulty or not at all. At the other end of the performance scale, a champion swimmer who floats high in the water is likely to encounter less resistance to his forward motion than one who cannot float as well. If all else is equal, such a swimmer clearly has a distinct advantage over his opponents who have less floating ability. (*Note:* While the ability to float is a decided advantage in swimming, its importance should not be overrated. Many people who cannot float have learned to swim without much difficulty and some of these have even broken world records and won Olympic titles in the sport.)

Buoyant Force. The swimmer in Fig. 97(a) is floating horizontally on the surface of the water. Because she is in a state of equilibrium in this situation, the sum of the forces acting on her in any direction must be equal to zero. In the vertical direction (and this is the important one in determining whether a body floats on the surface or sinks below it), the only forces acting are her weight and whatever vertical forces the water exerts on her. Clearly then, the resultant of these upward vertical forces (the so-called *buoyant force*) that the water exerts must be equal in magnitude to her body weight. This simple conclusion leads to the fundamental condition determining whether a given body will float or not. If the weight of the body is greater than the maximum buoyant force that the water can provide, the body will sink. If it is not, the body will float. Expressing this in mathematical terms, a body will float only if

$$\text{The weight of the body} \leqslant \text{the maximum buoyant force}$$

With the maximum buoyant force obviously of great importance in determining whether a body can float, it is relevant to examine the conditions that govern the magnitude of this force.

When the swimmer lies on the surface, she pushes aside (or displaces) a certain amount of water. Before this happens, this displaced water lies in equilibrium under the forces acting upon it. In the vertical direction, these forces are its own weight and the vertical upthrust, or buoyant force, exerted on it by the water below. And, since the water is in equilibrium, the two forces are necessarily equal in magnitude [Fig. 97(b)]. Now when the swimmer adopts her prone-lying position, she does nothing to effectively alter the state of the water below the water that she displaces. Consequently, the force that this water exerts on her is exactly the same as that which it previously exerted on the displaced water [Fig. 97(c)]. In other words, she experiences a buoyant force equal in magnitude to the weight of the water that she displaced. (This equality of the buoyant force and the weight of the displaced water was first discovered by Archimedes—while taking a bath!—and is widely known as *Archimedes' principle.*)

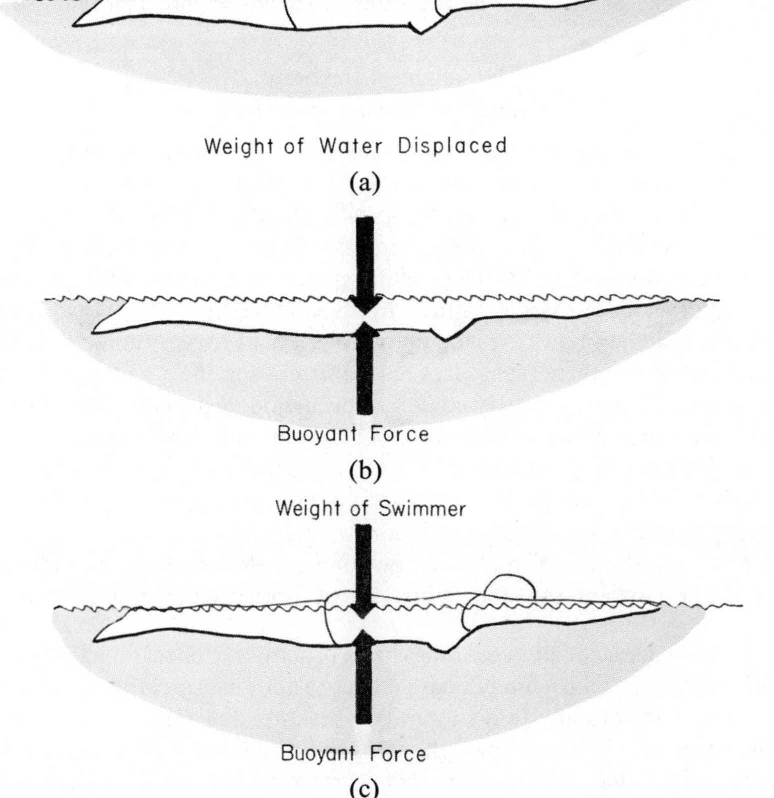

Fig. 97. The buoyant force is equal to the weight of the water displaced.

The maximum volume of water that a swimmer could displace would be a volume equal to that of her own body—such a maximal displacement occurring only if she were totally immersed in the water. This limit on the volume of water that can be displaced sets the upper limit on the magnitude of the buoyant force that can be exerted on the swimmer. The maximum buoyant force has the same magnitude as the weight of a volume of water equal to the volume of the swimmer's body.

Specific Gravity. The fundamental condition determining whether a body can float can now be restated as follows. A body will float only if:

The weight of the body $\leqslant$ the weight of an equal volume of water

In dealing with mathematical expressions of this kind it is permissible to divide both sides of the expression by an equal amount. Suppose, therefore, that both sides of the expression above were divided by "the weight of an equal volume of water." The resulting statement would then read—a body will float only if:

$$\frac{\text{The weight of the body}}{\text{The weight of an equal volume of water}} \leqslant 1$$

The fraction on the left-hand side is known as the *specific gravity* of a body and, as can be readily surmised, is a useful measure of its capacity to float.

The factors that determine the specific gravity of a human body shed a good deal of light on why some people can float with ease while others have no hope of doing so at all. The specific gravity of a human body (or, indeed, of any other body) is determined by its composition or physical makeup. Because a human body is made up of a variety of tissues (bone, muscle, fat, etc.) and because these themselves have different specific gravities, the amount of each that a person's body contains has a good deal to do with whether he can float or not. If his body contains a large amount of fat, which is relatively very light (specific gravity $\approx$ 0.8), he is much more likely to be able to float than if he is lean and heavily muscled (specific gravity of muscle $\approx$ 1.00) or "heavy boned" (specific gravity of bone $\approx$ 1.5–2.0).

The importance of body composition in determining an individual's specific gravity, and thus the ability to float, is reflected in a number of ways:

1. The volume of air in the lungs has a pronounced effect on an individual's ability to float. If a person inhales deeply, he adds considerably to the volume of air that is normally in his lungs (the residual air) and increases both the volume of his chest and the volume of his whole body. The increase in his body weight that accompanies this increase in volume is negligible. (The specific gravity of air is of the order of 0.0012, which is very small indeed.) With the numerator of the expression for specific gravity remaining virtually unaltered, and the denominator being markedly increased, the overall effect of this deep inhalation is to substantially reduce the body's specific gravity. The likelihood of the person being able to float is therefore enhanced. Conversely, if a person exhales forcefully, the specific gravity of his body is increased and his floating ability is correspondingly decreased.

The importance of this factor in terms of human flotation is clearly evident in the remarks of Whiting[2] who, after reviewing literature on this subject, stated that while most men and most women will float in water if they have taken a full inhalation, the majority of the men will sink unless they have more than just residual air in their lungs.

[2]H. T. A. Whiting, *Teaching the Persistent Non-Swimmer* (London: G. Bell & Sons, Ltd., 1970), p. 6.

2. As the relative proportions of the major body tissues change with age, so too does a person's specific gravity and ability to float. In general, the nearer a person is to the extremes in age (i.e., the nearer he is to being very young or very old) the more likely it is that his specific gravity will be low enough to permit him to float.

3. Women, because of their greater proportions of fat, tend to have lower specific gravities than men and are thus more likely to be able to float.

4. Studies[3,4] on the physiques of champion swimmers have shown that, in general, these people have slightly higher proportions of fat in their physiques than do champion athletes in most other sports.

Center of Buoyancy. In Fig. 97(b), the water that is subsequently displaced by the swimmer is in equilibrium under the action of two vertical forces —its weight and the buoyant force. The fact that this water is in a state of equilibrium means that these two forces not only must be equal in magnitude but also must act along the same straight line. Thus the buoyant force, like the weight, must act through the center of gravity of the about-to-be-displaced water. When the swimmer assumes the position of Fig. 97(c), the magnitude and line of action of the buoyant force are exactly as before, and the point through which the buoyant force acts (previously the center of gravity of the displaced water) is called the *center of buoyancy*.

The location of the center of buoyancy is of particular significance in determining what happens to a swimmer's body once a prone-lying position is assumed on the surface. If the center of gravity of the swimmer and the center of buoyancy coincide, or lie vertically one above the other, the body will retain its horizontal position. This, though, is a relatively uncommon situation, especially for males. (In tests of back-floating ability conducted by Whiting,[5,6] only one in six of his female subjects was able to maintain a horizontal floating position, and his male subjects fared even worse. In fact, of his 291 male subjects aged 15 yr and over, not one could maintain the position!) If the center of buoyancy does not coincide with the center of gravity or lie with it on the same vertical line, it is almost invariably found to be nearer the head than is the center of gravity. This then means that the

[3] J. E. Lindsay Carter, "The Somatotypes of Swimmers," *Swimming Technique*, III, October 1966, pp. 76–79.

[4] T. K. Cureton, *Physical Fitness of Champion Athletes* (Urbana, Ill.: University of Illinois Press, 1951).

[5] H. T. A. Whiting, "Variations in Floating Ability with Age in the Male," *Research Quarterly*, XXXIV, March 1963, pp. 84–90.

[6] H. T. A. Whiting, "Variations in Floating Ability with Age in the Female," *Research Quarterly*, XXXVI, May 1965, pp. 216–218.

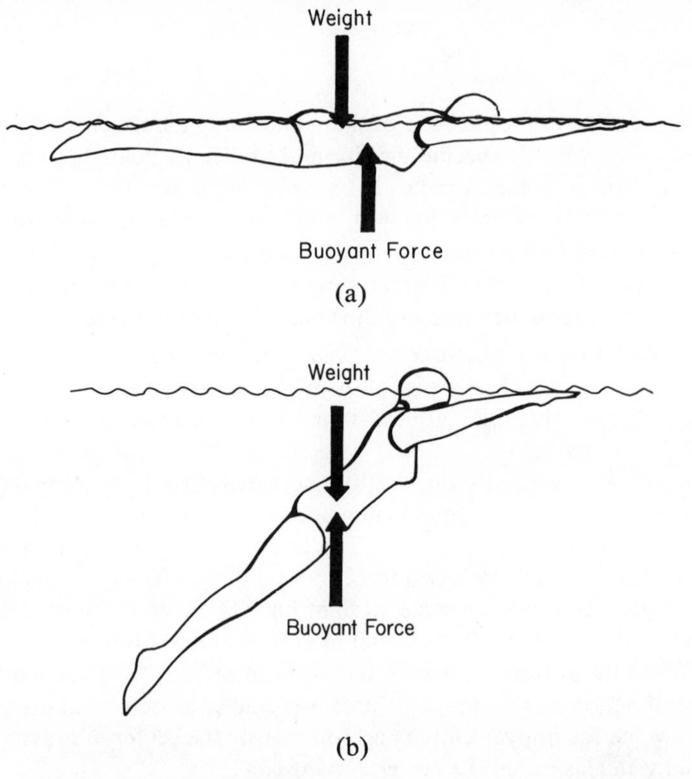

Fig. 98. Whether a body floats horizontally or rotates to some inclined position is governed by the relative positions of the lines of action of the weight and the buoyant force.

weight and the buoyant force act as a couple that tends to force the legs and feet downward [Fig. 98(a)].[7] In cases like this, the feet drop steadily until a point is reached at which the center of gravity and the center of buoyancy lie along the same vertical line. The body then floats in this position [Fig. 98(b)].

Relative Motion

When a water skier skims across a lake in the wake of a speedboat, to an observer on the shore it appears as if the water is still and the skier is moving across it at some speed—say 30 mph. However, if the skier looks down toward his feet, he can very easily get the reverse impression that he himself is stationary and that the water is rushing past him at this same 30 mph. No matter which of these two viewpoints is taken, the difference between the speeds of the water and the skis (i.e., the motion of one relative to the other) is exactly the same, namely, 30 mph.

Now it so happens that the effect that the water has on the skis depends on the relative motion between them rather than on the speed of either one.

[7]In Fig. 98(a) the magnitude of the weight is slightly greater than that of the buoyant force. Thus, the couple that produces the observed angular acceleration is actually made up of the buoyant force and a part of the weight equal in magnitude to the buoyant force. The remainder of the weight serves to accelerate the body downward.

For this reason either of the two viewpoints (that of the observer or that of the skier) can be taken in making an analysis of these effects.

In analyses of the influence of fluids on the motion of bodies involved in sports, it is generally more convenient to consider the body to be at rest and the fluid to be moving past it. Ganslen,[8,9] for example, has used this approach in his analyses of discus and javelin flight. To do this, he placed the implements in a wind tunnel and regulated the flow of air past them so that the relative motion of air and implement was the same as it would be if the implement were thrown in the normal way. Raine[10] has reported using a similar procedure in analyzing the air resistance encountered by a skier when he adopts various racing positions. [*Note:* It is perhaps of interest to note here that the resistance provided by air is fundamentally a "scaled-down" equivalent of the resistance provided by water. Both fluids (and many others) behave in essentially the same manner even though their effects on a body moving through them may appear to be markedly different. For this reason (provided due care is taken to get the scaling factors correct) it is possible to test how a supersonic aircraft, a glider, or even a discus will behave in flight, by noting its reaction to simulated flight in water.]

Fluid Resistance

When a discus is placed in a wind tunnel and air is made to flow past it, two simultaneous effects are produced. First, the direction of motion of the air, especially that nearest the discus, is altered so that it can pass around the obstruction in its path. Second, the air near the surface of the discus is slowed down as a result of coming into contact with the discus. These changes in the speed and direction of the airflow are brought about because the discus exerts forces on the air. And, of course, in reaction the air exerts equal and opposite forces on the discus.

The component of these latter forces that acts in the original direction of the airflow (i.e., before the air made its "detour" around the discus) is known as the *drag* (Fig. 99). When a body moves through a fluid, it is this drag, this component of the force exerted by the fluid against the body, that reduces the speed of the body along its path. When a swimmer pushes off at the completion of a turn, it is the drag that slows the forward motion of his glide and makes it necessary for him to resume kicking and stroking. It is the drag too that drastically reduces the speed of a badminton bird after it's been hit and causes it to follow a flight path that does not even approximate a parabola.

The component of force acting at right angles to the drag component is known as the *lift* (Fig. 99). An example of the importance of this lift com-

[8]Richard V. Ganslen, *Aerodynamic Factors Which Influence Discus Flight*, Research Report, University of Arkansas, 1958.

[9]Richard V. Ganslen and Kenneth G. Hall, *Aerodynamics of Javelin Flight* (Fayetteville, Ark.: University of Arkansas, 1960).

[10]A. E. Raine, "Aerodynamics of Skiing," *Science Journal*, VI, March 1970, pp. 26–30.

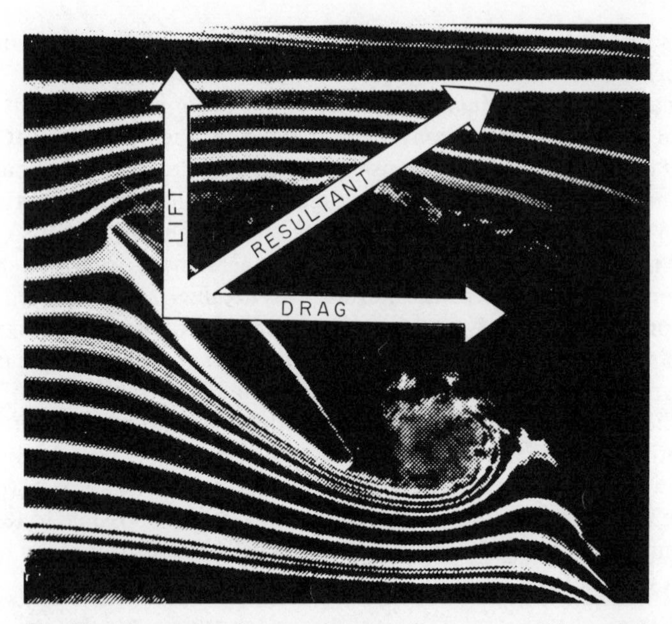

Fig. 99. Lift and drag forces acting on a discus mounted in a wind tunnel. The lines indicating the path followed by the air are obtained by injecting smoke through small jets positioned upstream (*Photograph courtesy of Richard V. Ganslen*).

ponent can be drawn from the sport of water skiing. Before a water skier starts his run, he assumes a predetermined position in the water, with his skis slanted forward and upward and the tips just out of the water. Then, when he is drawn forward by the pull of the rope, the lift component of the pressure exerted by the water on his skis causes him to be lifted up onto the surface of the water. The need to have the correct amount of lift sets definite limits on the starting position he can adopt and still be successful in getting up. If he has his skis near the vertical at this point, the drag will be very large indeed and the lift will be practically nonexistent. A similar lack of lift will be evident if he has his skis near the horizontal when he starts moving forward. In both these cases the skier will almost certainly fail to obtain a successful start.

Surface Drag. When air rushes past the discus in Fig. 100(a), the layer in contact with the discus is slowed down as a result of the forces that the discus exerts upon it. This layer of air tends to slow the layer next to it, which layer in turn tends to slow the next one, and so on. As a result of this progressive slowing of the air farther and farther away from the surface of the discus, the thickness of the air that is affected by this process gets progressively larger the farther the air travels along the body. After a certain distance, which varies according to the velocity and nature of the body, this layer of affected air (the so-called *boundary layer*) becomes unstable. Then, instead of neighboring

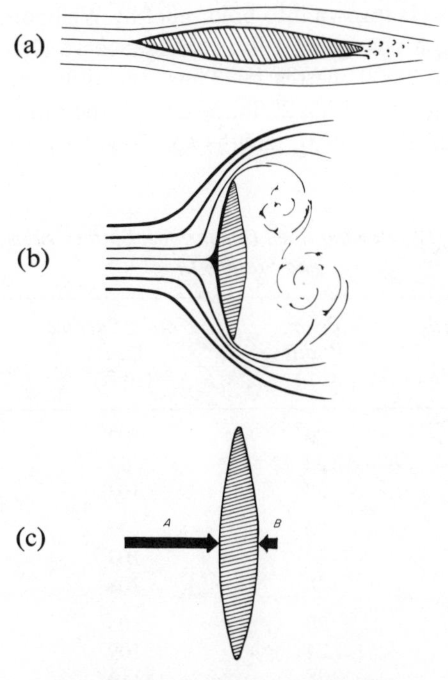

Fig. 100. The form drag experienced by a discus is a function of its orientation to the oncoming flow.

portions of the flow traveling along parallel paths as they did at the outset, they suddenly become violently mixed up. This transition from flow in parallel layers (called *laminar flow*) to flow with violent intermixing of the air (*turbulent flow*) results in the boundary layer becoming even thicker still. The process of slowing and mixing the air in close proximity to the surface of the discus requires that the discus exert force on it and, in reaction, that the air exert force on the discus. This latter force is known as the *surface drag*.

The magnitude of the surface drag that a given body experiences depends on a number of factors, including (1) the velocity of the flow relative to that of the body, (2) the surface area of the body, (3) the smoothness of this surface, and (4) the fluid involved. It is unlikely however, that a coach or an athlete would consider making adjustments in more than one or two of these four in an effort to reduce surface drag.

A rowing coach might take the surface area into account in making his choice of which shell his crew should use. On this subject, Wellicome[11] provides some interesting figures relating the surface drag of an eight-oared shell to its wetted surface area (Table 12). The importance of the observed differ-

[11]J. F. Wellicome, "Some Hydrodynamic Aspects of Rowing," in *Rowing, A Scientific Approach*, ed. J. G. P. Williams and A. C. Scott (London: Kaye & Ward, Ltd., 1967).

ences in surface drag is thrown into bold relief by Wellicome's statement that "Broadly speaking, a change of 1 lb in resistance at 17 ft/sec means a half length margin over 2,000 metres."[12] Thus, the difference in surface drag between a 60-ft long hull with a 20-in. beam and the same length hull with a 28-in. beam is equivalent to 3½ lengths (or 70 yd) over the normal racing distance!

*Table 12. Rowing Shell Dimensions, Surface Area, and Surface Drag**

Overall Length of Hull (ft)	*Beam Width* (in.)	*Wetted Surface Area*† (sq ft)	*Surface Drag*† (lb)
55	20	92	62
	24	97	66
	28	103	69
60	20	97	65
	24	103	68
	28	108	72
65	20	104	68
	24	109	71
	28	114	74

*Adapted from data in Wellicome, "Some Hydrodynamic Aspects of Rowing," pp. 32–33.
†The wetted surface area and surface drag were computed for a shell and crew weighing a total of 1900 lb and moving at a speed of 17 fps.

The second factor that an athlete or a coach might consider in an effort to reduce surface drag is the smoothness of the surface of the body. The hulls of rowing shells are usually highly polished with this in mind, and Wellicome has suggested that a hull that was very rough might cost a crew as much as three lengths in the course of a 2000-m race.[13] The rough surfaces presented in the large metal latches (or buckles) on a pair of ski boots have been shown to increase the drag on a downhill racer to such an extent that it adds about 0.3 sec to each minute of a race.[14] In an event that normally lasts about 2 min and in which the difference between winning and losing is often a mere few hundredths of a second, this factor of surface smoothness is obviously a very important one. Some interest in the relationship between surface smoothness and surface drag has also been shown in swimming. For instance, Karpovich,[15] in an early study of water resistance in swimming, found that

[12] *Ibid.*, p. 34.
[13] *Ibid.*, p. 28.
[14] Raine, "Aerodynamics of Skiing," p. 29.
[15] Peter V. Karpovich, "Water Resistance in Swimming," *Research Quarterly*, IV, October 1933, p. 26.

the use of a woolen swimsuit produced an increase in the resistance of a swimmer performing a glide compared with that found when the same swimmer used a silk suit, or no suit at all. More recently there has been some interest in the possibility of reducing a swimmer's surface drag by having him shave the hair from his limbs and torso. However, it seems most unlikely that any reduction in drag produced by this means would be large enough to have any noticeable effect on the outcome of a swimming race.

Form Drag. Because of its position, the discus in Fig. 100(a) has only a very small effect on the air flowing past it. However, if the discus is rotated through 90°, its effect on the airflow is appreciably increased [Fig. 100(b)].

As the oncoming air strikes the front face of the discus in Fig. 100(b), it is deflected outward from the center. When it reaches the rim, it is unable to make the sharp change in direction necessary to allow it to continue inward along the rear surface of the discus. (To actually achieve this would require that the air be acted upon by some large force that would accelerate it in the required direction. Since the only force even tending to produce such an effect is the pressure of the neighboring layers of air, there is little prospect of this happening.) Instead the flow breaks away (or separates) from the boundary formed by the discus. Later, and farther downstream, the pressure of the neighboring air forces the two diverging parts of the flow back together again.

In the course of this separation of the flow from the boundary, and the subsequent reuniting of the diverging parts, a "pocket" is formed behind the discus. In the state of turbulence that exists in this pocket, whirling currents of air (*eddy currents*) are formed. These currents ultimately detach themselves from behind the discus and flow downstream, where they eventually disintegrate.

An important characteristic of this turbulent pocket is the low pressure that prevails within it. This, together with the high pressure resulting from the oncoming airflow striking the front of the discus and being abruptly redirected outward, leads to a resultant pressure in the original direction of the flow. This is shown diagrammatically in Fig. 100(c), where the vector A represents the resultant pressure force (in the direction of the flow) on the front face of the discus and B the corresponding force on the rear face. The difference between these two vectors is the *form drag*.

Like surface drag, the magnitude of the form drag depends on a number of factors. Among these, and particularly relevant in a variety of sports, are (1) the cross-sectional area of the body perpendicular to the flow, (2) the shape of the body, and (3) the smoothness of its surface.

(1) In speed skating, competitors adopt a position in which their trunks are horizontal and their arms are held behind the body so that the cross-sectional (or frontal) area, and with it the form drag, can be reduced (Fig. 101). Racing cyclists crouch forward over the handlebars and jockeys huddle

behind the necks of their mounts with the same objective in mind. The descent of a skydiver is largely controlled by his frontal area. If he descends feet (or head!) first, his frontal area and form drag are both relatively small and, since the latter will not therefore slow his descent very greatly, he falls at a comparatively rapid rate. To avoid this, competent skydivers assume a position in which their frontal areas are as large as they can make them, thus slowing their rate of descent and prolonging their time of free fall. Then, when they finally resort to using their chutes, they further decrease their rate of descent by again increasing the frontal area.

Fig. 101. The speed skater adopts a near-horizontal trunk position in order to reduce his form drag, and to put his hip extensors in a position where they can be most effective in driving him forward.

(2) The influence that the shape of a body has on the magnitude of the form drag depends on the extent to which the body is streamlined. If the front of a body is shaped so that the direction of the flow is changed only gradually as it comes into contact with the body, the pressure on the front is markedly less than if the flow direction were abruptly altered. If, in addition, the rear of the body is tapered so that the flow is not required to make any sharp turns in order to remain in contact with it, the separation of the flow from the boundaries of the body, the low-pressure turbulent pocket, and the accompanying eddy currents are all greatly reduced. Thus, if all else is equal, a body with a gently rounded front and a tapered back (i.e., a streamlined body) will have less form drag than one that is not streamlined.

The relationship between streamlining and form drag is used in several sports. In yachting, canoeing, and rowing, consistent with other factors that must be taken into account, every effort is made to have the hull streamlined so that it will encounter as little form drag as possible. The positions adopted by speed skaters and downhill skiers (Fig. 102) are aimed not only at reducing the frontal area but also at making the athlete as streamlined as he can be under the circumstances.

In some sports there is a need to make the form drag as great as possible and therefore bodies that are not streamlined are used. In sailing downwind it is mainly the form drag of the sails that propels the boat forward and so the less streamlined these are, the greater is the force pushing the boat through the water. A somewhat similar situation exists in rowing, where the firmer the purchase that the blade of the oar makes on the water, the greater the force that goes into propelling the boat forward. For this reason the blades of oars are designed to create a large amount of form drag when they are moved through the water.

Fig. 102. The downhill skier adopts the so-called egg position to keep his form drag to a minimum.

(3) The smoothness of a body's surface influences its form drag because of the effect it has on the boundary layer. Consider a ball moving through the air. When air strikes the ball and is directed outward to pass around it, a boundary layer is formed close to the surface of the ball (see "Surface Drag," pp. 172–75). Then, as the flow continues around the ball, it separates from the boundary and creates the turbulent pocket already described. Now the point at which the flow separates from the boundaries of the ball very largely governs the magnitude of the form drag. If this separation point is near the front of the ball, a large turbulent pocket is created and the form drag is relatively high. If it is near the back of the ball, the reverse is true. The point at which the flow separates depends largely on the nature of the boundary layer around the front of the ball. If the boundary layer is laminar (i.e., adjoining layers of air flowing parallel to one another), the separation point will be farther toward the front of the ball than if it is turbulent.

The two principal factors that determine whether the boundary layer is laminar or turbulent—at least from a sports point of view—are the velocity of the flow and the smoothness of the surface of the ball. With respect to the first of these, once the flow reaches a certain critical speed, the boundary layer becomes turbulent and the form drag is drastically reduced. Lyttelton,[16] for example, reported that the drag on a cricket ball was reduced to approximately one-fourth of its previous value once this critical speed had been attained. The second of these factors, the smoothness of the surface of the ball, is important because it partly determines the magnitude of the critical speed—a magnitude that is greater if the surface of the ball is smooth than it would be if it were rough. The importance of this fact in the game of golf is reputed to have been first discovered by Scottish caddies who found that a well-worn or cut ball could be driven farther than a new one of the smooth type then in use. This finding ultimately led manufacturers of golf balls to dimple the surface of their product so that it would perform better. Incidentally, Bade[17] cites some interesting figures relating the depth of the dimple to the length of the carry, the airborne phase of the drive, and to the total length of the drive:

Depth of Dimple (in.)	*Carry* (yd)	*Total Length of Drive* (yd)
0.002	117	146
0.004	187	212
0.006	212	232
0.008	223	238
0.010	238	261
0.012	225	240

Wave Drag. When a swimmer does a racing dive he flings himself forward off the starting block in a near-horizontal direction. During the first part of the dive he is totally "immersed" in the air through which he is moving. Then, once he has entered the water (and before he returns to the surface), he is again totally "immersed," this time in the water through which he is gliding. During each of these successive immersions the swimmer's forward progress is opposed by the drag exerted on him by the surrounding fluid. When he is in the air, he is acted upon by the form and surface drags exerted by the air as he passes through it. Similarly, when he is underwater, he is subjected to the form and surface drags attributable to the water around him. In these respects he is no different from any other body that moves through a fluid in which it is totally immersed, for all such bodies experience form and surface drags.

The swimmer is different when he comes to the surface and settles into his normal stroking and kicking pattern, for now he is no longer completely immersed in one fluid but is instead operating at the interface between two

[16]R. A. Lyttleton, "The Swing of a Criket Ball," *Discovery*, XVIII, May 1957, p. 188.

[17]Edwin Bade, *The Mechanics of Sport* (Kingswood, Surrey, England: Andrew George Eliot, 1952), p. 54.

fluids—the water and the air. In cases like this—and there are a number of them in sports—the body involved is subjected to yet another kind of drag. And this arises because in its movement along the interface the body exerts forces that create waves. The reaction to these forces, called the *wave drag*, is a resistance force additional to those of the form and surface drags.

The events that led to the adoption of one of the present rules governing breaststroke swimming provide an example of the importance of wave drag. Up until the mid-1950s most breaststroke swimmers swam on the surface of the water in the usual way. Around this time, however, some of the world's leading exponents of the stroke realized that they could swim faster underwater than they could on the surface. This had a pronounced effect on the way breaststroke events were swum. For example, in the men's 200-meter breaststroke at the 1956 Olympic Games, several of the leading competitors swam almost the whole first lap underwater and then came to the surface during the remainder of the race only when forced to by their need for air. What these swimmers had apparently discovered was that, although one could expect an increase in both surface and form drags when the body was completely submerged, the concomitant decrease in wave drag was sufficient to more than offset these increases. The rules for breaststroke events were promptly changed, and now the number of strokes that can be executed underwater is limited to one per lap.

Lift. Many participants in sports are concerned with improving their performances by controlling the drag acting on a body. In addition, some of these people are also concerned with exerting a measure of control over the lift (the component of the air resistance force at right angles to the drag). Discus and javelin throwers, for example, try to throw the implements so that they will encounter a minimum amount of drag (and thus be slowed down as little as possible) and at the same time experience a maximum amount of lift (so that they can be "held up" in the air and their time of flight extended). Ski jumpers aim to achieve exactly the same kind of result during their flight through the air, and water skiers, as suggested earlier, seek the same effects at the start of a run.

At any given velocity of flow the magnitudes of the lift and the drag depend in part on how the body is oriented. If it is inclined perpendicular to the direction of the oncoming flow [as, for example, in the case of the discus in Fig. 100(b)], it will have a relatively large drag and little, if any, lift. On the other hand if it is parallel with the flow [as in Fig. 100(a)], it will have probably the smallest amount of drag possible for it but again little, or no, lift. It seems obvious therefore that if the body is to experience any marked lifting effect, the angle between the flow and the plane of the body (the so-called *angle of attack*) must be other than 0° or 90°—neither of which produce this effect. This in turn means that the drag must be somewhat greater than the minimum value that can be obtained. In short, a compromise is necessary between the conflicting objectives of maximum lift and minimum drag, since it is clearly impossible to obtain both at the same time.

For a few sports, efforts have been made to work out scientifically which angle of attack affords the best compromise between lift and drag. As an illustration of how this is done, consider the data in Table 13, which have been taken from Ganslen's study of the aerodynamics of discus flight. These data show how the lift and drag forces on a discus vary with changes in the angle of attack. The lift increases from 0 lb (at 0°) to 3.103 lb (at 27° and 28°) and then decreases to 0 lb again (at 90°). Concomitantly, the drag increases steadily from 0.264 lb (at 0;) to 3.985 lb (at 90°). Now the problem of working out which angle of attack provides the most appropriate compromise between the two components is overcome by expressing them as a ratio (lift/drag) and taking that angle for which the value of the ratio is the greatest. The final column of Table 13 contains the lift/drag ratios and reveals that, in this instance, the best angle of attack of those measured is 10°.

*Table 13. Lift and Drag Forces on a Discus**
(Wind velocity = 80 fps)

Angle of Attack (degrees)	*Lift* (lb)	*Drag* (lb)	$\frac{Lift}{Drag}$
0	0.000	0.264	0.000
10	0.974	0.337	2.890
20	2.393	0.928	2.579
25	2.885	1.301	2.218
27	3.103	1.547	2.006
28	3.103	1.665	1.864
29	2.475	1.785	1.387
30	2.520	1.838	1.371
35	2.275	1.965	1.158
40	1.911	2.147	0.890
45	2.002	2.502	0.800
50	1.938	2.748	0.705
60	1.547	3.367	0.459
70	1.073	3.694	0.290
80	0.573	3.794	0.151
90	0.000	3.985	0.000

*Adapted from data in Ganslen, "Aerodynamic Factors Which Influence Discus Flight."

The use of streamlined "wings" (or *airfoils*) mounted on Grand Prix racing cars provides an interesting application of the principles of fluid resistance. The idea is to increase the friction between the track and the tires as the car goes into a turn. This is done by mounting the airfoil upside down so that instead of being lifted—as an airplane is lifted by its wings as it roars down the runway—the car is forced down hard against the track. In other words, the lift component of the air resistance acts downward rather than upward. With the normal reaction increased in this manner, the objective of increasing the friction between the car and the track is thus achieved.

The Magnus Effect. The graceful (and maddening!) curved flight of a ball that has been sliced or hooked is well known to golfers. In tennis, baseball, soccer, volleyball, and many other sports it is also well known that a ball can be made to curve in flight if sufficient spin is applied to it before it becomes airborne.

This effect (known as the *Magnus effect*, after the German scientist who is credited with first noting it) can be explained in the following manner. When a body rotates, it tends to carry around with it the fluid that is in direct contact with its surface. This fluid, in turn, tends to similarly influence the neighboring fluid. In this manner, the body acquires a boundary layer that rotates with it. For the golf ball in Fig. 103, the arrows A and B indicate the direction in which the ball and its boundary layer are rotating. They also indicate that the air in that part of the boundary layer on the left-hand side of the ball (i.e., as shown in the figure) is moving in the same general direction as the oncoming flow, while that on the other side is moving in the opposite direction. These differences in the motion of the boundary layer, relative to the oncoming air, lead to differences in the pressures on either side of the ball. On the right-hand side, where the air in the boundary layer and that in the airflow meet "head-on," a zone of high pressure results. On the other side the similarity in the direction followed by the boundary layer and the airflow

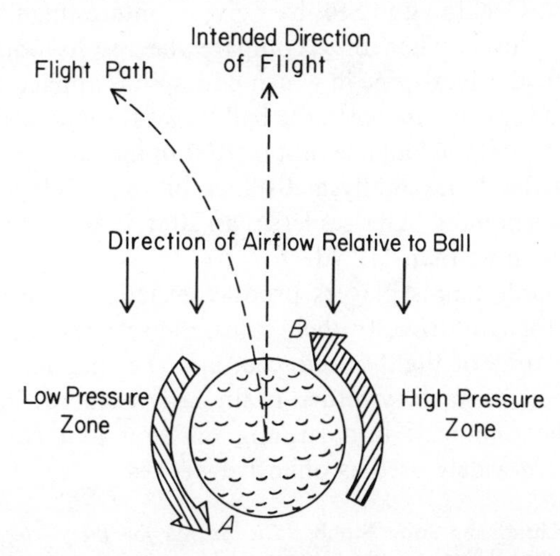

Fig. 103. The spin imparted to a golf ball causes an imbalance in the pressures exerted on it and a corresponding deviation from a straight-line path (Magnus effect).

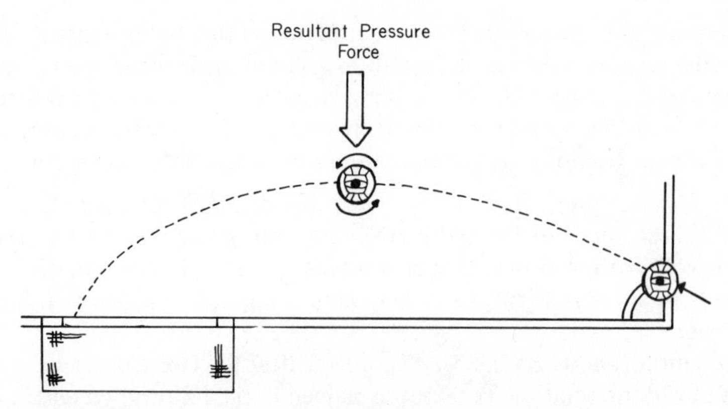

Fig. 104. A soccer player can score from a corner by imparting the appropriate spin to the ball as he kicks it.

leads to a zone of low pressure being created. The net result of this discrepancy in the pressures on either side of the ball is a resultant force acting on the ball from right to left [cf., the discus in Fig. 100(c)]. This resultant force causes the ball to deviate from a straight-line path.

While in some sports the Magnus effect produces results that are an embarrassment, more often than not it can be used to the performer's advantage. Even in golf, where slicing and hooking are very common faults, the Magnus effect is used to advantage in other ways. For example, when a golf ball is driven correctly, a certain amount of backspin is imparted to it. This backspin causes a high-pressure zone to develop beneath the ball, and this, in turn, results in the ball experiencing a force that lifts it and prolongs its time of flight. Cochran and Stobbs[18] give an interesting illustration of just how important this is when they compare the results obtained from two otherwise identical drives—one in which backspin is applied to the ball in the normal way and the other in which the ball leaves the club without spin of any kind. In the first case, the ball rises about 70 ft in the air, remains airborne for 5.5 sec, and carries horizontally a distance of some 200 yd. In very sharp contrast, the nonspinning ball rises less than 20 ft, is airborne for only 2.1 sec, and carries little more than 112 yd.

Tennis and table tennis players produce exactly the reverse effect when they execute a topspin drive. In their cases, the resultant pressure force acts downward, the time of flight is reduced, and the time an opponent has in which to make a satisfactory return is similarly reduced. Because this very often forces the opponent into rushing his shot and committing errors, topspin drives are widely used as offensive strokes.

[18]Alastair Cochran and John Stobbs, *The Search for the Perfect Swing* (London: Heinemann Educational Books, Ltd., 1968), p. 162.

The Magnus effect can also be helpful when a corner kick is taken in soccer. If the ball is kicked slightly off-center so that it acquires a rotation about a vertical axis, the resultant pressure can result in the ball being sufficiently deflected from its straight-line path that a goal can be scored directly from the kick (Fig. 104).

Part III

ANALYSIS OF SPORTS TECHNIQUES

Chapter 8

BASEBALL

Generally considered the invention of a Colonel Abner Doubleday who in 1839 laid out the first diamond in Cooperstown, N.Y., the game of baseball is today the national game of the United States, the principal summer game of Canada, and is also widely played in Cuba, Japan, Mexico, and Puerto Rico.

Basic Considerations

The four basic skills of greatest importance in baseball are throwing (which includes pitching), catching, batting, and base running.

Throwing

The objective in throwing is to move the ball from one point to some other specified point in the shortest possible time (in fielding) or with the optimum linear and angular velocities (in pitching).

Fielding. The time taken in moving the ball from one point to another is the sum of (1) the time taken to execute the throwing movement (i.e., from initiation of the movement until the ball is released) and (2) the time of flight. The first of these is governed principally by the magnitude and direction of the forces involved and the distances through which they act, while the second is governed by the velocity and height of release and, to some extent, by air resistance.

Pitching. Since the ball is initially at rest in the pitcher's glove, the linear velocity that it possesses as it leaves his hand on its way to the strike zone is directly related to the forces applied to it during the act of pitching. As the work-energy (p. 101) and impulse-momentum (p. 76) relationships suggest, the velocity at release is also related to the distance and time over which the various forces are applied. The pertinence of these two parameters is reflected, for example, in the long ($4\frac{1}{2}$ to 5 ft) stride that pitchers use to ensure that both the distance and time over which forces are applied are as large as they can be, consistent with other requirements and limitations.

The angular velocity with which the ball leaves the pitcher's hand depends on the torque applied to it during the release. Ignoring slight variations in the moment arm due to the seams, the magnitude of this torque is directly related to the magnitude of the forces exerted on the ball via the pitcher's fingers. In practice, these forces are a function of the grip employed and the actions of the pitcher's wrist and fingers during the release.

The magnitude and direction of the angular velocity of the ball largely determines the extent to which the Magnus effect (see pp. 181–83) operates and hence the amount to which the ball can be made to deviate from its normal parabolic flight path. The successful piching of curves, sliders, etc., is thus very dependent on the optimum angular velocity being obtained at the moment of release.

Catching

The objective here is to take hold of a ball flying through the air and to reduce its velocity to zero or near-zero. To achieve this objective a player must exert forces on the ball in a direction opposite to that in which it is traveling. In the process of exerting these forces, the player's hands, the means by which the forces are transmitted, are subjected to a certain amount of pressure. The magnitude of this pressure on the hands largely determines the player's ability to "hold" the ball. If he attempts to catch the ball in such a manner that the pressure becomes intolerably large, the player suffers acute discomfort and perhaps even an injury to his hands. In addition, he is very likely to drop the ball, thereby completely negating his efforts. The pressure exerted on any part of the player's hands is dependent on the magnitude of the force being exerted on the hands and by the area over which this force is distributed. Thus, to avoid the undesirable consequences of having too great a pressure on the hands, players strive to decrease the force exerted and to increase the area over which this force is spread.

Batting

The objectives in batting are to hit the ball at will and to impart to it that velocity necessary to carry it to some specific point within (or outside!) the ball park.

The path followed by a baseball once it has left the bat is governed by the height and velocity at which it leaves the bat and by the air resistance that it encounters in flight.

For all practical purposes the height of "release" (which is governed by the height of the ball as it enters or approaches the strike zone and by the height of pitch at which the batter prefers to swing) is of relatively minor importance.

The velocity of "release," on the other hand, is all-important. This velocity is governed by the respective masses of the bat and ball, their respective velocities before impact, and their mutual coefficient of restitution. In practice, however, the batter has a measure of control over only three of these factors: the velocities of bat and ball before impact and the mass of the bat (pp. 87–91).

Apart from the drag force, the principal effect that air resistance has on the flight of the ball is that due to the Magnus effect (pp. 181–83). Since this effect is due solely to the fact that the ball is spinning, the generally unwelcome deviations that result from it must be attributed to the factors that produce such spin.

The spin (or angular velocity) with which a ball leaves the bat can be regarded as the summed effect of (1) the spin imparted to the ball by the pitcher (and over which the batter has no real control) and (2) the spin imparted by the bat during its period of contact with the ball. This latter spin is the result of a frictional force that is created whenever the impact between bat and ball is an oblique one.

Base Running

A base runner's purpose is to traverse the distance from one base to the next without being tagged or thrown out. Apart from making the correct decision concerning whether or not he should run, the most important facet of his performance is the time he takes to cover the required distance. This time depends on the distance involved and on the average speed of the runner.

A base runner's average speed is a function of his speed at all points on his run—his speed in starting, in running, and in sliding or stopping. His speed in starting, running, and stopping (like that of a runner on a track, football field, or basketball court) is equal to the product of his stride length and his rate of striding, or stride frequency (see pp. 382–88). These in turn are determined very largely by the reaction to the forces that the runner exerts against the ground. A runner's speed in sliding is similarly governed by the reaction to the forces that he exerts against the ground.

Techniques

Throwing

Throwing is involved in two major aspects of the game of baseball—pitching (the more highly specialized aspect) and fielding. Since the techniques used in each of these have much in common, and to discuss them both in detail would be needlessly repetitive, the discussion that follows is largely restricted to a consideration of pitching. However, a number of the factors of specific importance in the throwing used in fielding the ball are also considered.

Fig. 105. An example of good technique in pitching from a wind-up stance

Pitching (Fig. 105). Having received the signal from the catcher, the pitcher takes up a position preparatory to beginning the sequence of movements that will culminate in the release of the ball in the direction of the strike zone. If the time involved in delivering the pitch is unimportant, the pitcher generally adopts a stance in which his back (or pivot) foot is placed across the rubber and his front (or striding) foot is somewhat behind and to the side [the wind-up stance, Fig. 106(a)].[1] However if there are men on base and, in particular, if there is a man on either first or second, the time taken to deliver the pitch becomes important. To reduce this time, and with it the chance of a runner stealing a base, the pitcher adopts a stance in which the first 90° or more of the angular motion involved in pitching from a wind-up stance is eliminated [the set stance, Fig. 106(b)].

From a wind-up stance[2] with his hands at waist height in front of his body and with the ball concealed in his glove, the pitcher initiates his pitching action by swinging his hands upward to a position overhead. This movement, accompanied by an approximately 90° turn of the pivot foot and by a shifting of weight from the striding foot to the pivot foot, is employed primarily to initiate the motion in a rhythmic manner, in the hope that this will relieve tension and assist in obtaining a correct timing of the movements that follow.

[1]The terms *front* and *back* are used here with reference to the position of the feet as the ball is released; the front foot is the one nearer the batter and the back foot is the one farther from the batter, at that instant.

[2]Because the actions in pitching from a set stance are basically an abbreviated form of those used in pitching from a wind-up stance, only the latter are discussed here.

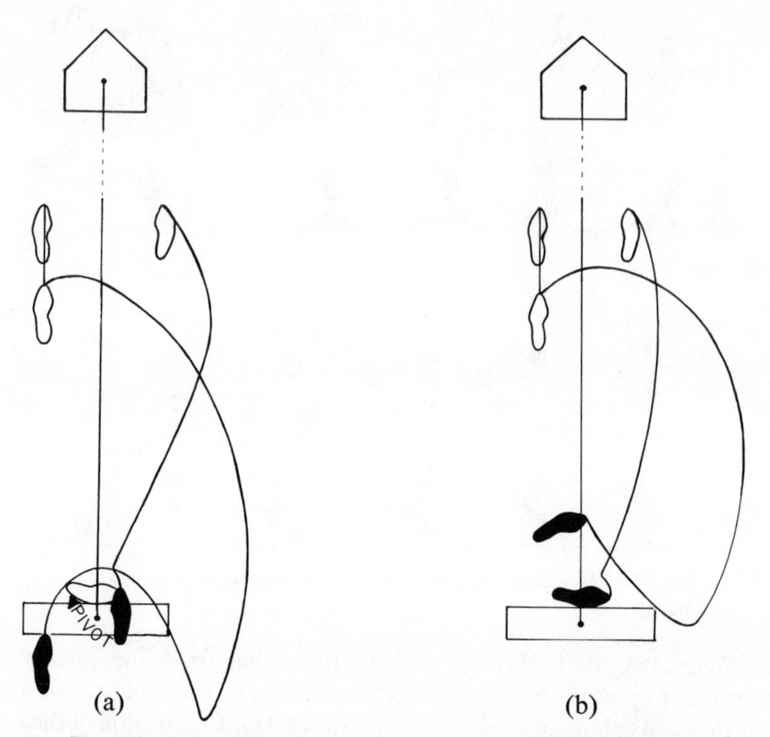

Fig. 106. Paths followed by the feet in pitching from (a) a wind-up stance, and (b) a set stance. (The paths depicted here were obtained from overhead, time-exposure photographs of small lights placed on a pitcher's feet.)

The winding-up actions (and pitchers vary considerably in the number and range of these that they employ) are followed by a coordinated series of movements involving

1. a backward push with the striding foot as it leaves the ground to pass forward over the rubber,
2. a swing of the striding leg, bent at the knee, forward and across the body,
3. an approximately 90° rotation of the hips and upper body,
4. a lowering downward and backward of the pitching hand, and
5. a backward inclination of the trunk.

The first of these movements (the backward push of the striding foot) has the effect of moving the pitcher's center of gravity forward of his pivot foot.

Thus when the pivot leg is forcefully extended moments later, the center of gravity is already moving forward and, more important, is in a position to be driven forward rather than upward. The other four movements listed serve to increase the distance through which the pitcher can apply force to the ball. Thus they contribute to an increase in the amount of work that can be done on the ball and to the magnitude of the velocity with which it can leave the pitcher's hand. The inclination of the trunk also enables the pitcher to bring additional muscular forces—those due to the contraction of the lateral muscles of his trunk—into effect.

Just as he is about to move forward into his delivery, the pitcher lowers his center of gravity by bending the knee of his pivot leg. Then, with his center of gravity some distance forward of his pivot foot, he vigorously extends this leg to drive himself forward into his stride. This movement is assisted by a simultaneous forward sweeping movement of the striding leg, which carries the striding foot across in front of his body to land to one side of an imaginary line joining the midpoint of home plate to the midpoint of the pitching rubber. (For a right-handed pitcher, the striding foot lands to the left of this line.) This off-line placement of the striding foot permits the hips to be more completely rotated to the front than if the foot were placed on line.

As the striding foot strikes the ground, the force involved in decreasing the body's vertical motion is reduced by allowing the knee to bend and increase the distance over which work is done on the body.

While the striding foot is being moved forward the some $4\frac{1}{2}$ to 5 ft of the average stride, the pitcher's hips and trunk are turned toward the front and his upper arm rotated outward to bring his elbow forward and upward. Then, once the striding foot has been grounded, the hip and trunk rotations are completed and the pitching arm is whipped forward to complete the throw. During this latter movement the elbow is pulled forward and around with such speed that the ball and the pitcher's forearm and hand are left well behind. (It is perhaps of interest to note that because of the sheer speed of the movements at this stage the pitcher's arm assumes positions that it is quite impossible for him to reproduce when his arm is stationary—Fig. 107.) From this position the forearm is whipped forward, the wrist is flexed (or "snapped") explosively, and the ball is released in the direction of the strike zone.

The angle between the pitcher's forearm and the horizontal, as the ball is released, varies from one pitcher to another. Some use an action involving an angle of approximately 90° (an overhand or overarm pitch); others, of approximately 180° (a sidearm pitch); and still others, probably the majority, an angle somewhere between these two extremes (a three-quarter pitch).

Since the ball will fly off at a tangent to the arc it is following immediately prior to release, the angle at which the ball should be released, and thus the

Fig. 107. The initial movements of the pitcher's delivery place his throwing arm in an extreme position (*Photograph courtesy of* Athletic Journal).

point on the arc at which this should occur, is clearly of considerable importance, This angle is governed by a number of factors:

1. the height at which the ball is released,
2. the amount the ball will drop under the influence of gravity before it reaches the plate,
3. the amount the ball will deviate upward or downward as a result of air resistance (and specifically, the Magnus effect) before it reaches the plate, and
4. the height at which it is desired to place the ball within the strike zone.

The height at which the ball is released is primarily a function of the physique of the pitcher and the length of stride and type of arm action (i.e., overhand, three-quarter, or sidearm) that he uses. The amount the ball drops due to gravity is governed by the length of time it takes to traverse the roughly 56 ft from the point of release to the plate. This, in turn, is governed by the horizontal velocity of the ball at the instant it is released—the smaller this velocity, the longer it will take the ball to reach the plate and the farther it will drop under the influence of gravity. The amount the ball deviates upward or downward as a result of air resistance has been studied by Selin[3] and been found to depend primarily on the direction in which the ball is spinning and the orientation of the axis about which it is spinning.

The data presented in Selin's study provide a basis for examining the extent to which each of the factors involved influences the angle at which the ball must be released. If each of the six types of pitch Selin studied were to be thrown horizontally at the maximum speed he recorded for them, the downward deviations due to gravity, while the ball traversed the 56 ft from the point of release to the plate, would be as shown in Table 14. The maximum values recorded by Selin for the upward deviations due to air resistance are also shown in Table 14. The summed effect of gravity and air resistance in this situation is obtained by adding algebraically the vertical deviations due to these two influences. (*Note:* For each type of pitch, the figure in the right-hand column of Table 14 represents the least downward deviation that could have been obtained—and then only if all the factors of maximum speed, horizontal direction of release, and maximum upward deviation had been combined in one throw.)

*Table 14. Vertical Deviations of a Pitched Baseball Due to the Effects of Gravity and Air Resistance**

Type of Pitch	*Maximum Speed Recorded* (fps)	*Vertical Deviation Due to Gravity*† (ft)	*Vertical Deviation Due to Air Resistance*† (ft)	*Total Vertical Deviation*† (ft)
Fast ball	121	−3.45	2.70	−0.75
Change-up	90	−6.23	2.40	−3.83
Knuckle ball	91	−6.09	2.25	−3.84
Curve ball	111	−4.11	0	−4.11
Slider	102	−4.85	0	−4.85
Sidearm-curve ball	92	−5.97	0	−5.97

*Based upon data in Selin, "An Analysis of the Aerodynamics of Pitched Baseballs."

†Positive values indicate upward deviations, and negative values indicate downward deviations.

[3] Carl Selin, "An Analysis of the Aerodynamics of Pitched Baseballs," *Research Quarterly*, XXX, May 1959, 232–40.

Now, if the pitcher releases the ball at a height of 7 ft (i.e., a height of 5 ft 9 in. above the 15-in.-high mound) and if the batter's armpits are 5 ft above the ground, the total downward deviation must be in the vicinity of 2 ft if the ball is to be placed near the upper limit of the strike zone. Except for the fast ball, the computed total deviations in Table 14 far exceed this value. It is clear, therefore, that pitchers of the caliber of Selin's subjects (pitchers in the Big 10 Conference) cannot throw other than their fast ball to the top of the strike zone by releasing it in a horizontal direction. Instead, they must release the ball at some angle above the horizontal.

If the batter's knees are 2 ft above the ground and the ball is released by the pitcher at a height of 7 ft, the maximum downward deviation that will permit the ball to pass above the lower limit of the strike zone is 5 ft. Remembering that the values shown in Table 14 represent the minimum downward deviations that Selin's subjects might have obtained, it appears that unless close to maximum speed and maximum upward deviation can be obtained, pitchers of a comparable standard are likely to have to throw all but their fast ball at angles above the horizontal in order to get them into the strike zone.

Thus it is apparent that for most overhand or three-quarter pitches the ball must be released a little before the pitcher's hand passes forward of his elbow (i.e., with his forearm inclined backward) so that as the ball flies off at a tangent it will be moving in a direction slightly above the horizontal.

Because adjustments in the length of stride and in the position of the pivot foot on the rubber have frequently been cited as means to correct consistent errors in a pitcher's placement of the ball over the plate, Edwards[4] conducted a study to examine carefully just what effect such changes produced. Using 47 pitchers from five Midwestern universities he found that

1. changing the length of the stride will not necessarily correct the fault of consistently pitching a fast ball high or low in relation to the strike zone, and
2. changing the position of the pivot foot on the pitching rubber will not necessarily correct the fault of consistently pitching a fast ball inside or outside home plate.

Once the ball has left the pitcher's hand, the angular motions of his arm and upper body are gradually reduced to zero as he "follows through." Although nothing that the pitcher does once the ball has been released can have any effect upon it, the follow-through is an important part of the pitcher's action. First, by allowing him to bring his various angular motions to a halt over a relatively large distance, it ensures that the internal muscular and ligamentous forces involved are much less than they would be if his body (and, most important, his pitching arm) were brought to a halt more abruptly.

[4]Donald K. Edwards, "Effects of Stride and Position on the Pitching Rubber on Control in Baseball Pitching," *Research Quarterly*, XXXIV, March 1963, pp. 9–14.

In this way the pitcher's follow-through helps him to avoid the risk of injury commonly associated with high internal forces. Second, if the angular motions of the pitcher's body were brought to a halt prior to, or during, the release, the forces responsible would almost certainly impair the application of appropriate forces to the ball. In this sense, therefore, the pitcher's follow-through is important not so much because of what it achieves but because of the undesirable effects that would almost certainly be produced if it were absent.

Immediately after his follow-through is completed, the pitcher must prepare himself to serve in his role of fielder. Since this requires that he assume a position with his weight supported evently on both feet (so that he can move with equal facility in either lateral direction) and since his follow-through will probably have been completed with one foot in the air and all his weight supported on the other, he must quickly readjust his position by thrusting against the ground with his supporting (striding) foot.

A listing of the fastest speeds of release ever recorded is presented in Table 15. A number of authors have reported the results of studies in which the speeds of release for various types of pitch, and for various subjects, have been determined. A summary of a selection of these is presented in Table 16.

*Table 15. The Fastest Speeds of Release Recorded for Pitched Baseballs**

Pitcher (club)	*Year*	*Speed*	
		mph	(fps)
Nolan Ryan (Angels)	1974	100.8	(148)
Bob Feller (Indians)	1946	98.6	(145)
Steve Barber (Orioles)	1960	95.5	(140)
Don Drysdale (Dodgers)	1960	95.3	(140)
Atley Donald (Yankees)	1939	94.7	(139)
Steve Dalkowski (Minors)	1958	93.5	(137)
Sandy Koufax (Dodgers)	1960	93.2	(137)
Joe Black (Dodgers)	1953	93.2	(137)
Dee Miles (Phila. A's)	1939	92.7	(136)
Johnny Podres (Dodgers)	1953	92.1	(135)
Christy Mathewson (Giants)	1914	91.3	(134)
Ryne Duren (Yankees)	1960	91.1	(134)
Herb Score (Indians)	1960	91.0	(133)
Mickey Lolich (Tigers)	1974	90.9	(133)
Bob Turley (Yankees)	1960	90.7	(127)
Walter Johnson (Senators)	1914	86.6	(114)

*Based on data in Don Merry, "Baseball's Fastest Pitchers," in *Street and Smith's Baseball Yearbook*. (New York: The Conde Nast Publications Inc., 1975), p. 64.

Table 16. The Speeds of Pitched Baseballs
(fps)

Type of Pitch	*Selin*[5] (College pitchers)			*Slater-Hammel and Andres*[6] (College pitchers)			*Kenny*[7] (High school, college, amateur, and professional pitchers)			*Bunn*[8] (Major league pitchers)	
	Min.	*Mean*	*Max.*	*Min.*	*Mean*	*Max.*	*Min.*	*Mean*	*Max.*	*Mean*	*Max.*
Fast ball	87.5	104	121	86	95–119	127	85	96	111	94	145 (Bob Feller)
Slider	92	98	102								
Curve ball	74	91	111	75	80–104	108	79	86	98	84	
Sidearm-curve ball	90	91	92				77	85	99		
Knuckle ball	75	84	91								
Change-up	62.5	79	90								

[5] Selin, "An Analysis of the Aerodynamics of Pitched Baseballs," p. 235.

[6] A. T. Slater-Hammel and E. H. Andres, "Velocity Measurement of Fast Balls and Curve Balls," *Research Quarterly*, XXIII, March 1952, p. 96.

[7] Adapted from data contained in James D. Kenny, "A Study of Relative Speeds of Different Types of Pitched Balls" (M.A. thesis, State University of lowa, 1938).

[8] John W. Bunn, *The Scientific Principles of Coaching* (Englewood Cliffs, N.J.: Prentice-Hall, Inc., 1962), p. 146.

It is interesting to note, and a cause for some confidence, that except for the mean speeds reported by Bunn (which are somewhat lower than might be expected for Major League pitchers when compared with the values obtained for college pitchers), the tabulated speeds reported by the various authors are in excellent agreement.

The behavior of the ball in flight—a source of considerable controversy over the years—has also been a subject of study.[9,10,11] In what has probably been the most exhaustive study of this question to date, Selin filmed over 200 pitches made by 14 pitchers from the Big 10 Conference and conducted a detailed analysis on a selected 30 of these pitches. For each pitch this detailed analysis included a determination of the velocity of the ball, the rate and direction of its rotation, the axis of rotation, and the amounts it deviated in horizontal and vertical directions from the parabolic path it would have followed in the absence of air resistance. Among the conclusions reached as a result of this analysis were

1. Every pitch showed either or both a vertical deviation or a horizontal deviation from the path it would have followed in the absence of air resistance. (The ranges of the deviations recorded are shown here in Table 17.)

*Table 17. Ranges of Horizontal and Vertical Deviations of 30 Selected Pitches**

Type of Pitch	*Range of Horizontal Deviations*† (ft)	*Range of Vertical Deviations*‡ (ft)
Fast ball	0.3 to 2.1	0.3 to 2.7
Curve ball	−1.2 to −0.1	−1.8 to 0
Slider	−0.8 to −0.4	−2.0 to 0
Sidearm-curve ball	−0.8 to −0.5	−1.6 to 0
Change-up§	−1.5 to 0.6	−2.0 to 2.4
Knuckle ball§	−2.5 to 0.8	−2.0 to 2.25

*Adapted from Selin, "An Analysis of the Aerodynamics of Pitched Baseballs."

†Deviation to the right (as a pitch is viewed by the pitcher) is regarded as positive, deviation to the left as negative.

‡Upward deviation is regarded as positive, downward deviation as negative.

§Some pitches rotated in a clockwise direction, others in a counterclockwise direction.

[9]Lyman J. Briggs, "Effect of Spin and Speed on the Lateral Deflection (Curve) of a Baseball; and the Magnus Effect for Smooth Spheres," *American Journal of Physics*, XXVII, November 1959, pp. 589–96.

[10]Selin, "An Analysis of the Aerodynamics of Pitched Baseballs," pp. 232–40.

[11]Frank L. Verwiebe, "Does a Baseball Curve?" *American Journal of Physics*, X, April 1942, pp. 119–20.

2. Within the ranges for the rates of rotation and velocities recorded in the study, the amount of deviation was determined primarily by the vertical angle—the obtuse angle between a vertical line drawn through the center of the ball and the axis of rotation, as viewed from the pitching rubber. In general, as the vertical angle approached 90°, the vertical deviation increased and the horizontal deviation decreased; and as the vertical angle approached 180°, the vertical deviation decreased and the horizontal deviation increased. (The mean vertical angle, and the direction of rotation for those types of pitch in which the ball rotated in a consistent direction, are shown in Fig. 108.)

3. The direction of deviation of the knuckle ball was apparently not related to the direction of rotation of the ball. With such low speeds of rotation (a mean of 5.3 rps compared with a range of means from 24.5–31.2 rps for the other types of pitch) the orientation of the seams relative to the axis of rotation and relative to the airflow may have determined the direction of the deviation.

The recent work of Watts and Sawyer[12] supports the contention that the orientation of the seams plays a major role in determining the flight path of a knuckle ball. After conducting a series of experiments in a wind tunnel, these investigators concluded that "the nonsymmetrical location of the roughness elements (strings) gives rise to a nonsymmetrical lift (lateral) force. A very slowly spinning knuckleball will have imposed upon it a lateral force that changes as the positions of the strings change."

It is of some interest to note that although Watts and Sawyer and Selin are in general accord on the role of the seams in producing lateral motion, their findings are at variance on the matter of speeds of rotation. While Selin[13] reports horizontal deviations of as much as 2.5 ft with speeds of rotation as high as 4.3 rps (equivalent to three complete revolutions between release and arrival over the plate), Watts and Sawyer conclude that, to be effective, a knuckleball should be thrown so that it rotates substantially less than once on its path to home plate.

4. The fast balls consistently displayed positive (upward) vertical deviation and this apparently created an illusion in which, although the ball fell toward the ground during the flight, its failure to fall as rapidly as expected made it appear to the batter to rise or hop as it approached the plate.

Fielding. As suggested earlier, when a fielder throws the ball, his objective is to complete its displacement from one point to another specified point in the shortest possible time. The time involved is the sum of the time spent in applying force to the ball and the time the ball spends in flight. The extent to which the time of flight depends on the time taken to execute the throwing

[12] Robert G. Watts and Eric Sawyer, "Aerodynamics of a Knuckleball," *American Journal of Physics*, XLIII, November 1975, pp. 960–63.

[13] Selin, "An Analysis of the Aerodynamics of Pitched Baseballs," pp. 89–90, 96.

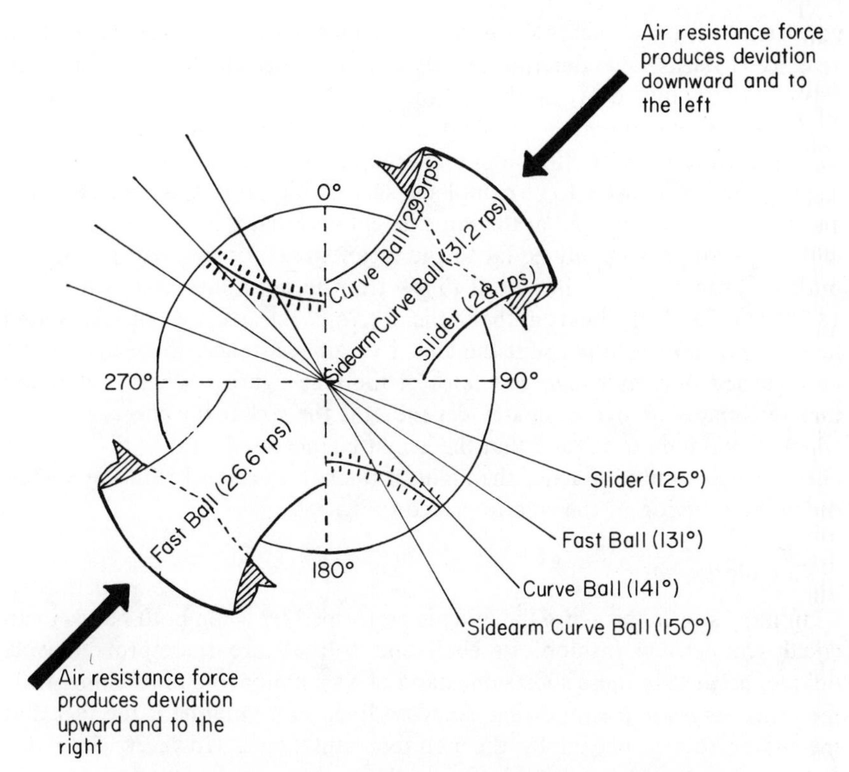

Fig. 108. Speeds, axes, and directions of rotation for various types of pitches. (*Based on data in Selin, "An Analysis of the Aerodynamics of Pitched Baseballs."*)

movement is apparent from the impulse-momentum relationship (p. 76), which indicates that if all else is equal, the velocity at which a ball is released is directly proportional to the time taken to make the throw. It is well to realize, however, that a reduction in the time of flight brought about by an increased velocity of release may not be greater than the increase in time that was needed to bring about this alteration in velocity. In other words, the total time involved may not have been decreased. For this reason an infielder may quite frequently use a short flick throw to, say, second base rather than use a throw with a more complete body and arm action that involves a shorter time of flight but a longer overall duration. On the other hand, an outfielder making a long throw to one of the bases or to home plate will almost certainly be most effective (i.e., will complete the throw in the least time) if he puts his body in such a position that he can apply maximum force to the ball.

If the height at which the ball is released and the height at which it lands, or is caught, are considered to be constant, the time of flight depends on the vertical velocity of the ball at the moment it is released. Therefore, in all throws, the ball should be released with as little vertical velocity as possible consistent with the successful completion of the throw. Since the vertical velocity at release also determines the extent to which the ball rises in flight, this is tantamount to saying that all throws should be as low as possible.

The equations presented in Chap. 3 to describe the motion of projectiles can be used to estimate the minimum amounts that a ball must be made to rise in order for a throw to be completed successfully. If it is assumed that the ball is released and caught at the same height and that it is thrown at a speed of 100 fps (approximately equal to the mean speed for fast-ball pitches), it must rise roughly 3.3 ft in a 90-ft throw from base to base and as much as 18.2 ft in a 200-ft throw from the outfield. (*Note:* Because the equations used here do not take into account the effects of air resistance, these figures can be regarded only as rough estimates of the true figures. Whether they are underestimates or overestimates depends on the unknown effects of drag, which would tend to require that the height be increased, and of the Magnus effect, which would influence the flight of the ball in accord with the magnitude and direction of the spin imparted to it.)

Catching

In most sports the act of catching is performed by using both hands in an equal symmetrical fashion. Baseball and softball are therefore relatively unique, because in these sports one hand plays a major role in catching while the other serves in a supporting capacity. Even more unique is the fact that the major role is played by the nondominant hand. However, since this frees the dominant hand and arm for the throw that invariably follows a catch, the reason for this use of the nondominant hand is not hard to find.

In the process of being caught, a ball gives up a certain amount of the kinetic energy that it possesses, to do work on the hands of the person making the catch. The amount of work done is equal to the loss in kinetic energy incurred in the process—i.e., assuming that the losses to nonmechanical forms of energy and the work done by gravity as the ball gives up potential energy are both sufficiently small that they can be disregarded. Thus, if the ball is brought to rest, the work done is equal to the kinetic energy it possessed immediately before it made contact with the player's glove. If the ball is merely slowed down, in preparation for a throw, the work done is equal to the difference between its kinetic energy at contact with the glove and the lowest value it attains.

Since the amount of work done is thus fixed by the circumstances in a particular case, the product of (1) the force that the ball exerts on the hands and (2) the distance through which the hands move is similarly fixed (work $=$ force $\times$ distance). Thus, if a player making a catch is to reduce the magnitude of the force exerted against his hands, he has just one way of achieving this—he must increase the distance through which this force acts.

For this reason—and, in some cases, to reduce the time of flight of the ball—a player making a catch will generally extend his arms so that the glove meets the ball at some distance from his body. Then, once the ball has struck his glove, his nongloved hand quickly smothers any possibility of it rebounding out of the glove, and his arms bend to allow the distance over which the ball is slowed down to be increased. (*Note:* This latter action also assists in putting the player in a position to execute a throw, once the catch has been completed.)

If a small amount of flexion is retained at the elbow as the player reaches his hands to meet the ball, the force it exerts as it strikes the glove initiates the further elbow flexion needed to cushion the impact. This effect is largely lost if the elbows are completely extended as the ball strikes the glove, for then the force exerted by the ball tends to act along the line of the extended arm and thus has little tendency to cause the arm to bend. As a result, there is every likelihood that the catch will be dropped, unless the player's conscious timing of the required elbow flexion is quite precise. Because it thus introduces an unnecessary problem in timing, the practice of fully extending the arms to meet the ball is to be avoided.

In addition to providing a thick protective layer and increasing the actual area that his extended hand can cover, a player's glove also serves the very important function of distributing the forces involved in catching the ball over a wider area than would be feasible with just the bare hand. In this way, the pressure on any one part of the hand can generally be kept within tolerable limits.

In preparing to catch the ball a player should adopt a body position that takes into account that the force exerted on him by the ball will generally have a tendency to rotate him about some axis through his feet. Thus, if the moment of this force is relatively large, it may be necessary for him to lean forward slightly in the direction of the oncoming ball so that his weight can effectively counter the rotary effect attributable to the ball.

Batting

Whether or not one agrees with the authority on batting who claims that "hitting a baseball is the single most difficult thing to do in sport,"[14] there can be no denying that, as the principal offensive weapon in baseball, batting is one of the most important skills in the game.

For the purpose of the analysis contained in this section, batting is divided into four phases or parts: (1) stance, (2) stride, (3) swing, and (4) follow-through (Fig. 109).

[14]Ted Williams and John Underwood, "Hitting Was My Life: Part V—Science of Batting," *Sports Illustrated*, XXIX, July 1968, p. 41.

Fig. 109. An example of good batting technique

Stance. The body position adopted by a batter as he waits for the pitcher to release the ball has a marked influence in determining the batter's subsequent actions.

One of the most important considerations is the placement of the batter's feet relative to the plate. The position of his feet in a forward and backward sense (toward or away from the pitcher) is a compromise between two conflicting factors. First, the farther the batter is away from the pitcher (i.e., the nearer he is to the back of the batter's box), the longer it will take for any given pitch to reach him and thus the more time he will have to decide if and how he will hit it. On the other hand, the farther back he is, the greater is the chance that a ball could be pitched to the outside of the strike zone and move away beyond the reach of his swinging bat. Thus the batter is obliged to place his feet in a position that will allow him the longest possible time to evaluate the pitch and yet still permit him to meet all legal pitches with the so-called "fat part" of the bat.

The distance of the batter's feet, relative to a midline through the plate, is also fixed to a large extent by his ability to bring this part of the bat to meet the ball at any point across the 17-in. width of the plate. If he stands too close to the plate, he is likely to be troubled by an inside pitch, while if he stands too far away, he is likely to experience a similar difficulty with an outside one.

The alignment of the batter's feet relative to one another is also a matter of some importance. In this matter the batter has essentially three choices. He can place his feet in a line parallel to that linking the midpoints of home plate and the pitching rubber (i.e., in a so-called *square* or *parallel* stance) or he can place them in a line directed toward either right or left field. For a right-handed batter the latter alternatives are known, respectively, as *closed* and *open* stances.

The choice between these stances depends among other things on two interrelated factors: (1) the distance through which the batter must move his various body parts in making his swing and (2) the time from the moment the swing is initiated until the bat is in a suitable position to make contact with the ball—a time often erroneously referred to as the batter's reaction time.

In using an open stance the batter's hips, shoulders, and arms are rotated somewhat more to the front than they would be if he used either a square or closed stance. As a direct consequence he has less distance to move them in order to bring the bat into position to meet the ball and, if all else is equal, requires less time in which to execute his swing. Thus, an open stance is likely to be particularly suitable for a batter who (through tardiness in making his decision to swing or through lack of strength) consistently has difficulty in getting his bat around early enough to make appropriate contact with the ball. In addition, a batter who does not have this problem but who would like a split second longer to evaluate the pitch before initiating his swing might also benefit from adopting an open stance rather than a square or close one.

The use of a closed stance enables a batter to exert the muscular forces responsible for rotating his hips, shoulders, and arms to the front, over a longer distance than is possible with either of the other two stances. Because the batter is thus able to do more work and to apply more force to the ball at impact, this type of stance is suitable for those who can execute their swings with sufficient speed that getting the bat around to a suitable position to meet the ball is not a problem.

The square stance, involving a position of the feet in between that of the other two stances, affords batters a compromise between the slower swing possible with an open stance and the more forceful hit possible with a closed one.

Another factor that may influence a batter's choice of stance is whether or not his dominant hand and his dominant eye are both on the same side. If they are, there is some evidence to suggest that he is likely to perform better with an open stance than with a closed one. Adams[15] compared the batting performances of six right-handed, right-eyed batters who used an open stance with those of six others of similar dominances who used a closed stance and found that the ones who used an open stance performed significantly better. In the course of a season, the batters who used an open stance struck out 91 times in 713 times at bat (12.8%) while those who used a closed stance struck out 105 times in 576 (18.2%).

[15]Gary L. Adams, "Effect of Eye Dominance on Baseball Batting," *Research Quarterly*, XXXVI, March 1965, pp. 3–9.

The distance between the batter's feet (the width of his stance) is logically dependent to a certain degree on his stature—a tall batter tending to place his feet a greater distance apart than a short one. However, irrespective of their heights, highly skilled batters generally adopt a stance in which the feet are slightly more than shoulder width apart.

The question of how the batter's weight should be distributed when he takes his stance is one of many on which there seems to be a lack of scientific evidence. Two factors that bear on this question are the ease with which the striding foot can be lifted and moved forward (the more weight there is on this foot, the more difficult is this action) and the position of the batter's center of gravity as he prepares to rotate his hips to the front at the beginning of his swing. (The farther forward his center of gravity, the more completely extended his rear leg is likely to be and thus the less forceful the hip rotation he can expect to produce.) These two factors alone would seem to support the contention that a greater proportion of the batter's weight should be borne by his back leg than by his front one.

Like several other aspects of his stance, the position in which the batter holds the bat is something of a compromise. If he holds it in a vertical position (i.e., with its center of gravity directly above his hands), he does not have to apply a torque to the bat to maintain it in position. Holding the bat in this way, therefore, involves less muscular work than holding it in any other position and, as such, affords the batter as easy and relaxed a carriage of the bat as is possible under the circumstances. While this is clearly one of the advantages of holding the bat in a vertical position, the need for the batter to rotate the bat downward through approximately 90° in order to make contact with the ball is something of a disadvantage, for this additional rotation (compared with what is necessary when the bat is held horizontally) provides additional sources of possible error. If the batter holds the bat in a horizontal position, the situation is essentially reversed—he avoids the possibility of making errors due to increasing the range of his motion but increases the likelihood of his becoming unnecessarily tense because of the additional muscular work he must perform. Which of these two extreme positions (or which intermediate position) a batter uses is likely to depend on his own judgement of the relative importance of the advantage and disadvantage associated with each. If he considers the additional rotation the more serious drawback, he is likely to hold the bat in a horizontal or nearhorizontal position. Conversely, if he regards the increased muscular work involved and the associated risk of unnecessary tension in the arms and shoulders as the more important consideration, he will likely hold the bat in or near a vertical position.

Stride. The purpose of the stride is to start the forward movement of the batter's weight in preparation for the swing that follows a split second later (approximately 0.04 sec after the striding foot is planted firmly, according to Hubbard and Seng[16]).

To execute the stride the batter lifts his front foot, advances it a short distance in the general direction of the pitcher, and then plants it firmly on the ground.

The height to which the batter lifts his striding foot as he brings it forward should be no greater than that necessary to ensure that his cleats clear the ground. If he lifts his foot higher than this he increases the time necessary to perform the stride and introduces additional room for error by increasing the range of the movement.

The direction in which the stride should be taken has been the subject of considerable discussion. Some authorities[17,18,19] suggest that skilled batters should modify the direction of their stride according to whether the ball is pitched inside or outside. Others, like Williams,[20] contend that because the stride is initiated as the ball is released by the pitcher (a point verified by Hubbard and Seng[21]) and because the batter cannot know exactly where it is going at this time, the direction of the stride should be consistent for all pitches. This latter contention is supported by the work of Breen,[22] who after "studying hundreds of major league batters and thousands of feet of motion picture film" concluded that one of the five "mechanical attributes" outstanding hitters have in common is consistency in the length and direction of their stride. Regardless of the type of pitch, the better hitters (and Major League players Aaron, Banks, Mantle, Mays, Musial, and Williams were among those studied) planted their striding foot in the same spot. The poorer hitters were not consistent in the placement of the foot.

It is generally recommended by batters and batting coaches that the length of the stride should be relatively short (of the order of 3 to 12 in.). The principal advantages in a stride length of this order, as compared to a longer one, would seem to be:

1. The batter's center of gravity remains sufficiently close to his back foot to allow a forceful hip rotation during the swing.

[16]Alfred W. Hubbard and Charles N. Seng, "Visual Movements of Batters," *Research Quarterly*, XXV, March 1954, p. 57.

[17]Bubba McCord, "The Physics in Hitting," *Athletic Journal*, L, December 1969, p. 46.

[18]Lew Watts, "Classroom Approach to Batting," *Scholastic Coach*, XXXV, March 1966, p. 20.

[19]Don Weiskopf, "Be a .300 Hitter!" *Athletic Journal*, XLIX, January 1969, p. 118.

[20]Williams and Underwood, "Hitting Was My Life: Part V—Science of Batting," pp. 43–44.

[21]Hubbard and Seng, "Visual Movements of Batters," p. 48.

[22]James L. Breen, "What Makes a Good Hitter?" *Journal of Health, Physical Education, Recreation*, XXXVIII, April 1967, pp. 36, 39.

2. The smaller range of motion leaves less scope for error and thus improves the batter's chance of obtaining a consistent action.

3. The forward and downward movement of the batter's head is reduced and with it the possibility that his sighting of the ball may be impaired by this movement.

Swing. From the instant the ball is released by the pitcher, the batter has roughly half a second in which to hit it before it passes behind him into the catcher's mitt. During this time his first task is to evaluate the pitch and make his decision as to whether or not he will swing at the ball. In this part of the process, the longer the batter can study the flight of the ball before making his decision (i.e., the longer his *decision time*), the more likely it is that he will make the correct choice. And, since consistently making the correct choice is an important prerequisite for success in batting, the longer the decision time a batter can allow himself, the better his performance is likely to be.

Having made a decision to swing, the batter's second task is to bring his bat around to meet the ball. The time he needs to do this (referred to here as his *swing time*) is also related to the standard of his batting performance, for if all else is equal, the shorter a batter's swing time, the longer his decision time and thus the higher the level of performance that he can produce.

The findings of Breen,[23] who reported the swing times of some proved Major League hitters and compared them with that of a Major League hitter with a batting average below .300, serve to illustrate this point. The respective lengths of the decision and swing times for a pitch traveling at 104 fps (the mean speed for fast balls as reported by Selin[24]) and the position of the ball at the latest moment the batter can still successfully commit himself to swinging at it are shown in Fig. 110. The advantage enjoyed by a top-class hitter, by virtue of his short swing time, is clearly evident.

The movements involved in a successfully executed swing proceed in a sequential fashion, with the hips, shoulders, arms, and finally the wrists and bat being driven forcefully around to the front.

The hip action, preceded by a "cocking" (or movement in the opposite direction) during the stride, is the result of the reaction to forces exerted against the ground by the batter's rear (nonstriding) leg and is accompanied by an inward rotation of this leg. Once the hip rotation is well advanced, the rotation of the shoulders begins. Then, when the shoulders have been brought around roughly parallel with the hips, the arm swing is initiated.

Because the linear velocity of the point on the bat at which the ball makes contact is of such critical importance in determining what subsequently happens to the ball (see pp. 87–91), good batters hold their leading arm straight, or nearly straight, during the swing. In this way the radius of the arc along

[23] *Ibid.*, p. 39.

[24] Selin, "An Analysis of the Aerodynamics of Pitched Baseballs," p. 235.

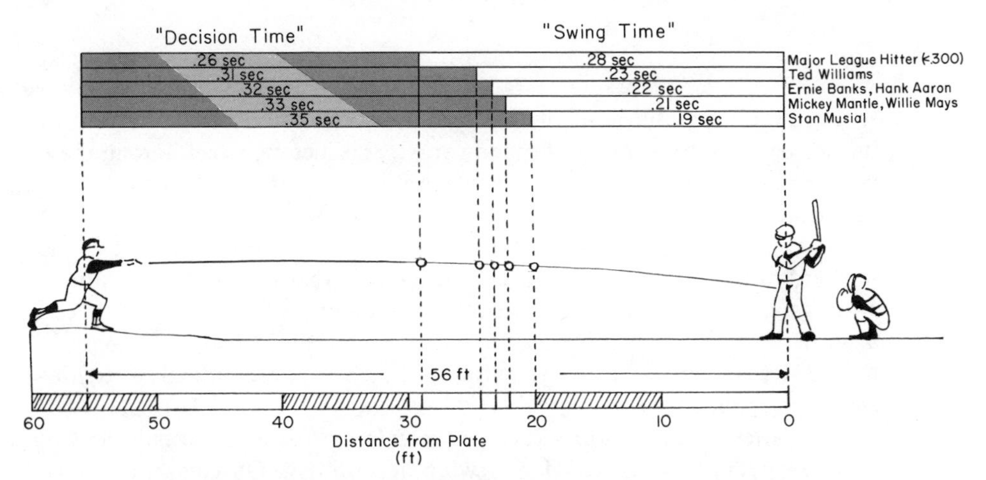

Fig. 110. Decision and swing times in batting (*Based on data in Breen, "What Makes a Good Hitter?"*)

which the "contact point" moves and the linear velocity of this point are kept as large as possible in this situation.

So that the bat can be appropriately aligned when it is brought forward to meet the ball and so that the forces exerted on the bat impart to it a velocity in the desired direction, the rotation of the hips, shoulders and arms should each take place in an approximately horizontal plane. With respect to this, Breen[25] found that the center of gravity of outstanding hitters "follows a fairly level plane throughout the swing," while that of lesser hitters showed a definite downward trend indicative of a "dipping of the shoulders."

The final contribution to the swing is made by the batter's wrists, just before the bat meets the ball. Some controversy surrounds the action of the wrists at this time, for while apparently no one questions that the wrists should be "uncocked" or straightened out (a movement known to anatomists as ulnar or medial deviation), some authorities contend that the wrists should be rolled (i.e., the wrist of the upper hand on the bat should be rolled over the wrist of the lower hand) and that this action should be initiated as the bat makes contact with the ball. However, since any additional muscular force that might be applied to the bat and thence to the ball as a result of such an action would likely be very small and possibly in an undesirable direction, and since an examination of motion pictures and still photographs of top-class batters fails to reveal any such action while the ball is in contact with the bat, the desirability of rolling the wrists in this way seems very much open to question. Because it is quite evident in a batter's follow-through that he has rolled his wrists, it could well be hypothesized that the controversy surrounding this question stems from an incorrect interpretation of the timing of the wrist-rolling action and the purpose that it serves. Instead of a protective movement initiated after the ball leaves the bat (and which it seems most likely to be), the action appears to have been looked upon as an additional source of muscular force to be used during the period of contact between bat and ball.

[25]Breen, "What Makes a Good Hitter?" pp. 36–39.

The required position of the bat and the batter's body at the instant of impact between bat and ball depends on several factors. Chief among these are the magnitude of the force with which it is desired to hit the ball and the direction in which it is intended the ball should go.

To apply maximum force to the ball, the batter must ensure that at the moment of impact he has a firm grip of the bat; that his wrists, arms, hips, and legs are braced one against the other; and that his feet are firmly braced against the ground. In this way the batter effectively increases the mass of the body that impacts with the ball and thereby correspondingly increases the force applied to it.

If a batter wishes to apply less than maximum force to the ball, he will generally reduce the speed of his swing accordingly. On some occasions, however, and specifically when bunting, even a reduction of the bat's velocity to zero is insufficient to produce the required reduction in the force applied to the ball. Therefore, at such times the batter holds the bat loosely in his hands and allows his arms to "give" as the ball strikes the bat, thereby keeping the effective mass of the bat relatively low throughout the period of impact.

The direction the ball follows after it leaves the bat is determined by a number of factors (see pp. 87–91). Of these, it is the angle of incidence that a batter generally uses as his means of fixing the direction in which he plans to hit the ball. By swinging early and making contact with the ball well forward of the plate or by swinging late and contacting the ball when it is over the plate, the batter obtains angles of incidence that enable him to send the ball to the same and opposite fields, respectively. Between these two extremes are hits in which the bat is parallel with the front edge of the plate as contact is made with the ball. Ignoring the effects produced by the spin on the ball, the zero angle of incidence in these cases leads to the ball being directed straight down the center of the field.

Follow-Through. As in the case of pitching, the follow-through in batting serves the dual purpose of reducing the risk of injury and preventing interference with the application of force to the ball.

Base Running

Running to First Base. Because there can be no question of whether or not a man should run, and because he is not required to stop as soon as he reaches the base, running to first base is generally simpler than running between bases.

As in running between any pair of bases, the time taken to reach first base depends on the distance covered by the runner and on his average speed over that distance. In regard to this, Garner's[26] study of the time taken by right-

[26] Charles W. Garner, "Difference in Time Taken to Reach First Base in Right and Left Hand Batting" (M.A. thesis, State University of Iowa, 1937).

handed and left-handed batters to reach first base produced some interesting results. When his 35 subjects batted left-handed, the mean distance they had to run to reach first base (88.71 ft) was less than the corresponding mean distance (92.32 ft) when they batted right-handed. However, this advantage was reversed when average speeds were considered, for the subjects' average speed when they batted right-handed (21.27 fps) was slightly faster than when they batted left-handed (21.20 fps). When both of these factors were considered together, it was found that the advantage of batting left-handed (i.e., a shorter distance to run) outweighed that of batting right-handed (i.e., a greater average speed). Thus, the subjects reached first base in a significantly shorter time when batting left-handed than they did when batting right-handed:

$$\text{Mean time to reach first base when batting left-handed} = \frac{88.71 \text{ ft}}{21.20 \text{ fps}} = 4.18 \text{ sec}$$

$$\text{Mean time to reach first base when batting right-handed} = \frac{92.32 \text{ ft}}{21.27 \text{ fps}} = 4.34 \text{ sec}$$

Running between Bases. Since the techniques used during the middle part of the run differ so little from those employed in sprinting on a track, the main interest here centers on methods of starting and stopping.

In taking a lead off the base there are two methods in common use—the standing, or stationary, lead and the walking lead. Each has its advantage. With a stationary lead the runner generally has his weight supported evenly on both feet and thus is in a position to move toward either base with equal facility. A walking lead, on the other hand, enables the runner to get a faster start because he is already in motion at the time he makes his decision to run.

The advantage that the walking lead has over the stationary lead, in terms of starting speed, is clearly evident in the results of Kasso,[27] who tested the time taken by 10 high school players to reach a line 25 ft from first base using four different leads. From fastest to slowest, the mean times recorded were ranked as follows:

1. 6-ft walking lead.

2 = { 4-ft walking lead.
8-ft stationary lead. }

4. 6-ft stationary lead.

All the differences between two means with different rankings were found to be statistically significant. (*Note:* To some extent the question of a walking lead versus a stationary lead is an academic one, for a good pitcher will

[27]Richard A. Kasso, "A Comparison of Two Methods of Leading Off First Base" (M.S. thesis, Springfield College, 1963).

prevent a runner from obtaining a walking lead by making the runner stop before he pitches the ball.[28])

To come to a stop at the end of his run, a runner must be subjected to forces in the opposite horizontal direction to that in which he is moving. If the runner simply runs onto the base, he evokes such forces from the ground by overstriding during the last few steps of his run (see p. 385). If he slides, the frictional forces exerted on him by the ground as he slides across its surface supplement whatever forces are exerted backward against his feet during the latter stages of his run.

For a runner to slide it is necessary for him to move his body from its sprinting position to a horizontal or near-horizontal position. This can be done in basically two ways: (1) by rotating his body forward some 70°–90° into a prone position (a head-first slide) or (2) by rotating it backward 90°–110° into a supine position (a feet-first slide). Although, at the time of writing, there appears to have been no published research concerning the relative merits of these two methods, and specifically how they compare in terms of speed, it would seem reasonable to expect that the head-first slide would be the faster of the two, for, as the runner's body is rotating forward, his legs are in position to drive strongly backward against the ground. Thus it is possible that the transition from a running position to a sliding position might well be effected with little if any loss of forward speed. The reverse of this argument applies in the case of the feet-first slide, for as the runner's body rotates into position, his legs move into a progressively less favorable position for the exerting of force backward against the ground. On other criteria, such as comfort, ease of learning, and safety, there can be little doubt that the advantages favor the feet-first method.

The techniques employed in changing direction to round one base and head for the next vary in the distance that the runner covers and in his average speed over that distance. If he runs straight to the base, makes a 90° turn to his left, and heads directly to the next base, he covers the least possible distance. His average speed, however, is likely to be less than if he followed a curved path as he made the change in direction. On the other hand, if he follows a curved path, the distance he runs is greater than it would be if he followed a straight one. (*Note:* This difference in distance obviously depends on how soon the runner begins to "round out" the turn. If he begins his curving run, say, 60 ft from the base, he will clearly run farther than if he began it at, say, 20 ft from the base.) Thus it can be seen that, in comparing any two methods, each one is likely to have an advantage over the other. The question is—which one has the greater advantage? Although it seems most unlikely that all the methods in common use are exactly equal in terms of effectiveness, there is as yet little evidence to support any one method in preference to another.

[28]Ron Square, *How to Develop a Successful Pitcher* (Englewood Cliffs, N.J.: Prentice-Hall, Inc., 1964), p. 173.

Chapter 9

BASKETBALL

Unlike baseball, which was first played somewhat earlier, basketball has developed to such an extent (since its invention by James Naismith in December, 1891) that it is now played in almost every country in the world. In fact, according to the *Encyclopaedia Britannica*,[1] it is now "... second only to soccer as an international sport."

Basic Considerations

The analysis in this chapter is subdivided into five sections involving three offensive techniques (passing, dribbling, and shooting) and two used on both offense and defense (footwork and jumping).

Passing

A player's objective in making a pass is to complete the displacement of the ball from his hand or hands to those of a predetermined teammate. Except in relatively rare instances (e.g., in the case of some hand-offs that a center, or pivot man, might use or in the even rarer case of a pass rolled along the floor), this task involves motion of the ball through the air.

The flight of the ball, like that of all such projectiles, is governed by its velocity and its height at release and by the air resistance it encounters en route. The passer's task, therefore, involves obtaining the combination of these three factors that will produce the optimum result.

Velocity at Release. The velocity with which the ball leaves the passer's hand or hands is determined by its velocity before he initiates his passing action and by the forces that he subsequently exerts upon it. In general, a player has at his disposal a large number of muscular forces that he may use

[1] "Basketball," *Encyclopaedia Britannica* (1967), III, p. 250.

to ensure that the ball obtains a release velocity of the desired magnitude and direction. However, since success depends in large measure on completing the actions involved before the defense can react and intercept the pass, it is important that those muscular forces that can be exerted most quickly be given priority. Thus the muscular forces producing flexion of the fingers, flexion of the wrist, and extension of the elbow should be the first to be called upon. Only when these forces are insufficient (as, for example, when a long pass is needed to initiate a fast break) should the less readily available forces of the trunk and legs be utilized.

For a pass to be completed successfully, the horizontal direction in which the ball is projected must be such as to take into account any displacement that the receiver undergoes while the ball is in flight. Thus, the ball must be thrown a sufficient distance forward of a moving receiver to allow his hands and the ball to arrive at the same point at the same time.

Height at Release. For any given case, the height at which the ball is released depends on the type of pass used and on the physical characteristics of the passer.

Air Resistance. While the cross-sectional area of a basketball is large compared to that of similar projectiles used in sports, the velocity at which it moves through the air is relatively small. Further, since the velocity is a more potent factor in determining the magnitude of the drag force than is the cross-sectional area, the drag on a basketball is also relatively small—so small, in fact, as to be of little practical significance.

When a player throws a pass, he almost invariably imparts spin to the ball. With the majority of passes this is a backspin that tends to slow the rate at which the ball falls under the influence of gravity (see Magnus effect, pp. 181–83). Provided the amount of backspin is not so great as to cause the receiver to have difficulty in catching the ball, its existence can only be regarded as desirable, for it enables the ball to follow a slightly more direct path than would otherwise be possible. On the other hand, if sidespin is imparted to the ball at release, the lateral deviation of the ball during flight will at least disconcert the receiver and at worst may cause him to miss the ball completely. In general, therefore, the application of sidespin to the ball should be avoided. The one major exception to this is bounce passing in which spins other than a backspin may be imparted in order to obtain a particular result once the ball strikes the floor.

To complete a pass, the ball must be caught by the receiver. The basic factors governing catching in basketball are essentially the same as those already discussed in connection with baseball except that no glove is used and the catch is generally made with both hands playing virtually identical roles.

Dribbling

Although not always clearly understood by players of the game—if their actions can be taken as any indication—the objective in dribbling is to advance the ball when passing is impossible or less desirable.

The act of dribbling consists of a number of repetitions of the following series of actions:

1. The dribbler applies forces to the ball and propels it (generally in a forward and downward direction) toward the floor. The forces applied by the dribbler, together with the force exerted on the ball by gravity, determine the velocity with which the ball leaves his hand. Ignoring the presumably negligible effect of air resistance, this velocity and the height of release determine the subsequent flight path of the ball.

2. The ball follows a curved flight path until it makes contact with the floor.

3. The floor exerts forces on the ball, altering both the magnitude and direction of its velocity. As in the many similar cases of impact in sports, the velocity of the ball as it leaves the floor depends principally on its velocity immediately before impact (assuming, as seems reasonable, that the other factors involved are essentially constant in any given situation).

4. The ball follows a curved path as it ascends from the floor to again make contact with the dribbler's hand. (*Note:* To ensure that his hand is in position to receive the ball as it rebounds from the floor, the dribbler must be moving with an average horizontal velocity approximately equal to that of the ball.)

5. The dribbler applies force to the ball to reduce its upward velocity to zero preparatory to propelling it back toward the floor. The amount of force the dribbler must exert to bring about this change in the velocity of the ball depends on the distance over which he applies the force and the velocity at which the ball is moving when his hand makes contact with it.

Shooting

The objective in shooting is very similar to that in passing, for it involves completing the displacement of the ball from one position (the hand or hands of the shooter) to another (the inside of the hoop). As Wooden[2] puts it—a shot is "a pass to the basket."

[2]John R. Wooden, *Practical Modern Basketball* (New York: The Ronald Press Co., 1966), p. 71.

Aside from the question of applying force to the ball (and in which process the factors of importance are identical with those already discussed in connection with passing), the principal factors governing the outcome of a shot are those that determine the flight path of any projectile,—its height, speed and angle of release, and the air resistance encountered in flight.

The height of release is determined by the physique of the shooter and by the type of shot he uses, the latter generally being influenced by the shooter's position on the court, by his preference for one type of shot over others that might be equally suitable, and by the need to get his shot away without having it blocked by an opponent.

Assuming a height of release that is fixed in some manner by these various factors, the combination of speed and angle of release most likely to result in a successful shot depends on the distance of the shot, the position and caliber of the man guarding the shooter, and the angle to the horizontal at which the ball should approach the basket (the *angle of entry*) to allow maximum tolerance or room for error.

Distance of the Shot. The distance of the shot directly influences the speed of release required—a tip-in or lay-up requiring markedly less speed than, say, a 20-ft jump shot. Further since the speed and angle of release required for a successful shot are interdependent, the distance of the shot indirectly influences the angle of release.

Because the effects of small deviations from the optimum speed and angle of projection become more pronounced the longer the ball is in the air, shooting accuracy also varies with the distance of the shot. While this in itself is little more than common knowledge, a study by Bunn[3] of the actual relationship between shooting accuracy and the distance of shots during competitive play has provided some interesting information (Fig. 111). His data suggest, for example, that on the average, one shot from 9 ft is worth more than two shots from 24 ft or that one from 3 ft is worth more than two from 15 ft or three from 24 ft. Incidentally, his 28% figure for shots from 15 ft compared with free-throw percentages that generally average in the neighborhood of 60–70% vividly demonstrates the effect that an active defense has in reducing shooting accuracy.

Position and Caliber of Defensive Man. The position, physique, and jumping ability of the man guarding the shooter all have a bearing on the speed and angle of release that can be successfully employed. The closer the defensive man is to the shooter, the taller he is, the longer arms he has, and the higher he

[3]John W. Bunn, *The Scientific Principles of Coaching* (Englewood Cliffs, N.J.: Prentice-Hall, Inc., 1972), p. 256.

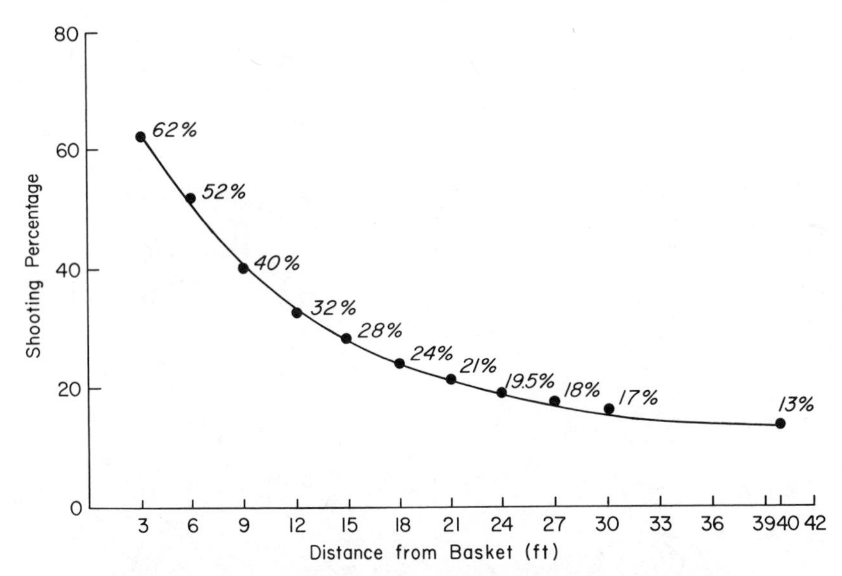

Fig. 111. Shooting accuracy varies with the distance of the shot (*Redrawn with permission from Bunn,* Scientific Principles of Coaching).

can jump, the more likely he is to be able to reach high enough to block the shot. Thus, the speed and angle at which the ball is released must be such as to allow it to pass beyond the reach of the defensive man.

Angle of Entry. The role of the angle of entry is a rather complex one and as such warrants detailed consideration.

If the ball approaches the hoop from directly above (i.e., its angle of entry is 90°), it is presented with an 18-in. diameter circular opening through which it can pass [Fig. 112(a)]. If it approaches from a lesser angle [Fig. 112(b), (c), and (d)], the opening at right angles to its path is elliptical with one diameter 18 in. in length and the other—the one that ultimately determines whether the ball can pass through the hoop without touching the rim—somewhat less than 18 in. The actual length of this second diameter can be computed using the equation:

$$d = (18 \sin \alpha) \text{ in.}$$

where d = the length of the diameter in question and α = the angle of entry.

When the angle of entry is such that d is equal to the diameter of the ball, the lower limit of possible angles of entry has been reached, for with any

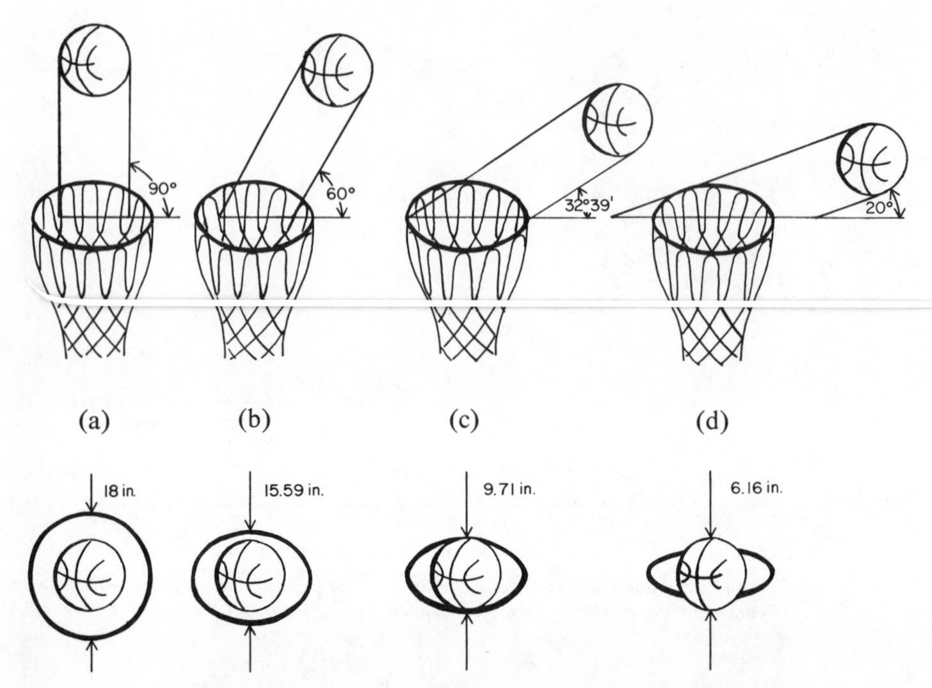

Fig. 112. The scope for error in the range of a basketball shot varies with the angle at which the ball approaches the hoop, i.e., the angle of entry (*Note:* The top half of the figure shows a series of oblique side views of the basket while the bottom half shows a corresponding series of oblique front views).

lesser angle the ball will inevitably strike the rim of the hoop.[4] With a ball of diameter 9.71 in. (the midpoint in the range of diameters permissible within the rules) this lower limit can be found by letting $d = 9.71$ in. and by solving for α:

$$9.71 = 18 \sin \alpha$$

$$\sin \alpha = \frac{9.71}{18}$$

$$= 0.5394$$

$$\alpha = 32°39'$$

The maximum angle of entry is obviously 90°, which angle is obtained if the ball is released from a point directly above the basket.

Because d increases as the angle of entry increases, the margin for errors in the range or distance of the shot is also related to the angle of entry—the nearer the angle of entry is to 90°, the greater is this margin. The horizontal

[4]Generalizations concerning the outcome when the ball hits the rim of the basket are very difficult to make because of the large number of variables involved. The discussion here is, therefore, confined to a consideration of angles of entry that permit the ball to pass through the hoop without touching the rim.

distance that the center of the ball can be away from the center of the hoop and still pass "cleanly" through the hoop (i.e., the margin for error) can be closely approximated using the equation:

$$E = \pm(9 \sin \alpha - r) \text{ in.}$$

where E = the margin for error, α = the angle of entry, and r = the radius of the ball. Using this equation, the relationship between E and α can readily be shown:

Angle of Entry, α	*Margin for Error, E (in.)*
90°	±4.15
80°	±4.01
70°	±3.60
60°	±2.94
50°	±2.04
40°	±0.93
32°39′	0.00

Therefore, from this standpoint alone, it would seem that the optimum angle of entry is one that is as close to 90° as possible. Several other aspects of the matter must be considered, however. One of these is the fact that for any given shot the higher the angle of entry, the higher must be both the speed and angle at which the ball is released. In the case of a 15-ft shot (e.g., a free throw) released from a height of 7 ft above the floor, the relationships between the angle of entry and the speed and angle of release, respectively, are shown in Fig. 113.

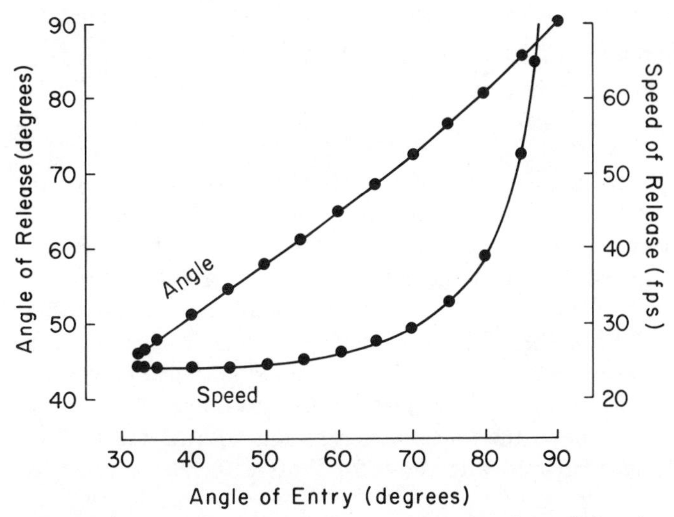

Fig. 113. For a given height of release and distance of shot (7 ft and 15 ft, respectively, in the present example) only one combination of speed and angle of release will yield a given angle of entry.

A careful consideration of Fig. 113 reveals that in this case, at least, angles of entry close to 90° are a practical impossibility. For example, an 87° angle of entry requires that the ball be released with a speed of approximately 65 fps (or 44 mph)—a speed well beyond the capabilities of an athlete using any orthodox shot. Even if an athlete could readily project the ball at such a speed, few indoor arenas would have ceilings high encough to "accommodate" the shot, for the ball would rise to a height of approximately 72 ft above the level of the court—a height roughly equal to that of a six-story building!

Another factor that should be taken into account is the effect that small errors in the angle of projection have on the distance of the shot. In this respect, Mortimer[5] has demonstrated that if all else is equal, "the seriousness of a 1° error increases with the angle of projection. As the angle increases, the same 1° error . . . causes the ball to deviate farther and farther from the center of the basket." Thus, while a high angle of projection is desirable in that it leads to a high angle of entry, it is less desirable than lower angles in the demands it puts on accuracy as the ball is released.

From all of this it can be seen that the question of the optimum angle of entry for a given shot is anything but a simple one. Probably the most difficult aspect of the whole question is how each of the various factors combine to determine the optimum angle of entry. In the case of a 15-ft shot released at a height of 7 ft, the extent to which each of the factors considered here influences the optimum angle of entry can be gauged from the data contained in Table 18. This table shows

1. a range of angles of entry between 32°39′ (the theoretical minimum) and 89°59′,
2. the angle of release necessary to attain each of these angles,
3. the margin for error in the length of each shot, and
4. the errors in the length of the shot resulting from deviations of 1° from the required angle of release.

Where the errors resulting from deviation in release angle exceed the margins for error that are available, the release angles concerned clearly require a higher degree of accuracy from the shooter than do the release angles for which this is not the case. The data in the latter cases (i.e., those that exhibit the greater "tolerance" of small errors on the part of the shooter) are shaded in Table 18. From these figures it seems reasonable to suggest that an angle of release between 49° and 55° is likely to provide the shooter with a greater likelihood of success than any angle outside this range.

Figure 114 shows the trajectories for shots with release angles within the range above compared with those for shots with release angles of 46° (the

[5]Elizabeth M. Mortimer, "Basketball Shooting," *Research Quarterly*, XXII, May 1951, p. 238.

Table 18. Optimum Angles of Entry and Release for a 15-ft Shot

Angle of Entry (degrees)	*Angle of Release* (degrees)	*Margin for Error* (in.)	*Error Due to 1° Error in Angle of Release** (in.)	
			+1°	−1°
32.65	46.14	0.00	1.43	−1.84
33	46.38	0.05	1.34	−1.74
34	47.06	0.18	1.09	−1.48
35	47.73	0.31	0.84	−1.22
36	48.41	0.44	0.60	−0.97
37	49.08	0.56	0.37	−0.73
38	49.75	0.69	0.14	−0.49
39	50.42	0.81	−0.09	−0.25
40	51.10	0.93	−0.31	−0.02
41	51.77	1.05	−0.54	0.21
42	52.44	1.17	−0.76	0.43
43	53.11	1.28	−0.97	0.66
44	53.79	1.40	−1.19	0.88
45	54.46	1.51	−1.41	1.10
46	55.14	1.62	−1.63	1.32
47	55.82	1.73	−1.84	1.55
48	56.50	1.83	−2.06	1.77
49	57.18	1.94	−2.28	1.99
50	57.86	2.04	−2.51	2.22
60	64.87	2.94	−5.01	4.75
70	72.37	3.60	−8.86	8.62
80	80.65	4.00	−18.63	18.40
89.98	89.98	4.15	−10,643.66	10,644.00

*A positive number indicates that the center of the ball passes beyond the center of the basket by the stated amount, while a negative number indicates that it falls short of the center of the basket by the amount stated.

minimum possible for a successful shot) and 73°. (The latter value has been chosen as roughly representative of the maximum angle of release that might conceivably be used in a game situation.) From the figure it is clear that the optimum angles of release yield trajectories with a "low arch" rather than the "medium arch" or "high arch" advocated by most writers on the subject.

When the ball strikes the rim or the backboard, instead of passing "cleanly" through the hoop, what happens to it next depends on the combined effect of the many factors that influence any elastic impact (see pp. 77–93). Assuming that the masses of the two bodies concerned and their coefficient of restitution are constant, the most important factors in determining the outcome are likely to be the point at which the ball makes contact with the rim or

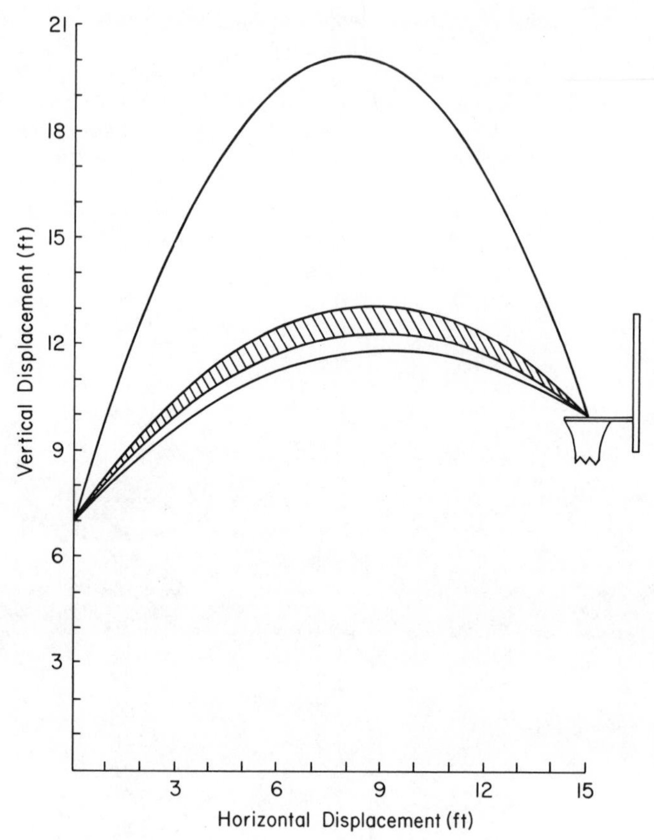

Fig. 114. Free-shooting: The paths followed by a ball released from a height of 7 ft at angles of 46°, 49°, 55°, and 73°. The optimum angles of release yield paths within the shaded area.

backboard, its velocity at that instant, and the extent to which it is spinning. Although it is very difficult to generalize in cases of such complexity, it would probably be true to say that contact nearer the back of the hoop than the front, a soft shot (i.e., a relatively low speed at contact), and the presence of some backspin on the ball would all be likely to improve the odds on the ball rebounding through the hoop.

Body Position and Footwork

To be effective on offense, a player must be able to outmaneuver a defensive man with rapid changes in his speed and direction of movement. Conversely, to be effective on defense he must be able to react readily to his opponent's attempts to outmaneuver him.

In attempting to achieve these objectives—whether he be starting from a stationary position, coming to a stop, or merely changing the speed or direction in which he is moving—an athlete must exert forces to accelerate (or decelerate) his body. The magnitude and direction of these forces determine the nature of the acceleration that he experiences (Newton's second law). It is, therefore, important that the player assume a position from which he can readily apply forces that are consistent with his needs. Because these forces result mainly from the action of muscles in his legs and are transmitted to the floor via his feet, the positions of his legs and feet are especially important in this regard.

Where starting from a stationary position is concerned, the greater the ease with which the player can disrupt his equilibrium state (i.e., the less the stability of his equilibrium), the more rapidly he can begin moving in a given direction and thus the more likely he is to be effective in eluding his opponent or in preventing his opponent eluding him. Therefore, when considering the optimum position to adopt in order to start quickly, a player should take into account the factors that influence his stability of equilibrium in addition to those that determine his ability to apply the necessary forces.

Jumping

In basketball, a player's purpose in jumping is to increase the height to which he can reach. In most cases (e.g., for rebounds and center jumps) the player endeavors to jump as high as he can and to so time his jump that at the peak of his flight he can place one or both hands on the ball. Then, if he is able to jump higher or time his jump more accurately than those opposing him, he is likely to be in a position to control or gain possession of the ball.

The maximum height to which a player can reach is the sum of (1) the maximum height to which his center of gravity can be lifted and (2) the distance to which he can reach beyond this height. As in the case of other projectiles, the first of these is governed by the height of the player's center of gravity and by his vertical velocity at the instant of takeoff. The second is a function of his body position in the air. (See p. 126 for a detailed consideration of the effects of variations in body position.)

In a few instances the actual height to which a player reaches is less important than obtaining an advantage over his opponent in terms of time. For example, in executing a jump shot it is more important that a player get his shooting hand to a height of, say, 8 ft before his opponent is able to reach to that height than it is for him to get his hand up to some greater height. In these cases, where speed in initiating a jump is more important than the maximum height to which the athlete can reach, the muscle actions (similar to those already discussed in connection with passing and shooting) must be such as require the least possible time consistent with the completion of the task.

Techniques

Passing

Of the two main methods of advancing the ball down the court (passing and dribbling), passing is generally the most effective and widely used. Although there exists an extensive variety of passes that might be used, in practice only a few of these are commonly employed. The others generally have inherent limitations that restrict their use to special situations and/or to specially gifted players.

The frequency and effectiveness with which the more common types of pass are used in competition have been studied by Allsen and Ruffner.[6] After analyzing the charts of 72 selected games (24 each at college, high school varsity, and college intramural level) they found that the two-hand chest pass was more frequently used than any other type of pass—38.6% of all passes recorded were of this type. Next most frequent were the one-hand baseball pass (18.9%), the two-hand overhead pass (16.6%), and the one- and two-hand bounce passes (7.3 and 7.2%, respectively). The one-hand baseball pass resulted in a lower percentage of completions than any other type of pass. On 9.3% of all baseball passes attempted, the ball was lost to the opposition. The next least successful were the one-hand bounce pass (9.1%), the two-hand shoulder pass (7.7%), and the two-hand bounce pass (7.3%). The best results were obtained with the two-hand chest pass and the little-used two-hand shovel pass, both of which resulted in the ball being lost only 2.5% of the time.

Chest pass. As indicated by the results of Allsen and Ruffner, the two-hand chest pass is both the most reliable and the most frequently used of the various passes in basketball.

To execute the pass the ball is held at approximately chest height with the thumbs pointed toward each other and the fingers comfortably spread behind the ball. From this position the ball is drawn backward a little as the wrists are "cocked" before applying force to the ball in the direction of the pass. This cocking action places the wrists in such a position that the muscles that flex the wrist can exert force over a longer distance and thus do more work on the ball than would otherwise be possible. The cocking of the wrists is immediately followed by a coordinated series of actions in which the elbows are extended and the wrists and fingers are rapidly flexed (or "snapped") to apply force to the ball. These actions are frequently accompanied by a step and a shifting of the body weight in the direction of the throw. Because the pass should normally be directed so that the ball can be caught by the receiver at some point between his waist and his shoulders, it must usually be released at an angle slightly above the horizontal in order to allow for the effects that

[6]P. E. Allsen and William Ruffner, "Relationship Between the Type of Pass and the Loss of the Ball in Basketball," *Athletic Journal*, XLIX, September 1969, pp. 94, 105–7.

gravity has on it during its flight. (*Note:* While it may be very effective in producing the kind of passing action required, it is as well to recognize that the common admonition of teachers and coaches to "throw the ball in a straight line parallel to the floor" is actually asking the impossible.)

For relatively short passes of less than 10–12 ft, all the force that needs to be applied to the ball can usually be provided by the muscles controlling the wrists and fingers, with those extending the elbow perhaps also making a small contribution. For somewhat longer passes a greater involvement of the elbow extensors and the muscles of the legs and trunk is normally required. The extent to which the various body parts contribute is reflected in the follow-through once the ball has been released—a short follow-through of wrists and fingers characterizing short passes and a complete follow-through with both arms characterizing longer ones.

Overhead Pass. Although most widely used as a pass into a post or pivot man, and as a pass to a cutter, the two-hand overhead pass can be particularly useful in other situations. Tall men often find it easy to protect the ball by holding it overhead. Thus, they are naturally in the ideal position from which to make an overhead pass. Similarly, a player receiving the ball overhead is probably in a better position for this kind of pass than he is for any other.

The two-hand overhead pass is made from a position in which the ball is held above and slightly forward of the head. (Apart from affording greater protection of the ball from defensive men behind him, this forward position makes it a little easier for the player to direct his pass forward and downward than would be the case if he held the ball directly above his head.) The ball is held in the hands with fingers spread and pointing upward and with the thumbs directed inward toward each other. To make it difficult for the defense to get to the ball the elbows are kept partially flexed and pointing outward.

The passing movement consists of a quick forceful flexion of the wrists and fingers, often accompanied by a forward step and a lifting of the body up onto the toes. Provided these latter movements are performed quickly enough not to prematurely reveal the passer's intention, they slightly increase his prospects of successfully completing the pass by allowing him to increase the speed and height at which the ball is released.

Except for those passes that are deliberately lofted beyond the reach of a defensive man, overhead passes are almost invariably directed forward and downward toward the chest of the receiver. The angle at which the ball is directed below the horizontal is governed by the distance of the pass and the speed of the ball at release—the longer the pass, or the less the speed of release, the nearer the angle of release must approach the horizontal.

Since most of the force applied to the ball derives from the action of the wrists and fingers, the follow-through after the ball has been released is naturally limited.

Bounce Pass. As a means of passing the ball through congested areas, or past a tall long-limbed defensive man, the bounce pass has few, if any, equals.

The two-hand bounce pass is executed in a manner identical to that of the two-hand chest pass except for the point at which the pass is aimed. Instead of being directed at the receiver's chest, a bounce pass is released in the direction of a point on the floor partway between the passer and where the receiver will be when he catches the ball. While to date there appear to have been no published reports of research which touch this aspect of the bounce pass, practical experience suggests that the ball should be bounced at a point about two-thirds of the distance between passer and receiver.

As in other similar cases, the result of the impact between the ball and the floor is influenced by whatever spin has previously been imparted to the ball. Topspin is frequently applied to increase the distance of a bounce pass and backspin to make the ball "come up" softly into the hands of a cutter. Sidespins are also used on occasion.

Because the distance that the ball must travel is greater for a bounce pass than for a direct one and because impact with the floor generally reduces the speed of the ball, a bounce pass will take a greater time to complete unless the force applied to the ball by the passer is correspondingly increased. For this reason, a bounce pass frequently requires a greater contribution of force from the muscles of the legs and trunk than is necessary for an equivalent chest pass. The increase in either the time to complete the pass or the force applied to the ball (which necessarily accompany the use of a bounce pass) increases the scope for error in the execution of the pass.

Baseball Pass. The baseball pass is used most as a means of initiating a fast break following a rebound or the scoring of a goal.

Consistent with the name, the passing action is similar to that used in a baseball throw. For a right-handed throw the ball is raised over and behind the right shoulder with the right hand behind the ball (fingers spread and pointing upward) and the left one guiding the ball back into position. The feet are comfortably spread and positioned so that a line joining the heel of the back foot to the toes of the front foot closely approximates the intended direction of the pass. (Such a position of the feet enables the hips to be fully rotated to the front during the throw and thus allows the muscles that produce this motion to add their contribution to the speed of the ball, should this be necessary.) From this initial side-facing position the throw is made with a rotation of the hips and trunk to the front followed by a coordinated action of the throwing shoulder, elbow, wrist, and fingers.

Dribbling

Although the height and speed of a dribble vary according to the situation, the technique used is basically the same in all cases.

The dribbling action is initiated by placing one hand on top of the ball

Fig. 115. Stroboscopic photograph of good technique in the execution of a high dribble. Note the length of time that the hand is in contact with the ball—consecutive photographs have been taken at equal intervals of time.

and, by extending the elbow and flexing the wrist and fingers, pushing it down toward the floor. Then, as the ball rises, one hand (generally the same hand as pushed it down) is placed on top of the ball and the procedure reversed. The wrist and elbow "give" as the hand accompanies the ball up to the high point of its bounce.

When a youngster first attempts to dribble a basketball, he almost invariably keeps his hand fairly rigid and hits the ball instead of pushing it. Because the time during which he is applying force is much less than it would be if he were to use the proper pushing technique, he must apply a much greater force to the ball to obtain the same result (impulse-momentum relationship, p. 76). Further, because his hand is in contact with the ball for only a very short time, he is less able to sense deviations from the required speed and path of the ball and to apply the forces necessary to correct them. The beginner's first attempts at dribbling are generally characterized, therefore, by a sharp stinging sensation in the hand (due to the relatively high force involved) and by an all-too-evident lack of control (due to the small period of contact between his hand and the ball).

The time during which a skilled dribbler's hand is in contact with the ball is probably a good deal longer than is commonly supposed. In a high dribble, the hand is normally placed on the ball when it is approximately 2 ft from the floor, accompanies the ball upward another 1–2 ft, and then pushes the ball down again through roughly the same distance. Thus, the hand may be in contact with the ball for as much as half of its journey. However, since the ball is moving relatively slowly as it nears the end of its upward motion—the farther it is from the floor, the smaller is its speed—the half of its journey in which the ball and hand are in contact is of longer duration than the other half. [*Note:* In the stroboscopic photograph of a basketball player performing a high dribble (Fig. 115) the player's hand is in contact with the ball in six out of eight (or 75%) of the exposures recording the dribbling cycle.] Because the main reason for a player electing to use a low dribble is to obtain a greater control over the ball and thus to reduce the chances of an opponent knocking it away, the proportion of time during which the player is in contact with the ball is even greater for a low dribble than it is for a high one.

Shooting

Shooting, the only means by which it is possible to score points, is probably the most important skill in the game of basketball. In light of this, it is little wonder that the techniques of shooting have received concentrated attention from coaches and players. It is hardly surprising, therefore, that the results of a statistical analysis conducted by Harvey[7] should reveal that the standard of shooting in major college games had improved steadily over an 18-yr period. Based on statistics from 73,400 major college games between 1949 and 1966, the data presented by Harvey (Fig. 116) revealed:

1. A more than 40% increase in the number of points scored per game. (The mean number increased from 109.5 in 1949 to 154.9 in 1966.)
2. An almost constant number of field goals attempted per game.
3. An almost constant number of free throws attempted per game—except for a 5-yr period (1953–1957) during which changes in the rules resulted in a substantial increase in the number.
4. A steady increase in both field-goal and free-throw shooting percentages.

In other words, these results showed conclusively that increases in the number of points scored per game during the period considered were the result of increases in the accuracy of the shooters rather than increases in the number of attempts made.

There are a wide variety of shots that a player might use in attempting to put the ball in the basket. Of these, some have been found by experience to offer better chances of success than others. Logically enough, these shots are used with greater frequency in games than those less likely to be successful. In a study of rebounding, Allsen[8] reported the frequency with which various types of shots were used in 39 selected games. His analysis, which unfortunately did not include lay-ups or tip-ins, showed that the one-hand jump shot was by far the most commonly used. This type of shot was used in 67.20% of the 3180 field-goal attempts recorded. Next in order were the one-hand set shot (21.01%), the right-hand hook shot (8.01%), and the left-hand hook shot (2.29%).

[7]John H. Harvey, "Statistical Trends in Basketball," *Scholastic Coach*, XXXVI, October 1966, pp. 22, 26.

[8]Phillip E. Allsen, "The Rebound Area," *Athletic Journal*, XLVIII, September 1967, pp. 34, 97–98.

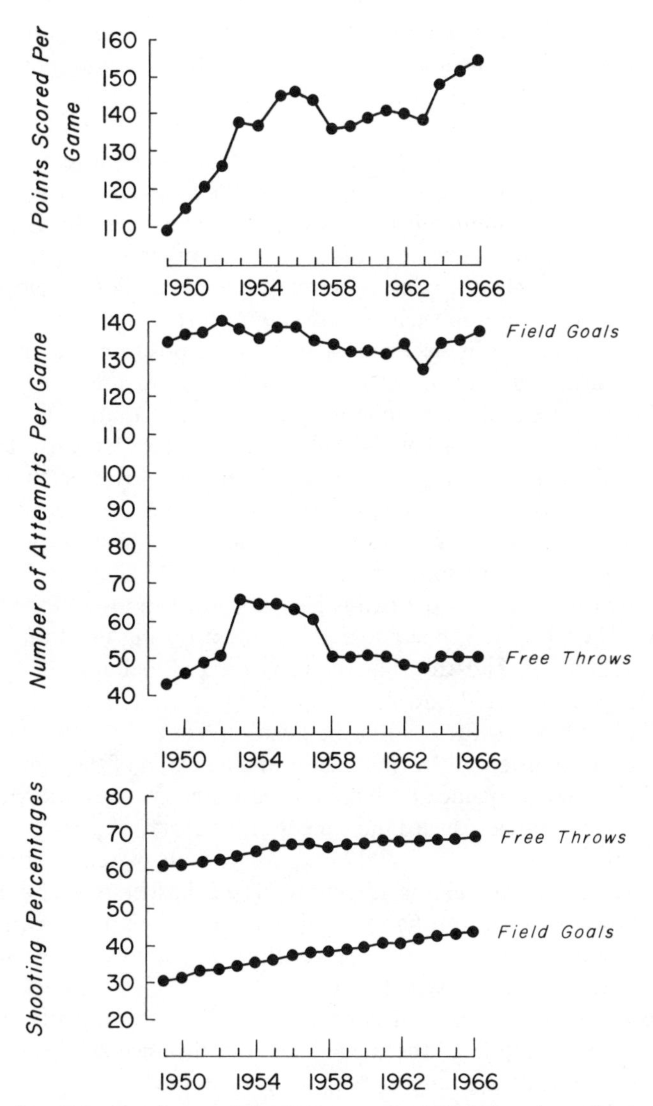

Fig. 116. Shooting statistics from a total of 73,400 major college games over an 18-yr period (*Based on data in Harvey, "Statistical Trends in Basketabll"*).

Set Shot. The similarity of the actions involved to those used in both lay-ups and jump shots has resulted in the set shot being frequently regarded as the foundation upon which these other shooting techniques are developed. Were this not enough reason in itself, the relative ease with which it can be learned has provided added support for the suggestion that the set shot is the logical one to be learned first.

For the execution of a set shot the foot on the same side as the shooting hand is placed a short distance forward of the other foot and pointed in the direction of the basket. The back foot is positioned slightly to the side and pointing forward and outward at approximately a 45° angle. This placement of the feet is the reverse of that used in almost all other sports techniques in which a ball is projected into the air, in that the forward foot is the one on the same side as the dominant hand. The main reason for this is that in set shooting, unlike projection techniques in many other sports, the need for accuracy far outweighs the need to develop high velocities at release. Thus, the opportunity to develop such velocities by incorporating large hip, trunk, and shoulder rotations is declined in favor of a position that permits the forces exerted on the ball to act almost exclusively in the direction of the basket. From this standpoint alone it would be appropriate if the back foot were also pointed at the basket, so that whatever forces were exerted by the back leg would be more likely to act in that direction. However, the desirability of increasing the shooter's stability by increasing the area of his base probably outweighs this consideration. Certainly most good shooters rotate the back foot outward to some extent.

For a right-handed shot the ball is held at chest height with the left hand underneath, in a position to support and guide it during the initial stages in the act of shooting, and the right hand is directly behind and somewhat below with palm forward and fingers pointing upward. The left hand then lifts the ball up past the face and into a position from which the shot is completed with an extension of the right elbow and a flexion of the right wrist and fingers. This whole sequence of actions is coordinated with an extension of the shooter's legs and a forward movement of his center of gravity.

So that the forces exerted on the ball as a result of these latter actions might more readily act in the direction of the basket, the shoulders are rotated a little to the left during the final stages leading up to the moment of release. This puts the shooter's eyes, right shoulder, elbow, and wrist and the ball in a direct line to the basket.

The extent to which the legs contribute to the force applied to the ball varies with the strength of the player and the distance of the shot. With youngsters who lack the strength to readily project the ball very far, the range and force of the leg action used is necessarily a good deal more than that required by stronger players for a shot of the same distance. Similarly, the farther the shot is from the basket, the more likely it is that the legs will have to contribute force to support that available from the arms and shoulders.

Jump Shot. A jump shot can be made at the end of a dribble, after moving to receive a pass, or simply from a stationary position. Whatever the case, the techniques involved are essentially the same.

The position of the shooter's feet as he begins to jump varies from player to player. Some prefer to have their feet in a position akin to that described for set shooting, while others—probably the majority—show preference for a position in which the feet are parallel to each other and equidistant from the basket.

The upward jump preceding the shot is obtained by a quick forceful extension of hip, knee, and ankle joints. The amount to which these joints are flexed prior to the jump should be just sufficient to allow the shooter to obtain a "lead" over the defensive man (see p. 221). While additional bending might well result in a higher jump, the extra time involved in the takeoff provides the defensive man with a better chance of being ready to block the shot. The truth of this is especially evident when a shooter increases the bend in his knees prior to starting upward into the jump. With the shooter's intention so clearly "telegraphed" in this way, it is little wonder that an alert defensive man can often block the resulting shot.

As the shooter extends his legs and pushes down against the floor with his feet, he raises the ball (held in the same manner as for a set shot) up to a position just forward of his head. This lifting of the ball continues until just prior to the release when the ball, still held by both hands, is high overhead. From this position the left hand is lowered as the right arm directs the ball into the air with an action similar to that used in a set shot.

Because, in general, the shoulders are not rotated in quite the same way as previously described for the set shot, it is usually necessary for the shooter to turn the palm of his shooting hand outward slightly (i.e., to pronate it), to ensure that the ball does not deviate from the straight-line path to the basket.

The instant at which the ball should be released relative to the peak of the shooter's jump is a topic on which there has been some disagreement. While Szymanski's study[9] of films of four N.B.A. guards executing the jump shot revealed that in all cases the ball was released "before the peak of the jump was reached," leading coaches (e.g., Auerbach,[10] Cousy,[11] and Sharman[12]) are almost unanimous in agreeing that the ball should be released at the peak of the jump.

This question of the timing of the release can be resolved by considering the forces that must be exerted on the ball if the shooter is to achieve his objective. To project the ball in the direction of the basket requires that both horizontal and vertical velocities be imparted to it as it is released. Regardless of when the ball is released, its vertical velocity at that instant is equal to the sum of the vertical velocity of the shooter and the vertical velocity of the ball

[9]Frank Szymanski, "A Clinical Analysis of the Jump Shot," *Scholastic Coach*, XXXVII, October 1967, pp. 8–9, 59–61.

[10]Arnold "Red" Auerbach, *Basketball for the Player, the Fan, and the Coach* (New York: Pocket Books, Inc. 1953), p. 20.

[11]Bob Cousy and Frank G. Power, *Basketball Concepts and Techniques* (Boston: Allyn and Bacon, Inc., 1970), pp. 47, 56.

[12]Bill Sharman, *Sharman on Basketball* (Englewood Cliffs, N.J.: Prentice-Hall, Inc., 1967).

relative to the shooter—the former resulting primarily from the forces exerted by the legs at takeoff and the latter from vertical forces exerted by his arms and shoulders. Now, since only a certain amount of vertical velocity is necessary in any given case, the amount obtained from each of these sources can be adjusted to meet the needs of the occasion. If the shooter has strong arms and shoulders or if the shot is a relatively short one, the ball may be released at the peak of the jump. In such cases, the vertical velocity of the shooter (and, thus, the contribution of the legs to the vertical velocity of the ball) is zero. For shots requiring more force than can reasonably be generated by the arms and shoulders alone, it is imperative that some of the force from the legs be used—or, in other words, that the ball be released while the shooter still retains some vertical velocity, i.e., before the peak of the jump is reached. (*Note:* On the basis of the results obtained by Szymanski it would appear that even highly skilled performers must utilize some of the force generated by their legs in order to impart the necessary velocity to the ball when shooting from a distance of 25 ft from the basket—the shooting distance used in the study.)

Lay-up Shot. A lay-up shot is one in which a player catches the ball on the move (either from a pass or at the end of a dribble), takes the one step permitted by the rules, leaps high into the air, and lays the ball up against the backboard so that it will rebound into the basket (Fig. 117).

To perform this movement legally and effectively, a player making a right-handed shot must gather the ball as he is in midair and about to land on his right foot or at the instant when his right foot contacts the ground. If he takes

Fig. 117. The lay-up shot

the ball earlier (i.e., when his left foot is in contact with the ground), he will either violate the rules governing "traveling" or make an unbalanced shot off the wrong foot. If he takes the ball later than described, he is likely to have to rush the final phases of the complete movement and make a hurried and inaccurate shot. Immediately following the grounding of the right foot, a normal running stride is taken on to the left foot from which the final jump and shot is made. During the transition from right to left foot the ball is held close to the body at about waist height. The takeoff for the final leap—directed in a vertical rather than a horizontal direction so that the shooter can gain the maximum height from which to release the ball—is made by extending the left leg vigorously in conjunction with a high lift of the right knee. During this movement the ball is being shifted from its waist-high position, past the face, to a position where at the maximum height of the jump it is as high as possible and about to be released.

The ball is normally released at the peak of the jump where it is most likely to be beyond the reach of the defensive man and from where it has the least distance to travel before hitting the backboard. (The less this latter distance, the less room there is for error in the displacement of the ball.) One of the most common faults in lay-up shooting, releasing the ball with too great a velocity, can also be guarded against to some extent if the ball is released at the peak of the jump. At this instant, the vertical velocity of the body is zero and thus the contribution of the body velocity to the ball velocity is lower than at any other time.

Two different positions of the shooting hand (and hence the arm) are in common use. In one, the hand is placed behind and below the ball with palm forward and fingers pointing upward. From this position, force is applied to the ball as the elbow extends and the wrist and fingers flex. In the other position, the shooting hand is under the ball, palm upward and fingers pointing forward. The release is effected with a lifting action of the arm followed by a flexion of the wrist just as the ball is about to leave the hand.

Because the first of these positions enables the shot to be made in a forward and upward direction (the direction in which the arm forces must be applied if the ball is released at the peak of the jump following a vertical takeoff) and because the arm action involved is similar (though of a much lesser range) to those used in jump and set shots, this method would appear to be the most appropriate for the majority of situations. However, where for one reason or another the velocity of takeoff for the final jump has a substantial horizontal component, it may well be best to use the hand-under-the-ball technique. For, assuming the ball is not moving backward relative to the shooter, it will have a horizontal velocity at least equal to that of the shooter himself as it is about to be released. If this velocity is in excess of what is needed—a situation that frequently arises—it becomes necessary for the shooter to apply a backward horizontal force to the ball. To do this when the hand is directly behind and below the ball is at best very difficult, and perhaps even impossible. However, when the hand is under the ball, as already described, the final wrist-flexing action provides just such a backward force as is required.

Body Position and Footwork

A prerequisite to success in executing most offensive and defensive skills is an ability to move quickly from one position on the court to another.

Starting. In starting from a stationary position the stance of the athlete has a direct bearing on the speed with which he can carry out a movement. With respect to this, an examination of those factors that govern the stability of equilibrium of a body—one of the two main considerations in starting—might suggest that the optimum starting position was one in which the athlete had his legs straight and his feet together and was balanced on his toes, for in this position he would have his center of gravity high and near the limits of a very small base. All of this, of course, is consistent with the requirements for moving quickly in any direction—*except in one major respect.* Because the athlete's legs are straight, they are not in a position to apply the force necessary to rapidly accelerate him. Thus a compromise must be reached between the conflicting demands of a near-minimum of stability and an optimum production of force. To achieve this compromise, the player adopts a stance in which his feet are spread, his heels are on the floor, his hip and knee joints are partially flexed (all of which contribute to the optimum production of force), and his center of gravity is over the midpoint of his base (overall the least stable position left open to it).

Slater-Hammel[13] studied the effect of four initial body positions upon the time that a player takes to move his whole body diagonally forward (to left or right) in response to a signal or stimulus. The four positions studied were:

1. knees straight with weight distributed over feet,
2. knees straight with weight on balls of feet,
3. knees bent with weight distributed over feet, and
4. knees bent with weight on balls of feet.

In each case the feet were parallel and approximately hip width apart, the body was bent forward, and the hands were placed in front of the body. Slater-Hammel found that the times for the starting positions with the body weight distributed over the feet were significantly shorter than those for the positions with the weight over the balls of the feet. (A subsequent analysis of changes in weight distribution revealed that most subjects consistently rocked back on their heels in getting started. Slater-Hammel therefore

[13] A. T. Slater-Hammel, "Initial Body Position and Total Body Reaction Time," *Research Quarterly*, XXIV, March 1953, pp. 91–96.

suggested that starting from positions with the weight on the balls of the feet took longer than when the weight was distributed over the feet because of the additional time spent in lowering the heels to the floor.) While times tended to be shorter for positions in which the knees were bent than for those with knees straight, the differences were not statistically significant and it was concluded that "the position of the knees had no marked effect upon the time taken."

The findings in a study by Cotten and Denning[14] are strongly supportive of those obtained by Slater-Hammel. In this study, the four starting positions used in the Slater-Hammel study were again used and the times taken to travel a short distance in each of three directions (left, right, and straight ahead) were recorded. The results revealed that the knees-bent and feet-flat positions were generally superior to the knees-straight and weight-on-balls-of-feet positions. The authors concluded that "[From] the information provided by both studies, it appears that a knees bent, feet flat stance is the best choice for optimum reaction-movement time. Hence, it would seem coaches and physical educators should consider abandoning the traditional 'weight on balls of feet' stance in favor of the feet flat stance."

Stopping. To stop, a player must exert horizontal forces against the floor in the direction in which he is traveling. If he does this correctly, the reaction to these forces reduces the motion of his center of gravity to zero before it can pass beyond the limits of the base formed by his feet. This reaction also causes the player to rotate in the direction in which he is moving. This rotational effect can be combated, however, if the player so positions his body that the vertical component of the reaction from the floor creates a torque in the opposite direction. (The sequence of an athlete performing a stride stop in Fig. 118 illustrates the application of these concepts in a specific situation.)

Fig. 118. Stroboscopic photograph of a player executing a stride stop

[14]Doyice J. Cotten and Donald Denning, "Comparison of Reaction-Movement Times from Four Variations of the Upright Stance," *Research Quarterly*, XLI, May 1970, pp. 196–99.

Jumping

The height to which a person can jump depends in part on whether the jump is made from a stationary position or at the end of an approach run. If all else is equal, jumps using an approach run result in higher heights being attained than jumps from a stationary position. The truth of this contention has been demonstrated by Enoka[15] in his study of the jump-and-reach scores obtained by volleyball spikers with jumps from a standing position and following one-, three-, and five-stride run-ups. The means for all three of the latter (respectively, 23.0 in., 25.3 in., and 25.5 in.) were significantly greater than the mean for standing jumps (22.4 in). Therefore, in situations where maximum height is required, and where time and other factors permit, a player should use a run-up preceding his jump. In rebounding and in jump-ball situations, for example, a player should endeavor to step into the position from which he takes off rather than execute the jump from a stationary position.

While using some form of run-up preceding the takeoff can increase the height to which a player is able to reach, it can also create problems. Because the player is moving horizontally as he places his feet in preparation for the jump, he has a tendency to continue moving in this direction. If left unchecked, this tendency causes him to take off in other than a vertical direction, thereby increasing the possibility of his committing a charging foul. With respect to this latter possibility, it is important to recognize that the horizontal distance that a player travels in reaching the peak of his jump is only about half of the total horizontal distance he will travel before landing. Thus, jumping to maximum height without committing a foul is only half the battle, for the player must still return to the floor without infringing—and this is a point that players sometimes overlook. To eliminate these problems, the player must reduce his horizontal velocity to zero by placing his feet forward of his center of gravity and exerting appropriate horizontal forces against the floor.

The actual jump itself results from a powerful extension of hip, knee, and ankle joints, supplemented by a forward and upward swing of the arms. The range and speed of the motion at each of the joints involved depends on a number of factors. Chief among these are the situation in the game and the proximity of opponents who may be disputing possession with the player concerned. Where possible, however, the position adopted by the jumper prior to beginning his upward movement should be such as to permit a range and speed of joint motion consistent with his obtaining the maximum vertical velocity at the instant of takeoff.

[15]Roger Enoka, "The Effect of Different Lengths of Run-up on the Height to Which a Spiker in Volleyball Can Reach," *New Zealand Journal of Health, Physical Education and Recreation*, IV, November 1971, pp. 5–15.

Chapter 10

FOOTBALL

Historically, the result of a progressive series of developments from the English game of rugby, the present-day game of football with its blocking, forward pass, platoon system, etc., is now so distinctively American that similarities between the two games are sometimes difficult to discern. The overwhelming popularity of football, however, is there for all to see. For such is the vast spectator appeal of the game that it now completely dominates the American sporting scene.

Basic Considerations

The six basic skills of greatest importance in football are passing, catching, running, blocking, tackling, and kicking.

Passing

The term *pass*, defined as the transfer of possession from one player to another of the same team, includes the "snapping" of the ball from the center to the quarterback, the punter, or the holder for a place-kick and the hand-off from one player to another, as well as the forward and lateral passes.

A player's objective in making a pass in football is exactly the same as that of a player executing a pass in basketball—to complete the displacement of the ball from his hand or hands to those of a predetermined teammate (p. 211).

Basically there are two types of pass:

1. those that involve motion of the ball through the air—the forward and lateral passes, and the center snap to the punter or to the holder for a field goal or point-after-touchdown attempt, and

2. those that do not involve motion of the ball through the air—the center snap to the quarterback and the hand-off from one player to another.

For passes in the first category the basic factors that influence the outcome are those that influence the outcome of any projectile motion, i.e., the speed, angle, and height of release and the air resistance encountered in flight.

The speed of release is determined by the magnitude and direction of the muscular forces exerted and the distance and time over which they act. Thus, a player executing a pass attempts to position his body so that he can apply appropriate muscular forces over such distance and time that the ball acquires the desired speed of release. For passes requiring a large speed of release (e.g., long forward passes) this means positioning himself so that the muscular forces of his legs and trunk may be utilized and those of his arms and shoulders can be applied over a near-maximum range (and for a near-maximum duration). For passes requiring less speed at the instant of release (e.g., most lateral passes) forces exerted by the muscles of the arms and shoulders over a relatively short range (and for a relatively short duration) will generally suffice.

The optimum direction in which to release the ball is governed by a number of factors. First, the ball should be directed toward that position in which the receiver's hands will be located when it arrives. (In practical terms this means that the ball should be directed toward a point some distance ahead of the intended receiver.) Second, the angle to the horizontal at which the ball is released must be such as to cause it to arrive in the required position at the appropriate time. Finally, the angle at which the ball is released must be such as to allow it to pass safely beyond the reach of any opposing players positioned between the passer and the receiver.

The height of release is fixed by the type of pass involved (an orthodox forward pass almost invariably being released at a greater height than a lateral pass, which in turn is almost always released at a greater height than a center snap to the punter), the physical characteristics of the player executing the pass, and the body position adopted for the purpose.

While no quantitative data appear to be available concerning the effects of air resistance on a football in flight, it seems likely that these effects are fairly pronounced and thus of considerable practical significance. Until adequate research evidence becomes available, however, the exact nature of the effects produced by air resistance must remain a matter of speculation.

The principal task of players involved in executing a "nonairborne" (snap or hand-off) pass is to ensure that the ball is always acted upon by an upward vertical force of a magnitude at least equal to the weight of the ball. For if it is not—and this is unquestionably the reason for most of the fumbles that occur in such situations—the ball experiences a downward acceleration that, unless countered, will ultimately result in it striking the ground.

Catching

Although there are differences in the magnitude of the forces exerted on the receiver's (or catcher's) hands, and consequently in the techniques employed, the basic factors of importance in catching a football are the same as those involved in catching a baseball (p. 187) or a basketball (p. 212).

Running

Because the techniques that yield the greatest speed differ from those that afford the best chance of avoiding an opposing player, the running done by a football player (who generally has both objectives in mind) is necessarily something of a compromise.

The speed at which a player runs is governed by the length of his stride and the number of strides he takes in a given time (speed = stride length × stride frequency, p. 382). Thus, if a player is running without any immediate danger of being obstructed or interfered with by a member of the opposing team, he is likely to attain the best results in exactly the same manner as does a sprinter—by using the combination of stride length and stride frequency that yields the greatest speed.

A player's ability to avoid an opponent by changing his speed and the direction in which he is moving depends on the nature of the forces he can exert against the ground via his feet and on the rapidity with which he can bring these forces into play. Thus, from the standpoint of being able to avoid an opponent, the optimum running technique is one that permits the athlete to exert large forces against the ground, in any given direction, and at the instant it first becomes apparent that such action is desirable.

Blocking

When one player throws a block on another, the outcome of the resulting impact is governed by essentially the same factors that determine the outcome when any two bodies collide, i.e., the masses and initial velocities[1] of the bodies involved and their mutual coefficient of restitution (pp. 77–93). Of these five factors, it is his mass and initial velocity over which a player has the greatest control.

If a man executes a block in which both his feet are off the ground at the instant he makes contact with his opponent, the mass involved in the impact is simply that of the man himself. However, if he executes a block with his feet firmly braced against the ground, the effective mass with which his opponent interacts is much greater than before due to the contribution exacted from the ground. In short, it might be said that his opponent is now interacting with the man-plus-the-earth rather than with the man alone.

[1]The term *initial velocity* is used here to mean the velocity immediately prior to impact.

The velocity at which a player is moving as he makes contact with his opponent is also of considerable importance in determining the outcome when a block is thrown—in general, the greater the speed of a blocker in the direction in which he desires to force his opponent, the greater the likelihood that he will achieve his objective.

Because he has little if any control over the mass and initial velocity of his opponent, a player preparing to throw a block must attempt to take these variables into account in the movements he makes. Thus, because of the difference in the speed at which the opponent is moving (and perhaps, too, because of a difference in mass), the body movements needed to effectively block a hard-running defensive man pursuing a punt will almost certainly be different from those used by an offensive lineman offering protection to the passer.

While the role of the coefficient of restitution must be acknowledged, it seems almost certain that the other variables already referred to are the principal ones determining the outcome when a block is thrown.

Tackling

Tackling may be regarded as a specialized form of blocking in which the objective is to reduce the ball carrier's motion to zero in the shortest possible time. Except for a few relatively minor differences, the basic factors involved in this process are the same as those discussed above with relation to blocking.

Kicking

There are four types of kick in common use. In three of these (the kickoff, the field-goal attempt, and the try for a point-after-touchdown) the ball is placed on a kicking tee appropriately positioned on the ground. In the fourth (the punt) the kicker releases the ball at near-waist height and then strikes it with his foot after it has fallen some distance toward the ground. In each case the ultimate success of the kick depends on the speed, height, and angle at which the ball leaves the kicker's foot and on the air resistance it encounters en route.

The speed and angle at which the ball leaves the foot is governed by the same factors that determine the result of any elastic impact—the masses and initial velocities of the bodies involved and their coefficient of restitution. Since it may be assumed that the mass of the ball and the coefficient of restitution are essentially constant, and since the initial velocity of the ball is zero in three of the four cases and very small in the fourth, it can readily be seen that the factors of greatest importance in fixing the speed and angle at which the ball is projected into the air are the effective mass and initial velocity of the kicker's foot. The former is determined by the extent to which muscular actions are used to bind the various body segments together as a unit and by the contribution exacted from the ground via the nonkicking foot (cf., "Blocking," pp. 237–38). The latter, the initial velocity of the kicker's foot, is determined by the forces exerted upon it and by the distance and time over which they act.

Except in the case of the punt, the height of the ball (or, more precisely, the height of the center of gravity of the ball) at the instant of release is subject to relatively little variation. (A ball kicked from a tee generally leaves the kicker's foot, i.e., is "released," at about 10–12 in. above the ground—a height approximately 5–6 in. higher than it had before impact.) The height of the ball as it leaves the punter's foot can vary from as little as 9 in. to as much as $2\frac{1}{2}$ ft, depending on the requirements of the situation. For example, if a long, low kick is required, the ball is contacted close to the ground. Conversely, if a high and somewhat shorter kick is required, the ball is contacted well above the ground. The reason for this relationship lies in the fact that as the height of release is altered, the angle at which the kicking foot is inclined at impact and thus the angle of release are also altered (Fig. 119).

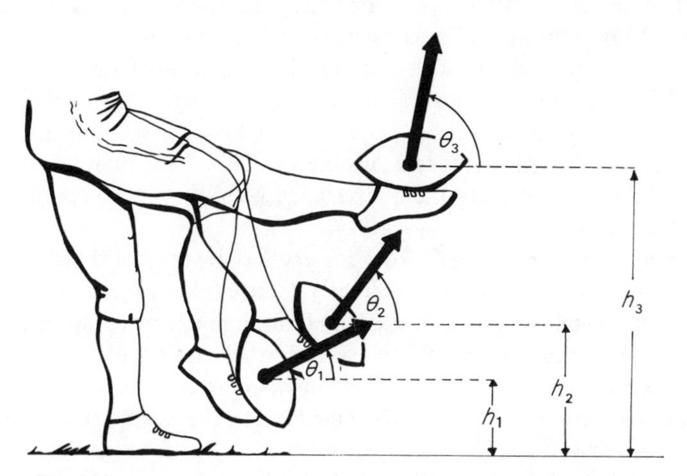

Fig. 119. In punting the height of release (h) and the angle of release (θ) are interdependent.

Techniques

For all the stress that football coaches seem to place on the importance of their players acquiring a mastery of the basic techniques of the game, these same techniques (with only one or two exceptions) have been virtually untouched by scientific analysis.

Passing[2]

Center Snap. The first pass in each play is that made by the center to either the quarterback, the punter, or the holder for a field-goal or point-after-touchdown attempt.

Center snap to quarterback. The center takes up his initial position with both hands on the ball, his head directly above the ball and inclined backward slightly so that he can see his defensive opponent, his back parallel with the ground, his knees well flexed, and his feet spread laterally at slightly more than shoulder width apart.

The positions of the feet relative to one another are subject to some variation—some centers taking a stance with one foot a few inches forward of the other (a so-called staggered stance), while others assume a position in which their feet are level with each other (a parallel stance). However, since the results of studies concerned with stances, reaction times, etc., repeatedly show advantages in favor of the staggered stance over the parallel one (p. 246), there would appear to be little justification for the continued use of the latter.

The snap is made with a rapid backward and upward sweeping motion of the right hand. During the first part of this motion the left hand is also brought backward and upward, guiding the ball toward the quarterback's hands. While coaches differ considerably in their opinions concerning the way in which the ball should be moved as it is brought upward, the method most widely advocated involves a turning of the ball through 90° to the left so that it can be presented to the quarterback with its long axis parallel to the line of scrimmage.[3]

As the center sweeps the ball from the ground, he takes a short, quick step forward with his back foot. This slightly reduces the distance that the ball must travel to reach the quarterback's hands and, much more important, gives the center more momentum than he would otherwise have. This latter is of importance if he is to complete his blocking assignment effectively.

Center snap to punter (Fig. 120). The snap to the punter differs from the snap to the quarterback in several respects. The need to get the ball into the hands of the punter as soon as possible after it is taken from the ground requires that the center impart a considerable speed to it in the short time at his disposal (the release speeds recorded in a study conducted by Henrici[4] ranged from 43.46–46.68 fps). In order to acquire this speed, the center usually takes up his position slightly farther behind the ball than for a snap to the quarterback—thereby increasing the distance over which he may exert

[2]Throughout the ensuing discussion it is assumed that the passer is right-handed.

[3]Harold L. McKain, "Techniques of Executing Fundamental Skills of Football" (M.A. thesis, State University of Iowa, 1963), pp. 73–74.

[4]Ronald C. Henrici, "A Cinemagraphical [sic] Analysis of the Center Snap in the Punt Formation" (M.S. thesis, University of Wisconsin, 1967).

Fig. 120. The two-handed technique for executing a center snap to the punter

force on the ball. In addition, some centers position their hands on the ball so that both can contribute to the speed at which the ball is swept backward (cf., the snap to the quarterback in which one hand propels the ball and the other serves in a guiding role). Although there appear to have been no studies conducted in which these two methods have been compared, it seems likely that the two-handed method is slightly faster than the one-handed—the increase in the torque available when two hands are used to accelerate the ball about the axis through the shoulders almost certainly being greater than the increase in the moment of inertia of the rotating system (arms-plus-ball) that tends to offset it ($T = I\alpha$, Eq. 58).

While the center invariably has his head inclined slightly upward and his eyes focused on the man in front of him when snapping the ball to the quarterback, a head-down position that enables the center to see back between his legs is frequently used when snapping the ball to the punter. The relative merits of these two methods (the visual and the nonvisual) have been studied by Slebos,[5] who found that the visual method was superior to the nonvisual in terms of accuracy but no different in terms of flight time (i.e., the time from the ball being lifted from the ground until it struck a target 12 yd behind the center).

Center snap to holder. Apart from the fact that the ball must be passed a lesser distance (7 yd as compared to 13–15 yd) and must be directed to a point closer to the ground (since the receiver is in a semikneeling position rather than standing erect), the snap to the holder for a place-kick is essentially the same as the snap to the punter.

Hand-off. The basic techniques involved in the execution of a hand-off are depicted in Fig. 121. The following points should be noted:

(a)[6] The quarterback's hand is wrapped around the ball with the fingers underneath supporting its weight; his eyes are focused on the abdomen of the player to whom he is giving the ball; and his shoulders are rotated in order

[5]Warren G. Slebos, "Football: A Comparison of the Visual and the Non-Visual Methods of the Spiral Center Pass" (M.A. thesis, The University of Iowa, 1968).

[6]The letters in parentheses refer to the corresponding positions in Fig. 121.

Fig. 121. The hand-off to a runner

to increase the length of his reach. (*Note:* A two-handed grip is generally preferred for hand-offs between players who are in close proximity to each other because of the greater security of possession that it entails. In the situation depicted, however, a two-handed grip would restrict the quarterback's reach and make it difficult for him to place the ball in the required position.)

The player receiving the ball has his right ("inside") elbow high—to provide an unobstructed path for the ball—and his left arm moving into position to provide support and protection. In order to observe the action taking place ahead of him, and for purposes of deception, the runner keeps his head up throughout the exchange.

(b) The ball has been pressed firmly into the runner's abdomen, his arms have closed around it—the left below the ball providing the necessary upward force to prevent the ball from falling—and the exchange is complete.

(c) A little more than one step later, the arm positions have been adjusted to conceal the ball and to place the hands over the ends of the ball, where they can act to prevent it from being dislodged sideways when the runner is hit.

Forward Pass. The techniques involved in executing a forward pass vary slightly according to the situation in the game and the type of forward pass (drop-back, roll-out, or jump) that is used. Since the similarities in the techniques far outweigh the differences, however, only the drop-back forward pass is considered here.

Once the ball has been received from the center, the quarterback's first task is to retreat to the position from which he intends to make the pass. For this purpose there are three basic techniques in common use:

1. The crossover, in which the quarterback has his body positioned at right angles to the line of scrimmage and retreats with a sideways running action. (*Note:* The distinctive crossing of one leg in front of the other, which is inherent in this action, gives this method its name.)

2. The sprint-back, in which the quarterback turns his back and sprints directly away from the line of scrimmage.

3. The backpedal, in which the quarterback faces the line of scrimmage and retreats by running backward.

While there are unquestionably numerous other factors that must be considered before deciding which of these methods is best employed by a given quarterback or in a given situation, it is of interest to note that a study by Pannes[7] revealed no significant differences in the length of time that it took quarterbacks to drop back to a specific distance using the three methods.

The technique used in the crossover method is shown in Fig. 122(b)–(g). After receiving the ball from the center, the quarterback shifts his weight to his left slightly—thereby facilitating the lifting of the right foot; pivots on the ball of the left foot—note that the heel of the left foot has been raised in (b) to facilitate this action; and drives vigorously against the ground by extending the hip, knee, and ankle joints of the left leg. His body is inclined at a near 60° angle to the horizontal at this latter instant—an orientation similar to that adopted in many sports when runners experience marked accelerations. As he completes his third retreating step, the quarterback grounds his right foot some distance ahead of his center of gravity (relative to the direction in which he is moving). This action is accompanied by a reaction from the ground that serves to brake his forward motion (p. 385). The process of

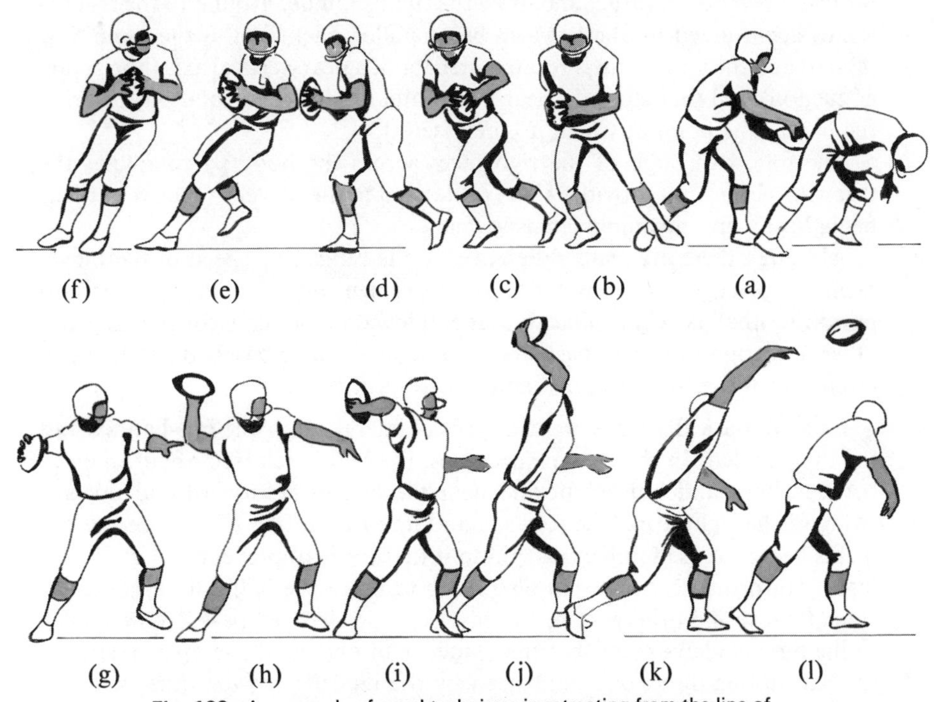

Fig. 122. An example of good technique in retreating from the line of scrimmage using the crossover method and throwing a forward pass

[7]Nicholas Pannes, "A Comparison of the Lengths of Time for Three Techniques of Quarterback Drop-back Passing to Three Different Depths" (M.S. thesis, Springfield College, 1967).

deceleration continues as the left and then the right foot are grounded at the end of the fourth and fifth steps, respectively. Note that the quarterback's body leans in one direction as he strives to build up speed (b)–(c) and then in the other as he slows to a halt (e). (A detailed consideration of the factors influencing the way in which the body leans as it is being accelerated, and decelerated, is presented in Chap. 15, pp. 298–400).

Once the quarterback has completed his fifth step and brought his horizontal motion momentarily to zero, his next task is to readjust his body position, or "set" himself, in preparation for the throw. To do this he pushes downward against the ground with his left foot, evoking a reaction that causes his center of gravity to be lifted and shifted toward his right. With his weight over his right foot and the right knee slightly bent, the quarterback is now in a position to step forward into the throw (f). The step of the left foot in the intended direction of the throw is accompanied by a raising of the ball to a position above the right shoulder and behind the head (g)–(h).

In the throw itself the hips are brought around to the front, as the right leg extends and drives the right hip forward (h)–(j); the muscles responsible for trunk rotation contract and draw the right shoulder around to the front—a movement aided by the left arm being pulled backward at the same time (h)–(j); and the upper arm is rotated first about a near-vertical axis (horizontal adduction) and then about a near-horizontal axis (medial rotation) to bring the ball to the point at which it is released (k).

The follow-through of the right arm across the body (l) completes the throwing action and provides some protection to the player in the event of his being hit by an oncoming defensive man.

Although there are some differences in the range and speed of the movements involved, the overall pattern of movement used in making a forward pass in football is very similar to that employed in baseball (for pitching and other overarm throws), in basketball (for a one-handed baseball pass), and in javelin throwing (pp. 188, 224, and 484, respectively).

Lateral. Basically there are two types of lateral pass—the one-handed and the two-handed. In the one-handed pass, the ball is released about midway between knee and hip height and near the end of a forward and upward swing of the right arm. The forward and upward motion of the hand as the ball is being released imparts a spin to it that tends to prevent it wobbling in flight. Unfortunately this spin also tends to make the ball a little harder to catch than would otherwise be the case. The two-handed pass is very similar to the one-handed except that the tendency of one hand to impart a spin to the ball in one direction is balanced by the tendency of the other hand to produce the reverse effect. As a result the ball generally leaves the passer's hands with no spin about its long axis. However, unless the wrists are "cocked" so that the ball is kept in a near-vertical position as it is swung forward and upward—an orientation that it then tends to retain once it has been released—the forward and upward swing of the arms tends to impart an end-over-end type of rotation about a transverse axis through the ball's center of gravity. This "tumbling rotation," like the "spiral rotation" of the one-handed pass, tends to make it more difficult to catch the ball than would otherwise be the case.

Catching

Although variations in the speed and direction at which the ball is traveling relative to the receiver necessitate variations in the techniques used to catch the ball, all such techniques have certain characteristics in common. Regardless of how the ball is moving as he attempts to make the catch, the receiver's first responsibility is to position his hands and body so that as he makes contact with the ball he is able to begin exerting the forces necessary to reduce its motion to zero (strictly speaking, to reduce the motion of the ball *relative to the receiver* to zero). In general this means moving so that the hands are positioned, palms toward the oncoming ball, across the flight path that it is following [Fig. 123(a) and (b)]. Then, as contact is made, the hands give slightly to reduce the force of the impact and therewith the possibility of the ball rebounding from the hands, and the fingers close around the ball providing both the couple necessary to reduce its angular motion to zero and the upward force to prevent it from falling toward the ground [Fig. 123(c)]. In those cases where the catch is made at some distance from the body the arms are flexed once the ball has been brought under control and the ball is thus drawn in close to the body [Fig. 123(d)]. The positions of the hands and arms are then adjusted quickly in an effort to reduce the possibility of the ball being dislodged if the receiver is hit.

Fig. 123. Catching the football

Running

Any discussion of the techniques used in running must consider the techniques used to get the body moving initially as well as those used once the player is in "full stride."

Starting. The question of the best body position (or stance) to adopt in order to get away quickly once the starting signal has been given is a complicated one. Among the factors that must be taken into account are:

1. The direction in which the athlete is required to move may vary considerably from one play to the next, and the initial stance that affords the best start on one occasion may be much less than satisfactory on another.
2. The need to conceal the direction in which the player intends to move—in order to avoid "tipping off" the opposition—makes it unwise to vary the stance adopted in an attempt to use the optimum one for each occasion.
3. The stance adopted must be consistent with the job that the player is expected to perform once he is in motion. For example, if a halfback is to take a hand-off from the quarterback after he has taken three strides, the stance he uses must permit him to be in a suitable position to receive the ball at that time.

In view of these various limiting conditions, it is clear that the optimum stance for any given player must necessarily be something of a compromise. It must give him the best all-round results in terms of speed (i.e., taking into account the various directions in which he might be called upon to move during the course of a game) and permit him to complete the specific task assigned to him.

Stances are generally classified according to the number of limbs in contact with the ground (thus, the two-, three-, and four-point stances) and according to whether the feet are equal or unequal distances from the line of scrimmage (parallel and staggered stances, respectively).

Which of the various combinations of foot position and points of contact affords the fastest start is a question that has attracted a considerable (and perhaps disproportionate) amount of attention from researchers.

The relative merits of the two- and three-point stances used by backfield players have been studied by several investigators:

Holtz[8] compared times taken to run 7 yd obliquely forward to the right, obliquely forward to the left, and straight ahead when using a two-point staggered stance with those obtained using a three-point staggered stance. He found that, for each of the three directions, the times obtained when the three-point stance was used were significantly faster than those obtained when the two-point stance was used.

[8]Louis Holtz, "Speed of Starting from Selected Stances Used in Football" (M.A. thesis, State University of Iowa, 1962).

Warshawsky[9] compared times to run 8 yd to the right, to the left, and straight ahead when using two- and three-point staggered stances and found that the two-point stance was significantly faster than the three-point stance when running to left and right. However, the three-point stance was significantly faster than the two-point stance when running straight ahead.

Robinson[10] compared times taken to run 10 ft in five different directions (90° to the right, 45° to the right, straight ahead, 45° to the left, and 90° to the left) using each of three different stances (a two-point parallel stance, a two-point staggered stance, and a three-point staggered stance). His results indicated that there were no significant differences in the times recorded when running 90° to the right, 45° to the right, or 45° to the left; the two-point stances were significantly faster than the three-point stance when running 90° to the left; and the two-point staggered stance was significantly faster than each of the other two stances when running straight ahead. On the basis of these results, Robinson contended that the best stance for all-around starting ability was the two-point staggered stance.

The obvious lack of agreement among the results obtained by Holtz, Warshawsky, and Robinson is almost certainly due to differences in experimental procedures used. For example, two of the investigators found that use of the three-point staggered stance yielded significantly better results than did use of the two-point staggered stance when the subjects ran straight ahead. The third investigator found exactly the opposite. This difference in result could have been due to differences in the distance over which times were recorded—7 and 8 yd on the one hand and 3.3 yd on the other. In other words, the two-point staggered stance may be faster than the three-point staggered stance over 3.3 yd and yet slower over 7 and 8 yd. [*Note:* A comparable situation exists in track where use of the bunch (or bullet) start has been shown to yield faster times off the blocks but slower times to 10 yd than those obtained when a medium start is used (pp. 390–91).] Another possibility is that differences in the weight distribution of the subjects while in the three-point stance may have led to the markedly different results. A study by Kadatz[11] is of particular interest in this respect. Kadatz studied the effect of variations in the amount of weight supported on the hand in a three-point stance and found that the time between the starting signal and the first movement of the subject (the *reaction time*) was decreased significantly with each successive increase in the amount of weight on the hand. He also found that the time from the first movement of the subject until he had traveled forward 1 yd (the *movement time*) was significantly less when 35% of the

[9]Lawrence Warshawsky, "A Comparative Time Study of Two Backfield Stances in Football" (M.S. thesis, University of Illinois, 1963).

[10]Franklin H. Robinson, "A Comparison of Starting Times from Three Different Backfield Stances" (M.Ed. thesis, Springfield College, 1949).

[11]Dennis M. Kadatz, "The Relationship of Weight Distribution and Charging Time for Football Linemen" (M.A. thesis, University of Alberta, 1965).

weight was supported on the hand than when only 5% was thus supported. In short, it may be that the three-point staggered stance is superior to the two-point staggered stance only if sufficient weight is placed on the supporting hand. Another possible explanation—and one which is related to the matter of weight distribution—is that in some cases the actions used by the subjects were not typical of those that a running back would use in an actual game situation. Holtz, for example, states, "Most of the subjects when running from a three-point stance stayed as close to the ground as possible and hence would have had difficulty in getting the ball from the quarterback."[12] It is conceivable, therefore, that the superiority of the three-point stance found in the studies of Holtz and Warshawsky was due to the use of positions and actions atypical of the game situation.

Differences in the results obtained when the subjects ran in directions other than straight ahead may similarly be accounted for by differences in the procedures used by the several investigators.

It should be obvious from the preceding discussion that, despite extensive research efforts, the relative merits of the two- and three-point stances have yet to be clearly and convincingly demonstrated.

The relative merits of the three- and four-point stances used by linemen have also been subjected to examination:

Fitch[13] compared the "starting speeds" (movement times over a distance of 59 in.) of linemen using various combinations of initial stance and weight distribution and concluded that

1. The three stances used in the study—the three-point staggered stance, the three-point parallel stance, and the four-point parallel stance—were equally effective in terms of starting speed in each of the six directions for which times were taken. (These six directions were 90° to the right, 45° to the right, straight ahead, 45° to the left, 90° to the left, and straight back.)

2. For moving straight ahead and at an angle of 45° to the right, the times recorded when the subjects had their weight forward were significantly less than those recorded when they had their weight back. There were no significant differences associated with differences in weight distribution when the subjects moved in any of the other four directions.

(*Note:* Fitch's findings that a position in which the weight is forward decreases the time when charging forward—compared with that obtained when the weight is back—is in close agreement with the findings of Kadatz referred to previously.)

[12]Holtz, "Speed of Starting from Selected Stances Used in Football," p. 10.

[13]Robert E. Fitch, "A Study of Linemen Stances and Body Alignments and Their Relation to Starting Speed in Football" (P.E.D. dissertation, Indiana University, 1956).

Bolt[14] conducted a cinematographical study in which he compared the times taken by 10 experienced linemen to (1) pull out and run 8 ft to the left, (2) pull out and run 8 ft to the right, and (3) charge forward 2 ft, using three- and four-point staggered stances. The results obtained from this study are tabulated below.

Direction	*Result*
Moving to the right	8 of the subjects were faster using the three-point stance. 2 of the subjects were faster using the four-point stance.
Moving to the left	4 of the subjects were faster using the three-point stance. 5 of the subjects were faster using the four-point stance.
Moving straight ahead	1 of the subjects was faster using the three-point stance. 7 of the subjects were faster using the four-point stance.

On the basis of these results Bolt concluded that there "seemed to [be] a definite advantage gained by the use of the four-point stance for the line charge"; the three-point stance was superior for pulling out to the right; and neither stance was superior for pulling out to the left. He also made an interesting point concerning the relationship between which hand is on the ground in the three-point stance and the ability of a lineman to move to left or right:

> It should be mentioned here that all the men participating in the experiment used the three-point stance with the right hand down. This method seems to help them pull to the right but to hinder them in pulling to the left. If it can be assumed that placing the left hand on the ground would help the individual to pull out to the left, but hinder him in pulling to the right, there is a logical basis for the practice of some coaches who have their men on the right side of the line assume a stance with their right hand down and pull only to the right and the men on the left side of the line assume a stance with the left hand down and pull only to the left.[15]

Numerous other characteristics of the stance and of the initial movements involved in starting have been the subject of study:

Miles and Graves[16] examined the effects of variations in the way the starting signal was called on the "speed of charge" (actually the time from the starting signal until a recording device was triggered as the subject began to

[14]Don Bolt, "Four-Point Stance in Football," *Athletic Journal*, XXIX, April 1949, pp. 38, 68–69.

[15]Bolt, "Four-Point Stance," p. 69.

[16]W. R. Miles and B. C. Graves, "Studies in Physical Exertion: III. Effect of Signal Variation on Football Charging," *Research Quarterly*, II, October 1931, pp. 14–31.

move forward), the "unison of charge" (the extent to which the times for the seven subjects tested together varied from each other), and the number of "off-sides" charged against the subjects. Anticipatory (e.g., charging on the fourth number called) and nonanticipatory (charging when a specified digit is called) signal systems were compared. The signals were called at 40, 60, 100, and 120 digits per minute or, in other words, with 1.5, 1.0, 0.6, and 0.5 sec, respectively, between digits.

The results indicated that

> (1) Reaction to anticipatory signals is . . . much more prompt than to non-anticipatory. . . .
>
> (2) Off-sides are rather more numerous with anticipatory signals.
>
> (3) Unison of play is somewhat better with anticipatory signals. . . .
>
> (4) The rate at which the signals are called makes a very great difference. The favorable rate for anticipatory signals is 100 single digits per minute, while for non-anticipatory signals the best rate appears to be about 60 single digits per minute.
>
> (5) The best position for a starting signal within a series of digits appears to be from about two seconds to five seconds after the men have set. This ordinarily falls on the fourth or fifth digit called. Offsides are considerably less when charging occurs early in the series of digits. This applies to both kinds of signals.
>
> (6) Calling digits in non-rhythm converts the ordinary anticipatory series into one producing much slower speed. . . . Non-rhythm does not appear to increase offsides. . . . However, it does produce poorer unison of charge.
>
> (7) In anticipatory signals the players tend to start the charge at the same moment when the ball starts in action; whereas in non-anticipatory signals they get away almost one-tenth of a second after the ball has moved. This fact gives their opponents a real advantage, and counts strongly against non-anticipatory signals in general.[17]

Thompson, Nagle, and Dobias[18] also studied the response of football players to various types of starting signals. They found that rhythmic digit starting signals resulted in significantly faster times (to react and take an 18-in. step forward with the back foot) than did either nonrhythmic word-digit or nonrhythmic color signals.

The force with which a charging lineman makes contact with his opponent is another topic that has interested researchers:

Owens[19] examined the effects of variations in the front-to-rear and lateral spacings of the feet and in the hand-to-toe anterior-posterior spacing on (1)

[17] *Ibid.*, p. 31.

[18] Clem W. Thompson, Francis J. Nagle, and Robert Dobias, "Football Starting Signals and Movement Times of High School and College Football Players," *Research Quarterly*, XXIX, May 1958, pp. 222–30.

[19] Jack A. Owens, "Effect of Variations in Hand and Foot Spacing on Movement Time and on Force of Charge," *Research Quarterly*, XXXI, March 1960, pp. 66–76.

movement time and (2) force of shoulder impact at the end of movement through a 36-in. horizontal distance. Among his findings were the following:

1. No one of the 40 different stances used in the study was significantly superior to another in terms of the force exerted at impact.
2. The force exerted at impact appeared to be closely related to the movement time and to the weight of the individual.

[*Note:* Rosenfield[20] and Elbel, Wilson, and French[21] also studied factors that influenced the force at impact and in both cases found a significant relationship between body weight and the force exerted at impact—findings in keeping with that of Owens. In the studies of Rosenfield and Elbel, Wilson, and French no significant relationship was found between the *response time* (i.e., the time from the instant the starting signal was given until the instant the subject's shoulder struck a dummy and triggered a recording device) and the force exerted at impact. Unfortunately the authors of each of these studies referred to the response time as the "speed of charge" and, since this might well convey the totally false impression that the speed of a player at the instant of impact is not a factor in determining the force exerted, it should be carefully noted that *the speed at which subjects were traveling at the instant of impact was not measured in either of these studies.*]

Owens[22] also studied the effects of rhythmical and nonrhythmical preparatory and starting signals on the force of impact but found no significant differences. He did find, however, that the response time was shortest and the number of off-sides was greatest when a rhythmical count was used in giving the preparatory and starting signals.

Running. Were it not for the need to change speed and direction in an attempt to avoid would-be tacklers and blockers, the basic running techniques used in football would be practically identical to those used by track athletes (pp. 389–400). However, when a good player is in imminent danger of being tackled or blocked, he changes quickly from the normal running action of a sprinter to one that affords him a better chance of avoiding his opponent. This process generally involves a marked reduction in his stride length and a concomitant increase in his rate of striding—the combined effect of which is to increase the time during which he is in contact with the ground and thus to minimize the delay in initiating evasive action once the opponent "commits himself" (i.e., makes his intentions apparent). The evasive action itself generally takes the form of applying force in a lateral direction by thrusting down-

[20]Richard J. Rosenfield, "Measuring Reaction Time and Force Extended by Football Players" (M.A. thesis, University of Kansas, 1950).

[21]Edwin R. Elbel, Donald Wilson, and Clarence French, "Measuring Speed and Force of Charge of Football Players," *Research Quarterly*, XXIII, October 1952, pp. 295–300.

[22]Owens, "Effect of Variations in Hand and Foot Spacing on Movement Time and on Force of Charge," pp. 66–76.

ward and sideways against the ground and/or against the opponent. Successfully executed, such action carries the runner beyond the grasp of his opponent and into position to continue on his way.

Blocking

Although writers on the subject commonly list as many as 10 to 15 types of blocking maneuver, the majority of these are simply variants of the two basic blocks—the shoulder block and the cross-body block.

Shoulder Block. In executing a shoulder block the blocker drives his head directly forward toward his opponent's abdomen. Then, having concealed his intention until the very last moment, he slides his head to one side or the other so that hard contact is made with the shoulder. To increase the area in contact and thus his ability to control his opponent, the blocker holds his upper arm in a position so that it forms an almost straight-line extension of the blocking shoulder. The hand on the same side is held close to the chest (perhaps even gripping the jersey) in order to reduce any tendency to use it illegally.

In driving forward into his opponent and once contact has been made, the blocker takes short, fast steps that permit him to remain in almost continuous contact with the ground—a situation that permits an almost uninterrupted application of large forces against the opponent. In addition, these short driving steps are made with the feet spread well apart in order to reduce the possibility of the blocker being knocked off-balance by the force of the impact or by his opponent's subsequent attempts to break clear.

Cross-Body Block. The initial movements in a well-executed cross-body block look very much like those used in executing a shoulder block—the blocker drives forward with short, fast steps, his feet well apart and his head directed at his opponent's abdomen. Then, just before contact is made, he quickly turns his arms and head to one side and swings his body around so that he contacts his opponent at waist level with his hip and side.

The results of a study by Klumpar[23] suggest that delaying the sideways turning of the body until the last moment before contact is made is a factor of considerable importance in determining the effectiveness of a cross-body block. In 92% of the successful cross-body blocks that he analyzed, the blocker turned his head and shoulders when he was approximately 1 ft away from his opponent. In only 46% of the unsuccessful cross-body blocks analyzed did the blocker get that close to his opponent before beginning to turn sideways. Thus it would seem that on many occasions a blocker becomes

[23]Emil Klumpar, "An Analysis of Blocking in Football" (M.A. thesis, State University of Iowa, 1939).

so concerned with the body position required that he overlooks or underestimates the importance of timing.

The results obtained in another study of blocking techniques might very well be attributable to this same failing. Nordman[24] conducted a study to determine the relative effectiveness of the shoulder block and the cross-body block used against defensive backfield players (i.e., linebackers, defensive half backs, and safety men). After analyzing 26 game films (11 high school games and 15 university games), he concluded that for both high school and university teams the shoulder block is superior to the cross-body block in downfield blocking situations—225 (or 77%) of the 293 shoulder blocks observed were effective in preventing the defensive man from tackling the ball carrier compared to 104 (or 56%) of the 185 of the cross-body blocks that were similarly effective.

Tackling

Head-On Tackle (Fig. 124). The tackler's approach toward the ballcarrier is almost identical to the approach he would use if he were going to execute a shoulder block. The one major exception is the low and widespread position of his arms—a position adopted to thwart any last-minute evasive action on the part of the ballcarrier and to enable the tackler to get a firm grasp of his opponent at the earliest possible moment. Once contact has been made, the tackler wraps his arms tightly around his opponent's thighs, vigorously extends his hip, knee, and ankle joints, and lifts his opponent clear of the ground—the latter action depriving the ballcarrier of the opportunity to exert the forces necessary to sustain his forward motion. The tackle is completed with both players falling to the ground under the influence of gravity. (*Note:* The continued driving action of the tackler's legs acts to hasten the completion of the tackle by moving the center of gravity of the combined ballcarrier-plus-tackler system farther away from the transverse axis through the feet about which the system tends to rotate.)

Fig. 124. Executing a head-on tackle

[24]Gerald R. Nordman, "A Comparison of the Effectiveness of the Shoulder Block and the Body Block in Football" (M.A. thesis, State University of Iowa, 1959).

Fig. 125. Executing a side-on tackle

Side-On Tackle (Fig. 125). Except that the angle at which the tackler approaches the ballcarrier is different, the technique used in making a side-on tackle is practically identical to that employed in executing a head-on tackle.

Kicking[25]

Although they have a great deal in common, it is convenient to consider the techniques of place-kicking and punting under separate headings.

Place-Kicking. A place-kick may be considered to consist of four consecutive phases or parts: (1) the initial stance, (2) the approach, (3) the swing and kick, and (4) the follow-through.

Initial stance. The kicker takes up his initial position at some predetermined distance behind the kicking tee—in general, approximately 2 yd behind the tee for field-goal and point-after-touchdown attempts and some 5–10 yd for kickoffs. (To minimize the undesirable effects of variations in the length of his approach, the good kicker takes care to see that this distance is always exactly the same for kicks of the same type.) The position adopted is one in which the athlete stands with his right foot slightly in advance of his left; his knees bent a little in readiness for the extension of the legs that will drive him forward into his approach; his trunk erect or inclined slightly forward—the latter requiring slightly more muscular effort than the former but putting the body in a slightly better position from which to move forward; his arms hanging in a relaxed, comfortable position; and his head inclined forward with eyes focused on the spot where the ball will be at the instant he kicks it.

Approach. The approach in field-goal and point-after-touchdown attempts begins with a short step of the right foot. This is followed by a much longer step that brings the left foot into the position it will occupy as the right foot is swung downward, forward, and upward into the ball.

The opinions of writers on the subject are in general agreement as to the optimum position of the left foot relative to a line through the ball in the intended direction of the kick—most recommend a position some 2–4 in. to the left of this imaginary line, a position consistent with the normal lateral spacing of the feet as a person brings one foot past the other in such activities as walking and running. Opinions differ markedly, however, when it comes

[25]Throughout the ensuing discussion it is assumed that the athlete kicks with his right foot.

to the position of the left foot in a forward-and-backward direction. Dodd,[26] for example, recommends a distance of "about two and a half inches," while Bunn,[27] to cite just one of the many other opinions available, suggests that "the non-kicking foot should be placed about 6 inches back of the point of the ball." The only quantitative data available at the time of writing do little to resolve the question. In a study of the place-kicking technique of professional Jim Bakken, Becker[28] found that the left foot was placed behind the ball at distances ranging from 10.4–11.1 in.—distances considerably in excess of those normally recommended. Since it is very likely that the position of the left foot has a considerable bearing on the results obtained by a place-kicker, the conflicting opinions of experts and the current lack of quantitative data suggest that this is a topic worthy of further investigation.

There has been a good deal of interest in recent years in the so-called soccer-style place-kick. In this style the kicker takes up his position some distance to the left of the intended line of the kick and then moves in an arc as he takes his two approach steps to bring him up to the ball.

The various claims made in favor of the soccer-style kick have attracted the attention of a number of researchers. Plagenhoef[29] conducted an analysis of the techniques employed by a former professional soccer and football player and found that the side approach (soccer-style) kick produced a greater ball velocity than did the regular straight approach:

Type of Kick	*Ball Velocity* (Average of five kicks)
Side approach (instep kick)	95.5 fps
Straight approach (instep kick)	82.0 fps
Straight approach (toe kick)	82.0 fps

He then examined the two factors of greatest importance in determining the velocity with which the ball leaves a kicker's foot—the effective mass of the foot and its velocity immediately before impact—and concluded that since the side approach was no better than the straight approach for obtaining "maximum foot velocity," the effective mass of the foot was the more important factor. Manzi[30] compared the soccer-style approach and the straight

[26]Robert L. Dodd, *Bobby Dodd on Football* (Englewood Cliffs, N.J.: Prentice-Hall, Inc., 1954), p. 274.

[27]John W. Bunn, *Scientific Principles of Coaching* (Englewood Cliffs, N.J.: Prentice-Hall, Inc., 1972), p. 191.

[28]John W. Becker, "The Mechanical Analysis of a Football Place Kick" (M.A. thesis, University of Wisconsin, 1963).

[29]Stanley Plagenhoef, *Patterns of Human Motion: A Cinematographic Analysis* (Englewood Cliffs, N.J.: Prentice-Hall, Inc., 1971), pp. 98–103.

[30]Robert Manzi, "A Comparison of Two Methods of Place Kicking a Football for Distance and Accuracy" (M.S. thesis, Springfield College, 1967).

approach in terms of the distance and accuracy of the resulting place-kicks and concluded that

1. The soccer-style approach leads to greater accuracy at each of the three distances (20, 26, and 32 yd from the goal post) at which trials were conducted.

2. Either method of approaching the ball, where the number of steps is limited to two, will produce the same or similar distance kicked.

(*Note:* The length of the approach run used by the subject in Plagenhoef's study was not reported.)

Swing and kick. As his left foot is grounded, the kicker begins the vigorous forward swing of his right leg that will ultimately bring the right foot into contact with the ball. The swing begins with the right leg bent at a near 90° angle and the right foot at about hip height and well behind the line of the body. (The extent to which the foot is behind the body—probably a function of the length of the last stride of the approach—influences the distances through which forces may be exerted upon it.) From this position the flexors of the right hip contract forcefully to swing the thigh forward and downward about a transverse axis. During this initial part of the swing the thigh, lower leg, and foot maintain an essentially constant relationship to one another and rotate as a unit. Then, as the thigh nears that point at which it will pass vertically beneath the hip, its angular velocity begins to decrease and angular momentum possessed by the thigh is added to that already possessed by the lower leg and foot. The latter experiences a marked angular acceleration; the knee extends rapidly—probably aided by a contraction of the knee extensor muscles of the thigh; and the foot swings at ever-increasing speed downward, forward, and finally upward into the ball.

The position of the kicker at the instant his foot strikes the ball and his actions during that brief interval in which his foot is in contact with the ball—approximately 0.015 sec according to Roberts and Metcalfe[31]—are obviously of critical importance in determining the final outcome. The following points deserve mention:

1. The foot is moving in an upward and forward direction at the instant contact is made with the ball.

2. The knee is still slightly flexed as the foot strikes the ball and the last few degrees of extension take place while the foot is in contact with the ball.

[31]Elizabeth M. Roberts and A. Metcalfe, "Mechanical Analysis of Kicking" in *Biomechanics*, ed. by J. Wartenweiler, E. Jokl, and M. Hebbelinck (Basel, Switzerland: S. Karger, 1968), p. 317.

The rate of knee extension, 15 msec [0.015 sec], before contact is of the order of 1,500 to 2,000 degrees per sec, and it is faster or nearly the same throughout contact.[32]

3. The right hip is flexed to approximately 140° as contact is made and throughout the period of contact, indicating that the angular motion of the thigh has been reduced to zero.

4. The head is forward and the back is somewhat rounded to allow the eyes to be focused on the ball.

5. The arms (the left forward and the right to the side and slightly backward) act to "balance" the action of the legs.

Marshall[33] constructed a mechanical kicking machine in order to examine the effect of various factors on the distance achieved in place-kicking and, on the basis of an extensive series of tests, concluded that

1. The optimum point of contact for the toe on the ball was approximately $5\frac{1}{2}$ in. up on the seam when the ball was tilted 15° toward the kicker and the tee was set 15 in. in front of the point directly below the axle (a point roughly equivalent to the knee joint of the kicking leg). When the ball was placed in this position, the angle between the kicking leg and the vertical was approximately 10°. If the ball was moved away from the kicker and the tilt increased, it was possible to locate other positions from which the ball could be kicked almost as far.

2. There were no significant differences in distances obtained in kicking the leather football compared with those obtained in kicking the rubber football.

3. The use of a detachable rubber kicking toe produced no significant changes in the distance that the ball could be kicked.

4. The position of the laces made very little difference in the direction of the flight or the distance that the ball traveled when kicked under the optimum conditions [i.e., those specified in (1) above].

5. Varying the air pressure from 9 to 15 lb [sic] produced no significant effect on the distance that the ball could be kicked.

6. A medium-high, slowly revolving, end-over-end kick results from kicking the ball at the optimum height and with the optimum tilt.

7. Height may be obtained either by tilting the ball toward the kicker or by striking it lower with the toe.

[32]*Ibid.*, p. 317.

[33]Stan Marshall, "Factors Affecting Place-Kicking in Football," *Research Quarterly*, XXIX, October 1958, pp. 302–8.

Bona[34] trained two groups of beginning kickers for a period of 6 weeks (one group using a toe kick and the other an instep kick) and then tested them for accuracy in point-after-touchdown and field-goal kicks and for distance when kicking off. He found that

1. The group that used the toe kick was significantly more accurate than the group that used the instep kick when attempting points-after-touchdown.
2. There was no significant difference between the two groups in terms of accuracy in kicking field goals or in distance achieved in kicking off.
3. There were no significant differences between the groups in terms of either accuracy or distance when kicks with rubber and leather footballs were compared.

Follow-through. The follow-through in place-kicking serves to protect the body from injury and to ensure that the forces necessary to stop the leg swing do not interfere with the act of projecting the ball into the air.

Although the discussion thus far has been concerned primarily with the place-kicks used when field goals and extra points are attempted, much of what has been said applies equally well to the place-kick used to start the game after touchdowns and at the beginning of each half. However, there are also some differences between the two types of kick—a man taking a kickoff having some distinct advantages over one attempting a field goal or a point-after-touchdown. In the first place, he has a virtually unlimited amount of time in which to make his kick. This places him at an advantage compared with a man attempting to kick a field goal or an extra point, for the accuracy and/or distance that the latter can obtain is almost certainly limited by a need to get the ball into the air quickly. Second, the path that the ball must follow in order to pass beyond the reach of the oncoming defensive men increases the demand for accuracy on the part of a kicker attempting field goals and points-after-touchdown. The man taking the kickoff has considerable latitude in this respect and the possibility of the ball's being deflected early in its flight is a relatively remote one. The possibility that it may have to be kicked so high as to cause a reduction in the total distance of the kick is even more remote.

Punting (Fig. 126). The basic techniques used in punting are very similar to those used in kicking field goals and extra points. The principal differences between the two arise as a result of the punter having first to catch the ball and then, as he is moving forward onto his left foot, to release it in a downward and forward direction where it will eventually be hit as the right foot swings through. (*Note:* Although the punter imparts no forward velocity to

[34]Richard C. Bona, "A Comparison of the Instep Kick to the Toe Kick in Football" (M.S. thesis, Washington State University, 1963).

the ball at the time he releases it—he simply lets it drop or guides it downward with his hands—the ball retains the forward velocity it had as a result of the kicker's forward motion. Fortunately for the kicker who would otherwise overrun it, the ball thus falls in a forward and downward direction once it has been released.)

Fig. 126. Punting

The importance that various factors have in determining the success of a punt have been investigated by Smith.[35] From a cinematographic analysis of three punters (an expert, an average punter, and a beginner), he arrived at the figures shown in Table 19.

Since there was very little difference between the three subjects on each of the other measurements taken, Smith concluded that the five parameters listed "made up the difference between good and poor punters, and that these points should be stressed in coaching the punting of a football."

*Table 19. Characteristics of the Technique Used by Punters of Different Ability**

	Expert	*Average*	*Beginner*
Distance ball dropped	6.4 in.	18 in.	30 in.
Velocity of ball at release	92 fps	87 fps	74 fps
Angle of release	47.5°	31.5°	32°
Distance kicked	61 yd	57 yd	46 yd
Time per kick (catch to release from foot)	1.385 sec	1.548 sec	1.697 sec

*Based on data presented in Smith, "A Cinematographic Analysis of Football Punting."

The factors influencing success in punting were also the subject of a study by Alexander and Holt.[36] These two researchers filmed a series of punts

[35] William H. Smith, "A Cinematographic Analysis of Football Punting" (M.S. thesis, University of Illinois, 1949).

[36] Alan Alexander and Laurence E. Holt, "Punting, A Cinema-Computer Analysis," *Scholastic Coach*, XLIII, June 1974, pp. 14–16, 44.

executed by "two exceptional Canadian punters" and selected for analysis an average and an exceptional punt by each subject. The results obtained (Table 20) shed considerable light on several long-standing questions concerned with punting techniques. It was found, for example, that there was virtually no difference (less than 3%) in the speeds of the kicking foot immediately prior to contact for average and exceptional punts. This finding, which "contradicted the generally accepted theory that the speed of the foot . . . is the critical factor in punting performance," led Alexander and Holt to suggest that the critical factor in punting was not the speed of the kicking foot itself but the force that this foot imparted to the ball—the latter a function of both the speed of the foot and the manner in which contact was made between foot and ball. With respect to the nature of the contact between foot and ball, Alexander and Holt observed that "Whenever the ball is met on the instep across the laces—as is commonly advocated—the metatarsals are brought into play" and "the foot 'gives' and causes a loss of energy in the same way that a boxer 'gives' with a blow. When, however, the ball is contacted higher on the ankle and more across the foot, the foot does not 'give'." They concluded, therefore, that the most efficient application of force is obtained if the ball is contacted high on the ankle and with its long axis across the line of the foot. They also noted that the extent to which the punter is lifted in the follow-through—a process in which the vertical momentum of the punter's leg is transferred to his body as a whole—is an indication of the efficiency of the ball-foot contact. "Good contact produces a slower moving leg in the follow-through; hence, less or no lift of the body."

*Table 20. Characteristics of the Techniques Used to Produce Average and Exceptional Punts**

	Distance (yd)	*Speed of Foot before Contact* (fps)	*Speed of Ball after Contact* (fps)	*Angle of Ball across Foot* (deg)	*Body C.G. Lift in Follow-Through* (ft)
Subject K. C.					
Exceptional	67	79.9	108.2	24	.21
Average	45	77.6	98.9	10	.56
Subject D. D.					
Exceptional	60	68.6	90.4	25	.56
Average	40	67.9	86.3	15	.70

*Based on data presented in Alexander and Holt, "Punting, A Cinema-Computer Analysis."

Chapter 11

GOLF

The basic techniques of golf have probably been subject to more widespread examination than have those of any other sport. One manifestation of this interest in the techniques of the game is the large number of instructional books and articles written on the subject by players, coaches, teachers, and others. Unfortunately, within this considerable volume of material there exist numerous areas of disagreement as to what constitutes the optimum technique for making each of the various shots. In addition, most of the arguments presented in favor of one technique over another are based upon nothing more secure than personal opinion and experience. The outstanding exception to this is a book by Cochran and Stobbs.[1] This book, the end product of a 6-yr research project sponsored by the Golf Society of Great Britain (G.S.G.B.) and aimed at examining objectively some of the basic techniques in the game of golf, provides scientifically established answers to many of the questions about which there has been so much discussion and disagreement hitherto. Many of the results obtained in the G.S.G.B. project are referred to in this chapter.

Basic Considerations

The objective in golf is simply to effect the displacement of the ball from one position to another with the least number of shots possible. The initial position of the ball is that in which the golfer places it on the tee, and the desired final position is within the hole some prescribed number of yards away.

[1] Alastair Cochran and John Stobbs, *The Search for the Perfect Swing* (London: Heinemann Educational Books, 1968).

The displacement of the ball from one position to the other is usually effected by a shot, or sequence of shots, in which the ball travels some distance through the air and then bounces and rolls a farther distance. With the exception of putting, the initial passage of the ball through the air (the *carry*) is generally responsible for a greater proportion of the displacement achieved than is the bouncing and rolling (the *run*) that follows. The contributions of the carry and the run in a series of typical drives are shown in Table 21. (The figures in this table have been obtained using equations derived by the G.S.G.B. research team.[2])

Table 21. The Relative Contributions of the Carry and the Run to the Total Length of a Drive

Total Length of Drive (yd)	*Carry* (yd)	*Percentage of Total Length*	*Run* (yd)	*Percentage of Total Length*
160	121	76	39	24
180	145	81	35	19
200	169	85	31	15
220	193	88	27	12
240	217	91	23	9
260	241	93	19	7
280	265	95	15	5

[*Note:* (1) These results are for "squarely struck *drives*" with a British 1.62-in. diameter ball. Differences between these figures and the equivalent figures for an American 1.68-in. diameter ball are almost certainly negligible. (2) The computation of the total length of drive assumes "some sort of average ground conditions."]

The Carry

For any given case, the length of carry obtained depends on (1) the speed, and (2) the direction at which the ball leaves the face of the club, (3) the height of the ball at that instant, and (4) the nature of the air resistance that it encounters in flight.

(1) The speed of any body immediately following an elastic impact is governed by the masses and initial velocities of the bodies involved and by their mutual coefficient of restitution (pp. 77–93). In the case of a golf ball struck by a club, three of these five factors either are fixed by the rules or are subject to such little variation that they may safely be regarded as constant. These are the mass of the ball (0.003 slug, equivalent to a weight of 1.62 oz); the initial velocity of the ball (0 fps); and the coefficient of restitution which,

[2] *Ibid.*, p. 229.

although it varies slightly for differing speeds of impact, is essentially a function of the materials of which the clubhead and ball are constructed.

Variations in the mass of the clubhead, the second body involved in the impact, effect the ease with which the club can be swung and thus the speed of the clubhead at impact—increasing the mass of the clubhead reduces its impact speed, while decreasing the mass produces the reverse effect. Since, if all else is equal, an increased clubhead mass will produce an increase in the speed of the ball at release, and a decreased clubhead speed will have the opposite effect, which of these two factors is the dominant one is of some importance.

The effects that changes in the mass of the clubhead have on the initial velocity of the ball, and thus on the length of a shot, have been reported by Daish.[3,4] Following an analysis in which four golfers of "varying ages. . . and a fair range of golfing ability" swung a club whose weight was varied from 3.5 to 12.3 oz, Daish concluded that "varying the mass of the clubhead over the wide range from 5 to 11 oz has little or no significant effect on the initial velocity imparted to the ball" and "should produce no difference of any consequence in the length of shot obtained."

While it might well be thought that the "body" interacting with the ball immediately before the latter is released is a combined unit of club-plus-golfer-plus-earth, experiments have shown that this is not the case. In fact, at impact the clubhead behaves essentially as if it were not directly connected to the golfer. The truth of this has been dramatically demonstrated by the G.S.G.B. research team,[5] who found that the length of drives hit with a No. 2 wood with a freely moving hinge between the clubhead and the clubshaft varied little from those hit with a normal club—30 drives with each club yielded averages of 215 and 220 yd, respectively. Even then, it was considered likely that this small difference was probably due to factors other than those operating at impact.

Since each of the factors so far considered has been shown to allow little effective variation, it is apparent that observed differences in the speed at which the ball leaves the club must be mainly attributable to differences in the one remaining factor—the speed of the clubhead at impact. Now since the clubhead is momentarily at rest at the peak of the backswing, its speed at the instant of impact must be determined by the forces exerted on the club during the downswing (Newton's second law). Of these forces (gravity, air resistance, and the muscular forces applied to the grip), clearly the last is capable of the widest variation. Differences in the speed at which the ball

[3] C. B. Daish, "The Influence of Clubhead Mass on the Effectiveness of a Golf Club," *Institute of Physics and the Physical Society*, Bulletin XVI, September 1965, pp. 347–49.

[4] C. B. Daish, *The Physics of Ball Games* (London: The English Universities Press Ltd., 1972), pp. 14–15, 102–109.

[5] Cochran and Stobbs, *The Search for the Perfect Swing*, pp. 145–47.

leaves the club are therefore more likely to be due to differences in the muscular forces applied to the grip than to differences in any other single factor.

(2) The direction in which the ball is moving at "release" must be considered in terms of both the angle that the vector representing the velocity of the ball makes with the horizontal (the angle of projection or release) and the angle it makes with the intended line of the shot. These angles are measured in vertical and horizontal planes, respectively.

The angle of projection of the ball is governed primarily by the inclination of the clubface (i.e., its loft). Consider the velocity of the clubhead immediately before impact to be resolved into two components—one acting perpendicular (or normal) to the clubface and the other acting parallel with the clubface [Fig. 127(a)]. At impact the natural tendency of the clubhead to keep moving as it was an instant earlier (Newton's first law) results in the ball experiencing forces in the same direction as these components [Fig. 127(b)]. The normal force causes the ball to be accelerated in that direction.

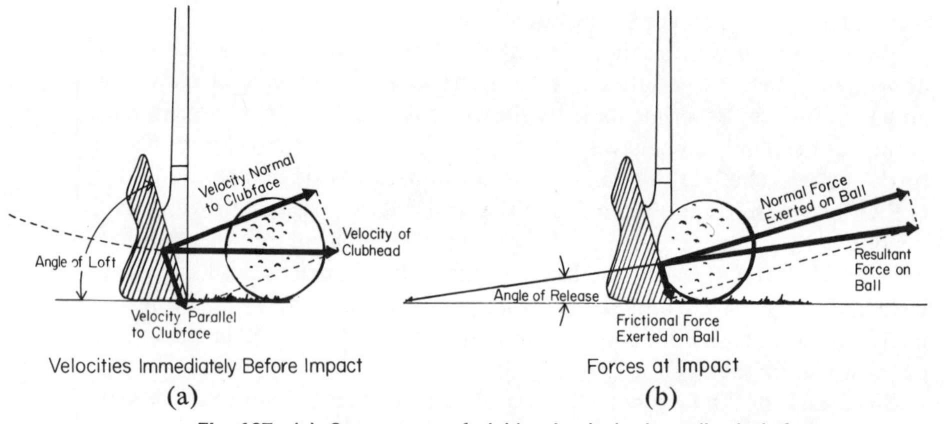

Fig. 127. (a) Components of clubhead velocity immediately before impact. (b) Components of the force exerted on the ball at impact.

The end product of this acceleration (the normal component of the velocity of the ball as it leaves the club) is influenced by the elasticity of both bodies—the less the elasticity, the less the magnitude of this component. The effect of the force along the clubface is governed by the limiting friction between the surfaces in contact—those of the clubface and the ball. If the force exerted exceeds that of the limiting friction, some slippage occurs and the component velocity of the ball in this direction following impact is correspondingly reduced. In fact, if the surfaces were perfectly smooth, the release velocity of the ball would have a zero component along the clubface and the ball would fly off in a direction perpendicular to this surface. However, since in general the surface of the clubface and that of the ball are both "rough," this situation is unlikely to apply in practice. Instead, the ball is generally acted upon by a friction force that not only causes it to be released in a direction slightly below the perpendicular to the clubface but also causes it to acquire a certain amount of backspin.

With respect to motion in the horizontal plane—strictly, a horizontal plane upon which the path of the ball is considered to be projected—the success of any given shot depends on bringing the clubhead to meet the ball so that the horizontal directions in which (a) the clubhead is moving and (b) the clubface is "pointing" at impact both coincide with the direction in which it is intended the ball should go [Fig. 128(a)]. Failure to achieve this consistency in directions causes the path of the ball to deviate laterally from that intended [Fig. 128(b)].

(3) While the height of the ball at release, relative to the height of that point at which it will land, is a significant factor in determining the length of the carry, it is one over which the golfer has very limited control.

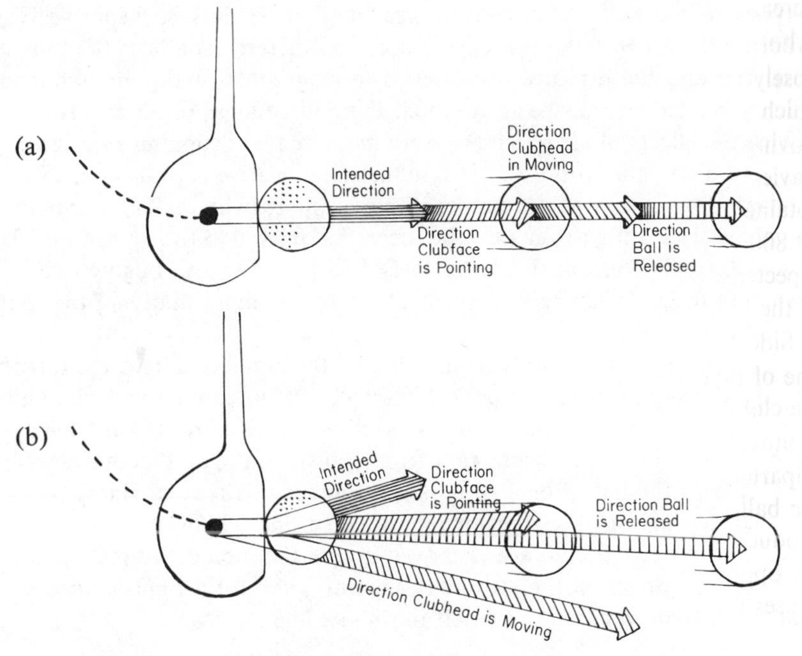

Fig. 128. (a) To obtain the intended direction of release the directions in which the clubhead is moving and "pointing" must both coincide with that intended direction. (b) Failure to obtain the required coincidence of the three directions results in the ball deviating laterally from the desired direction.

(4) A golf ball in flight is subject to forces exerted upon it by the air through which it passes. These forces may be regarded as the summed effects of a

resistance to the linear motion of the ball through the air (drag) and a resistance to the angular motion of the ball as it spins about an axis through its center of gravity. Of these two it is only the latter over which the golfer could be said to exert any real measure of control.

The resistance to the angular motion of the ball serves to modify the drag and also causes the ball to deviate vertically and/or laterally from the path it would otherwise travel. There are basically four types of angular motion that may be imparted to a ball—topspin, backspin, "slicing" sidespin, and "hooking" sidespin.

Topspin, which markedly reduces the lift component and therewith the time of flight and the distance of the carry, is normally applied only in error. This most frequently occurs when the ball is "topped" (i.e., the clubhead hits only the top part of the ball).

Backspin, on the other hand, is directly attributable to the "down-the-clubface" (or frictional) component of force that a lofted club normally applies to the ball. As already indicated (p. 182), this backspin serves to increase the lift that the ball experiences and thereby lengthens the time it is airborne and the distance it carries. The magnitude of the lift obtained is closely related to the speed at which the ball rotates. In an experiment in which the effects of air resistance were measured by dropping a ball into the moving airstream within a wind tunnel (see relative motion, pp. 170–71), Davies[6] found that at an airspeed equivalent to what might normally be obtained with a high-iron shot the lift varied from 0 lb (no spin) to 0.055 lb (at 8000 rpm). Thus, at this latter rate of rotation (again, roughly that to be expected from a high-iron shot) the lift force was more than half the weight of the ball (0.101 lb).

Sidespin generally results from bringing the clubface across the intended line of the shot, during the period in which it is in contact with the ball. If the clubface moves across the intended direction line and toward the golfer (a movement known as bringing it from "outside-in"), a "slicing sidespin" is imparted to the ball. For a right-handed golfer this type of sidespin causes the ball to curve to the right of the intended line. A "hooking sidespin" is produced when the clubface is moved across the intended direction line in the opposite direction (i.e., from "inside-out") and for a right-handed golfer causes the ball to curve to the left during its flight.

The Run

Of all those factors that have a part in determining what happens once the ball hits the ground, the only one that is capable of much variation and

[6]John M. Davies, "The Aerodynamics of Golf Balls," *Journal of Applied Physics*, XX, September 1949, pp. 821–28.

is not essentially fixed by the initial impact between club and ball is the coefficient of restitution. If the coefficient of restitution is zero (i.e., the impact is an inelastic one), the ball simply imbeds itself in the ground at the point at which it lands. However, if the coefficient of restitution is greater than zero, as normally is the case, the ball bounces following impact with the ground. In such instances the distance covered during the bounce (and during each successive bounce) depends very largely on the magnitude of the coefficient. For example, if the ground is hard and the coefficient is therefore high, the ball is likely to cover greater distances with each bounce and, too, to bounce more often than it would if the ground were soft.

After a number of bounces the vertical velocity with which the ball strikes the ground decreases to the point where the ground-reaction force is insufficient to carry it once more into the air. At this point the ball begins to roll, the distance it rolls being governed by its horizontal velocity at the time and by the forces (e.g., gravity and rolling friction) that subsequently act upon it.

Putting

In putting the golfer has three distinct tasks to perform:

1. He must determine the direction in which to hit the ball in order to have it fall into the hole. If a straight line joining the ball and the hole runs directly uphill or downhill or goes across a part of the green that is smooth and flat, choosing the best direction in which to hit the ball is a very simple matter indeed—it should be hit directly along this line. More common, though, are cases in which the surface between ball and hole slopes to one side or perhaps slopes in different directions at different points. In such instances the golfer must take heed of the effect that gravity will produce on the path the ball will take. For example, if the green slopes down to the right of the straight line between the ball and the hole and the golfer directs the ball along this line, that component of the ball's weight that acts downhill will accelerate it in that direction. As a result the ball will follow a curved path below the direct line between ball and hole and, unless the distance of the putt is very short or the slope of the green very small, this downward deviation will cause the ball to pass some distance below the hole. An experienced golfer would therefore choose to direct his putt above the line between ball and hole, knowing that gravity would tend to bring it down and around toward the hole.

2. He must determine how hard the ball must be hit in order to impart to it the speed necessary for it to cover the required distance. In this assessing of the "strength" of the putt, the experienced golfer takes particular note of the distance of the putt, the extent to which it is an uphill or downhill shot, whether the green is soft or hard, or dry or wet, and the length of the grass and the direction in which it lies.

3. Finally, having thus determined the velocity that he intends to impart to the ball, the golfer's task is to execute the putting stroke in a manner consistent with this intention. Basically, this means that the must have the face of the putter at right angles to the chosen line throughout the short period of impact and moving at a speed that will result in the desired speed being imparted to the ball.

Techniques

Grip

There are three principal methods of gripping the club for driving strokes—the overlapping (or Vardon) grip, the interlocking grip, and the baseball (or two-handed) grip. With respect to the relative merits of these three, the results of an experiment by Walker[7] are of interest. After comparing the performances of 24 male golfers who used each of the three grips in turn, Walker found that no one of the grips was statistically superior to either of the others, in terms of greater distance or accuracy.

Stance

The placement of the feet relative to one another and relative to the intended direction of the shot are of some importance in determining the velocity with which the clubhead meets the ball.

If a golfer places his feet together, the narrowness of his base inevitably makes him conscious of the need to maintain his balance and thus precludes him from obtaining a maximum contribution of force from the muscles of his legs and hips. A stance with the feet wide apart also hampers the production of force by these muscles. Logically, therefore, most skilled players use a spread of their feet somewhere intermediate between these two extremes—in general, slightly more than shoulder width apart—for shots requiring maximum or near-maximum effort and a position with the feet closer together for shorter shots requiring only a limited contribution from the legs and hips.

The placement of the feet relative to the intended direction of the shot (as in baseball, the terms *open*, *square*, and *closed* are used to describe the basic variations) also depends to some extent on the length of shot required. For shots requiring maximum or near-maximum effort there appears to be very little evidence to suggest that one alternative is superior to any other. However, for those shots (and particularly short pitches, etc.) where accuracy rather than maximum distance is the prime consideration, a somewhat open stance tends to restrict the range of the backswing, thereby decreasing the scope for errors in the execution of the swing, without preventing the required force from being obtained.

[7]Alan Walker, "The Relationship of Distance and Accuracy to Three Golf Grips" (M.S. thesis, Springfield College, 1964).

The Swing[8]

The swing, that succession of movements that culminates in the clubhead striking the ball, may be regarded as the central element about which the whole game of golf is built.

Although, from the first movement following the address (the position adopted by the player before he begins his swing), all actions of the golfer's body and the club must be coordinated into one smooth sequence, for the purposes of analysis it is convenient to consider the full swing as consisting of four major parts—*backswing* (or *upswing*), *downswing*, *impact*, and *follow-through*.

In both backswing and downswing the motion is essentially rotary and for simplicity may be considered in terms of two levers rotated about their respective axes. These levers are the club itself, rotating about an axis passing through the golfer's hands, and a combined shoulders-arms-hands lever, rotating about an axis inclined to the horizontal and passing through the golfer's chest (Fig. 129).

Fig. 129. The golf swing may be analyzed in terms of a shoulders-arms-hands lever rotating about an axis through the upper chest and a club lever rotating about an axis through the hands.

Backswing. The purpose of the backswing is to put the golfer and the club into the optimum position from which to start the downswing.

[8]In the analysis of the swing that follows, it is assumed that the golfer is right-handed and that he is executing a drive for maximum distance. The terms *forward* and *backward* are used with reference to the direction in which the golfer intends to hit the ball. Forward is in that direction, and backward is in the opposite one.

The backswing begins with a simultaneous backward movement of the clubhead and the hands and a rotation of the trunk to the right.

These first movements are sometimes preceded by a "pressing" or "cocking" action in which the golfer gently pushes his right knee in toward his left one and then, as the knee returns to its original position, begins the withdrawal of hands and club. This movement (evident in the results of Carlsöö's study[9] of forces and muscular actions in the golf swing, and occasionally referred to by golf writers[10,11,12]) may serve to help the golfer initiate his backswing in a systematic and relaxed manner, but apart from this would appear to have no particular merit, for it merely adds to an already wide range of possible sources of error.

As the combined backward movement of the hands and club and the rotation of the trunk continue, the left arm is raised and swung across the trunk, the wrists are "cocked" (or bent sideways toward the thumbs), and the left forearm is rolled so that the back of the left hand lies in an approximately vertical plane.

The end of the backswing is reached with the hands at or slightly above head height, the trunk rotated approximately 90° from its original position, and the wrists cocked so that the club shaft lies over and behind the head at some 0—45° above the horizontal.

According to Williams[13] an objective examination of the paths followed by the hands and the clubhead during the upswings of top-class players reveals that the path of the hands varies hardly at all from player to player while the path of the clubhead varies considerably. He therefore concluded "that the path followed by the clubhead in the upswing has little significance and is a matter of personal preference."

Carlsöö[14] also reached some interesting conclusions regarding the backswing. He concluded, for example, that the backswing could be divided into two consecutive parts—an accelerating movement backward and upward lasting about 0.3 sec and a retarding or braking movement lasting until the top of the swing about 0.35 sec later. This retardation was characterized by a change in the direction of the horizontal couple that the feet exerted against the ground and by marked changes in muscular activity; the activity of those muscles that had initiated the backswing diminished and that of their

[9]S. Carlsöö, "A Kinetic Analysis of the Golf Swing," *The Journal of Sports Medicine and Physical Fitness*, VII, June 1967, pp. 80–81.

[10]Julius Boros, *Swing Easy, Hit Hard* (New York: Cornerstone Library Inc., 1968), p. 68.

[11]Robert McGurn and S. A. Williams, *Golf Power in Motion* (New York: Cornerstone Library Inc., 1969), p. 29.

[12]Sam Snead, *The Driver Book* (New York: Cornerstone Library Inc., 1965), p. 57.

[13]David Williams, *The Science of the Golf Swing* (London: Pelham Books Ltd., 1969), pp. 46–55.

[14]Carlsöö, "A Kinetic Analysis of the Golf Swing," pp. 81–82.

Fig. 130. A forward movement of the hips initiates the downswing.

antagonists (i.e., those muscles that perform the opposite function) increased. Furthermore, the muscles that produced this retarding or braking effect on the backswing continued to be active in the downswing during which they acted as "very essential movement-promoting muscles."

Downswing. The objective on the downswing is to have the clubhead arrive at the point of impact moving at maximum speed in the required direction and with the face of the club "pointing" in that same direction.

The downswing begins with a forward movement of the hips that, with good golfers, actually begins approximately 0.1 sec before the clubhead reaches the limit of its backswing.[15] This moving forward of the hips rotates the whole upper body (Fig. 130) and moves both levers through the first part of the downswing. The forces responsible for this forward movement of the hips and the lesser forces exerted by the same hip and leg muscles later in the downswing have been estimated to account for $2\frac{1}{2}$ of the total 3–4 hp generated in a good drive. Thus it can readily be seen that "the muscles of the hips and legs constitute the main source of power in long driving."[16]

The positions of the shoulders, arms, hands, and club relative to one another are unchanged as the hips are driven forward. Then, when the left arm reaches an approximately horizontal position, this first stage (the "one-piece stage" as Williams[17] calls it) comes to an end and the angle between club-

[15]Cochran and Stobbs, *The Search for the Perfect Swing*, p. 82.

[16]*Ibid.*, p. 81.

[17]Williams, *The Science of the Golf Swing*, pp. 17–20.

shaft and left arm, previously about 60°–70°, becomes progressively larger. From this point onward, the hands continue to move along their circular arc at a fairly constant speed while the clubhead's speed increases dramatically as the angle between the left arm and the clubshaft straightens out.

To understand what happens during this second stage of the downswing, it is necessary to consider the forces that the golfer's hands apply to the grip of the club (Fig. 131). The resultant of these forces may be resolved into two components:

1. *A radial component* (or centripetal force) acting toward the axis about which the shoulders-arms-hands lever rotates. This component serves to constrain the motion of the handgrip of the club to a circular arc and, because its line of action does not pass through the center of gravity of the club, also tends to cause the club to be rotated (or, more precisely, to be angularly accelerated) relative to an axis through this point. The direction of this rotation is consistent with an "uncocking" of the wrists or, to put it another way, with an increasing of the angle formed by the lines of the club shaft and the left arm.

2. *A tangential component* acting, as the name suggests, in a direction tangential to the path followed by the handgrip of the club. This component serves to accelerate the handgrip, and the club as a whole, in the direction in which it acts. When its line of action does not pass through the center of gravity of the club, this component also tends to angularly accelerate the club in a direction opposite to that of the radial component. The net effect of these two components is to accelerate the club along a circular path and to cause it to rotate relative to an axis through its center of gravity. The direction of this rotation is governed by how the opposing tendencies of the two components compare.

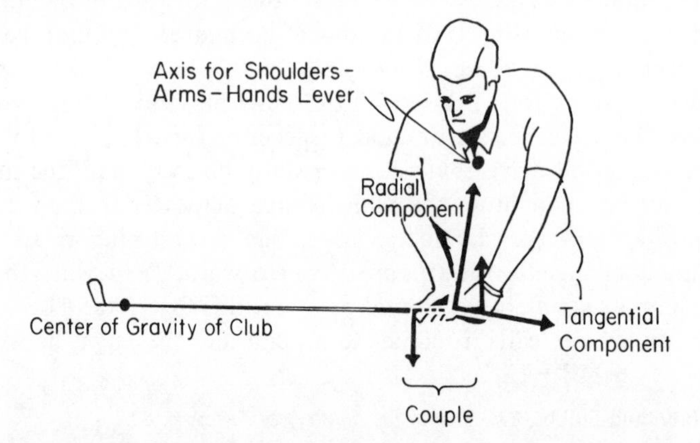

Fig. 131. The forces applied to the handgrip of the club, considered relative to the axis through the chest

In addition to the resultant force that the hands exert on the grip, there exists the possibility that the combined actions of the hands also result in a couple being applied to the grip. For, if the right hand is pressed down forcefully against a resistance of equal magnitude provided by the left hand, a couple that tends to uncock the wrists comes into existence. The question of whether such a couple exists is a source of some disagreement in the literature. Following an analysis of the swing of Bobby Jones, Williams[18] concluded somewhat emphatically that "The mathematics... proves beyond any argument that hand (or wrist-uncocking) leverage has nothing to do with accelerating the clubhead in what is usually referred to as the 'hitting area'." Cochran and Stobbs,[19] on the other hand, profess the more widely held view. Without presenting any evidence in support of their contention—their book is intended for the lay reader and therefore does not include all the detailed scientific evidence upon which it is based—they state that "The obvious way to add speed to the clubhead is by applying some effort at the hinge [the hands] . . . . To do this, the right arm has simply... to *push:* that is to try to straighten out at the elbow."

Until this question is satisfactorily resolved, any analysis of the second stage of the downswing should probably consider both possibilities. If the golfer does not apply a couple that contributes to the uncocking action of the wrists, this action must result solely because the torque due to the radial component exceeds that due to the tangential component. Alternatively, if a couple is applied via the hands, this couple, together with the radial component of the resultant force, must produce the characteristic "uncocking" action despite the contrary tendency of the tangential component.

The "uncocking" of the wrists during the second part of the downswing is commonly attributed to a centrifugal force acting on the clubhead to pull it outward, away from the axis about which it is rotating. While this explanation might have some superficial appeal, it is inconsistent with the facts of the matter, as a consideration of the forces acting on the club readily reveals.

Apart from gravity and air resistance (both of which can be instantly dismissed as having no direct bearing on the question), the only external forces acting on the club are those applied at the grip. Now, for the purposes of examining this particular question, consider the resultant of these forces to be resolved into a radial (or centripetal) component that acts *inward* along the line of the clubshaft and *toward the axis through the hands,* and a tangential component acting perpendicular to the line of the clubshaft. (*Note:* These are *not* the same radial and tangential components referred to earlier, *relative to the axis through the golfer's chest*—cf. Fig. 131 and 132). The reaction to the centripetal force exerted by the hands *on the club* is a centrifu-

[18]Williams, *The Science of the Golf Swing,* p. 23.

[19]Cochran and Stobbs, *The Search for the Perfect Swing,* p. 66.

gal force that the club exerts *on the hands.* Thus, since the only centrifugal force involved, in the rotation of the club about an axis through the hands, acts not on the club but on the hands, it cannot possibly be responsible for pulling the clubhead outward.

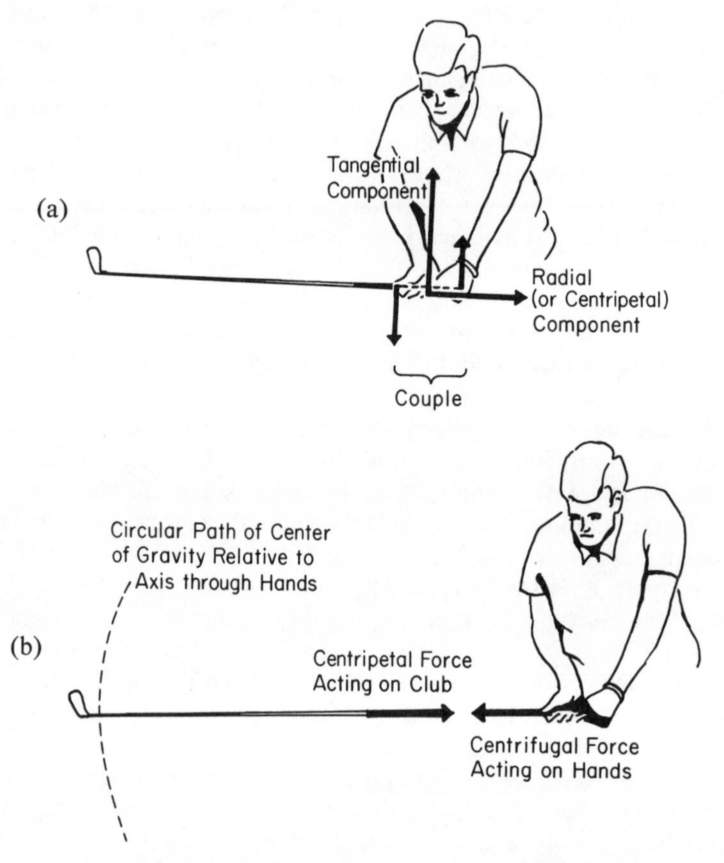

Fig. 132. Forces applied to the handgrip of the club, considered relative to the axis through the hands

As the downswing is executed, the right elbow is brought down and close to the right side of the body, in the process transmitting to the grip the force produced by the contraction of muscles on the right side of the body. During the final few inches that the hands travel in the downswing, the wrists are "rolled" through approximately 90° so that the back of the left hand and the clubface (until this time in a near-vertical plane parallel to the intended direction of the shot) are brought around perpendicular to that direction.

Impact. The critical features of the swing at the instant of impact are the orientation of the clubface, the position of the clubhead, and the velocity at which it is moving. Theoretically—provided the clubface is at right angles to the required direction, the center of gravity of the clubhead is directly behind the center of the ball, and the clubhead is moving forward with the maximum speed possible, under the circumstances—the position of the golfer is of little consequence. Experience suggests, however, that the need to satisfy these three conditions allows only minor variations in the position of the golfer at impact and that the optimum position has the following characteristics (Fig. 133).

Fig. 133. The position at impact

1. The clubhead is in line with or just behind the hands. (Computer studies and observations of good golfers, referred to by Cochran and Stobbs,[20] suggest that the clubhead is moving fastest at, or possibly just before, the point where it catches up with the hands).

2. The back of the left wrist and hand is in a vertical (or near-vertical) plane perpendicular to the intended direction of the shot.

3. The golfer's center of gravity is forward of a midline between his feet, thus placing a greater proportion of his body weight on his left foot than on his right one. In his study of a Swedish champion, Carlsöö[21] found that at impact the left foot transmitted a vertical force of 74 kg (163 lb) to the ground, while the corresponding figure for the right foot (24 kg or 53 lb) was less than one-third of this. (*Note:* While some of this vertical force resulted from the actions taking place at impact—the golfer himself weighed only

[20]Cochran and Stobbs, *Ibid.*, p. 58.

[21]Carlsöö, "A Kinetic Analysis of the Golf Swing," pp. 79, 81.

84 kg (185 lb)—it is clear that by far the greater proportion of the golfer's weight was supported on his left foot.)

4. The axis of the shoulders-arms-hands lever, passing through the golfer's chest, is in the same position it has been in throughout the downswing. The golfer brings this about by inclining his trunk, the lower end of which has been driven forward earlier by leg and hip action, slightly backward and inclining his head forward with his eyes firmly focused on the ball. This latter serves to ensure that the axis is not lifted and thus tends to eliminate any risk of the ball being "topped."

Follow-Through. The follow-through, which serves the same purposes as it does in other similar activities, consists of a gradual slowing down of the body and club movements that led up to the moment of impact.

Putting

Unlike driving, in which the need for maximum clubhead speed at impact largely determines the body actions that can be successfully employed, success in putting can be achieved using a wide variety of techniques (Fig. 134). For, aside from the obvious need to meet each of the three requirements mentioned earlier (pp. 267–68), present knowledge of putting techniques sheds very little light on what methods are most suitable. The truth of this statement was well borne out by the results of a cinematographical analysis of 16 "first-class" professionals in which it was found that "The only features where the professionals showed a measure of constancy were the ball position and the head position. Most of them had the ball placed opposite the left foot and their eyes almost directly above the ball."[22]

Cochran and Stobbs[23] also reported the results of a number of other valuable experiments on putting. Among their many findings were the following:

1. For all practical purposes it is impossible with a normal putter to put any useful spin on the ball.

2. If the ball is hit off-center, the clubhead tends to rotate and the length and direction of the putt are affected. For example, an otherwise 20-ft putt stops 4–6 ft short and about 7 in. to one side if it is hit 1 in. off-center. [If the reaction that the ball exerts on the club at impact does not pass through the center of gravity of the clubhead, it will tend to rotate the clubhead relative to an axis through this point—see eccentric force, p. 105. In this event the direction in which the ball is hit deviates laterally from that in which it would otherwise have moved. And, because the part of the clubface with which

[22]Cochran and Stobbs, *The Search for the Perfect Swing*, p. 136.
[23]*Ibid.*, pp. 128–42.

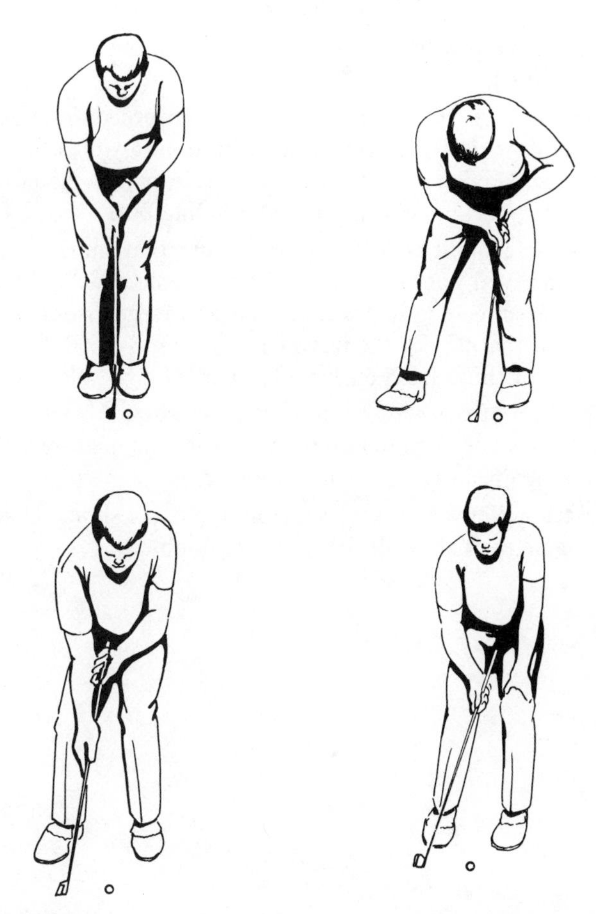

Fig. 134. Success in putting has been achieved using a wide variety of techniques.

the ball is in contact is moving at a lesser speed than it would have been if the impact had not been off-center, the speed imparted to the ball is also reduced.]

3. Distances lost and lateral deviations from the desired direction due to the balls being hit off-center could be reduced if some of the weight of the clubhead could be shifted to the heel and some to the toe, still leaving the center of gravity in the middle. [An eccentric force exerted by the ball at impact causes the clubhead to be angularly accelerated. For any given case, the magnitude of this angular acceleration, and consequently the effect produced on the ball's subsequent motion, is inversely proportional to the moment of inertia of the clubhead—see angular analogue of Newton's second law, pp. 151–52. Thus, since the suggested redistribution of the weight of the clubhead increases its moment of inertia, the effects of an off-center

hit will be less marked with such a putter than with one in which the weight is not so distributed.]

4. The direction in which the ball sets off is governed more by where the face of the putter is pointing than by the direction in which the head of the putter is moving. Having the blade "square" at impact is therefore the most important single point to concentrate on in holing out.

5. Random irregularities in the green ensure that putts hit in precisely the same manner will not necessarily yield the same results. An experiment designed to determine the importance of such irregularities on success in putting revealed that 2% of the missed putts from 6 ft, 50% of those from 20 ft, and 80% of those from 60 ft could be attributed solely to this factor.

6. Comparison of the performances of professionals using blade, center-shafted, and mallet-style putters during tournament play revealed that no one type was significantly better than another.

7. A scratch golfer would save something like six shots per round if the regulation diameter of the hole ($4\frac{1}{2}$ in.) were doubled.

Chapter 12

GYMNASTICS

The sport of gymnastics encompasses a vast field of activity. It includes not only the 10 Olympic events (floor exercise, vault, rings, horizontal bar, parallel bars, and side horse for men, and floor exercise, vault, beam, and uneven bars for women) but also such varied activities as trampolining, tumbling, flying rings, rope climbing, dual balancing, and pyramid building.

Because the field is so vast, and the amount of space that can be devoted to it here is limited, this chapter is confined to a consideration of just three of the men's Olympic events (floor exercise, vault, and horizontal bar) and trampolining.[1]

The discussion of each of these events includes (1) a consideration of the basic concepts underlying the techniques used in the event, and (2) a series of brief analyses of selected techniques to show how these concepts apply in specific cases. With respect to the latter, elementary techniques are given preference over more advanced ones, although (where possible) these too receive some attention.

FLOOR EXERCISE

The gymnastic movements used in floor exercises are widely regarded as forming the foundation upon which the rest of gymnastics is built. For this reason the basic concepts underlying these movements deserve careful consideration.

[1]In general, the basic concepts underlying the techniques in the other Olympic events are similar to those for the events considered here. The biomechanical bases for the various techniques in the women's floor and beam events, for instance, are almost identical to those described in connection with the men's floor exercise; similar close relationships exist between the men's and women's vault and between the horizontal bar on the one hand and the parallel bars, rings, and uneven bars on the other hand.

Basic Considerations

Floor exercises are generally composed of leaps; springing and tumbling movements (including necksprings, headsprings, rolls, cartwheels, somersaults, etc.); and held positions exhibiting balance, suppleness, and strength.

In leaps and in those springing and tumbling movements in which the gymnast projects himself into the air, his success depends on his ability to do the following three things in an aesthetically pleasing way:

1. acquire lift and rotation at takeoff,
2. control his rotation while in the air, and
3. control his motion as he lands.

Once the gymnast leaves the ground, he becomes a projectile, the flight path of which (like all other projectiles) is determined by the velocity and height of his center of gravity at the instant of takeoff. Thus any attempt to improve his performance by acquiring more lift must involve modifying either or both of these quantities if it is to have the desired effect. The rotation that a gymnast acquires at takeoff derives from one or both of two sources—a couple and/or an eccentric force.

When the gymnast becomes airborne, the only way in which he can control his rotation is by adjusting his moment of inertia. By bringing the masses of his various body parts closer to the axis of rotation through his center of gravity, he can increase his angular velocity; by moving them farther away, he can decrease it (conservation of angular momentum). The extent to which he can make such adjustments is generally limited, however, by the nature of the aesthetic and technical requirements of the activity. He could, for example, tuck toward the end of a neckspring and thereby increase his angular velocity. There is no doubt, though, that such a maneuver would be much less aesthetically pleasing than one in which his body was kept fully extended throughout the flight. Furthermore, it fails to meet the technical demands of the movement, which require that such springs be completed in an essentially straight position. Under such circumstances it should be clear that while the gymnast can exert a small measure of control over his rotation when he is in the air, the most important determining factor as far as his rotation is concerned is *not* what he does in the air but what he does on the ground during the takeoff. For it is here that is decided what angular momentum he will have during the flight and how long this angular momentum will have to take its effect (i.e., how long he will be airborne).

At the instant at which the gymnast lands (the instant of touchdown), his body is rotating about an axis through this center of gravity, which is itself undergoing translation. If he wishes to come to a stop (i.e., if he doesn't wish to immediately move off into another movement), he must evoke such forces from the landing surface as will cancel this translation and rotation. The logical (and most used) method of doing this is for the gymnast to land so that the reaction forces that he evokes act eccentrically. These eccentric forces provide a rotating effect of such magnitude and direction as to reduce his angular velocity to zero. Their translatory effect similarly reduces the linear motion of his center of gravity to zero. If the gymnast wishes to move straight into another stunt as soon as he has landed, he tries to evoke a reaction consistent with his needs. If these include a continuing rotation in the same direction (as in, say, a flip-flop followed by a back somersault), he will endeavor to obtain a reaction that rather than canceling his rotation at least permits some of it to be retained and perhaps even enhances it. On the other hand, if he wants to follow on with a movement involving a rotation in the opposite direction (and this, it might be said, is much less likely), he tries to more than cancel his rotation so that he will actually acquire some angular momentum in the direction in which he now plans to move.

Another important factor concerning landing is the distance over which the body moves while being brought to rest or while having its motion redirected. Whichever of these two is involved, a certain amount of work is done on the body, and the magnitude of the forces that act on it and that it must be able to withstand depends on the distance involved (work-energy relationship). In gymnastics the need to reduce the magnitude of the force by increasing the distance is recognized and all landings are accompanied by some flexion at the joints supporting the body—e.g., the hip, knee, and ankle joints, for landings on the feet. Good form in floor exercises requires that the distance involved be kept within fairly narrow limits, however, and anything approaching maximum bending of the limbs involved renders the gymnast liable to be penalized.

The basic concepts involved in rolling movements (forward and backward rolls, cartwheels, etc.) are exactly the same as those described above in connection with the springs and the other tumbling movements, except that here the gymnast does not become airborne and is therefore not concerned with those factors that create lift.

For held or static positions the gymnast endeavors to assume as stable an equilibrium position as he can in keeping with aesthetic and technical demands. Thus he seeks to have

1. his gravity line passing through the midpoint of his base—or, to express it another way, to have the distance from his gravity line to the limits of his base as large as possible—and
2. his center of gravity as low as possible.

In practice, however, it is only rarely that the requirements of the position permit him to exercise any substantial control over his stability except via the first of these.

Techniques

Front Scale,[2] Standing Scale Frontways (F.I.G.), Front Single Leg Lever, Arabesque

This held position, like most others, can be moved into in a number of different ways. For the purpose of this analysis it is assumed that the gymnast moves into it in what is probably the simplest possible manner—from an erect standing position. From this position, the gymnast first shifts his center of gravity slightly to one side so that it is directly over the midpoint of the base provided by what will be his supporting leg. He then lowers his head and trunk forward; raises his arms forward and upward; and, keeping both legs straight, simultaneously raises his free (or nonsupporting) leg up behind him.

In executing these movements the gymnast's task is to assume the desired final position without moving his gravity line relative to the base or, in other words, to keep his body as stable as he can, consistent with the requirements of the position. However, since the movement of his free leg does not tend to shift his center of gravity backward as far as it tends to be moved forward by the movement of his upper body, these actions have the effect of shifting his gravity line away from its central position. To prevent this, the gymnast moves his hips backward slightly as he lowers his upper body and raises his free leg. This action serves to reduce the forward shift of the center of gravity due to the upper body, add to the backward shift due to the free leg, and add still farther to the backward shift by moving the supporting leg in that direction. The net result is that the gymnast's center of gravity retains its position directly over the midpoint of the base (Fig. 135).

Headstand

To perform a headstand from a crouch position the gymnast places both hands on the floor about shoulder width apart and with his fingers comfortably spread. He then places his forehead in position well forward of a line joining his hands—it is generally considered that lines joining the three "points" of contact, head and hands, should form an approximately equal-sided triangle. Then, having ensured that the area of his base will be as large as is practicable (by spreading his fingers and by putting his head well forward of his hands), the gymnast pushes off both feet and moves his hips upward and forward over his base. Having thus reached an angled headstand position (hips high, legs straight, and feet low), all that remains is for the gymnast to rotate his legs about an axis through his hip joints until they reach the vertical.

[2]The terms used here to describe the various gymnastic positions and movements are those believed to be in most widespread use in the United States. However, since gymnastics terminology is anything but standardized, a number of synonymous terms are also listed. In each case where the first term given is different from that of the International Gymnastics Federation (F.I.G.), the term used by that body is indicated.

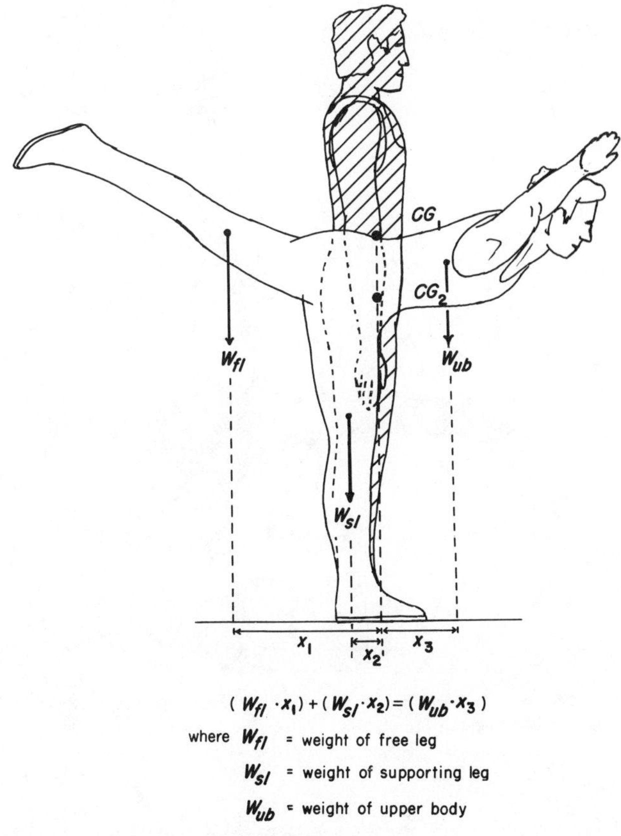

Fig. 135. Front scale

In reaction to this final upward motion of the legs, the gymnast's upper body tends to rotate in the opposite direction (Fig. 136). While most of this reaction is "absorbed" by the floor, its effects are evident in the slightly arched position in which the gymnast finishes. This arched position serves a useful function, for it assists in keeping the gymnast's center of gravity over the midpoint of his base—in this case, the centroid of the triangle formed by his "points" of support.

To overcome any tendency of his center of gravity to move away from this position of maximum stability, the gymnast applies additional forces to the floor via his hands and head. If his center of gravity tends to move toward the forward limit of his base, the gymnast contracts the muscles that flex his neck and this increases the force that his head exerts against the floor. The resulting increase in the eccentric force, which the floor exerts in reaction to this, tends to overcome the body's tendency to rotate forward. Movement of the gymnast's center of gravity toward the backward limit of his base is similarly countered by contracting the wrist flexor muscles and increasing the force exerted by the hands against the floor.

Fig. 136. Angular action and reaction in moving from an angled to a full headstand position

Handstand

Although very similar in many respects, the handstand is considerably more difficult than the headstand because the higher center of gravity and the smaller base that the gymnast has in this position make it inherently less stable.

From a standing position the gymnast steps forward and places his hands on the floor shoulder width apart and with fingers spread. He then swings his rear leg upward and backward. As this leg approaches the limit of its swing and is slowed down, the angular momentum that it possesses is transferred to the body, to aid in rotating it toward the handstand position. Meanwhile the gymnast's forward leg is also contributing by extending at the hip, knee, and ankle joints and exerting a force downward and backward against the floor. The reaction to this force has a relatively large moment about a horizontal axis through the gymnast's wrists and causes the body to rotate about that axis in the desired direction.

Once the gymnast's feet leave the floor, his body weight, acting through his center of gravity at some distance from the axis of rotation, serves to reduce the angular momentum that he acquired earlier. Recognizing that this will inevitably happen, the gymnast tries to acquire just as much angular momentum as his weight can effectively cancel by the time his body arrives in a position vertically above his hands. If he pushes off too hard and acquires more angular momentum than his weight can effectively overcome, the gymnast risks overbalancing because his body is moving too fast as it reaches the vertical position. Similarly, too little angular momentum seriously reduces his chances of getting his body to the vertical.

Fortunately, the situation does not require quite the degree of precision that it might appear to, for the gymnast can make adjustments to allow him to reach the desired equilibrium position despite *small* errors in the amount of angular momentum imparted to his body at the outset. If too much angular momentum has been acquired, he can contract his wrist flexors, increase the pressure exerted against the floor by his fingers, and thus increase the moment of the reaction opposing his angular motion. If he has too little angular momentum, he can increase his angular velocity by bending his arms or legs or by arching his back slightly (i.e., by decreasing his moment of inertia). It should be pointed out here that while each of these latter adjustments may allow the gymnast to "save" the situation, they are all regarded as bad form in gymnastics and could incur some penalty in a competitive situation.

Once in a balance position, the gymnast works to maintain this in much the same way as he would in a headstand. To prevent overbalancing forward, he contracts his wrist flexors and increases the pressure exerted against the floor by his fingers; to prevent overbalancing in the opposite direction, he increases the pressure exerted by the heels of his hands by contracting his wrist extensor muscles.

The gymnast's body should be in a straight rather than arched handstand position. For, apart from aesthetic considerations, it requires less effort on the part of the gymnast to maintain such a position. Consider the simple wooden blocks in Fig. 137. In (a), each of the two top blocks rests squarely upon the one below it and no forces other than their weights (and the reactions they elicit) are necessary to maintain them in position. In (b), the addition of the third block causes the superstructure to topple because, although the line of gravity of each block passes within the limits of its base, the resultant weight of the top two blocks falls outside the base supplied by the bottom one. These three blocks can be maintained in equilibrium, however, by the addition of an external force. If a strap or tape is fastened to them along the side indicated [Fig. 137(c)], the forces it exerts are sufficient to keep the system in equilibrium. Put in very simple terms, this is the essential difference between the straight and arched handstands. In the straight handstand, each "block" (or body part) rests upon the one below and the effort required of the gymnast to maintain this position is relatively small [Fig. 138(a)]. In an arched handstand, on the other hand, the gymnast's muscles and ligaments must be additionally active to "tie" the precariously balanced body parts together [Fig. 138(b)]. In short, it requires more effort to maintain such a position.

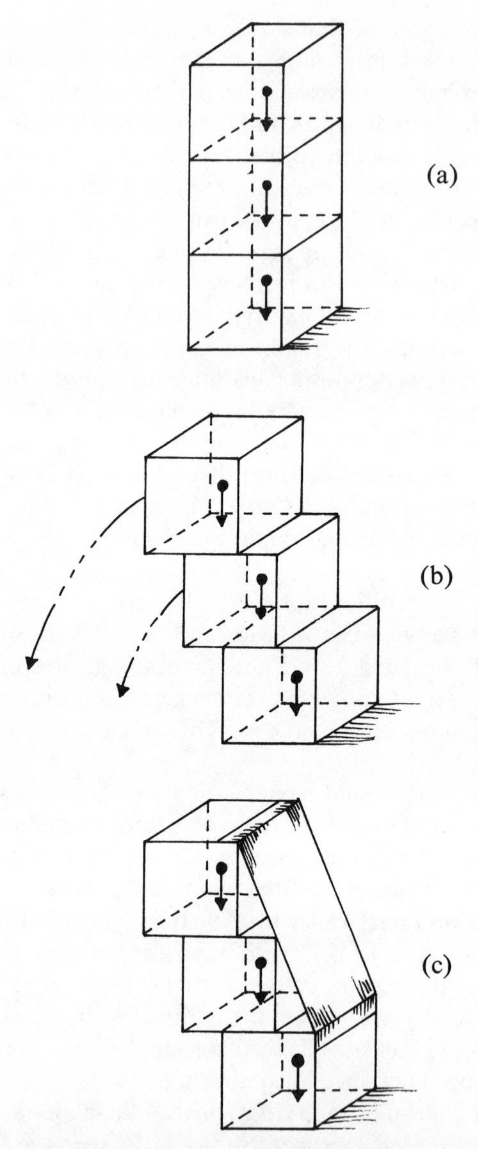

Fig. 137. The resultant weight of the top two blocks in (a) falls within the limits of the base provided by the third block, and the structure is in equilibrium. In (b), although the line of gravity of each of the top two blocks passes within the limits of the base provided by the block below, the line of action of their resultant weight is such as to cause the structure to topple. If a strap is attached to the blocks as shown in (c), the forces exerted by the strap serve to maintain the structure in equilibrium.

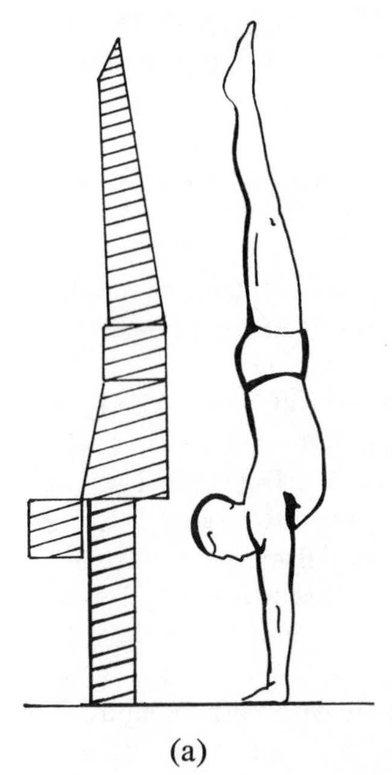
(a)

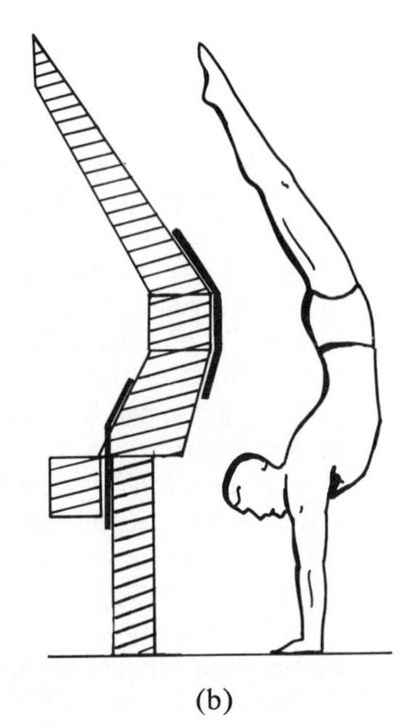
(b)

Fig. 138. Straight and arched handstands

Forward Roll,[3] Front Roll

The forward roll (Fig. 139) is one of the most basic of gymnastic movements and as such is one of the first stunts attempted by most beginning gymnasts.

Possibly the most interesting thing about the biomechanics of the forward roll is the fundamental difference that seems to exist between the way in which beginners are often taught this movement and the way in which advanced gymnasts perform it. The beginner is commonly exhorted to "tuck up tight" or to "roll up into a ball" once her feet have left the floor. Her rolling rotation is thus mainly controlled by the forces that she exerts while her feet are on the floor and only a very little by subsequent adjustments of her body position (and thus, of course, of her moment of inertia). In this respect this type of forward roll is very like those springs and tumbling movements in which the gymnast becomes airborne—once the gymnast's feet leave the floor, her control over her angular motion is relatively limited. The advanced gym-

[3]Although this section of Chap. 12 is devoted to a consideration of techniques in the men's floor exercise, the forward roll, the backward roll, and the cartwheel are discussed in terms of a woman gymnast in order to be consistent with the stroboscopic photographs depicting these techniques (Fig. 139–141).

nast (Fig. 139), on the other hand, initiates the roll and then, with her legs straight and her back only slightly rounded, controls the rotation with her legs. For instance, if she requires a fast rolling action (e.g., so that she can come to a standing position with straight legs), she extends her legs, reaches high with her feet, and brings them down and forward relatively fast. In this way her legs acquire a large amount of angular momentum that she then transfers to her body as a whole as they slow down—normally when her heels are about 2–3 ft from the floor. This additional angular momentum supplements that which the body possesses as a result of forces exerted before the gymnast's feet left the floor.

Whichever of the two techniques of controlling the roll is used (and the latter would certainly appear to have the greater merit), the rolls are initiated in basically the same manner (see the first two positions, Fig. 139). From a position in which both her feet are on the floor, the gymnast extends her legs and evokes a reaction from the floor that imparts angular momentum to her body. She then places her hands on the floor or mat (if they are not already there), bends her arms, tucks her head, and lowers her shoulders to the floor to initiate the rolling action.

The completion of the roll is also basically the same with both techniques. The advanced gymnast flexes at the hips and knees, thereby bringing her heels close to her buttocks and adopting a position similar to that used by the beginner. When her feet make contact with the floor and her buttocks start to lift off it, the moment of her weight tends to reduce her angular momentum (cf., kicking up into a handstand in previous section). If the gymnast is in any

Fig. 139. Forward roll

danger of having insufficient angular momentum to carry her forward over her feet, she stays as tightly tucked as she can, thereby reducing her moment of inertia, increasing her angular velocity, and improving her chances of completing the movement satisfactorily.

Backward Roll

Although most beginning gymnasts find it a little harder to learn, a backward roll (Fig. 140) is, in essence, merely the reverse of a forward roll. Starting from a crouch position, the gymnast pushes herself backward so that her line of gravity passes outside the backward limit of her base. This movement, which results in backward angular momentum being imparted to the body, initiates the backward roll. In the middle part of the roll the gymnast can use the same two alternatives described in conjunction with the forward roll. She can adopt a tightly tucked position and roll "like a ball" or she can roll in a more open position. If she chooses to do the latter, the parts played by the legs and upper body are the reverse of those they have in the forward roll, for when the gymnast's buttocks reach the floor, she thrusts her upper body backward. When this backward angular motion of her upper body is slowed, its angular momentum is transferred to the legs, which are then lifted up and over her head. Finally, by ensuring that her body has a small moment of inertia, (i.e., is piked if the legs are straight or tightly tucked if they are not), the gymnast speeds the passage of her center of gravity over her hands. Once her center of gravity has passed beyond her hands, she extends her arms and lifts her head to arrive in the required finishing position.

Fig. 140. Backward roll

Cartwheel

The cartwheel (Fig. 141) is another elementary movement, generally taught fairly early in any sequence of gymnastic instruction. From a standing position with her arms stretched upward and her body facing the direction in which she will travel, the gymnast takes a step forward and shifts her weight forward over her front foot. She then rotates her body sideways, swings her rear leg upward, and lowers her trunk until the hand on the same side as the front foot has been brought to floor level. Then, with this hand turned and placed on the floor, she pushes off strongly with her front foot. The reaction to this push substantially increases the angular momentum of the body and assists in lifting it into and beyond the vertical side-handstand position. During her descent from this position the gymnast supplements the angular momentum that she already has by pushing against the ground with each hand as it passes below and behind her center of gravity. The movement is completed with the gymnast in a side-facing astride position.

Fig. 141. Cartwheel

Round off, Arab Spring

A round off is a cartwheel in which the gymnast brings both legs together as he passes through the side-handstand position and then, with his moment of inertia reduced to a near-minimum, executes a quarter turn about the long axis of his body before snapping his legs down toward the floor. The quarter turn he makes as he places his hands on the floor together with the quarter turn he executes as his feet pass overhead result in his landing facing back in the direction from which he came. And herein lies the importance of this movement—it is a simple and yet effective way for the gymnast to change from forward-rotating to backward-rotating movements while moving in one direction along a straight line.

Neckspring, Neck Kip (F.I.G.), Backspring, Mat Kip, Snap up

The neckspring is one of a whole family of gymnastic movements known as *kips* or *upstarts*. These movements are used to lift the gymnast's body and to rotate it forward from a horizontal position to a vertical (or near-vertical) one. To execute a neckspring the gymnast moves into a back-lying position, with his legs straight and raised overhead so that his feet point back behind him. His weight is supported on his shoulders, and his palms are placed on the floor behind them. From this position the gymnast whips his legs strongly in a forward and upward direction. Shortly after they have passed beyond the vertical position, their motion is abruptly decreased and angular momentum is thus transferred to the rest of the body. As the gymnast's upper body is being lifted from the floor by this transfer of momentum, he forcefully extends his neck and arms and drives his body forward and upward into the air. If the reaction of the floor to these forces passes behind his center of gravity, it will not only lift him forward and upward but will also tend to increase his forward angular momentum at the same time. Once the gymnast becomes airborne, he maintains an arched-body position that keeps his moment of inertia as small as possible, without violating the tenets of good form. Then, when his feet touch the floor and if he has had sufficient lift and angular momentum at the instant of takeoff, his body will rotate upward and over his feet to bring him into a standing position.

The angle of the legs to the horizontal at the moment their angular motion is arrested has been shown by Spencer[4] to be a critical factor in determining success in performing a neckspring. In successful trials he found that the mean angle of "leg thrust" was 52° and that this was significantly different from the mean angle of 64° for unsuccessful trials (i.e., those in which the gymnast was unable to complete the movement by coming forward to a position "on or beyond a balance on the feet"). Thus it would appear that those who fail to successfully complete this movement tend to stop (or slow) the angular motion of their legs prematurely.

The fairly common practice of teaching the whipping action of the legs by having the gymnast place the palms of his hands on his thighs so that he can then push them through the desired movement is deserving of comment. To perform a neckspring a gymnast needs an appropriate blending of lift and angular momentum. If he sacrifices one to improve upon the other, his prospects of achieving this appropriate blending are likely to be reduced. In the present instance, the pushing of the palms against the thighs may lead to an increase in the angular momentum acquired by his legs. However, because this means that the muscles of the neck must now supply virtually all the force to project him forward and upward, this gain in angular momentum is

[4]Richard R. Spencer, "Ballistics in the Mat Kip," *Research Quarterly*, XXXIV, May 1963, pp. 213–18.

obtained only at the expense of a marked decrease in lift. As a result, necksprings performed in this way almost invariably fail to carry the gymnast forward over his feet unless he assumes a tucked position during the latter stages of his flight. In effect, therefore, the gymnast is merely exchanging one problem for another.

Handspring

Following a short run and a hop, which brings him into the appropriate position, the gymnast performs a handspring by rotating through a handstand into the air, and over to land in a standing position. The hopping movement that leads into the handspring is made with the body inclined forward and the nonhopping leg extended behind. Then, when the hopping foot lands slightly behind the gymnast's center of gravity, the body is in a position to enable it to rotate forward without delay. As the body begins this forward rotation, the gymnast brings his nonhopping foot forward and places it on the floor in front of the other one. A simultaneous lowering of his arms and trunk, together with a thrust from his rear leg, moves his center of gravity forward, over, and beyond his front foot. The angular momentum that has already developed is added to by the strong upward swing of his rear leg and by the moment of his weight about a horizontal axis through the ankle of his front leg. As the gymnast reaches forward and places his hands on the floor, a forceful extension of his front leg substantially increases his angular momentum and his body is carried upward toward the handstand position. To ensure that the moment of the reaction of this leg drive (about an axis through his wrists) is as large as is practicable, the gymnast reaches his hands well forward as he places them on the floor. (A side view of a well-executed handspring reveals that hip, shoulder, and wrist joints are in an approximately straight line at this moment.) Assuming that the gymnast has now acquired all the angular momentum he is likely to need, his next task is to develop the necessary lift to project him upward into the airborne phase of his handspring. To do this, he contracts the appropriate muscles of his arms and shoulders and thrusts forcefully downward against the floor, as his center of gravity passes forward and over his hands. (A study[5] of the muscular action of a skilled performer has verified that a number of muscles that produce this kind of effect are particularly active at this time.) The reaction to this downward thrust tends to lift the gymnast's body into the air. The factors governing the final phases of the movement (i.e., from the moment the gymnast becomes airborne until he completes the handspring) are identical to those outlined relative to the neckspring.

[5]M. Hebbelinck and J. Borms, "Cinematographic and Electromyographic Study of the Front Handspring," in *Biomechanics: Technique of Drawings of Movement and Movement Analysis*, ed. by J. Wartenweiler, E. Jokl, and M. Hebbelinck (Basel, Switzerland: S. Karger, 1968).

Forward Somersault, Salto Forward (F.I.G.)

The forward somersault, a tightly tucked forward roll performed while the gymnast is airborne, is normally preceded by a short run and a low jump to bring both feet together ready for takeoff. When his feet contact the floor at the end of this low jump, the gymnast cushions the shock of their landing by bending slightly at hip, knee, and ankle joints. This action also places these joints in positions from which, moments later, they can be vigorously extended to drive the body upward into the air. The reaction to this forceful extension of the legs acts in an upward and backward direction and passes behind the gymnast's center of gravity. Thus, the reaction accounts not only for the lift that the gymnast obtains but also for the greater proportion of his subsequent angular momentum. The latter may be added to by a forward and downward movement of the arms from a position overhead or by an upward and backward swing of the elbows from a position in front of the body. Once the gymnast leaves the ground, he quickly moves into a tucked position, thereby decreasing his moment of inertia and speeding his rotation. Then, toward the end of his flight (and in accord with the visual cues he obtains by looking down in front of his body for the floor), he comes out of his tuck, slows his rotation, and prepares to land in an erect standing position.

Flip-Flop, Flic-Flac (F.I.G.), Back Handspring, Backflip, Flip-Flap

Basically a backward handspring from a two-footed takeoff, a flip-flop can be performed either from a standing position or following a roundoff.

For a standing flip-flop, the gymnast adopts an erect position with arms by his sides (or extended behind him) and his feet slightly apart. Then, with his trunk still kept fairly erect, he bends at hip and knee joints and lowers his buttocks downward and backward [Fig. 142(a)–(b)]. This places his center of gravity beyond the backward limit of his base. With his body thus beginning to rotate backward, he swings his arms vigorously in a downward, forward and upward direction [Fig. 142(a)–(f)] and drives hard against the floor by extending his hip, knee, and ankle joints [Fig. 142(d)–(g)]. The reaction to the swing of the arms and to the drive of the legs projects the gymnast upward and backward into the air and, since its line of action passes behind his center of gravity (relative to the direction in which he is moving), imparts to him the required backward angular momentum. Once airborne, the gymnast maintains his body in as straight a position as possible consistent with its angular momentum and the length of time he will have in the air. In this respect a markedly arched back is generally regarded as an indication that the gymnast's takeoff was deficient and that as a result he is having to compensate (by decreasing his moment of inertia) in order to successfully complete the movement. When the gymnast's hands reach the floor, his angular momentum should be sufficiently large to allow his body to pass through a hand-

Fig. 142. Flip-Flop. The ground reaction forces exerted on the gymnast are indicated by the superimposed vectors (*Based on data in Payne and Barker, "Comparison of the Take-Off Forces in the Flic-Flac and the Back Somersault." Photographs courtesy of Howard Payne*).

stand position and (aided by the same kind of upthrust used in a handspring) into the air to an erect standing position. If the gymnast does not have sufficient angular momentum at this time, he will almost instinctively bend his arms and legs (decreasing his moment of inertia about an axis through his wrists) in an attempt to rotate safely over his hands to land on his feet. Since this is an unattractive way to complete the movement, such last-minute adjustments can incur penalties in competition.

In a roundoff flip-flop the gymnast already possesses a certain amount of backward angular momentum when his feet land at the end of the roundoff. Thus, provided this is not entirely dissipated, the need for angular momentum to be generated during the takeoff is generally not so great as it is for a standing flip-flop. As a consequence, a gymnast will normally use a much less vigorous and extensive arm swing in a roundoff flip-flop than he would in a standing one.

Backward Somersault, Salto Backward (F.I.G.)

This movement requires actions at takeoff that are somewhat similar to those required for a flip-flop and actions in the air that are akin to those for a forward somersault.

The actions at takeoff in a well-executed standing backward somersault and the ground-reaction forces associated with them are shown in Fig. 143.

Aside from the obvious difference in direction, the in-the-air actions required for a backward somersault differ from those used in a forward somersault in one major respect, the position of the head during the flight. Because of its position relative to the rest of his body, the gymnast performing a backward somersault can see the floor much earlier in his flight than when performing a forward somersault. (In the latter, his body obscures his view of the floor until relatively late in the flight.) As a direct result of this difference between the two movements, it is much easier for the gymnast to achieve the correct timing of his movements as he prepares to land from a backward somersault than from a forward one.

Some confusion exists concerning the role of the head in initiating the backward angular momentum required to complete the somersault. Some coaches tell the gymnast to drive vertically upward at takeoff and then, as he reaches the peak of his flight, to throw his head back to start the backward somersaulting action. While this advice often produces the desired results, it is quite incorrect from a mechanical standpoint. If the gymnast did project himself vertically into the air with zero angular momentum (as implied in such instructions), the action of throwing his head back would produce a contrary reaction in the rest of his body. In other words, the rest of his body would start a forward somersaulting motion—albeit a very small one, considering the moments of inertia of the bodies involved.

The reasons why this advice often produces the required results are not difficult to find. First, if the gymnast keeps his head in an erect position throughout the takeoff he is likely to obtain a greater vertical velocity than if he allows it to start dropping back in anticipation of the somersault to follow. This greater vertical velocity at takeoff ensures that he has correspondingly more time in the air and thus more time for his angular momentum to take effect. Second, the sharp throwing-back of the head at the peak of the flight coincides with a quick tucking action in which the knees are brought to the chest and the hands are brought down to meet them. These actions simultaneously decrease the gymnast's moment of inertia and increase his angular velocity relative to the transverse axis through his center of gravity. Since the throwing-back of the head and the sharp increase

Fig. 143. Backward somersault. The ground reaction forces exerted on the gymnast are indicated by the superimposed vectors (*Based on data in Payne and Barker, "Comparison of the Take-Off Forces in the Flic-Flac and the Back Somersault." Photographs courtesy of Howard Payne*).

in the body's angular velocity occur at the same time, it has generally been assumed that the former is the cause of the latter. This is incorrect. The head is thrown back so that the gymnast can see where he is going; his angular velocity increases because his moment of inertia has been decreased; and the fact that these two actions occur simultaneously has little, if any, significance. Coaches who wish to give advice that is both helpful and correct should therefore omit any reference to the head action causing or initiating the required backward rotation.

In teaching the flip-flop and the backward somersault, coaches generally give quite different instructions for the two movements—e.g., "Fall backward, leaving the feet behind while reaching and looking for the floor" and "Strive for height jumping forward and upward before throwing the head back and tucking," respectively. The differences in takeoff actions were the subject of a recent study by Payne and Barker.[6] In this study, four skilled gymnasts performed each of the two movements from a force-platform—a device that recorded the forces exerted against it by the feet of the gymnasts. After examining force-time records and motion-picture films for the best trials by each subject, Payne and Barker concluded that the backward lean of the body is much more pronounced in the flip-flop (about 48° to the horizontal as the feet lose contact) than in the backward somersault (70°); that the more upright body position in the somersault and the "harder drive" associated with it yield additional vertical force; and that the time of flight for the somersault is almost twice that for the flip-flop. On the basis of these several findings they suggested that "the usual instructions from the coach in teaching these movements have a sound mechanical basis."

LONG HORSE VAULT

The long horse vault differs somewhat from the other Olympic gymnastic events in that it consists of one short-duration movement ($1\frac{1}{2}$ to 2 sec from takeoff to landing) rather than a linked series of movements executed over a comparatively extended period (e.g., 50–70 sec for the floor exercises).

Basic Considerations

The vault, performed over a horse 5 ft 3 in. long and 4 ft 5 in. high, is judged on four basic criteria—difficulty (if the vault is optional) or interpretation (if compulsory), where the hands are placed relative to specific zones painted on the horse, the flight onto and off the horse (*preflight* and *flight*, respectively), and form in the execution of the vault.

For the purpose of analysis, it is convenient to consider a vault as being composed of seven consecutive parts or phases: (1) run-up, (2) hurdle step, (3) takeoff, (4) preflight, (5) support, (6) flight, and (7) landing.

Run-Up

The purpose of the run-up is to get the gymnast to the optimum point for the takeoff into his hurdle step, with as much horizontal velocity as is possible consistent with the movements he must subsequently perform. There are, therefore, two quite distinct requirements—speed and accuracy.

With respect to the first of these, the factors to be considered are those that govern the speed of any runner, regardless of the sport in which he participates—that is, his rate of striding (or stride frequency) and his stride

[6]A. H. Payne and P. Barker, "Comparison of the Take-Off Forces in the Flic Flac and the Back Somersault in Gymnastics" in *Biomechanics V-B*, ed. by P. V. Komi (Baltimore: University Park Press, 1976), pp. 314–21.

length. Any improvements in this facet of a gymnast's performance can come about only by virtue of an improvement in one or both of these factors. He must either take more strides per second or cover a greater distance with each one. (For a more complete discussion of the factors governing speed in running, refer to Chap. 15.)

Accuracy is primarily a function of intelligent practice. The wise gymnast makes use of the same methods for standardizing his run-up as do competitors in the jumping events on track and field programs. Paramount among these is the careful measurement of the length of the run-up—a practice by no means as widespread as it should be. It might be noted by way of comparison that because of the restriction that the rules impose on the length of the gymnast's run-up ("it must not be longer than 20 meters [64 ft 7¼ in.] including the springboard"), it should be possible for gymnasts to strike the correct spot for their takeoff into the hurdle step with more consistency than, say, long jumpers or pole-vaulters can strike their optimum takeoff spot, for these latter customarily use run-ups 2 to 2½ times as long as the gymnast is permitted.

Hurdle Step

The hurdle step is the transitional phase between the run-up and the takeoff. Its purpose is to enable the gymnast to adjust the body position he uses in sprinting, with as little loss in speed as possible, to one from which he can best initiate his takeoff.

To execute the hurdle step the gymnast lowers both arms during the last step before the hurdle. Then, as his center of gravity moves over the foot from which he will spring into his hurdle step, he sweeps his arms and free leg in a forward and upward direction. Coordinated with the reaction to the drive from his other leg, these movements lift him into the air and impart backward angular momentum to his body. This angular momentum rotates his body into the backward inclined position he needs as he lands on the board.

During the flight phase of the hurdle step, the gymnast brings both feet together and his hands down behind him in preparation for the landing on the board and the subsequent vigorous upward drive into the takeoff.

Takeoff

In this, the most critical phase of his vault, the gymnast's objective is to obtain the velocity and height of takeoff and the amount of angular momentum that is optimum for the vault in question. Of these three quantities, the first and last are the most important. The second, the height of the gymnast's center of gravity at the instant of takeoff, is not amenable to very pronounced change.

Provided he has his arms extended forward and upward and his hip, knee, and ankle joints extended at the moment he leaves the board, there is little else that he can reasonably do to improve his performance from this standpoint.

The magnitude of a gymnast's velocity at the instant of takeoff is governed by the velocity of his run and the changes in this velocity produced during the takeoff phase. These latter generally take the form of a sharp increase in the vertical velocity and a decrease in both horizontal and resultant velocities. The following figures, obtained from a motion-picture analysis of a champion gymnast performing a hecht vault,[7] might be regarded as fairly typical.

	Instant of Takeoff for Hurdle Step (fps)	*Instant of Takeoff from Board* (fps)
Horizontal velocity	21.9	16.3
Vertical velocity	4.6	11.1
Resultant velocity	22.4	19.7

Of particular importance in these changes is the pronounced increase in the vertical velocity. This is brought about mainly by the reaction to the forceful extension of the gymnast's hip, knee, and ankle joints and the vigorous forward and upward swing of his arms.

While there may be a few cases in which the gymnast imparts a backward angular momentum to his body during the takeoff, for the vast majority of vaults he needs angular momentum in the opposite direction. For example, all 45 vaults listed in the *F.I.G. Code of Points*[8] require the gymnast to take off with a certain amount of forward angular momentum. To acquire this forward angular momentum, the gymnast ensures that the reaction forces that the board exerts upon him in the latter stages of the takeoff pass at some distance behind his center of gravity. To ensure that this happens, he will often flex at the hips so that his center of gravity moves forward of the line of action of the reaction force—a line approximating that of his extended legs.

Preflight

The motion of the gymnast during this phase of the vault is governed by his velocity, height, and angular momentum at the instant of takeoff and by his body position in the air. Since the tenets of good form set fairly narrow limits on what this latter should be, those factors associated with the takeoff afford the only means of effecting substantial changes in the preflight.

[7]Edward A. Gombos, "Gym Forum," *Modern Gymnast*, IV, No. 7, September–October 1962, pp. 23–25.

[8]*F.I.G. Code of Points*, 1975, pp. 36–42.

Support

During the support phase the gymnast modifies his angular motion and acquires lift for his flight off the horse. The modifications in his angular motion are dictated by the requirements of the particular vault. If he is required to continue rotating forward once his hands leave the horse, his concern is with maintaining and perhaps adding to the forward angular momentum that he already possesses. On the other hand, if the latter phases of the vault require him to rotate backward, he must reverse the direction of his angular momentum during the support phase. In the first case, where the direction of rotation is the same throughout the vault, the inertia of the gymnast's body tends to carry his center of gravity forward over his hands, once these land on the horse. The forward rotation involved in this is enhanced if the reaction from the horse acts along a line passing behind the gymnast's center of gravity. As the center of gravity passes forward, over, and beyond the hands, a similar muscular action to that used in a handspring on the floor evokes from the horse a reaction that both lifts the gymnast forward and upward into the air, and if it acts eccentrically, increases his forward angular momentum. In those cases where the direction of rotation must be reversed during the support phase, the gymnast contracts the extensors of his shoulder joints and thrusts backward against the horse via his hands. The resultant of the reaction to this backward thrust and the vertical reaction from the horse passes forward of the gymnast's center of gravity. Acting in this way, it first causes his angular momentum to be reduced to zero and then to be increased in the opposite direction. Just before the gymnast's hands leave the horse, and while his center of gravity is still some distance behind them, he thrusts down hard against the horse. As in the previous, similar case, the reaction to this thrust projects him upward and forward into the air and provides additional angular momentum.

Flight

This phase of the vault is very similar to the preflight phase in that the same factors of velocity, height, and angular momentum at takeoff and body position in the air govern what happens to the gymnast. The only major difference lies in the fact that many vaults involve substantial changes in the gymnast's body position during the flight, while relatively few require comparable changes during the preflight. Consequently, adjustments in the body position of the gymnast can be used to control the flight to a much greater extent than is normally possible during the preflight.

Landing

Aside from the greater velocities that are involved, and the fact that there is no question of preparing for subsequent movements, the important factors in the landing phase of a vault are exactly the same as those involved in landing in floor exercises.

Techniques[9]

Straddle, Long Astride Vault, Long Fly

Generally the first vault that a gymnast learns to perform over the long horse, the straddle (Fig. 144) is one of a small group of vaults (including squat, stoop, and hecht vaults) requiring forward rotation during the preflight and backward rotation during the flight.

After rotating forward to place his hands on the horse, the gymnast pushes forcefully downward and backward, to project his body into the air and to give it the required backward rotation. As his legs begin their rotation downward, the gymnast separates (or straddles) them so that his feet pass on either side of (and, hopefully, above) the horse. Then, as he nears the end of his flight, he brings his legs together in preparation for the landing. Because this latter action is a symmetrical one, the angular action of one leg is effectively canceled by the equal and opposite action of the other, and the motion of the rest of his body is unaffected.

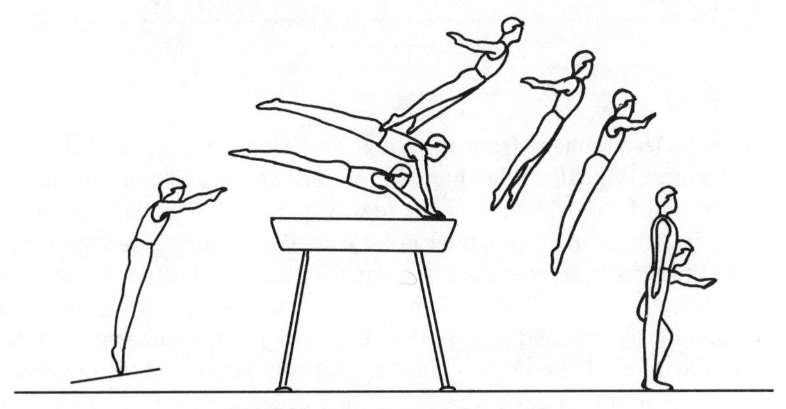

Fig. 144. Long horse vault: straddle

Squat, Through Vault

The squat (Fig. 145) is very similar in many respects to the straddle. It differs mainly in that once the legs begin their rotation downward and forward after the hands touch the horse, the gymnast flexes his hip and knee joints so that his legs can pass forward and, seemingly, between his arms. Because his moment of inertia is less at this time than during the corresponding part of a

[9]For the vaults described here it is assumed that the gymnast places his hands on the far end of the horse.

straddle vault, he needs to impart slightly less angular momentum to his body in a squat vault than he does in a straddle—that is, to obtain the same angular displacement by the time he lands on the mat. Once his feet have passed forward over the end of the horse (by which time his hands have thrust downward and backward and then broken contact with the horse), the gymnast fully extends his body. He then retains this extended position until his feet have rotated underneath him to a position somewhat forward of his center of gravity, at which time he is ready to land.

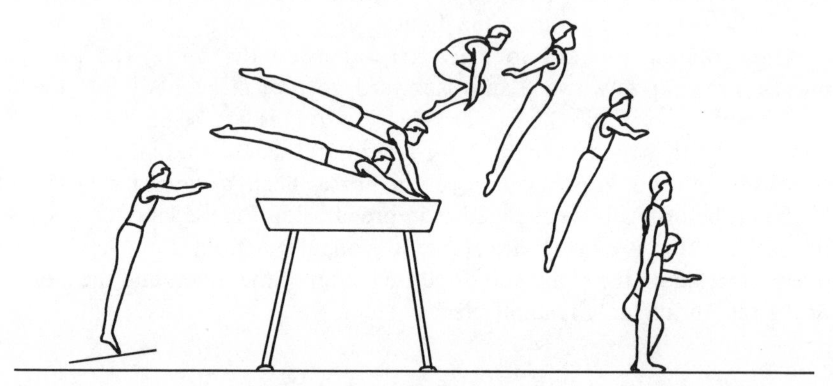

Fig. 145. Long horse vault: squat

Hecht

A logical development from the squat and stoop vaults, the hecht (Fig. 146) is the most difficult of the three. In the performance of a hecht vault, the downward and forward swing of the legs, which begins during the support phase, is executed with the whole body extended. Because his moment of inertia is thus much greater than in a squat (where his body is tucked) or a stoop (in which it is piked), the gymnast needs either to acquire more angular momentum or, by increasing his time of flight, to give the same amount more time to take effect. If he does acquire a relatively large angular momentum during the support phase, he runs the danger of having his feet rotate downward into the horse and ruin the vault. In practice, therefore, the second alternative receives the greater attention. In order to increase his time of flight the gymnast adjusts his angle of takeoff from the board so that his preflight is relatively low and his center of gravity is still moving forward and upward at the moment his hands touch the horse. Because the reaction to the forceful downward and backward thrust that he then applies to the horse does not have to markedly change the direction in which he is moving, the velocity with which he leaves the horse is much greater than it would be if his center of gravity had been moving downward as his hands landed. This greater velocity, and specifically its greater vertical component, leads to an increase in the time of his flight as compared with that in squat or stoop vaults.

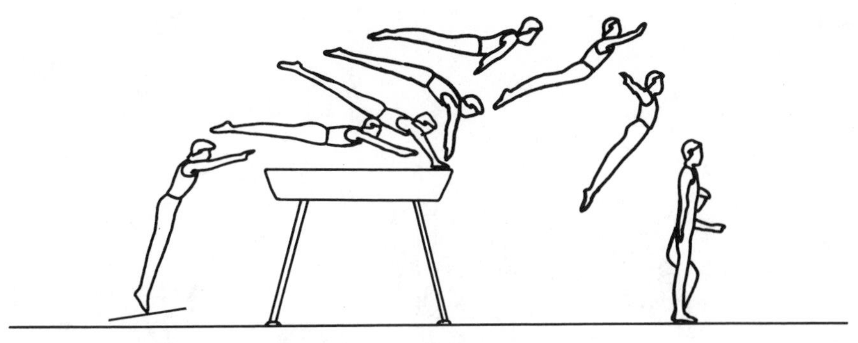

Fig. 146. Long horse vault: hecht

Hecht—Forward Somersault

At the very highest levels of competition the hecht is performed with a turn or a full turn during the flight and with the hands placed on either the near or the far end of the horse. In addition, the *F.I.G. Code of Points*[10] lists, among the most difficult of vaults, a hecht followed by a forward somersault in a tucked position. A close examination of Fig. 147—a redrawing of the diagram that accompanies the listing in the *F.I.G. Code of Points*—reveals that to perform this vault in the approved manner a gymnast must rotate backward as his hands leave the horse (cf., positions 3 and 4) and then initiate a tucked *forward* somersault. Since this implies a change in the direction of rotation during the flight (a physical impossibility in the absence of an external torque, p. 148), it seems unlikely that this vault will ever be executed in the manner prescribed! A vault that might well be substituted for the "impossible" hecht forward somersault and one that is not currently listed by the F.I.G. is the hecht followed by a tucked *backward* somersault. Because this vault follows the pattern of other hecht vaults in requiring forward rotation during the preflight and backward rotation during the flight and because it would appear to be within the physical capabilities of a highly skilled gymnast, it would seem that a strong case could be made for its inclusion in future F.I.G. lists.

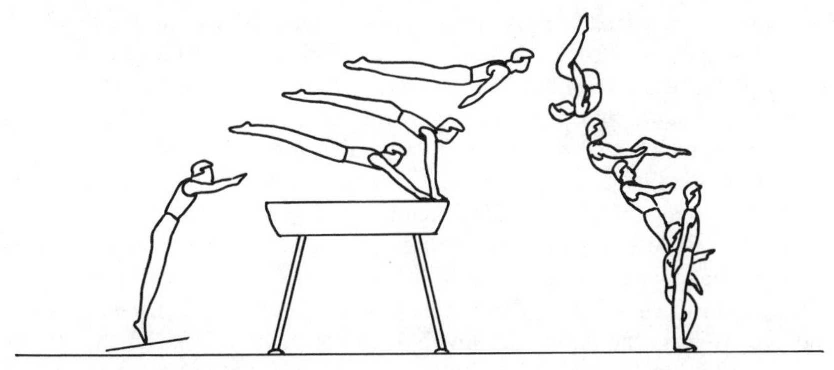

Fig. 147. Long horse vault: hecht followed by a forward somersault

[10]*F.I.G. Code of Points*, 1975, p. 41.

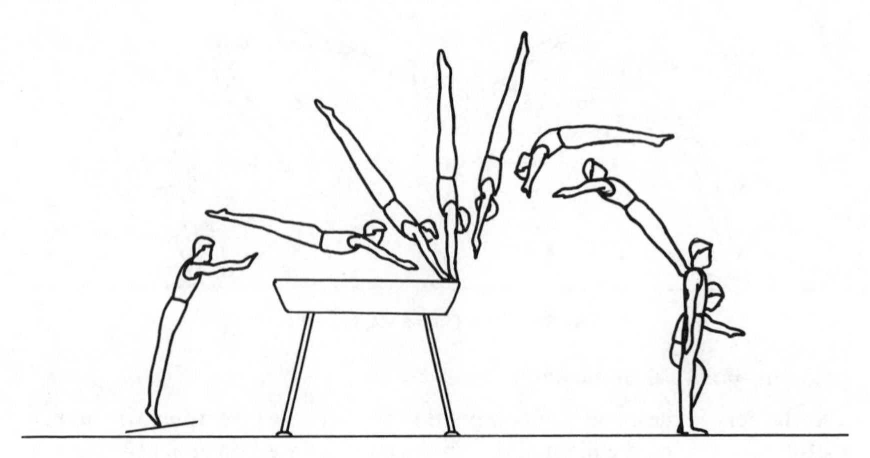

Fig. 148. Long horse vault: handspring

Handspring

The handspring (Fig. 148) is probably the simplest of the group of vaults that require forward rotation during both the preflight and the flight.

In performing a handspring, a gymnast tries to minimize the angular distance through which his inertia must rotate him once his hands land on the horse. Thus he endeavors to have his body inclined at a relatively large angle to the horse at this instant. Then, once his center of gravity has passed forward of his hands, he drives downward against the horse, thereby evoking the reaction required to lift him into the air and increase his angular momentum. Throughout the flight phase, he endeavors to maintain his body in a fully extended position. Excessive arching of the back or other deviations from a straight body position are generally indicative of too little angular momentum and/or too low a flight. This latter, of course, leads to less time in the air than is necessary to complete the vault in the approved fashion.

Among the advanced variations of the handspring vault is that in which the gymnast is required to pass through a piked position during his flight off the horse—the so-called Yamashita vault. In effect, this vault requires that instead of executing the final "half somersault" (from the handstand on the horse to erect standing on the mat) in a straight body position, he does so in a position in which his body has a much smaller moment of inertia. Because his body rotates faster in such a position that it does when fully extended, and because the required angular displacement is approximately the same, the gymnast needs less angular momentum to perform a Yamashita than to perform a handspring. While in itself this suggests that the Yamashita is less difficult than the handspring, empirical evidence indicates that the reverse is true. Apparently, the difficulties involved in correctly timing the changes in body position are of such magnitude as to more than offset the effects of differences in the amount of angular momentum required.

HORIZONTAL BAR

Apart from the vaulting horse, the horizontal (or high) bar is generally the first piece of apparatus to which beginning gymnasts are introduced. It is on this that they learn the basic swinging movements that are of such importance not only on the horizontal bar but on the rings and parallel bars as well.

Basic Considerations

The rules governing competitive work on the horizontal bar state that "exercises must consist exclusively of swinging without stops."[11] Thus, any consideration of the basic factors underlying horizontal bar exercises must be concerned with those factors applicable to swinging movements.

When a gymnast swings on a horizontal bar, he is acted upon by three forces and a couple (Fig. 149):

W, his weight acting vertically downward through his center of gravity.

R, a reaction force exerted by the bar against his hands. This force is here considered to consist of two components—a normal (or centripetal) component (R_N), which acts along a line from the center of gravity of the gymnast to the center of the bar, and a tangential component (R_T).

A, a small, and generally insignificant, air-resistance force.

M, the resultant moment of all the small frictional forces (or couples) exerted on his hands as they slide over the surface of the bar.

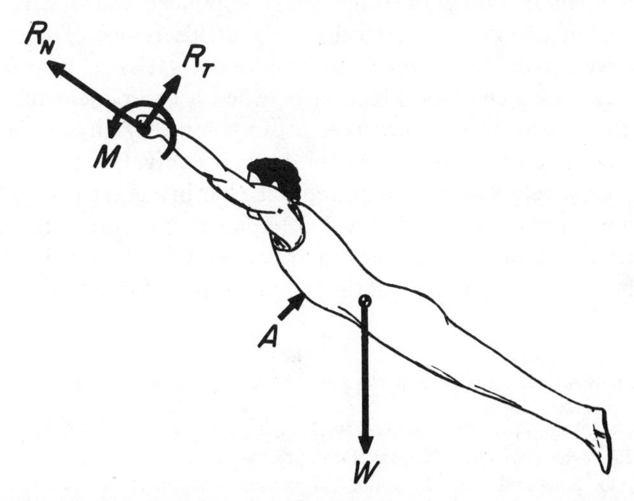

Fig. 149. A gymnast performing a horizontal bar exercise is acted upon by three forces (*W*, *R*, and *A*) and a couple (*M*).

[11]*F.I.G. Code of Points*, 1975, p. 18.

Weight

While the magnitude and direction of this force are constant, its line of action changes continually as the gymnast swings. These changes are reflected in the moment of the gymnast's weight about an axis through the bar. When the line of action passes through the bar (i.e., the gymnast's center of gravity is directly above or below the axis), the moment is zero. Alternatively, when the line of action is displaced horizontally away from the bar as far as it can be without the gymnast releasing his grip, the moment of the weight is as large as he can make it. This latter condition is obtained when the gymnast's body is fully extended in a horizontal position (e.g., as in the downswing for a forward or backward giant swing). For other positions of the center of gravity, the moment of the weight is intermediate between these maximum and minimum values.

The direction of the moment of the weight also changes during the swing. When the gymnast's body is swinging downward, the moment of his weight is in the same direction as that in which he is moving. Then, as he swings beneath the bar and up the other side, the moment decreases to zero and then increases in the opposite direction. If his swing continues so that his center of gravity passes over the bar, the direction of the moment of his weight is again reversed. Thus, the gymnast's weight serves to accelerate his downward motion and to retard his upward motion.

Centripetal Component

The principal function of the centripetal component is to repeatedly change the direction of the gymnast's motion so that his center of gravity moves along a curved path. The centripetal component varies in magnitude from zero (when a swing changes direction or when a component of the weight provides all the centripetal force needed) to some very large values indeed. Kunzle[12] has stated that the centripetal force is equivalent to four times the gymnast's body weight as he swings under the bar in a giant swing; Cureton[13] has reported obtaining a higher value (equal to five times the gymnast's weight) during the same movement; and Sale and Judd[14] have reported still higher values (in the range 4.8–5.4 times body weight) for a similar exercise on the rings.

[12] G. C. Kunzle, *Olympic Gymnastics: Horizontal Bar* (London: James Barrie Books, Ltd., 1957), p. 142.

[13] Thomas K. Cureton, "Elementary Principles and Techniques of Cinematographic Analysis," *Research Quarterly*, X, May 1939, pp. 15–17.

[14] Digby G. Sale and Ross L. Judd, "Dynamometric Instrumentation of the Rings for Analysis of Gymnastic Movements," *Medicine and Science in Sports*, VI, Fall 1974, pp. 209–16.

Tangential Component

The tangential component of the bar reaction force is an eccentric force that serves to accelerate the gymnast in the direction in which it acts and to angularly accelerate him about an axis (most frequently a transverse axis) through his center of gravity. The magnitude of the tangential component is especially important during the final phases of circling and kipping movements that end with the gymnast on or above the bar.

Moment

When a gymnast swings on a horizontal bar, his hands tend to rotate about it in accord with the rest of his body. This tendency is opposed by frictional forces that the bar exerts on the gymnast's hands. If the gymnast is swinging forward with what is commonly referred to as an *overgrasp* of the bar (forearms pronated, thumbs between the hands), these frictional forces are exerted along the curved length of his fingers toward the fingertips [Fig. 150(a)]. Thus, they have the effect of elongating the fingers slightly and of wrapping them around the bar. For this reason such a grip is generally regarded as a fairly secure one when the gymnast is swinging forward. When he reverses the direction of his swing, however, the effects are similarly reversed and the friction forces tend to loosen or break his grip. Thus swinging "against the grip" in this way calls for considerable caution.

To partially overcome this problem, the gymnast adjusts his grip at carefully chosen moments during the course of his exercise. The optimum moments are those at which the bar reaction force and the gymnast's weight

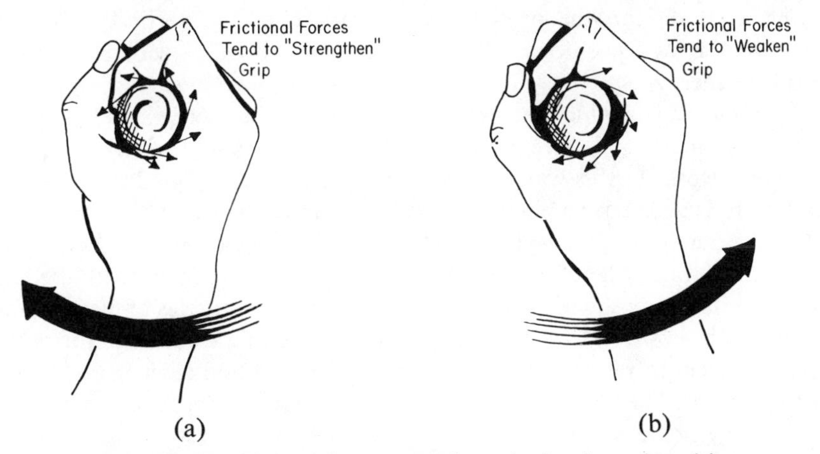

Fig. 150. The frictional forces exerted on the hands tend to (a) strengthen the gymnast's grip when he swings forward and (b) weaken it when he swings backward.

act so that the pressure on the hands is minimal. As a specific example, consider a gymnast who reaches the peak of a backward swing with his body horizontal. Because his angular velocity is zero at this instant, the centripetal force applied to him via his hands is also zero (cf. Eq. 60). In addition, since his weight is acting downward at right angles to the line of his arms, little (or none) of this force is transmitted via his hands to the bar. Thus there is a relative absence of pressure on his hands at this instant, which is, therefore, an opportune moment to adjust their position to obtain a more secure grip.

The body position just described is a desirable starting point in the discussion of another important concept in horizontal bar exercises. At the instant mentioned (i.e., the moment the peak of a backswing is reached) the gymnast has a certain amount of potential energy. Then, as he swings down, some of this potential energy is transformed into kinetic energy. However, because of the friction between his hands and the bar (and, to a much lesser extent, because of the air resistance), not all the potential energy that is lost is converted into kinetic form. Some of it is converted into other nonmechanical forms, of which probably the most important is the heat generated as the hands move about the bar. (Incidentally, this conversion of mechanical energy to heat energy is reflected in the high incidence of blisters—"blisters are friction burns"[15]—among horizontal bar exponents.) The net result of this process is that the gymnast passes through the low point of his swing with less mechanical energy than he had at the outset. Then, as he moves upward in the forward part of the swing, the conversion from kinetic energy back to potential energy is accompanied by similar losses. As a direct consequence of these losses the gymnast's center of gravity does not rise as far on the forward swing as it did on the preceding backward one—the vertical difference between the two positions reflecting the losses to other (nonmechanical) forms of energy.

The only way in which the gymnast can offset these energy losses—and, of course, he must if his swing is not to be "damped out" entirely—is to do muscular work. In this way, the losses in mechanical energy due to friction and air resistance can be balanced, or more than balanced, by the gains from the conversion of chemical energy (involved in muscular contraction) to mechanical form. Therefore, as the gymnast swings forward beneath the bar, he arches his body slightly. This puts the appropriate muscles of his hips and shoulders in position so that as he starts to swing upward, he can forcefully contract them to raise his legs and to press forward and slightly downward with his arms. These actions have two effects. First, they produce an increase in the body's potential energy by lifting it a little higher; second, they increase its kinetic energy (and its angular velocity) by decreasing the moment of inertia. Correctly executed, these movements therefore serve to maintain

[15]Joe Brown and Perry Childers, "Blister Prevention: An Experimental Method," *Research Quarterly*, XXXVII, May 1966, p. 187.

(or increase) the body's mechanical energy. And this is, of course, reflected in the range and speed of the swing.

These two characteristics of the swing are also influenced by the gymnast's body position during the downswing. Apart from losses due to friction and air resistance, the kinetic energy possessed by a gymnast as he passes beneath the bar is equal to the work done on him by gravity during the downswing.[16] The work done by gravity depends on the moment of the gymnast's weight and the angle through which the body is moved (i.e., the left-hand side of the equation below. Because the angle is usually determined by other factors (and, in particular, the height of the previous swing), the only way in which the gymnast can exert any measure of control over the kinetic energy he obtains is by adjusting his body position and thereby altering the moment of his weight. Therefore, if the gymnast wants a large amount of kinetic energy as he passes beneath the bar—and, in general, this would be the case—he endeavors to keep his center of gravity as far from the bar as he can during the downswing.

Air Resistance

The magnitude of this force is generally so small that its effects can be ignored.

Techniques

Underswing, Cast

An underswing (Fig. 151) is the usual method of initiating a large swing at the start of an exercise. From a standing position slightly behind the line of the bar, the gymnast springs forward and upward to take his grip. Because his center of gravity is behind the bar as he takes hold of it, his body swings gently forward. Then, as he approaches the end of his subsequent backward swing, the gymnast pulls up on his arms, brings his chest close to the bar, and simultaneously forces his body into an arched position. These adjustments in body position increase the extent of his backward swing and put his hip and trunk flexor muscles into position for the next phase of the movement. As the gymnast starts to move forward from this position, he swings his legs forward and upward to bring them close to his hands. At the same time he allows his arms to straighten and his chest to fall away from the bar. Then, as his body approaches the forward limit of its swing, he extends his hip joints and thrusts his legs upward and outward. This action, accompanied by a forceful

[16]The angular analogue of the work-energy relationship can be expressed in algebraic form as

$$T\theta = \tfrac{1}{2}I\omega^2 + mgh$$

where T = torque, θ = the angle through which this torque is applied, I = the moment of inertia, ω = the angular velocity, m = the mass, and h = the height of the center of gravity of the body involved.

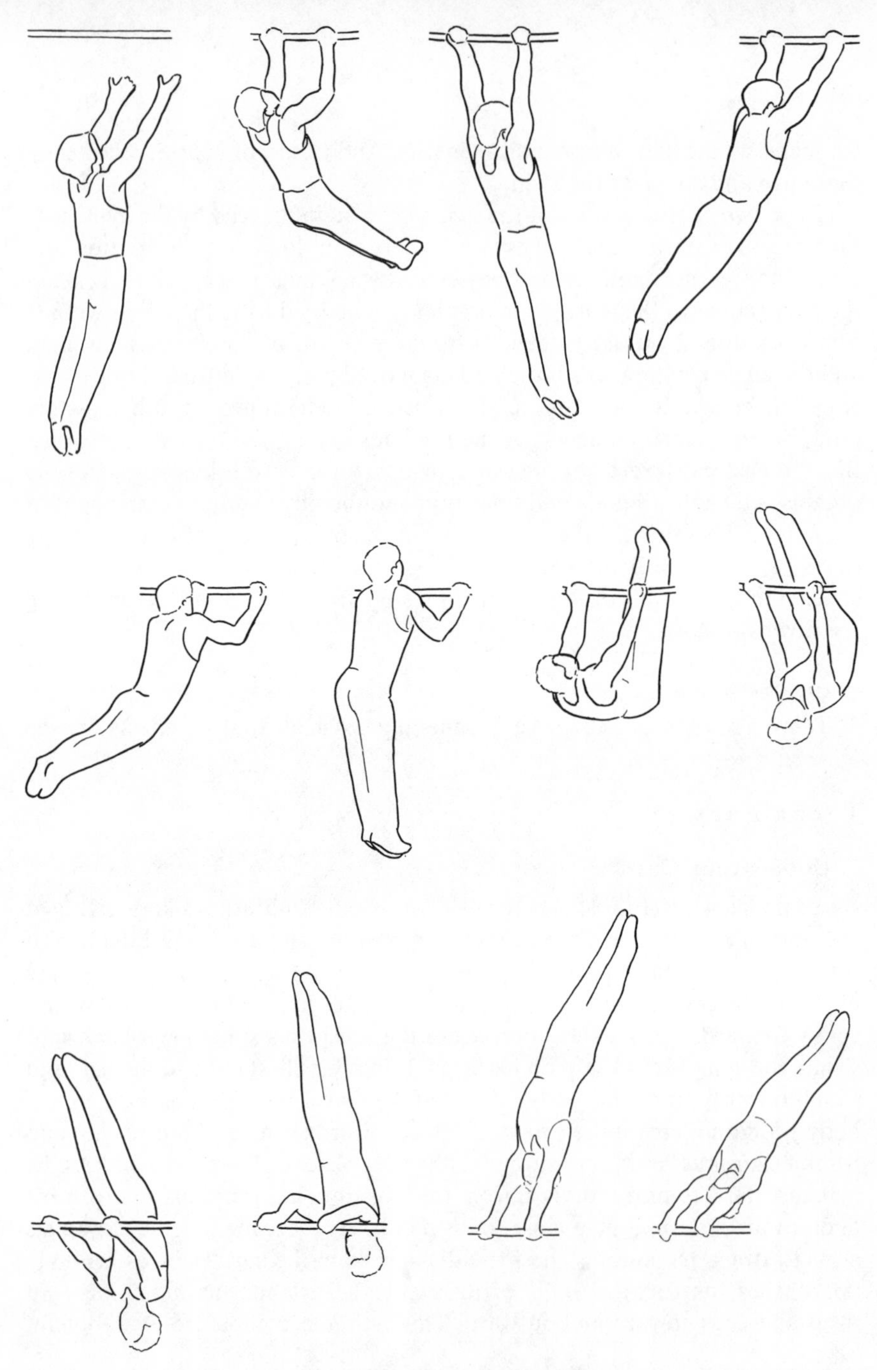

Fig. 151. Underswing (or cast)

flexion of his shoulder joints, lifts his center of gravity up and away from the bar, giving his body a large amount of potential energy and putting it in such a position that the moment of his weight will be near-maximum as he begins his return swing.

Single Leg Upstart (Leg Acting), Single Leg Rise

The single leg upstart is probably the simplest way for a beginning gymnast to get on top of the bar. The movement begins with the gymnast swinging upward and forward with his body in an arched position. Then, near the forward limit of his swing he pikes sharply at the hips and brings his legs close to the bar, with one leg passing beneath it and between his arms. This action produces a marked reduction in the gymnast's moment of inertia and causes him to swing back faster than he swings forward. As he nears the limit of his backward swing, he presses down with his arms and extends his hips. This serves to raise the height of his center of gravity and to bring his hips in closer to the bar, thereby decreasing his moment of inertia even further. Correctly timed, these movements lead him to a straddle-sitting position atop the bar.

Forward Kip, Upstart, Kip

The forward kip (Fig. 152) is an important element in horizontal bar exercises at practically all levels of competition. It starts in the same way as a single leg upstart—with a forward and upward swing with the body arched,

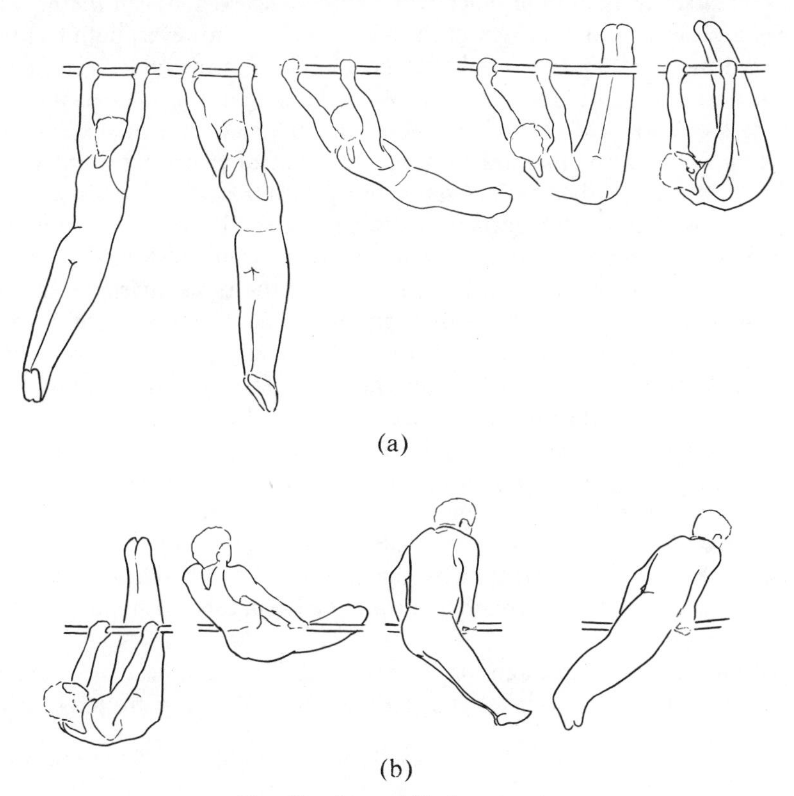

Fig. 152. Forward kip (or upstart)

followed by a sharp piking at the hips. (Following a biomechanical analysis of the upstart, Bevan and Corser[17] suggested that "the timing of this piking movement is probably a crucial element in the skill." While coaching manuals frequently refer to the piking movement being executed at the start of the backward swing, they found that all three of their subjects invariably began to pike much earlier than this—approximately 65–67% of the piking movement being completed before the end of the forward swing.) This piking movement brings the gymnast's ankles close to the bar and once again decreases his moment of inertia and increases his angular velocity. Then, as his center of gravity is about to pass under the bar in the course of his backward swing, the gymnast presses down hard with his arms and extends his hips. Correctly executed, these movements lift his center of gravity close to the bar and decrease his moment of inertia still further in the process. The net result is that he continues to rotate upward and backward into a front support position.

Reverse Kip, Back Kip Backward (F.I.G.), Reverse or Back Upstart, Back Kip

Another in the family of kipping (or upstart) movements, the reverse kip (Fig. 153) starts as did the other two already discussed. When the hips are piked and the legs are brought back toward the bar, however, both feet pass between the arms, and the thighs are held in close to the body. This tightly piked position ensures that the moment of the weight is as large as it can be during the downward part of the swing and also places the hip extensors in position for a forceful contraction near the end of the backswing. At this point the gymnast drives his legs vertically upward (by extending his hips) and presses down hard with his arms. These movements lift his center of gravity above the bar, thereby putting him in a position from which he can acquire the angular momentum necessary to complete the three-quarter backward seat circle that follows. As his body begins to swing downward from the peak of his backswing, the gymnast returns to his tightly piked position with a vigorous downward motion of his hips and legs. The angular reaction to this markedly increases the downward force acting on the bar and causes it to be deflected downward. As his body passes beneath the bar and his shoulders begin to rise, the gymnast extends at the hips, once again lifting his center of gravity and decreasing his moment of inertia. Correctly timed, this lifting action is facilitated by the recoil of the bar that thus gives up its strain energy to do work on the gymnast. With his center of gravity thus held in close to the bar, the gymnast rotates around and up to arrive in a back support position atop the bar.

During the final hip extension it appears as if the gymnast's legs remain stationary (or nearly so) while his upper body is lifted up and around into line

[17]Randall Bevan and Tom Corser, "The Biomechanical Study of Gymnastic Movements" *The Gymnast*, March 1969, p. 30.

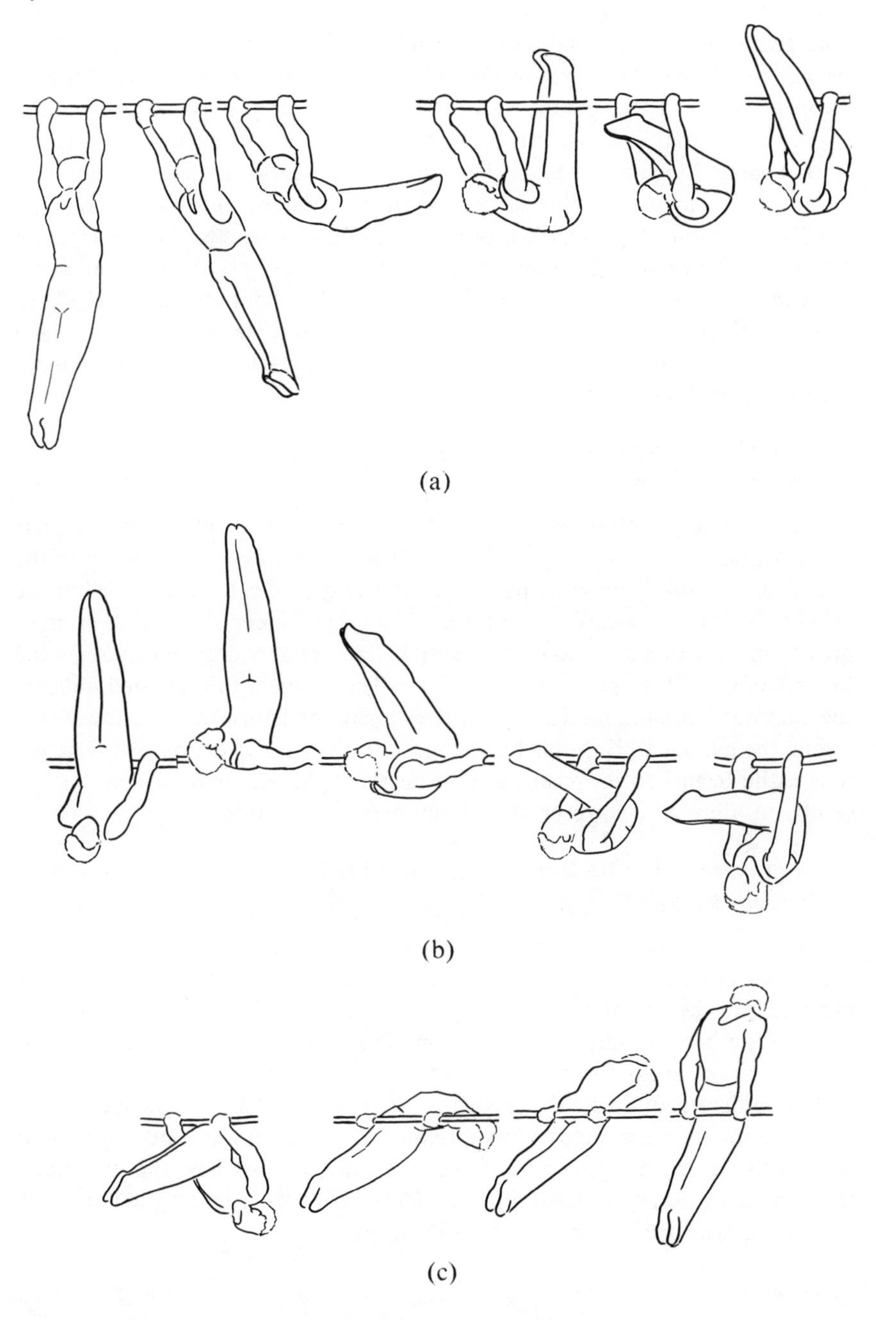

Fig. 153. Reverse kip

with them. This occasionally leads coaches and observers into drawing the wrong conclusions concerning the nature of the movement. What actually happens is this: as the gymnast swings under the bar with his body rotating in, say, a counterclockwise direction, both his trunk and his legs are rotating in that same direction. Then, as he extends at the hips, he increases the rate at which his trunk is rotating counterclockwise. This additional counterclockwise action of his trunk produces a clockwise reaction of his legs. At this point these latter have two contrary tendencies: (1) to move counterclockwise due to the rotation of the whole body about the bar and (2) to move clockwise in reaction to the trunk movement. The result is that these two effectively cancel each other, and the gymnast's legs appear to remain stationary (or almost stationary).

Back Hip Circle, Hip Circle Backward (F.I.G.), Backward Circle

The back hip circle, one of the most elementary of all horizontal bar movements, starts with the gymnast in a front support position on top of the bar. From this position with his center of gravity vertically above the bar, he swings his legs backward and upward. This movement shifts his center of gravity in the same direction, increasing his potential energy and moving the line of action of his weight behind the bar. The moment thus created initiates the backward circling motion. As his legs make their forward and downward swing, the gymnast brings his hips in close to the bar. Then, by piking at the hips as he begins to move upward, he decreases his moment of inertia sufficiently to allow him to return to a front support position.

Front Hip Circle, Hip Circle Forward (F.I.G.), Forward Circle

Although basically very similar to the back hip circle, the front hip circle is somewhat more difficult to perform well. From a front support position, the gymnast presses downward and backward to move his center of gravity forward of the bar and initiate his rotation. Then, as he rotates downward and his head passes under the bar, he flexes forcefully at the hips, driving his head and shoulders around in the direction of the turn. This strong hip action serves to increase his angular velocity and to carry him around once again toward the front support position. Then, finally, to complete the full circle, the gymnast pushes his head and shoulders forward and over the bar while pressing downward and backward with his arms.

Back Uprise, Back Up

This movement is very commonly performed in sequence with two of the other movements analyzed here—the underswing and the back hip circle. As the gymnast swings backward under the bar following a high forward swing (or an underswing), he presses down strongly with his arms. This action increases his mechanical energy and causes him to move higher on his backswing than he would otherwise. In addition, the forward and downward movement of the arms (relative to the body) produces a forward and upward reaction of the legs. This latter slows their backward swing and keeps them some distance below the gymnast's shoulders throughout the remainder of the backswing. At the peak of the backswing the gymnast presses backward and downward with his arms. The angular reaction to this pressing movement pulls his hips into close proximity to the bar in preparation for the next movement—generally some form of backward circle. The movement of the hips into the bar is facilitated by the restrained backward swing of the legs that places the center of gravity closer to the bar than it would have been had a normal backswing been made.

Giant Swing Backward,[18] Backward Grand Circle

The giant swings, of which the backward giant is generally the first learned, form the basis of most of the advanced work on the horizontal bar. From a momentary handstand position on top of the bar, the gymnast presses with his arms to move his center of gravity backward and away from the bar. This movement is accompanied by a conscious stretching of his body aimed at moving the center of gravity as far as possible from the bar. Because it ensures that the moment of his weight is as great as it can be (and, consequently, that his kinetic energy at the bottom of the swing will also be as large as he can make it), the gymnast maintains this fully extended position throughout his downward swing. As he is about to pass under the bar, the gymnast arches his back slightly in preparation for the movements to follow. Then, as he begins his upward swing, he flexes sharply at the hips and presses downward and forward with his arms. These movements increase his potential and kinetic energies and offset the effects of friction and air resistance that would otherwise make it impossible for him to complete the full giant swing in good form (and, in particular, with straight arms). Passing through the horizontal once again and into the fourth quadrant of his circle, the gymnast arches his back and does further work against gravity by pressing downward with his arms. This downward pressing movement is generally preceded by a shift in his grip to place his hands more nearly on top of the bar than they were before—a shift that makes it easier for him to press downward a moment

[18]Gymnastics terminology is often confusing when it comes to describing the direction of a rotation. For swinging movements beneath the bar the words *forward* and *backward* have their usual meaning. Thus, in forward swinging, the front of the body precedes the back, and in backward swinging the back of the body precedes the front. For movements in which the gymnast completely circles the bar, however, these directions are reversed. Thus, in a giant swing backward, the front of the gymnast leads the back as he rotates around the bar.

later. Ideally this hand shift takes place at a time when the centripetal force required to maintain the angular motion is equal to the component of the gymnast's weight acting along a line joining his center of gravity to the axis. Under such circumstances, the pressure between the gymnast's hands and the bar is at a minimum, and movements of his hands around the bar are relatively easy to make. The gymnast completes his circle by rotating upward into a momentary handstand on top of the bar. If all the preceding movements have been carried out effectively, he arrives in this position with his body in a straight line.

Giant Swing Forward, Forward Grand Circle

The basic factors influencing the performance of a forward giant swing (Fig. 26) are similar to those for the backward giant. From a momentary handstand position the gymnast stretches upward, tucks his head, and allows his center of gravity to fall forward of the bar. During the downswing he retains this fully extended position and, in order to acquire as much kinetic energy as possible, approaching the bottom of the downswing he flexes his hips slightly and then, as he begins his upward swing, he reverses this action, extending his hips and pressing downward and forward against the bar. As before, this increases his potential and kinetic energies and enables him to continue upward to finish atop the bar in a handstand position.

If the gymnast wishes to finish his forward giant with a dismount over the bar, as is frequently the case, he increases the range and force of his hip extension and arm pressing movements in the third quadrant of the circle. The additional work he does in this way is reflected in the potential energy (and height) he has at the high point of his flight.

Dismounts

The final phase of a horizontal bar exercise, the dismount, can be subdivided for the purpose of analysis into three parts—(1) the release or takeoff, (2) the flight, and (3) the landing. The basic factors governing performance in all three of these are the same as those already described in the preceding sections, for corresponding parts of vaulting and floor exercises.

TRAMPOLINING (OR REBOUND TUMBLING)

While the trampoline is without doubt the most popular of all the pieces of gymnastic apparatus from a purely recreational standpoint, it has yet to become accepted as an integral part of competitive gymnastics. The first World Trampoline Championships were held as recently as 1964 and the event has yet to make an appearance on an Olympic Games program.

Basic Considerations

When a gymnast stands in the center of a trampoline bed, he is in equilibrium under the action of two forces, his weight and the vertical reaction exerted on him by the bed. To begin bouncing, he must somehow exert such forces as will evoke from the bed a vertical reaction greater in magnitude than his weight. He has two main methods of achieving this:

1. He can drive hard downward against the bed by vigorously extending his hip, knee, and ankle joints. (This is the more obvious and important way.)
2. He can swing his arms strongly in an upward and forward direction. (That this action does have the required effect can readily be demonstrated by a gymnast standing on a trampoline with his hip, knee, and ankle joints "locked" or tightly tensed. If he swings his arms vigorously forward and upward, returns them gently behind him, and repeats this cycle several times, he can build up a moderate bounce without any direct assistance from the legs.)

In practice, of course, the good gymnast makes use of both of these methods.

Because his vertical velocity at takeoff largely determines his time of flight, which in turn largely determines the complexity of the stunts he may perform while in the air, the acquisition of vertical velocity is clearly of some importance to a trampolinist. In contrast, there are several reasons why the acquisition of horizontal velocity at takeoff is undesirable:

1. Safety. If a trampolinist acquires a relatively large amount of horizontal velocity at takeoff, he runs the risk of incurring serious injury by landing beyond the limits of the bed.
2. Control. If he acquires a relatively small amount of horizontal velocity at takeoff, the trampolinist must make due allowance for the additional horizontal forces to which he will be subjected when he next lands on the bed. Unless he makes adequate allowance for them, he is likely to be faced with increasing problems in control.
3. Form. Horizontal motion (or "travel") is generally regarded as bad form and as such is penalized in competitions.

For these various reasons, trampolinists attempt to take off and land in the center of the bed—or, in other words, with a zero horizontal velocity at takeoff.

To perform all but the very simplest of stunts (tuck jump, pike jump, etc.), a trampolinist needs to initiate rotation about one or more of his three principal axes (p. 147). Rotations about his transverse and frontal axes (somersaulting and side-somersaulting rotations, respectively) are almost invariably initiated by the reaction to an eccentric thrust that he exerts against the bed.

Where the takeoff is from the feet (and possibly in a few other instances), it is generally necessary for the gymnast to displace his center of gravity horizontally so that the vertical reaction from the bed can act eccentrically. In this process, usually accomplished by pushing the hips forward or backward, the gymnast's center of gravity acquires a certain amount of horizontal velocity that, if unchecked, causes him to travel during the airborne phase of the stunt. To prevent this, the good gymnast pushes horizontally against the bed with his feet just prior to takeoff (and after the required horizontal displacement has been obtained). The reaction to this thrust reduces the horizontal velocity of his center of gravity to zero and thus ensures that the stunt is executed without travel. Rotations about the gymnast's longitudinal axis (twisting rotations) are generally initiated by the reaction to a couple exerted via the feet against the bed or by a "trading" of somersaulting angular momentum for twisting angular momentum (pp. 156–57).

When a gymnast lands on a trampoline bed, the trampoline does work on him to bring him momentarily to rest—at that point where the bed is maximally depressed. Because of the way in which trampolines are constructed, the distance through which the gymnast moves during this slowing-down process is relatively large and this ensures that the forces exerted upon him by the bed are relatively small. For this reason trampolinists have a wide range of acceptable landing positions compared with other gymnasts and compared with athletes in almost all other sports.

Techniques

Feet Bounce

From a standing position, with feet approximately shoulder width apart, the gymnast projects himself into the air by bending his knees, driving with his legs, and swinging his arms forcefully forward and upward. In the air, he "points his toes" and brings his legs together (in keeping with the requirements of good form) and continues the circling movements of his arms so that they will be appropriately positioned for a further forward and upward swing at the next takeoff. Finally, as he lands on the bed, he substantially increases the area of his base by flexing his ankle joints (so that he lands flat-footed) and by moving his feet so that they are once more about shoulder width apart. In much the same way as increasing the area of the base effects the stability of a body in equilibrium, these adjustments increase the gymnast's ability to control his landing and the movements that follow it. In fact, a failure to make these adjustments is undoubtedly one of the major reasons why complete beginners at trampolining invariably travel back and forth across the bed instead of bouncing up and down on the one spot.

Fig. 154. Knee drop. The eccentric reaction evoked if the gymnast tries to straighten his legs too early causes him to be projected in an upward and backward direction and rotated forward.

Knee Drop

Because it requires only a very small adjustment in body position from that adopted for feet bounces, the knee drop is generally one of the first of the six basic drops[19] taught to beginners. To perform a knee drop the gymnast does a feet bounce and, after reaching the peak of his flight, flexes his knees so that his shins are approximately horizontal. (The contrary reaction of the remainder of the body is so small as to be virtually imperceptible.) Except for moving his knees apart slightly to improve his ability to control the landing, and bringing his arms down behind him in preparation for the next takeoff, the gymnast retains this position for his landing on the bed. Then, as the bed is recoiling preparatory to projecting him into the air, he swings his arms forward and upward, thereby adding to the lift obtained. Once safely into the air again, the gymnast drops his lower legs into line with the rest of his body, ready for a landing on the feet. With respect to this straightening of the legs, one of the more common faults in performing a knee drop is trying to straighten the legs too soon. If this movement is begun before the performer has broken contact with the bed, an eccentric reaction is evoked that tends to send him traveling backward and rotating forward onto his face (Fig. 154).

[19]Disregarding the feet bounce (which is listed as a basic drop by some authors), there are six basic drops: the knee drop, the seat drop, the hands and knees drop, the front drop, the back drop, and the hands drop.

To avoid the undesirable consequences of this kind of "levering off the toes," beginning trampolinists should be instructed to delay the straightening of their legs until they are high above the bed.

Seat Drop

Another of the first drops taught to beginners, the seat drop, is the first discussed here that requires rotation to be initiated at takeoff. As the gymnast takes off to perform a seat drop, he leans backward slightly. This moves his center of gravity backward so that the vertical reaction from the bed acts eccentrically and imparts a slight backward rotation to his body. To combat the traveling tendency that this shifting of his center of gravity produces, the gymnast exerts a small backward force against the bed as he takes off. The gymnast retains an extended body position until relatively late in his downward flight when he flexes at the hips. This action lifts his legs to the horizontal and brings his trunk forward to a near-vertical position ready for landing. During these movements, and "absorbing" part of the reaction to the lifting of his legs, the gymnast moves his hands downward and backward into position beside, or slightly behind, his hips. To initiate the forward rotation necessary to bring him back to an erect position during the next airborne phase, he pushes against the bed with his hands. The reaction to this downward thrust passes behind his center of gravity and creates the required rotation.

Hands and Knees Drop

The hands and knees drop is the first of the basic drops considered here that requires forward rotation at takeoff. In taking off to execute a hands and knee drop the gymnast leans forward slightly, putting his center of gravity forward of the line of action of the thrust of the bed. Any tendency for forward travel to result from moving the center of gravity forward in this way is overcome by the gymnast pushing gently forward against the bed as he takes off. Nearing the peak of his flight he brings his arms and knees toward each other, each of these movements effectively balancing the other, and prepares for a landing in which his knees and hands land simultaneously. The backward rotation he needs in order to return to an erect position is obtained from the reaction to a forceful thrust he exerts via his hands against the bed.

Front Drop, Stomach Drop, Belly Drop

The front drop is little more than a somewhat advanced version of the hands and knees drop. The forward rotation required is initiated in exactly the same manner as for the hands and knees drop except that, since the body's moment of inertia is greater in an extended position than in a partially tucked hands-and-knees position, slightly more angular momentum is required at takeoff. As an indirect result, there is generally a greater tendency for beginning trampolinists to travel forward when executing a front drop. Because this almost invariably results in a rough landing and is also potentially very dangerous, instructors are well advised to place some emphasis on pushing forward with the feet at takeoff or, to use other words, on remaining in the center of bed. (With respect to the latter, an instruction to "place your belly where your feet were" generally produces the desired result.) Once in the air the gymnast rotates gently forward so that by the time he again contacts the bed he has his trunk horizontal, his legs together and extended horizontally behind him, and his arms bent with his hands, palms down, slightly forward of his head. Contact is then made simultaneously with legs, trunk, forearms, and palms. (*Note:* The average pressure on the gymnast when he lands in a correctly executed front drop is much less than that for any of the other basic drops because of the relatively large area involved.) To complete the stunt the gymnast pushes down against the bed with his hands and obtains the eccentric reaction necessary to rotate him once more into an erect position.

Back Drop

The only basic drop in which the performer cannot see where he is going, the back drop would probably be rated by most beginners as the most difficult of the five considered here. Backward rotation is initiated at takeoff by an eccentric thrust, and travel is controlled by a gentle backward pressure against the bed. In the air, the gymnast retains an extended, sometimes arched, position until he begins his descent, when he pikes at the hips to bring his legs and trunk to a position almost at right angles to each other, ready for landing. The landing is made with the whole back, from buttocks to shoulders, contacting the mat simultaneously. As the bed recoils, the gymnast partially extends his hips, thereby imparting some angular momentum to his legs. Then, when the motion of his legs is reduced a split second later, this angular momentum is transferred to the whole body and serves to rotate him back to an erect position. This kipping action (akin to that used in a neckspring) is necessary because the gymnast does not have other suitable means of obtaining an eccentric reaction from the bed as he does in most of the other basic drops.

Swivel Hips

Frequently regarded as the first major challenge to the skill of a beginning trampolinist, the swivel hips movement consists of a seat drop followed by a half twist about a vertical axis to land in a second seat drop (Fig. 87). As the bed is recoiling from his first seat drop landing, the gymnast thrusts forcefully downward against it with his hands. The reaction to this thrust acts eccentrically and, as in a regular seat drop, causes the gymnast to rotate forward about his transverse axis. Once this thrust has been delivered, and before he leaves the bed, the gymnast throws his arms and head upward and to one side. As the motion of these body parts decreases, the angular momentum they possess is transferred to the body as a whole and causes it to rotate about a vertical axis through the center of gravity. Thus, as the gymnast leaves the bed, he is rotating about both a transverse axis and a vertical axis. Once he becomes airborne, he extends at the hips to bring his legs into line with his trunk. This decreases the moment of inertia and increases the angular velocity relative to the vertical axis (and it is generally the rotation about this axis that presents the greater difficulty) at the same time as it produces exactly the reverse effects relative to the transverse axis. The moment of inertia about the vertical axis can be further decreased by the gymnast raising both arms and bringing them together overhead. (Beginners often find that concentrating on this action leads to a more effective throw of the arms at takeoff and a straighter body position in the air than they can achieve otherwise.) As the half twist is nearing completion and the gymnast is descending toward the bed, he flexes again at the hips to bring his legs and trunk into position for the final seat drop landing.

Barrel Roll, Roller

Similar to the swivel hips in that it starts and ends with a seat drop, the barrel roll incorporates a full twist about the body's long axis in between successive seat drop landings. As he is being lifted by the bed following his first seat drop landing, the gymnast leans backward slightly. This brings his

Fig. 155. Full turntable

center of gravity and his hands roughly into the same transverse line. Thus when he thrusts vertically downward with one hand a moment later, the reaction to this thrust causes him to rotate about his long axis, without also causing a rotation about his transverse axis. The rotational effect of the reaction from the bed (relative to the long axis) is supplemented by a simultaneous turning of the head and shoulders in the direction of the impending twist. Once in the air the gymnast extends at the hips to bring his trunk and legs into line with his long axis. This decreases his moment of inertia and increases his rate of rotation. Then, nearing the completion of the full twist, he reverses this process to bring his legs and trunk into position for landing.

Turntables

The turntable stunts (half turntable, full turntable (Fig. 155), one-and-a-half turntable, etc.) are some of the very few trampoline movements involving rotation about the gymnast's frontal axis. From a front drop landing, the gymnast pushes downward and sideways with his hands and downward with his knees against the bed. The reactions to the eccentric downward thrusts from the hands and knees add to the lift that the gymnast acquires and, if correctly executed, balance each other so that there is no resulting rotation about his transverse axis. The reaction to the sideways component of the push with the hands acts eccentrically to a vertical axis through the gymnast's center of gravity (frontal axis) and causes him to rotate about this axis. In the air he assumes a more or less tightly tucked or piked position until he has rotated almost completely around into position for his next landing. At this stage he straightens his body for the front drop landing that completes the stunt.

The basic factors governing the performance of a turntable movement are exactly the same irrespective of how much rotation is involved. The progression from a simple to a more difficult turntable is accomplished merely by making one or more of the following modifications:

1. Increasing the height from which the gymnast goes into his front drop.
2. Increasing the downward thrust of the hands and knees.

[*Note:* Both (1) and (2) result in an increase in the height of the gymnast's rebound from the bed and a greater time in which to obtain the required rotation.]

3. Increasing the sideways thrust of the hands against the bed.
4. Decreasing the moment of inertia of the gymnast (by tucking or piking more tightly) when he is in the air.
5. Delaying the extension of the body until much nearer the instant of landing on the bed.

Somersaults

The absence of a run-up and the need to take off and land on the same spot constitute the only major differences between somersaults executed in a tumbling or floor exercise and those executed on a trampoline.

Forward Somersault. As the gymnast starts to rise with the bed, he leans forward slightly so that the line of action of the thrust from the bed passes behind his center of gravity and causes him to rotate forward. Then, just as he is about to leave the bed, he pushes his hips back, brings his trunk forward, and thrusts his feet forward against the bed. The reaction to the horizontal component of this thrust reduces the forward horizontal motion of the gymnast's center of gravity and ensures that his takeoff is in a vertical direction. [It might be noted here that coaches and performers put differing emphasis on these movements preceding the takeoff. Some stress the movement of the hips; some, the thrust of the feet; and some, the forward movement of the upper body. Since each of these movements evokes a horizontal reaction from the bed, it matters little (from a biomechanical standpoint) which of them is emphasized. What is important is that the body's horizontal motion is effectively checked before the gymnast leaves the bed.] Once airborne, the gymnast's actions are identical to those already described for a forward somersault on the floor except that since he has a good deal more time at his disposal, he can perform these actions at a more leisurely rate.

Backward Somersault. The techniques employed in the correct execution of a backward somersault are fundamentally the same as those used in a forward somersault, except for the obvious differences in direction.

Codys

A cody (Fig. 156, pp. 325–26) is a forward or backward somersaulting movement performed from a takeoff in the front drop position to a landing on the feet.

Front Cody. This movement involves a three-quarter forward somersault from front drop to feet. After a high bounce, which is the only really effective way of gaining height for this stunt, the gymnast lands on the bed in a front drop position. Then, as it recoils, he pushes down strongly into the bed with his knees. The eccentric reaction to this push initiates the forward rotation that subsequently carries him around to a landing on the feet.

Fig. 156. Codys (a) front cody

Back Cody. Although requiring a one-and-one-quarter somersault as compared to a three-quarter somersault, the back cody is easier in some respects than a front cody. From a front drop landing, the gymnast thrusts strongly downward with his hands as the bed recoils. The eccentric reaction to this push (which, incidentally, is a good deal easier to exert than a push with the knees) initiates the backward rotation needed for the somersault to follow.

Because the front drop position is ill-suited to the purpose, gymnasts performing codys frequently use a special technique to aid them in developing the required somersaulting angular momentum. This technique, known as a *kaboom* because of the two-beat rhythm involved, consists of executing the front drop so that one end of the body strikes the bed slightly before the other. The eccentric reaction force exerted on the end that lands first causes the other end to be driven down into the bed even more forcefully than would normally be the case. The eccentric reaction force exerted on the end that lands second then serves to impart angular momentum to the body in the desired direction. For example, in executing a front cody, the gymnast endeavors to land so that his chest contacts the bed first. The reaction force exerted by the bed on his chest tends to rotate him backward and, thus, to drive his knees deeper into the bed. This action increases the eccentric reaction force exerted by the bed on the knees and this, in turn, increases the forward angular momentum imparted to his body.

Fig. 156. Codys (b) back cody

Ball Out, Cannonball Somersault

Little more than an advanced version of the back drop, the ball out [Fig. 157 (pp. 328–29)] involves a back drop followed by a forward one-and-one-quarter somersault to a landing on the feet. From a landing in the back drop position, the gymnast extends his hips vigorously as the bed recoils. Then, as he slows this hip extension (or kipping) movement a moment later, the angular momentum of his legs is transferred to his body as a whole. (*Note:* Because the gymnast requires a good deal more angular momentum to complete this stunt than he does to complete a simple back drop, this hip extension movement must be correspondingly more vigorous than it is in the latter case.) As soon as he leaves the bed, the gymnast decreases his moment of inertia by assuming a tightly tucked position. Then, nearing the end of his one-and-one-quarter somersault, he comes out of the tuck into a fully extended position for landing.

Fig. 157. Ball out (or cannonball somersault)

Chapter 13

SOFTBALL

Developed from baseball via the indoor version of that game, softball is a comparatively young sport that, perhaps suprisingly, enjoys an even greater international following than its parent.

While the rules of the two games differ considerably, the only major difference in techniques lies in the method of pitching. Unlike baseball, where the ball is pitched with anything from an overarm to a sidearm action, in softball the rules require that the ball must be delivered with an underarm action.

Since the other techniques used in softball have already been analyzed in the chapter on baseball (Chap. 8), the discussion in this present chapter is confined to an analysis of pitching.

Basic Considerations

The basic factors governing pitching in softball are identical with those already discussed relative to pitching in baseball.

Techniques

Pitching

Stance. In taking his initial position on the rubber, the pitcher places his feet approximately shoulder width apart with the heel of his right shoe in contact with the front half of the rubber and the toe of his left one in contact with the back half.[1] This position of the feet provides a relatively broad base and therefore reasonable stability; it affords the opportunity to obtain

[1]For the sake of simplicity it is assumed throughout this analysis that the pitcher is right-handed.

the maximum distance through which to exert force on the ball; and last, but not least, it conforms with the rules.

Having carefully placed his feet in the required position, the pitcher assumes an erect stance with his center of gravity directly over his left foot. Both hands are held in front of him at, or slightly below, waist height with the ball concealed in the glove, Fig. 158(a)].

Delivery The two main methods of delivery are:

1. the *windmill* (Fig. 158), in which the pitcher's arm rotates through approximately 360° in a vertical or near-vertical plane before the ball is released, and
2. the *figure eight* (Fig. 159), in which the pitcher's arm is moved horizontally to a position behind his back before being brought forward again to the point where the ball is released.

While only the former is considered in this chapter, it should be noted that many of the points made apply equally well to both methods of delivery.

The wind-up preceding the release of the ball begins with the pitcher moving both hands downward and forward to a position near his knees. This movement draws his shoulders forward and, aided by a slight flexing of the knees, serves to set his center of gravity moving in a forward and downward direction. Once his hands have reached the low point of their downward motion, the pitcher pushes against the rubber with his left foot and then brings his left leg forward and upward in unison with a forward and upward swing of the arms [Fig. 158(c)]. As a result of these actions the pitcher's center of gravity moves forward of his right foot and into a position from which moments later a forceful extension of the right leg drives it still farther in a forward and upward direction—Fig. 158(d) and (e). (*Note:* The flexing of the knees during the earlier downward swing of the hands not only decreases the body's moment of inertia and therefore facilitates the passage of the center of gravity over the right foot but also puts the right leg into a position from which it can later drive strongly downward and backward against the ground.)

During the course of these striding movements the arms continue forward and upward until they are roughly horizontal [Fig. 158(c)], at which stage the right arm continues its backward rotation while the left one remains where it is and assists in maintaining the required balance. As the right arm swings overhead and begins to descend [Fig. 158(d) and (e)], the pitcher's body turns so that his hips and trunk face sideways. This action serves two main purposes:

1. It places the body in a position where the muscles responsible for hip and trunk rotation can make a contribution to the speed of the ball at release.
2. It increases the distance through which the ball may be accelerated.

Fig. 158. The windmill pitch. [*Note:* The subject of this photo sequence—a pitcher who has represented his country in international competition on several occasions—shows two minor deviations from the "ideal" form described in the text: (1) his right foot is further back on the pitching rubber than it needs to be and this slightly reduces the range of his pitching action; (2) his left foot lands slightly to the right of a direct line to the plate, thereby limiting the contribution that hip rotation can make to the speed of the ball.]

(c) (d) (g) (h) (i)

Fig. 159. The figure-eight pitch

The stride is completed when the heel of the left foot strikes the ground some 5–6 ft from the rubber and slightly to the left of the intended line of the pitch. This off-line placement of the foot permits the hips to be fully rotated to the front and thus make a maximal contribution to the speed of the ball at release. After the left heel has landed, the rest of the foot is quickly grounded and the hip, knee, and ankle joints of the left leg flex to reduce the force of the impact.

Once the left foot has been grounded, the pitcher's body rotates to the front, partly due to the eccentric ground reaction evoked and partly as a result of the internal muscular forces exerted at that time. This rotation of the body brings the right shoulder forward and causes the path followed by the ball to be "flattened out" as it approaches the point of release. [*Note:* The path followed by the ball is determined by summing the effects produced by(1) the rotation of the right arm about an axis through the right shoulder and (2) any displacement which that axis undergoes. With respect to the latter, if the axis is moved horizontally, the path followed by the ball is flattened or elongated in a horizontal direction. Similarly, if the axis is moved vertically—as it is earlier in the pitching sequence (Fig. 160)—the elongation occurs in a vertical direction.]

Release. Once the point of release has been reached, the centripetal force exerted on the ball by the pitcher's hand (and which has hitherto kept the ball moving along a curved path) is removed. Then, no longer restrained in this manner, the ball flies off in the direction it was moving at the instant of release (i.e., in a direction tangential to its path at the point of release).

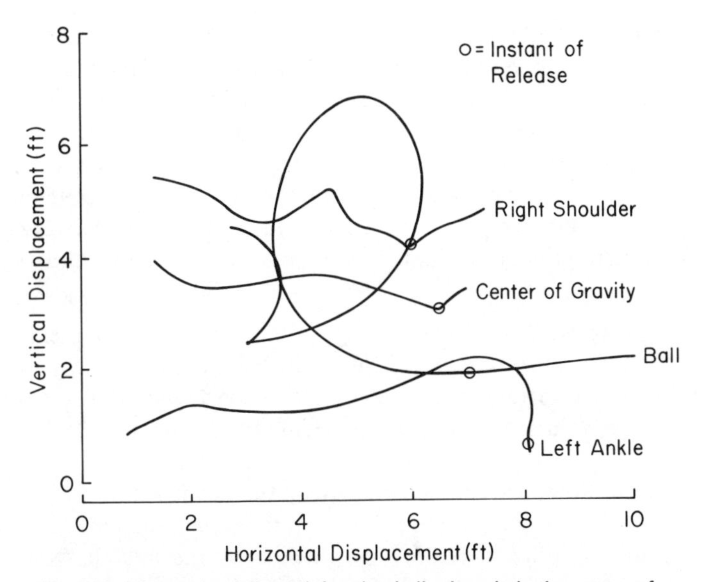

Fig. 160. The paths followed by the ball, the pitcher's center of gravity, right shoulder, and left ankle during the execution of a windmill pitch (*Adapted from data in David J. East, "A Cinematographical Analysis of a Softball Pitch," unpublished term paper, University of Otago, New Zealand, 1969. The pitcher in Fig. 158 was the subject in the cinematographic analysis conducted by East.*)

Because the direction in which the ball is moving as it leaves the pitcher's hand largely determines the ultimate success of a pitch, the point at which the ball is released is obviously of critical importance. If it is released before it reaches the correct (or optimum) point, it is likely to be lower than intended as it passes over the plate. Similarly, if the release takes place at some point beyond the optimum, the ball is likely to be higher than intended.

The flattening of the path followed by the ball during the latter stages of the delivery affords the pitcher some room for error in his point of release. For, with such a flattened path, a point of release that deviates slightly from the optimum produces a less marked difference in the direction in which the ball is released than would be the case if the ball's path were a true circular arc. Thus, the flattening of the path followed by the ball is not only an indication that the hip and trunk rotations have been correctly executed but, because it permits some latitude in the point of release, is also of value in itself.

The height at which the ball should be released, relative to the pitcher's body, depends on a number of factors, including the speed of the ball as it is released and that point in the strike zone (high or low) at which it is desired to place it. In general, however, the optimum point of release lies somewhere between knee and hip height.

The speed of the ball at release has received attention from a number of investigators. Lieber reported[2] that Bill Massey "ace of the world champion Clearwater, Fla., team" attained a speed of 98.8 mph (or 145 fps)—a speed, incidentally, that is exactly the equal of that reported for Bob Feller's fast ball (p. 196). Other reported values [the 108 fps and 109 fps of the two male pitchers reported by Cooper and Glassow[3] and the mean of 59.95 mph (87.92 fps) for the nine pitchers tested by Miller and Shay[4]] were probably obtained using less talented subjects than Massey and are therefore understandably somewhat lower.

Follow-Through. The follow-through after the release of the ball serves exactly the same purposes in softball as it does in baseball, viz., to reduce the speed of the various body parts (and in particular, the pitching arm) without risk of injury and without impairing the application of forces to the ball.

Once his follow-through is completed, the pitcher moves quickly into position to field the ball should it be hit in his direction. This need to move quickly into a fielding position is heightened by the fact that the pitching rubber is only 46 ft from the plate (cf. 60 ft, 6 in. in baseball) and the pitcher thus has even less time before the ball can be returned to him than does his baseball counterpart.

[2]Leslie Lieber, "The Big Hardball vs. Softball Duel," *This Week Magazine*, May 14, 1961.

[3]John M. Cooper and Ruth B. Glassow, *Kinesiology* (St. Louis: The C. V. Mosby Co., 1963), p. 66.

[4]Robert G. Miller and Clayton T. Shay, "Relationship of Reaction Time to the Speed of a Softball," *Research Quarterly*, XXXV, October 1964, p. 436.

Chapter 14

SWIMMING

The techniques of swimming have undergone profound changes over the years. Chief among these have been the evolution of the front crawl, back crawl, and butterfly strokes and the concomitant relegation of breaststroke from its role as the premier stroke—a role it played in the early days of modern competitive swimming.

Basic Considerations

A competitive swimmer's objective is to swim the full distance of his race in the prescribed manner (i.e., in accord with the rules governing starting, turning, finishing, and the execution of the stroke) in the least time possible. Or, to state it in terms that take both the distance of the race and the time involved into account, his objective is to obtain the maximum average speed of which he is capable.

Now the average speed obtained is equal to the product of two factors:

1. *The average stroke length*, i.e., the average horizontal distance traveled during the completion of one complete cycle of the swimmer's arms. Thus,

$$\text{Average stroke length, } \overline{SL} = \frac{\text{distance stroked}}{\text{number of complete arm cycles}}$$

2. *The average stroke frequency*, i.e., the average number of complete arm cycles executed in a given time. Thus,

$$\text{Average stroke frequency, } \overline{SF} = \frac{\text{number of complete arm cycles}}{\text{time spent stroking}}$$

As an example, consider the backstroke swimmer who takes 10 strokes (or complete arm cycles) to cover 60 ft in 12 sec. His average stroke length is

$$\overline{SL} = \frac{60}{10} = 6 \text{ ft/cycle}$$

and his average stroke frequency is

$$\overline{SF} = \frac{10}{12} = 0.83 \text{ cycle/sec}$$

His average speed $\bar{S}$, the thing he is most concerned about, is equal to the product of these two factors:

$$\begin{aligned} \bar{S} &= \overline{SL} \times \overline{SF} \\ &= 6 \text{ ft/cycle} \times 0.83 \text{ cycle/sec} \\ &= 4.98 \text{ fps} \end{aligned}$$

Since the speed at which a swimmer moves through the water (at least while he is stroking) is wholly dependent on his stroke length and stroke frequency, it is appropriate that those factors that determine the magnitude of each of these parameters should be considered next.

Stroke Length

The stroke length is governed by the forces exerted on the swimmer—the *propulsive forces*, which drive him forward through the water in reaction to the movements he makes, and the *resistive forces*, which the water exerts on him to oppose that motion.

Propulsive Forces. While the forces exerted on the swimmer in reaction to the movements of his arms are generally regarded as the prime source of his forward propulsion, opinions differ concerning the exact magnitude of the contribution from the arms. Karpovich,[1] for example, determined the speeds that front crawl swimmers could develop using the arms alone, the legs alone, and the arms and legs together and concluded that good crawl swimmers derived about 70 percent of their forward speed from their arms and 30 percent from their legs. Armbruster, Allen, and Billingsley[2] accord the arms even greater credit when they state that "the arms provide about 85% of the total power of the sprint crawl stroke" and Counsilman[3] goes still further with his statement that "the arm stroke in the crawl is the main source of propulsion and, in the case of most swimmers, the only source of propulsion."

[1] Peter V. Karpovich, "Analysis of the Propelling Force in the Crawl Stroke," *Research Quarterly*, VI (Supplement), May 1935, pp. 49–58.

[2] David A. Armbruster, Robert H. Allen, and Hobert S. Billingsley, *Swimming and Diving* (London: Kaye and Ward, Ltd., 1970), p. 71.

[3] James E. Counsilman, *The Science of Swimming* (Englewood Cliffs, N.J.: Prentice-Hall, Inc., 1968), p. 25.

Two studies in which swimmers were tethered to devices that recorded the forces they exerted while swimming "on the spot" shed some light on the relative contribution of the arms in strokes other than the front crawl. In the first of these Mosterd and Jongbloed[4] found that the forces exerted by the arms and legs in the butterfly stroke were of approximately the same magnitude, while in the breaststroke "the work of the legs dominates a bit." Magel[5] reached essentially the same conclusions: "The work of the legs makes a much larger contribution to the total propulsive force in this stroke [breaststroke] than in the front and back crawl, where the arms provide the major portion of the propulsive force. In butterfly swimming, the propelling forces delivered by the arms and legs appear to be approximately the same." (*Note:* Although the extent to which the results of studies like these can be considered to apply in the case of a swimmer free to move forward in the water is unknown, it would seem likely that they provide a reasonable indication of what takes place under such conditions.)

While the importance of the arms as a source of propulsion is well recognized, the exact manner in which this propulsive force is obtained remains a matter of open debate. In the past it has generally been considered that the drag force exerted against the swimmer's arms, in reaction to his efforts to move them directly backward, is responsible for his being propelled forward [Fig. 161(a)]. Thus articles and texts on swimming have routinely emphasized the importance of the arms' being pushed horizontally backward through the greatest distance possible. However, it has been suggested in two relatively

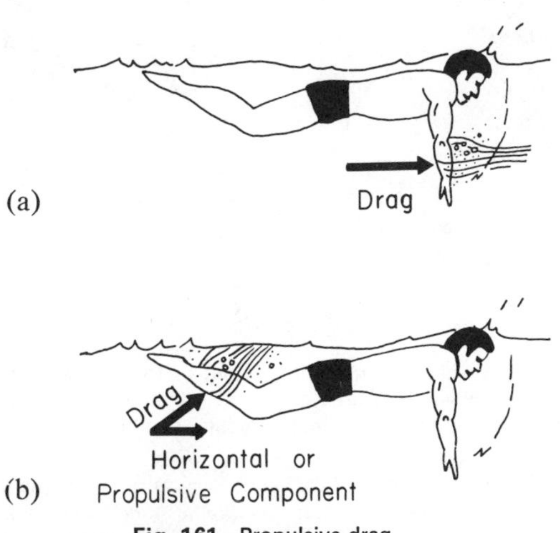

Fig. 161. Propulsive drag

[4]W. L. Mosterd and J. Jongbloed, "Analysis of the Stroke of Highly Trained Swimmers," *Arbeitsphysiologie*, XX, 1963, p. 291.

[5]John R. Magel, "Propelling Force Measured During Tethered Swimming in the Four Competitive Swimming Styles," *Research Quarterly*, XLI, March 1970, p. 72.

recent papers[6,7] that lift rather than drag may be the principal source of the propulsion generated by the arms.

An examination of the path followed by a swimmer's hands during one arm cycle of breaststroke provides a basis for considering the concepts involved here. In the course of such an arm cycle, the swimmer's hands move forward, outward, inward, and forward again (Fig. 162). They do not, as is generally supposed, move backward to any appreciable extent. (*Note:* While the hands move backward *relative to the shoulders*, the shoulders themselves are moving forward at approximately the same rate. The general impression, that the hands move backward through the water, is thus more apparent than real.) Now, if the swimmer's palms face outward and backward as his hands move outward in the pulling phase of the stroke, the resultant force exerted by the water on the hands will be in an inward and forward direction (Fig. 162). Further, if the hands are moving in a direction at a right angle to the direction of the swimmer's motion, the lift component of this resultant force acts in the direction in which he is moving and thus serves to propel him

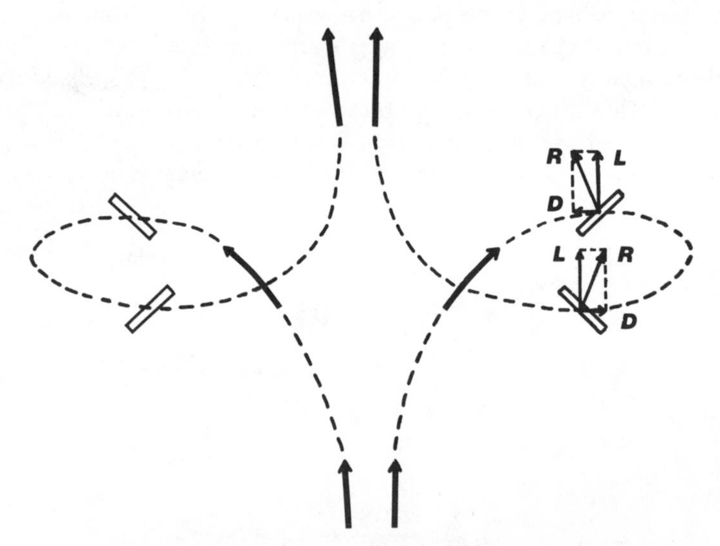

Fig. 162. With appropriate positioning of the hands relative to the direction in which they are moving (i.e., with an appropriate angle of attack), lift may be the dominant propulsive influence.

[6]James E. Counsilman, "The Application of Bernoulli's Principle to Human Propulsion in Water," *First International Symposium on Biomechanics in Swimming, Waterpolo and Diving Proceedings*, ed. by L. Lewillie and J. P. Clarys, Université Libre de Bruxelles Laboratoire de L'effort, 1971.

[7]Ronald M. Brown and James E. Counsilman, "The Role of Lift in Propelling the Swimmer," *Selected Topics on Biomechanics: Proceedings of the C.I.C. Symposium on Biomechanics*, ed. by John M. Cooper (Chicago, Ill., The Athletic Institute, 1971), pp. 179–88.

forward. The drag component acts at right angles to the direction in which the swimmer is moving and thus has no effect on his forward motion. During the recovery phase of the stroke the palms of the swimmer's hands are directed inward and backward and the lift components of the resultant forces on the hands once again act to propel him forward. Under the circumstances described—and similar circumstances can be shown to exist to some extent in all four competitive strokes—it seems evident that lift plays a prominent role in propelling the swimmer through the water.

Since these general ideas were first put forward, several researchers[8,9,10,11] have conducted studies aimed at examining the premise that lift is the principal source of propulsion in swimming and, while the validity of this premise has yet to be conclusively demonstrated, the evidence points strongly in that direction. In keeping with these developments, there has been a considerable shift in the emphasis that swimmers and their coaches place on the various aspects of the arm pull and, in particular, on the orientation (or "pitch") of the hand and arm relative to the direction in which these parts are moving.

The effect that the position of the hand and fingers have on the propulsive force that a swimmer is capable of producing has long been a subject of discussion. In order to examine the relative merits of various hand positions, Counsilman[12] made plaster casts of the same hand held in five different positions, placed them one at a time in a wind tunnel, and determined the drag force exerted against each of them. He found that the drag recorded decreased in the following order:

1. hand flat, fingers and thumb together;
2. hand flat, fingers together, thumb out at the side;
3. hand flat, fingers held apart;
4. hand cupped, fingers held together;
5. hand flat, wrist and fingers extended slightly.

There was very little difference in the results obtained for the first three positions and significant differences between these three and the other two. He therefore concluded that a swimmer should not cup his hands or extend his wrist since in so doing he would lose some propulsive force.

[8]K. M. Barthels and M. J. Adrian, "Three-dimensional Spatial Hand Patterns of Skilled Butterfly Swimmers," in *Swimming II*, ed. by L. Lewillie and J. P. Clarys (Baltimore: University Park Press, 1975), pp. 154–60.

[9]G. W. Rackham, "An Analysis of Arm Propulsion in Swimming," in *Swimming II*, pp. 174–79.

[10]Robert E. Schleihauf, "A Biomechanical Analysis of Freestyle," *Swimming Technique*, XL, Fall 1974, pp. 89–96.

[11]Bob Schleihauf, "A Hydrodynamic Analysis of Breaststroke Pulling Proficiency," *Swimming Technique*, XII, Winter 1976, pp. 100–105.

[12]Counsilman, *The Science of Swimming*, pp. 9–12.

With the advent of the idea that lift may be the principal source of propulsion, the optimum position in which to hold the hand once again becomes an open question. If it is ultimately established that the bulk of the propulsive force exerted on a swimmer derives from the lift component of the resistance his arms encounter rather than from the drag component, it would occasion little surprise if the optimum hand position were found to be one in which the hand was cupped to approximate the geometry of a hydrofoil (Fig. 163).

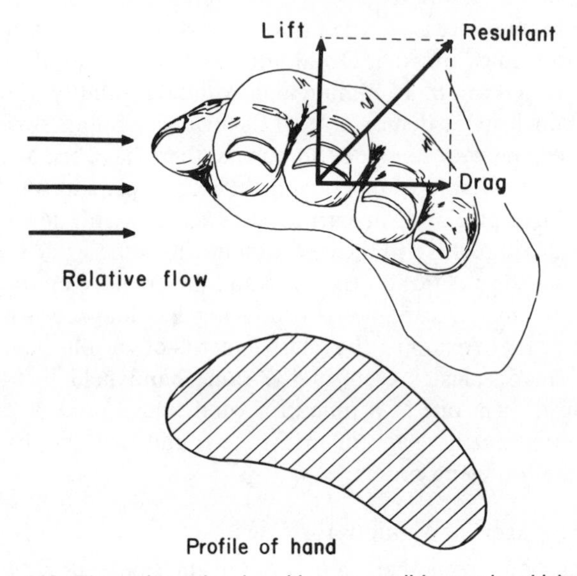

Fig. 163. The optimum hand position may well be one in which the hand approximates the geometry of a hydrofoil.

There are three useful functions that the kicking actions of the legs may serve:

1. They may aid in the production of propulsive forces—a function they serve in breaststroke, and probably, too, in the butterfly stroke.[13]
2. They may serve to decrease the resistive forces that oppose the motion of the body through the water—a role now generally accepted as that which the kick serves in the front and back crawl.[14]
3. They may serve to increase the propulsive forces and decrease the resistive forces simultaneously.

The propulsive function of the legs (where they serve such a function) is generally accounted for in terms of drag. Thus, for example, when the swim-

[13] *Ibid.*, p. 69.
[14] *Ibid.*, pp. 25–30.

mer in Fig. 161(b) thrusts his legs downward and backward as he executes the first beat of a dolphin kick, the horizontal component of the drag force acting on his legs tends to accelerate him horizontally.

The possibility that lift may be the principal source of the propulsion derived from the legs has also been proposed:

> We may suggest. . . that the explanation given for the hand motion that lift is responsible for the propulsive force on the hand also applies to the foot motion. In fact if the feet were being used as paddles [i.e., if drag were the propulsive force] it would be difficult to explain why they are not pushed straight back, since that would be so much easier.[15]

Obviously there is an urgent need for further research aimed at resolving the extent to which lift and drag are involved in propelling a swimmer forward. For, until such research is conducted and the issues involved are clarified, it is virtually impossible to reach concrete conclusions concerning the optimum propulsive force. While continued experimentation in practice will undoubtedly suggest some answers to the many questions involved, completely satisfactory answers can be obtained only through a series of carefully controlled scientific experiments. The problems involved are much too complex to be amenable to complete solution by any other means.

Resistive Forces. Three types of resistive force act on a swimmer to decrease his stroke length—form drag, surface drag, and wave drag (pp. 172–79).

Form drag. The magnitude of the resistive form drag that a swimmer encounters is governed by the speed at which he is traveling forward through the water and by the cross-sectional area that he presents to the "oncoming flow." Since any attempt to reduce a swimmer's form drag is aimed at increasing his forward speed, any reduction gained by decreasing the forward speed clearly defeats the whole purpose of the exercise. Thus a decrease in his cross-sectional area offers the most likely means by which a reduction in form drag can lead to an increase in a swimmer's forward speed. In practical terms this generally means an adjustment in the swimmer's body position in the water and/or in the range of his stroking or kicking actions.

The effect that variations in body position and limb movements have on the resistance that a swimmer encounters has been studied by a number of researchers.

Counsilman[16] reported the results of a study in which a swimmer was towed in a prone position with his head held in (1) a "normal" position with the water at hairline level and (2) a high position with the water at eyebrow level. These results suggested that when the head is held in the higher position there is a significant increase in the drag. This is almost certainly attributable to an increase in the cross-sectional area that the swimmer presents to the

[15]Brown and Counsilman, "The Role of Lift in Propelling the Swimmer," p. 187.

[16]Counsilman, *The Science of Swimming*, pp. 20–21.

flow—an increase due to the legs and feet dropping as the head is raised.

Alley[17] noted differences in the drag force associated with differences in the type of front crawl (or "flutter") kick used by a swimmer as he was towed through water at various speeds. When his subject was towed at speeds less than 4.3 fps, yet greater than he could normally attain with his leg kick alone, the drag recorded was less with a "normal kick" (feet approximately 12 in. apart at maximum spread) than it was with a "short kick" (feet approximately 6 in. apart at maximum spread). In addition, the drag recorded when the subject kicked was considerably less than that recorded when his legs were idle. All of this suggests that, within the range of speeds involved, the leg kick is capable of decreasing resistance and that a "normal kick" is more effective than a "short kick" in this regard. Although the means by which the leg kick reduces resistance has not been established, it seems likely that the summed effect of the various vertical reactions to the movements of the legs and feet causes them to be raised higher in the water than they would otherwise be and that the associated reduction in cross-sectional area results in a corresponding decrease in the form drag encountered by the swimmer.

Kruchoski[18] obtained similar results in a study of the back crawl kick—he concluded that although the kick acted as a retarding force when a swimmer was towed at speeds greater than he could attain using his legs alone, the continuation of the kick resulted in less resistance being encountered than when the legs were not kicking.

Surface drag. It seems unlikely that the surface drag exerted on a swimmer is of sufficient magnitude to be of much practical significance.

Wave drag. The wave drag depends, among other things, on the swimmer's speed, his body shape, and the movements he makes in proximity to the water surface.

The effect of the swimmer's speed on the magnitude of the wave drag has been suggested by the results of Alley[19] and Counsilman,[20] both of whom reported (1) the formation of a pronounced bow wave as the speed at which they towed their subjects increased and (2) a corresponding sharp increase in the rate at which the drag increased relative to the speed.

Among those movements that are particularly effective in creating waves—and thus are particularly undesirable because of the increased drag that they produce—are large up-and-down movements of the swimmer's body. For

[17] Louis E. Alley, "An Analysis of Water Resistance and Propulsion in Swimming the Crawl Stroke," *Research Quarterly*, XXIII, October 1952, p. 269.

[18] Eugene P. Kruchoski, "A Performance Analysis of Drag and Propulsion in Swimming Three Selected Forms of the Back Crawl Stroke" (Ph.D. dissertation, State University of Iowa, 1954).

[19] Alley, "An Analysis of Water Resistance and Propulsion in Swimming the Crawl Stroke," p. 261.

[20] James E. Counsilman, "Forces in Swimming Two Types of Crawl Stroke," *Research Quarterly*, XXVL, May 1955, p. 133.

this reason the practice of pressing vertically with the arms (downward near entry and upward near "release") should be avoided, since it serves merely to produce the kind of up-and-down motion mentioned.

Stroke Frequency

The stroke frequency that a swimmer attains depends on the time he spends in executing each of the two recognized phases of his arm stroke—the pull and the recovery. In the three strokes in which the arms are recovered out of the water the recovery phase tends to be considerably shorter than the pull phase—presumably because of the decreased resistance encountered when the arm(s) move through air rather than water. [Ringer and Adrian[21] found that the pull phase of a group of Yale Varsity swimmers averaged 0.758–0.759 sec (or 65–66% of the total time taken to execute one front crawl stroke) and the recovery phase, 0.391–0.400 sec (or 34–35%]. In the breaststroke there tends to be rather less difference between the times spent in executing each phase, although because the resistance to the forward motion of the hands during the recovery is less than that experienced during the pull, the former is still of a somewhat shorter duration.

The durations of the pull and recovery phases are functions of:

1. The positions of the hand, forearm, and arm relative to an axis through the shoulder(s). If all else is equal, the less the moment of inertia of the arm, the less the time necessary to move it through a given range.
2. The range of motion through which the limb moves. If all else is equal, the greater the range of motion, the greater the duration of the phase in question.
3. The torque applied about the axis through the shoulder(s). Again if all else is equal, the greater the torque applied, the shorter the duration of the phase.

Thus a swimmer has at least three ways in which he can modify the duration of the pull and recovery phases in order to obtain his optimum stroke frequency—he can modify the "shape" of his arm actions, he can adjust the range of these actions, or he can alter the muscular torques that he applies to produce them.

Interrelationship of Stroke Length and Stroke Frequency

Stroke length and stroke frequency are to a very large extent interdependent. As a swimmer increases his stroke length, he generally finds it necessary to increase the time over which he applies forces during the pull phase of the

[21]Lewis B. Ringer and Marlene J. Adrian, "An Electrogoniometric Study of the Wrist and Elbow in the Crawl Arm Stroke," *Research Quarterly*, XL, May 1969, p. 361.

stroke. Thus, while his stroke length increases, his stroke frequency tends to decrease. Similarly, to increase his stroke frequency he generally tends to reduce the time he spends pulling, and this usually leads to a reduction in his stroke length. Thus when a swimmer increases one of these two factors, he must ensure that the other does not suffer a comparable (or more than comparable) decrease if he is to gain any advantage in terms of speed.

In view of the obvious importance of obtaining the optimum combination of stroke length and stroke frequency it is rather surprising that the question of how to determine this optimum combination has received so little attention from those concerned with swimming techniques. Among the few published studies on the subject is that of East[22] who analyzed the performances of swimmers in the 110-yd events at a national championship in an attempt to determine the relationships among stroke length, stroke frequency, and performance. He found significant relationships in the men's freestyle, backstroke, and butterfly events and in the women's butterfly.

> In the men's freestyle, improvements in performance from one end of the time range (89 sec) to the other (69 sec) were characterized by a marked increase in stroke frequency (34 percent) and a slight decrease in stroke length (6 percent). . . . Improvements in performance in men's backstroke from one end of the time range (75 sec) to the other (62.5 sec) were characterized by negligible changes in stroke frequency (1 percent) and a marked increase in stroke length (18 percent). In other words a better performance resulted almost entirely from an increase in stroke length. Similar results were obtained in men's butterfly, with stroke frequency [decreasing] 3 percent and stroke length increasing 24 percent . . . Differences in performance between men and women competing in the same swimming style seem to be directly attributed to the differences in stroke length as the stroke frequencies are similar. . . . This is in direct contrast with sprint running where differences in performance between men and women sprinters seem to be the result of differences in stride frequency [see p. 388][23]

For those events where significant relationships were found, East presented optimum values for the stroke length and stroke frequency (Table 22). These values were obtained using equations derived in the statistical analysis of his data.

Techniques

The techniques of each of the competitive swimming strokes are considered here under four headings—body position, leg action, arm action, and breathing.

[22]David J. East, "Swimming: An Analysis of Stroke Frequency, Stroke Length and Performance," *New Zealand Journal of Health, Physical Education and Recreation*, III, November 1970, pp. 16–27.

[23]*Ibid.*, pp. 22–23.

Table 22. Optimum Stroke Frequencies and Stroke Lengths for Selected Performances*

Time for 110yd (sec)	Men's Freestyle		Men's Backstroke		Men's Butterfly		Women's Butterfly	
	SF (cycles/sec)	SL (ft)	SF (cycles/sec)	SL (ft)	SF (cycles/sec)	SL (ft)	SF (cycles/sec)	SL (ft)
55	0.94	6.1						
56	0.93	6.1						
57	0.91	6.1						
58	0.89	6.1						
59	0.87	6.1						
60	0.85	6.2			0.92	5.9		
61	0.84	6.2			0.91	5.8		
62	0.82	6.2	0.77	6.8	0.91	5.7		
63	0.80	6.2	0.77	6.7	0.90	5.6		
64	0.78	6.3	0.77	6.6	0.90	5.6		
65	0.76	6.3	0.77	6.5	0.90	5.5		
66	0.75	6.4	0.77	6.4	0.89	5.4		
67	0.73	6.4	0.77	6.3	0.89	5.3	0.97	4.9
68	0.71	6.5	0.77	6.2	0.89	5.3	0.96	4.9
69	0.69	6.6	0.77	6.1	0.89	5.2	0.96	4.9
70	0.67	6.6	0.77	6.0	0.89	5.1	0.95	4.8
71			0.77	5.9	0.89	5.0	0.94	4.8
72			0.77	5.8	0.89	5.0	0.94	4.7
73			0.77	5.7	0.89	4.9	0.93	4.7
74			0.77	5.6	0.89	4.8	0.93	4.6
75			0.78	5.5	0.89	4.7	0.92	4.6
76							0.92	4.6
77							0.91	4.5
78							0.91	4.5
79							0.90	4.4
80							0.90	4.4
81							0.90	4.4
82							0.89	4.3
83							0.89	4.3
84							0.89	4.2
85							0.88	4.2

*Adapted from data in East, "Swimming: An Analysis of Stroke Frequency, Stroke Length and Performance," pp. 25–26.

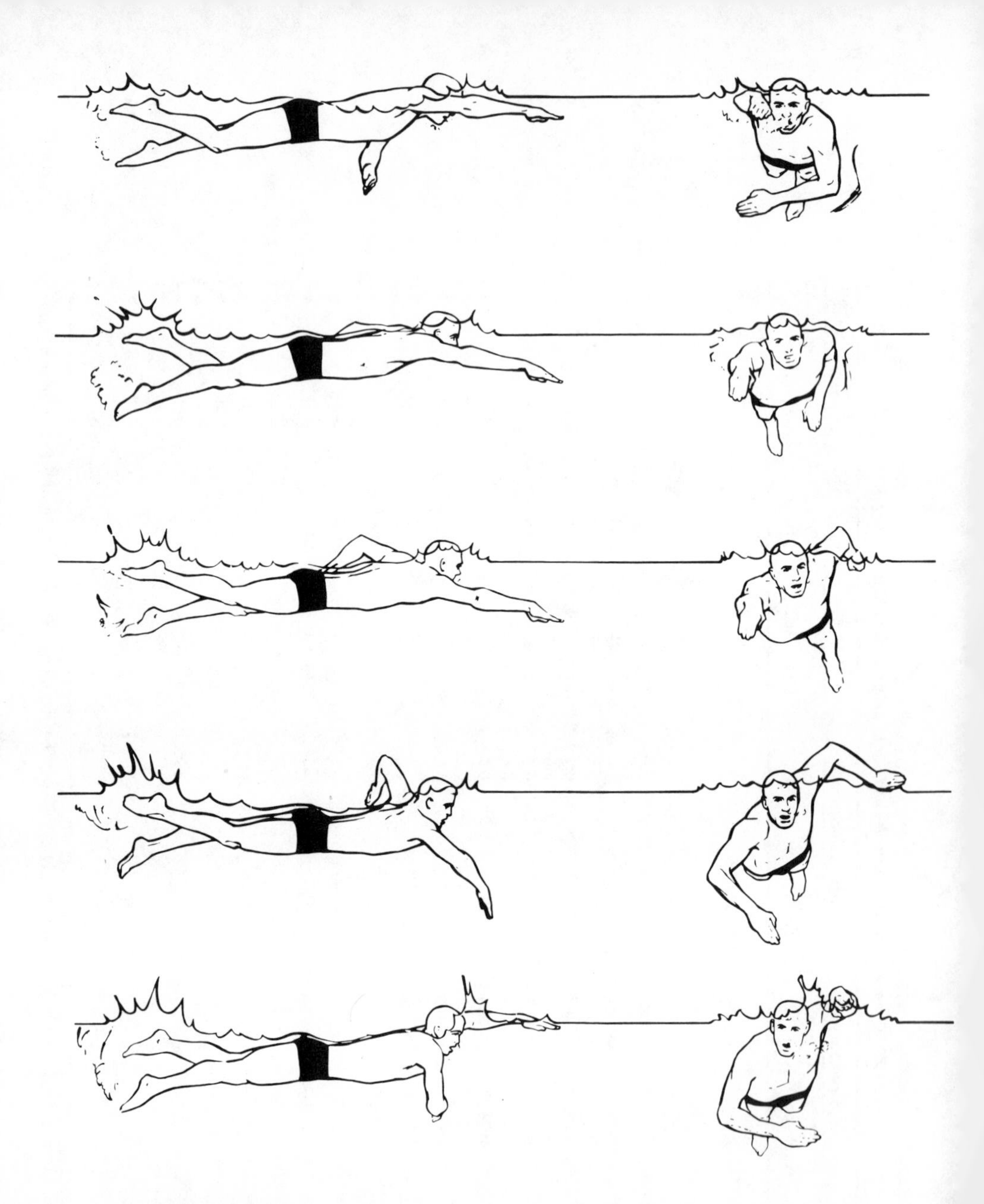

FRONT CRAWL

Body Position

The ideal body position for the front crawl (Fig. 164) would allow the swimmer to both maximize the propulsive forces that he exerted and minimize the resistive forces that he encountered.

In attempting to satisfy these requirements, the good crawl swimmer assumes a prone position in which his head is relatively low in the water

Fig. 164. Front crawl (*Reproduced with permission from Counsilman, The Science of Swimming.*)

(waterline at or above the hairline), his hips slightly lower than his shoulders, and his legs relaxed and extended to the rear.

The elevation of the body relative to the waterline depends on a swimmer's buoyancy (a swimmer with a favorable specific gravity will "ride higher" than one who has a less favorable specific gravity) and on the speed at which he moves through the water (the greater his speed, the higher he will "ride" in the water). While a high body position in the water serves to reduce resistive drag, the swimmer has little real control over the factors that produce such a high position. His buoyancy is fixed—barring drastic changes in his body composition—and the speed at which he swims should already be the fastest he can manage. One thing that will certainly not produce a high body position and a concomitant reduction in drag is a conscious effort to achieve it. Where a favorable specific gravity is not the explanation, a high body position in the water is the result of a swimmer's speed and not the cause of it. To consciously attempt to get "on top of the water" by lifting the head or by pressing vertically downward with the arms in the early part of the stroke will inevitably more than offset any advantage normally associated with a high position. Either the legs will tend to drop as the head is raised (thereby increasing the swimmer's cross-sectional area and the form drag he encounters) or the body, lifted by one arm and then dropped before the next arm can lift it again, will begin to oscillate vertically (increasing both form and wave drags).

Another aspect of body position that is often overlooked and yet is definitely worthy of attention is lateral alignment. If the head, trunk, or legs—those parts of the body that are generally immersed in the water—are allowed to deviate laterally from the straight-line direction in which the swimmer should be moving, the cross-sectional area of the body and the form drag are almost certain to be increased. Such deviations—generally caused by faulty breathing or arm actions—are therefore to be avoided.

The results of a study by Allen[24] provide an interesting sidelight on this question. Allen determined the time it took a group of swimmers to swim 10 yd using their arms alone under two sets of conditions:

1. with the legs unsupported and free to respond to the movements of the upper body,
2. with the legs supported on interconnected balsa floats and thus unable to "counterbalance" the actions of the upper body.

Despite a presumed reduction in drag due to the elevation of the feet, Allen found that the times were faster when the feet were unsupported. Apparently the inability of the legs to counterbalance the arm action produced the "pronounced lateral hip action" observed by Allen, and this in turn produced an increase in the drag sufficient to more than offset whatever advantage was gained by having the feet elevated.

Leg Action

Although the precise function of the leg action—to increase propulsion and/or to decrease resistance—has yet to be satisfactorily resolved, there appears to be fairly substantial agreement concerning which form of leg action produces the best results in practice. Most leading swimmers use an action (the so-called "flutter kick") in which the legs alternate in a vertical, or near-vertical, up-and-down motion, first thrusting upward and backward as the leg is brought toward the surface and then downward and backward as it descends to complete the cycle. With the knee extended and the ankle

[24]Robert H. Allen, "A Study of the Leg Stroke in Swimming the Crawl Stroke" (M.A. thesis, State University of Iowa, 1948).

plantar flexed, the upward action is primarily one of hip extension—the complete upward action being executed with the leg straight. The downward action, on the other hand, incorporates hip flexion, together with knee flexion and extension, in that order.

Several studies have been conducted in an attempt to resolve various questions that have been raised concerning the optimum leg action.

One of the earliest and probably the most comprehensive of these was a study by Cureton.[25] Among the many conclusions reached in this study were the following:

1. The propulsive effects produced by the flutter kick do not derive "from the action of the two legs acting as a wedge on the V-shaped patch of water enclosed" but rather from an action akin to that used by fish. In this action the swimmer's body is pushed forward in reaction to the forces that he exerts to push water backward with his legs and feet.

2. "A conservative statement would say that the up-kick is certainly as valuable, if not more so, for propulsion than the down-kick, although the evidence given indicates that it is *more* effective."

3. Swimmers with the best kicks get a much greater percentage of their leg power from the hips (a mean of 51% was obtained for "four good sprinters of varsity caliber") than do swimmers with relatively poor kicks. One of Cureton's subjects in this latter category recorded 20.5%, while three others could not progress at all when they had to rely on hip action. [The figures were obtained by comparing times recorded when (a) knees, (b) ankles, and (c) knees-and-ankles were immobilized in turn, with those obtained when the subject used his normal flutter kick.]

4. Expert swimmers with better kicks have greater flexibility in the ankles than average swimmers with poorer kicks. (Robertson[26] has since found significant relationships between ankle flexibility and propulsive force.)

5. For the majority of the subjects the best performances were recorded when the knees were allowed to bend to approximately 15°.

6. For best results a person should kick as wide a kick as his physique and strength will permit, up to a maximum of 24 in. (A number of other investigators[27,28] have also endeavored to determine the optimum width of kick. However, since their conclusions, like those of Cureton, are based

[25]Thomas K. Cureton, "Mechanics and Kinesiology of Swimming (The Crawl Flutter Kick)," *Research Quarterly*, I, December 1930, pp. 87–121.

[26]David F. Robertson, "Relationship of Strength of Selected Muscle Groups and Ankle Flexibility to the Flutter Kick in Swimming" (M.A. thesis, State University of Iowa, 1960).

[27]Allen, "A Study of the Leg Stroke in Swimming the Crawl Stroke."

[28]George L. Poulos, "An Analysis of the Propulsion Factors in the American Crawl Stroke" (M.A. thesis, State University of Iowa, 1949).

upon tests in which the subjects were not using their arms, and thus were traveling at much lesser speeds than they would when swimming the full stroke, the applicability of their results to this latter situation is open to question.)

7. The width and rate of kick are intimately related, a wide kick usually calling for a slower rate, and vice versa.

The effects produced by variations in the width of kick have also been studied by Alley,[29] who found that a normal kick (approximately 12 in. between feet at maximum spread) was superior to a short kick (approximately 6 in. between feet at maximum spread) on practically all the tests conducted. When his subject swam using only his legs, he produced greater propulsive forces with the normal kick than with the short one. When he was towed at speeds greater than he could normally attain with his legs alone, use of the normal kick resulted in less drag being created than did use of the short kick. Finally, when he swam using both arms and legs, a given arm action combined with a normal kick almost invariably produced greater propulsive forces than the same arm action combined with a short kick.

Attempts have also been made to determine the optimum number of kicks per complete arm cycle. Thrall,[30] for example, compared the performances of three varsity swimmers when they used a "normal" six-beat kick and when they used a "feathered" or two-beat kick. (*Note:* The actions of the legs are commonly referred to in terms of the number of downward beats that are made during a complete arm cycle.) He found that the average speed of his three subjects increased by 0.6, 0.8, and 0.7 fps when they used a six-beat kick as compared to when they swam using their arms only. When they used a two-beat kick, the corresponding changes in speed were 0.1, 0.1, and 0 fps. He therefore concluded that, at least for the stroke frequency used in the study (1 cycle/sec), the six-beat kick makes a greater contribution to forward speed than does the two-beat kick.

More recently, Eaves[31] has suggested that the widespread popularity of the six-beat crawl—most good sprint swimmers use a leg action of this kind—can be accounted for in terms of the swimmer's need to conserve angular momentum about his long axis. Thus he contends that as the swimmer's right arm and shoulder go down and his left arm and shoulder come up (near the beginning of the pull with the right arm), his left leg and hip must go down and his right leg and hip come up, in order to provide the necessary

[29]Alley, "An Analysis of Water Resistance and Propulsion in Swimming the Crawl Stroke," pp. 253–70.

[30]William R. Thrall, "A Performance Analysis of the Propulsive Force of the Flutter Kick" (Ph.D. dissertation, State University of Iowa, 1960).

[31]George Eaves, "Angular Momentum and the Popularity of the Six-Beat Crawl," *First International Symposium on Biomechanics in Swimming, Waterpolo and Diving Proceedings,* ed. by L. Lewillie and J. P. Clarys, Université Libre de Bruxelles Laboratoire de L'effort, 1971.

conservation of angular momentum. When the left arm and shoulder go down near the beginning of the next pull (with the left arm), the reaction from the lower body must be in the opposite direction—the right leg and hip going down and the left leg and hip coming up. This, he maintains, can only occur if the number of leg beats per half arm cycle (or per pull) is an odd number, i.e., is of the form $(2n + 1)$ where $n = 0, 1, 2, \ldots$, etc. The number of beats per complete arm cycle must therefore be of the form $2(2n + 1)$ where $n = 0, 1, 2, \ldots$, etc.

Of the various alternatives this theory permits, only the first three—2-beat, 6-beat and 10-beat—are at all feasible. With respect to these, he concludes that "... the 2-beat crawl is too slow [cf. Thrall's conclusion] and the 10-beat crawl is too fast a leg-beat for most swimmers, leaving the 6-beat crawl as the only crawl-stroke comfortable as regards the speed of the leg-beat and the conservation of angular momentum."

Arm Action

For the sake of analysis the arm action is generally considered to be divided into two parts:

1. a *pull phase* that begins as the hand enters the water and ends as it leaves,
2. a *recovery phase* during which it is moved forward above the water in preparation for the next pull phase.

Pull Phase. As the hand enters the water some distance in front of the corresponding shoulder, it travels forward and slightly downward with the elbow—bent and somewhat higher than the hand at entry—extending in the process. Although a swimmer may gain some propulsive lift in the course of this initial forward and downward motion of his hand, it seems likely that its overall effect is to retard his forward progress. However, since this forward and downward motion also places the hand almost as far forward as it can go relative to the body and into a position from which it can then be pulled backward beneath the body, it seems likely that whatever the swimmer loses as a result of the resistance created by the initial forward motion is more than offset by advantages he gains in terms of the added range through which he can exert propulsive forces. Were this not the case, swimmers might be well advised to try to have their hands moving backward at a speed at least equal to their forward motion at the time of entry—cf., recovery leg action in running, pp. 395–96. Instead, however, the hand travels some 12–28 in. before reaching a speed equal to the forward speed of the swimmer's body. (*Note:* In the case of poor swimmers this means that the arm may actually be at right angles to the trunk before any forward propulsion is obtained!)[32]

[32]Counsilman, *The Science of Swimming*, p. 45.

Having reached the point at which his arm begins to evoke a propulsive reaction from the water—a point commonly referred to as the *catch*—the swimmer's task is to ensure that (consistent with anatomical and physiological limitations) this propulsive reaction is of the greatest magnitude possible, that it acts over the longest possible distance, and that its line of action is such as to provide the best possible results in terms of forward speed.

To ensure that the propulsive reaction is as large as he can reasonably make it, the swimmer bends his elbow as he begins to bring his arm downward and backward to pass beneath his body. This reduces the distance between (1) the point at which the resultant force he exerts on the water is applied and (2) the shoulder joint axis about which his arm is rotating. This in turn ensures that, for any given torque applied to the upper arm, the force exerted against the water is greater than it would be if the arm were kept straight. In addition, by keeping his elbow high relative to his hand during the early stages of the pull, the swimmer permits the muscles that medially rotate his upper arm to make a contribution to the force that he exerts backward and downward against the water.

The bending of the arm and the maintenance of the elbow in a relatively high position also has the effect of keeping the hand moving horizontally backward over a longer distance (and for a longer time) than would be the case if the arm remained straight throughout. (*Note:* Over much of the distance that the hand traverses in the latter case, its motion is primarily in a vertical direction—downward during the early stages of the pull and upward during the final stages. As a direct consequence, the distance over which the arm can make a major contribution to the swimmer's forward speed is probably fairly small.)

The forces that the water exerts on a swimmer—like the forces acting on any body—serve to accelerate him in the direction in which they act and, if they act eccentrically, to rotate him about an axis through his center of gravity. Since a swimmer's prime concern is to get down the pool as fast as he can, he obviously wants the forces exerted on him to act in a forward horizontal direction. However, because the resistance that he encounters also has a bearing on the results of his efforts, it is important that these horizontal forces do not set up rotations that will add materially to the resistive forces acting on him. One way of achieving this is for the swimmer to pull along a line vertically below and parallel to his long axis. Not only does this encourage the bent-arm, high-elbow action already referred to, but it also reduces the likelihood of his initiating unwanted rotations about his frontal (anteroposterior) axis—rotations that would tend to move his body out of alignment in a lateral direction.

In practice, however, deviations from such an action are necessary. In the first place the hand is usually placed in the water in front of the corresponding shoulder—presumably to minimize the resistive drag during the first part of the pull by ensuring that the arm is in the most streamlined position possible and perhaps also to eliminate any unnecessary drag associated with increasing the body roll to get an "on-line" entry. Second, once the swimmer's arm has passed beneath the line of his shoulders and the most productive part of the pull phase has been completed, it must be moved outward in order that it can pass around the body and up toward the surface. (Incidentally, in order to continue evoking propulsive force during these latter stages of the pull phase, most good swimmers keep the palm of the hand in a vertical or near-vertical plane and "facing" backward as they bring the arm from beneath the body and then back and up in preparation for the recovery). These two deviations result in the hand following a path that, when viewed from above or below, looks like an inverted question mark.

The path followed by the swimmer's hand relative to the midline of his body has frequently been the subject of comment, much of it concerned with whether the hand should cross the midline during the course of the pull. However, if one accepts the basic proposition, that the resultant propulsive forces exerted on the swimmer as a result of his pull should act along a line vertically below his longitudinal axis, this question effectively answers itself. For, since the resultant propulsive force at any instant is the summed effect of the propulsive forces evoked by the arm, forearm, and hand as they move backward through the water, it is clear that this resultant will act at a point somewhere between the fingertips and the shoulder joint—probably somewhere between the wrist and the elbow. Under these cricumstances, the hand or some part of it *must* cross the midline of the body if the point at which the resultant propulsive force is applied is to lie on a line vertically below the swimmer's longitudinal axis.

Recovery Phase. At the completion of the pull, the swimmer lifts his arm from the water preparatory to swinging it forward for entry and the start of its next pull phase. The act of withdrawing the hand from the water is effected with a high lifting action of the arm (elbow high), coordinated with a slight rolling of the trunk, which elevates the shoulder and thus contributes to the lifting action of the arm. Once the hand is clear of the water, the arm is swung forward as those muscles that abduct and rotate the shoulder and flex the elbow come into play. To minimize the lateral angular reaction of the legs to the recovery action of the arm, the arm is brought forward in as near a vertical plane as the swimmer's body position will permit.

Breathing

In addition to being of obvious importance physiologically, the manner in which the swimmer breathes is of some consequence from a biomechanical standpoint. Unless he incorporates his respiratory movements into the whole stroke so that they neither interfere with his ability to produce propulsive force nor add to the resistance that he meets, his performance will be materially affected. In order to see that this does not occur or, if it must, to minimize the effects produced, the good swimmer generally tries to take the least number of breaths consistent with his physiological needs and to take them in such a manner that his body position is changed as little as possible in the process. Thus when he needs a breath, he rotates his head about its long axis until his mouth, deep in the trough of the bow wave created by his head, is just clear of the water. He then inhales and returns his head, rotating once again about its long axis, to its original position.

Time trials conducted by Cureton[33] have shown "conclusively that swimmers can swim faster for short distances without breathing regularly and that they become slower in direct proportion to the number of breaths taken." The explanation for this might very well be accounted for in terms of resistance measures reported by Karpovich.[34] He found that ordinary turning of the head for breathing increased the resistance about 0.5 lb at a speed of 3 fps and about 1.5 lb at a speed of 5 fps.

BUTTERFLY STROKE

The dolphin butterfly stroke evolved from the orthodox breaststroke during the early 1930's. The first stage in this evolutionary process was the discovery that an out-of-the-water recovery action not only conformed with the then-existing rules governing breaststroke swimming but also markedly increased a swimmer's speed by decreasing the resistance that he encountered. This development was closely followed by the appearance of the so-called dolphin kick in which both legs move simultaneously in a vertical plane instead of in the roughly horizontal plane of the then-orthodox breaststroke kick. While this type of kick has since been shown to be vastly superior to previous types of breaststroke kick in terms of speed, it was many years before its use gained official sanction—it was 1955 before the dolphin kick was officially acceptable in intercollegiate competition and 1956 before it made its debut in an Olympic Games event.

Body Position

In some respects the body position adopted by a good butterfly stroke swimmer is similar to that adopted by his counterpart in a front crawl event—he assumes a prone position in the water; he endeavors to minimize the drag opposing his motion by keeping his head relatively low, his legs relatively

[33]Thomas K. Cureton, "Relationship of Respiration to Speed Efficiency in Swimming," *Research Quarterly*, I, March 1930, p. 66.

[34]Peter V. Karpovich, "Water Resistance in Swimming," *Research Quarterly*, IV, October 1933, p. 26.

high, and the up-and-down movements of his body to a minimum (insofar as is consistent with the other things he must do). In other respects there are differences. Because both arms pull and then recover together, the use of a body roll to facilitate either or both of these actions is not feasible—the swimmer's body therefore retains its prone position throughout each stroke. The simultaneous action of the arms also acts to the swimmer's advantage during the recovery, for since the action of one arm effectively balances (or reacts with) the opposite action of the other arm, the swimmer need have no concern that his legs will become misaligned and increase the resistance with which he has to contend.

Leg Action

While the technique widely used in the years between the discovery of the butterfly arm action and the bestowing of official recognition on the dolphin kick involved a combination of the "new" arm action and the "old," orthodox kick, this particular combination (the so-called butterfly-breaststroke) has now all but disappeared. Its place has been taken by the dolphin butterfly stroke that is now used almost exclusively at all levels of competitive swimming.

While perhaps not apparent to the casual observer, the leg action in the dolphin butterfly stroke is practically identical with that used in the front crawl except that both legs move together rather than in opposition and the number of beats per arm cycle is generally considerably less. This is well illustrated by the sequences of Fig. 165. The top sequence depicts one complete cycle of the action of the right leg of a front crawl swimmer. (The actions of the left leg can be seen in the background.) The bottom sequence shows the actions of both legs during one cycle of a dolphin kick. Now, while there are *slight* differences in the inclination of the trunk and in the extent to which the hip, knee, and ankle joints are flexed during the downbeat, the similarity between the two sequences is obvious. (*Note:* These sequences have been taken from photographs of a swimmer kicking on a kickboard and thus do not exactly replicate the actions used when the whole stroke is employed. Instead, since the legs are the sole source of propulsion in this case, the range and force of the kicks in Fig. 165 are somewhat greater than they would be in a whole stroke.)

In front crawl events the number of downward beats of the legs per complete arm cycle usually varies with the length of the race—a six-beat crawl generally being used for sprint events and some lesser number for longer events. In butterfly events the swimmer normally executes two leg beats per arm cycle, the first, and generally the more forceful, starting as the hands enter the water, the second starting as the hands pass beneath the line of the shoulders and ending shortly before they leave the water.

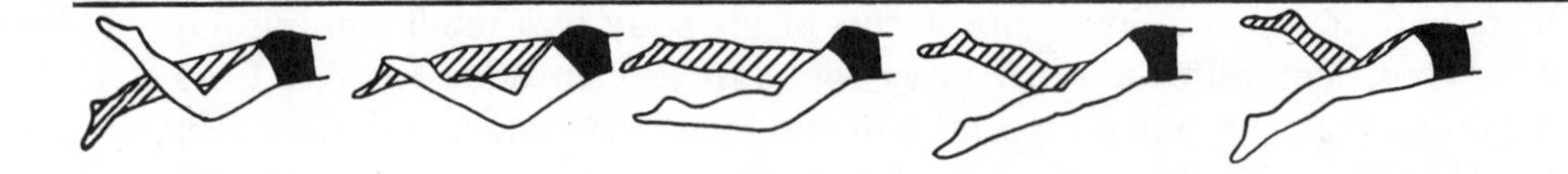

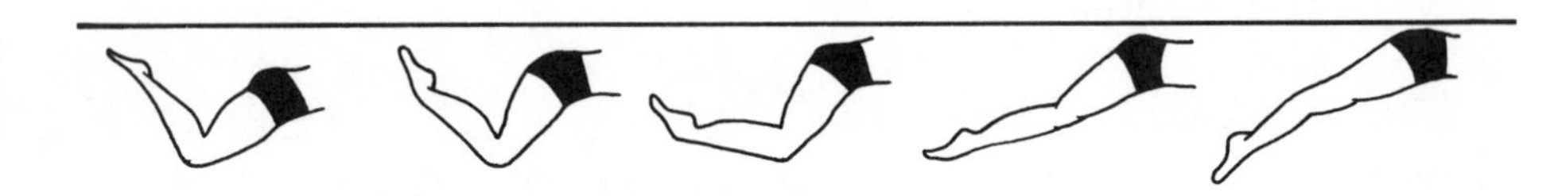

Various aspects of the dolphin kick have been the subject of study; Barthels and Adrian[35] studied the muscular activity, the ranges of motion at various joints, and the timing of the kick of four intercollegiate swimmers and found:

1. When only the legs were used for propulsion, no major-minor kick pattern was observed. For each subject, each kick in a sequence was identical in character to the others. When a full stroke was used, alternating major and minor kicks (i.e., major and minor in terms of time and/or range of joint motion) were noted in all cases. On the basis of these findings Barthels and Adrian questioned the value of practicing series of identical kicks using the legs alone if refinement of the timing of the leg action is the desired objective.

2. Coordinated contractions of the rectus abdominis and erector spinae muscles revealed their roles in producing, respectively, flexion and extension of the spine. When these two muscles contracted together, they served to stabilize the trunk. Barthels and Adrian concluded that "The activity of these musles indicated an active participation by the trunk as an inherent part of the total kicking movement and would suggest the need for the study of spinal movement during the kick in future research."

3. The muscles of the lower leg apparently became stretched due to the pressure of the water on the foot, and contracted reflexly to prevent farther stretching. Barthels and Adrian concluded that "the development of flexibil-

[35]Katharine M. Barthels and Marlene J. Adrian, "Variability in the Dolphin Kick Under Four Conditions," *First International Symposium on Biomechanics in Swimming, Waterpolo and Diving Proceedings*, ed. by L. Lewillie and J. P. Clarys, Université Libre de Bruxelles Laboratoire de L'effort, 1971.

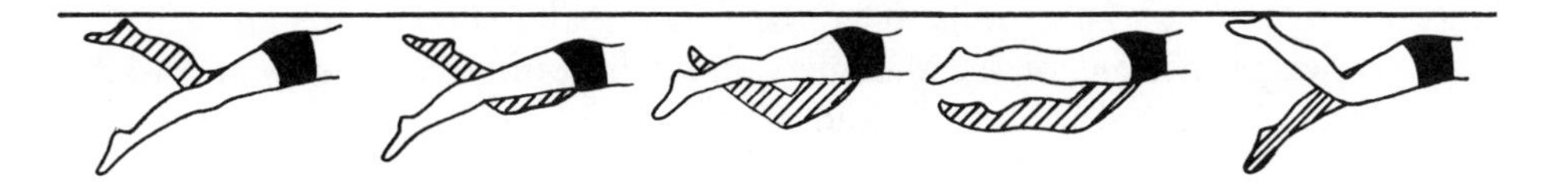

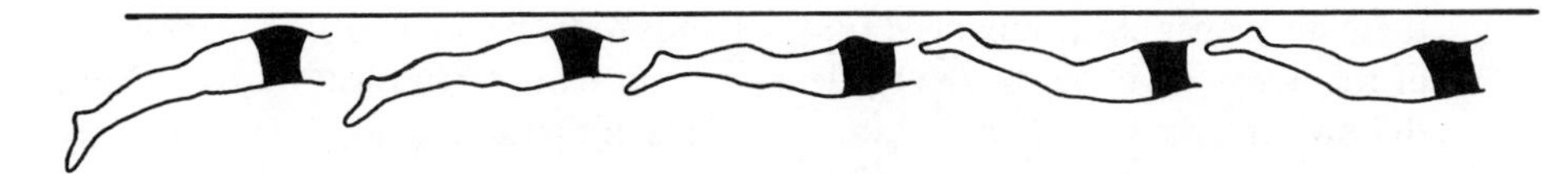

Fig. 165. The leg action in the dolphin kick (lower sequence) is very similar to that in the front crawl (upper sequence).

ity for greater plantar-flexion of the foot would be more worthwhile than concentration on strength development in the lower leg."

4. Some interrelationships between the ranges of motion at hip and knee joints were noted—a large range of motion at the hips being associated with a small range at the knees, and vice versa.

Kersten[36] examined the effects of two types of dolphin kick—kick A (a maximum knee action-minimum hip action kick) and kick B (a maximum hip action-minimum knee action kick)—on the time taken to swim 10 yd. He found that kick A was significantly faster than kick B when only the legs were used and that the whole stroke using kick A was significantly faster than the whole stroke using kick B. (In the latter case the difference between the mean times was 0.3 sec for 26 highly trained competitive swimmers and 0.8 sec for 8 less experienced swimmers.) He suggested therefore that his findings "appear to warrant the conclusion that the Dolphin leg drive when the subject utilizes maximum knee action and minimum hip action is more effective than the Dolphin leg drive when the subject utilizes maximum hip action and minimum knee action."

Arm Action

The arm action used in the butterfly stroke (like the body position and leg action) bears a strong resemblance to the corresponding action in the front crawl. Since both arms pull and then recover simultaneously, however, some modifications of the basic crawl stroke action are necessary.

[36]Orville A. Kersten, "Propulsion Factors in Swimming the Dolphin Butterfly" (M.A. thesis, State University of Iowa, 1960).

Pull Phase. The hands enter the water (palms downward and slightly outward) in front of the shoulders and a little more than shoulder width apart. The arms are relaxed and almost straight and the elbows are a little higher than the hands. Then, as the hands move forward, downward, and backward to the catch position, they are brought slightly wider apart thereby tracing out the first part of the so-called double-S pull (also known as the hourglass or keyhole pull) currently used by virtually all good butterfly swimmers. From this position the hands are pulled backward and inward toward the midline of the body, thereby following essentially the same path as in the front crawl. However, because both arms are pulling simultaneously, there no longer exists any need for the hands to be brought across the midline of the body (cf. p. 355)—the tendency of one arm to produce rotation about the swimmer's frontal axis (and, too, his longitudinal axis) is effectively canceled by the opposing tendency of his other arm. Instead, with arms bent and elbows high, the hands are drawn in toward the midline, pass beneath the shoulders, and finally go outward and backward to finish at (or near) the upper thighs.

Recovery Phase. As the pull phase ends, the arms are well-nigh straight and the palms of the hands are facing upward. From this position the swimmer rotates his arms outward and then swings them forward and around, close to the surface of the water, and toward the point at which the next entry will be made. [*Note:* The bent-arm, high-elbow action characteristic of the recovery phase in the front crawl is not used in the butterfly stroke for at least two reasons: (1) The anatomical structure of the shoulder joint makes it a virtual impossibility to perform such an action unless the body rolls and (2) the balancing of one arm by the other eliminates the need for such an arm action anyway.]

Breathing

While raising the head to take a breath almost inevitably increases the resistance and thus adversely effects a swimmer's speed, a good swimmer minimizes this effect by appropriate timing of his breathing pattern. In the first place he breathes only as often as necessary to meet his physiological needs—perhaps once every two to three arm cycles in sprint events and once every one to two arm cycles in the longer events. Second, he times his movements so that he lifts his head when it has the least distance to move to achieve the desired result. This occurs in the latter half of the pull phase when the shoulders have been elevated by the action of the arms and by the second beat of the leg action taking place at that time. Once he has lifted his head up and back until his mouth is just clear of the water—any unnecessary elevating of the head will tend to force the legs lower and create added resistance—the swimmer inhales and lowers his face back between his arms, which by this time are swinging past the line of his shoulders on their way forward to the entry.

BACK CRAWL

Originally a form of inverted breaststroke, the back crawl (Fig. 166) has since evolved into what might more aptly be described as an inverted front crawl—the simultaneous action of the arms and the so-called "frog kick" of the early technique having given way to the alternating arm action and the "flutter" kick of the modern stroke.

Body Position

The back crawl is the only competitive stroke in which the swimmer adopts other than a prone position in the water. In the back crawl he assumes a near-horizontal position on his back with his chin close to his chest (yet far enough away to allow the bow wave to break across the top of his head), his trunk and legs loosely extended, and his hips just low enough in the water to ensure that his kick will be beneath the surface. The relationship between the positions of the head and hips is of particular importance. If the swimmer has his head so far back that he is unable to see the water being disturbed as he kicks, he is likely to cause his hips to rise to the point where his kick becomes ineffective. Thus, although the resistance encountered will almost certainly be less with the body in a horizontal position, the loss of propulsion from the legs almost certainly outweighs any gains from that source. Conversely, if the head is brought too far forward, the hips are likely to be lowered more than is necessary to obtain an effective kick. The added resistance evoked in this manner and the lack of any offsetting advantage to such an alignment make it an unnecessary liability.

Leg Action

The leg action in the back crawl is essentially the same as that in the front crawl except for the obvious differences due to the change in the swimmer's position.

Arm Action

While there has long been substantial agreement among swimming authorities concerning the body position and leg action most likely to produce success in the back crawl, only in recent years has there been any suggestion of agreement concerning the optimum arm action. The main source of disagreement has been with regard to the relative merits of the straight-arm and the bent-arm actions.

The 1936 Olympic champion Adolph Kiefer (U.S.A.) is generally credited with popularizing the straight-arm action. According to Armbruster, Allen, and Billingsley:

> The technique developed by Kiefer contained three features that distinguished his back crawl stroke from the generally used form. First, the recovery of the arms was made in a very low, lateral fashion, with the arms held straight. Second, the arms entered the water just above the line opposite the shoulders instead of straight up from the shoulders, alongside the head. Third, the arms were drawn through the water just below the surface as contrasted with the deep pull then employed generally and were held straight.[37]

[37]Armbruster, Allen, and Billingsley, *Swimming and Diving*, p. 10.

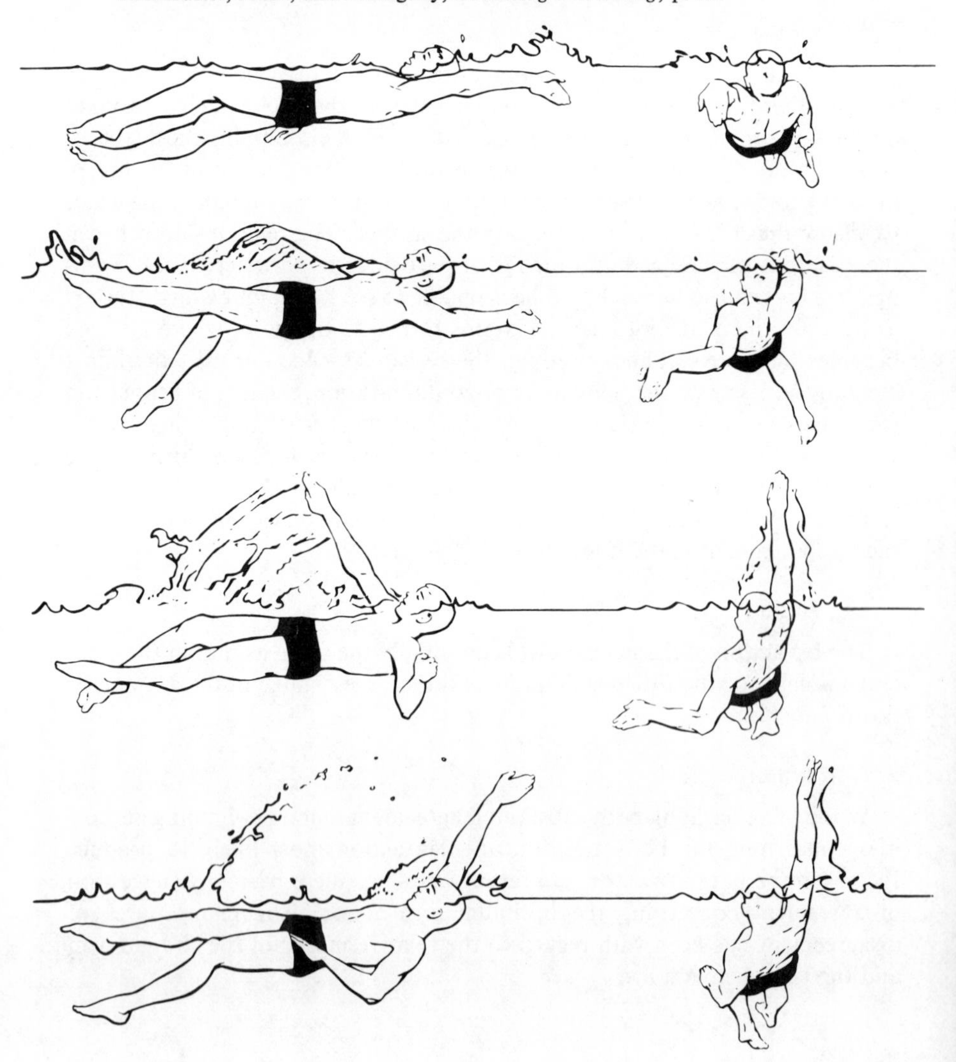

Fig. 166. Back crawl (*Reproduced with permission from Counsilman,* The Science of Swimming.)

Following his successes the technique used by Kiefer rapidly became accepted as the optimum for back crawl swimmers and remained so for probably 15–20 yr. Around this time it began to become apparent that many of the top swimmers were in fact bending their arms during the pull phase of the stroke rather than keeping them straight in the "Kiefer style." Today another 20–25 yr later, good back crawl swimmers use a technique (Fig. 166) that retains none of the three distinguishing features of the Kiefer technique referred to by Armbrusteı, Allen, and Billingsley. The recovery of the arm proceeds in an essentially vertical plane with entry being made at near-full reach, above and only slightly to the side of the shoulder. The pull begins with the arm nearly straight, continues with the arm bending to a near 90° angle, and finishes with the arm nearly straight again.

Two studies of the back crawl—among the very few studies that have been concerned with this stroke—are of interest.

Mindheim[38] compared the times that 10 experienced competitive swimmers took to swim 30 yd using straight- and bent-arm actions. He found no significant difference between the times recorded when his subjects used one arm action and those recorded when they used the other. Since most of his subjects normally used the straight-arm action and since all of them apparently had some difficulty mastering the bent-arm method—both circumstances tending to bias the results in favor of the straight-arm action—the results obtained would appear to be capable of being interpreted in two different ways. Either there is no difference betewen the two actions or the superiority of the bent-arm action was masked by the subjects having had extensive practice in the use of the straight-arm method. The results currently being obtained in practice suggest rather strongly that the latter of these two is the more likely to be correct.

Kruchoski[39] investigated the effectiveness of three depths of arm stroke—shallow (approximately 6 in.), medium (approximately 12–16 in.), and deep (maximum depth). Although he did not specifically state which type of arm action was used, it would seem reasonably safe to assume that the straight-arm action, very much in vogue at the time, would have been used by each of the three subjects. While the results obtained by Kruchoski can hardly be regarded as unexpected—his subjects consistently performed better when using the medium-depth action, the one they normally used in competition, than when using either of the other two—his observations are of some interest. He noted that:

> (1) While using the shallow-depth arm stroke, the subjects produced considerable lateral movement which could be expected to increase resistance and therefore decrease surplus propulsive-force.

[38] Arthur D. Mindheim, "Speed and Propulsive Force of Two Types of Back-Crawl Stroke" (M.A. thesis, State University of Iowa, 1960).

[39] Kruchoski, "A Performance Analysis of Drag and Propulsion in Swimming Three Selected Forms of the Back Crawl Stroke."

(2) The deep arm stroke caused excessive rolling of the body and caused the hips to be moved out of line horizontally.

(3) The use of the medium-depth arm stroke tended to minimize these factors.

(4) The degree of flexibility each subject possessed in the shoulder region limited the maximum depth of arm stroke.

(5) While using the deep arm stroke, each subject felt as if his legs were being forced into a deeper position. This increased body inclination and therefore increased resistance.

The arm action currently used by good back crawl swimmers eliminates, or minimizes, many of these and other similar problems. Once the hand has moved downward and slightly backward to the point at which the catch is made, the swimmer's arm begins to bend. In this way the arm can not only push backward forcefully (probably more forcefully than if it were kept straight)[40] but, by passing neither too far to the side nor too deep, can avoid evoking the horizontal or vertical reactions that will cause the resistance to be increased [cf., No. (1) and (5)]. The pull is completed with the arm straightening, rotating, and thrusting down beside the swimmer's thigh. This final downward thrust provides the impetus to roll the body about its long axis and facilitates the recovery of the arm by elevating the shoulder on the same side. In addition, the lowering of the opposite shoulder puts the arm on that side in a stronger anatomical position for the pull that is just beginning.[41] The recovery is executed with the arm, straight and relaxed, being lifted from the water and swung overhead in a vertical or near-vertical plane. The recovering of the arm in this manner eliminates the possibility that the feet may move laterally in reaction to the recovery—a very real possibility if a Kiefer-type lateral recovery action is used.

Breathing

Since the swimmer's head is held in a constant position and his mouth and nose are clear of the water at all times other than during starts and turns, the optimum pattern of breathing is a physiological question rather than a biomechanical one.

BREASTSTROKE

Probably the first swimming stroke to be developed, the breaststroke is ill-suited to propel a man through water at speed—a fact attested to by its secure rating as the slowest of the four competitive strokes. Its principal

[40]Forbes Carlile, *Forbes Carlile on Swimming* (London: Pelham Books, 1964), p. 191.

[41]Kruchoski, "A Performance Analysis of Drag and Propulsion in Swimming Three Selected Forms of the Back Crawl Stroke," pp. 47–48.

value would appear to lie outside the realm of competition. It may be used for long periods with a minimal output of energy, it allows the swimmer a clear view ahead, and it makes use of a very simple breathing technique—features that account at least in part for the important role the stroke plays in lifesaving and survival swimming.

Body Position

The ideal body position is similar to that for swimming the front crawl and the butterfly stroke. The swimmer should have his body as near horizontal and as streamlined as he can, consistent with his need to have (1) his head close enough to the surface to permit him to breathe with relative ease and (2) his legs sufficiently below the surface to ensure that they can exert the maximum propulsive force possible.

In breaststroke the changes in the positions of the swimmer's limbs probably have a greater influence on the resistance he encounters than do the corresponding changes in any other stroke. A study by Kent and Atha[42] provided some very interesting results concerning the magnitude of the resistance encountered at different stages in the breaststroke action (Fig. 167). When they towed their subjects through the water at velocities up to 0.5 m/sec (equivalent to the average speed for a 2-min 13.3-sec, 200-m), they found that the resistance increased in the following order: glide, post-thrust, breathing, pre-thrust, and recovered. Relative to the glide position, the resistance at 1.5 m/sec increased by a multiple of 1.91 for breathing, 2.01 for post-thrust, 2.28 for pre-thrust, and 2.37 for recovered. (For example, with one subject who encountered resistance of 21.4 lb when gliding at 1.5 m/sec, the resistance values for the other four positions were 40.8 lb for breathing, 43 lb for post-thrust, 48.7 lb for pre-thrust, and 50.7 lb for recovered. These results are depicted in Fig. 167. Similar results were recorded for the other two subjects.)

Two aspects of these results stand out. First, the resistance increased tremendously from that recorded for the glide to that recorded for the breathing position—an increase that almost doubled the resistance! Second, although the towing speed of 1.5 m/sec is admittedly very fast—close to the average speed necessary to break the world record for 100 m—the magnitudes of the resistive forces acting on the swimmer are almost certainly much larger than is commonly supposed. (It is surely no wonder that the swimmer slows down during the recovery of his arms and legs!)

[42] M. R. Kent and J. Atha, "Selected Critical Transient Body Positions in Breaststroke and Their Influence Upon Water Resistance," *First International Symposium on Biomechanics in Swimming, Waterpolo and Diving Proceedings*, ed. by L. Lewillie and J. P. Clarys, Université Libre de Bruxelles Laboratoire de L'effort, 1971.

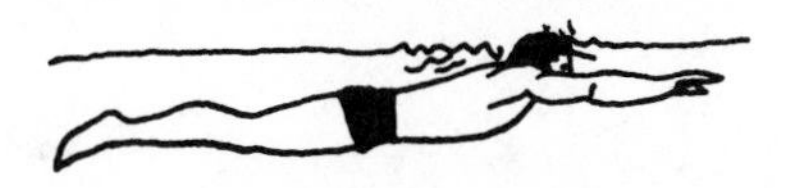

Glide (21.4 lb)

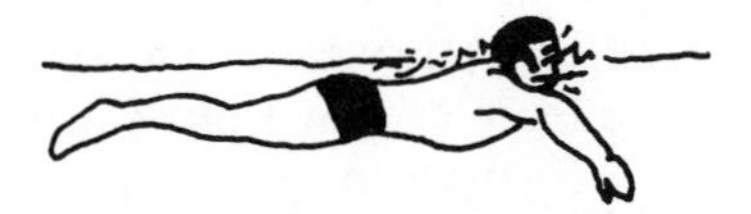

Breathing (40.8 lb)

Recovered (50.7 lb)

Pre-thrust (48.7 lb)

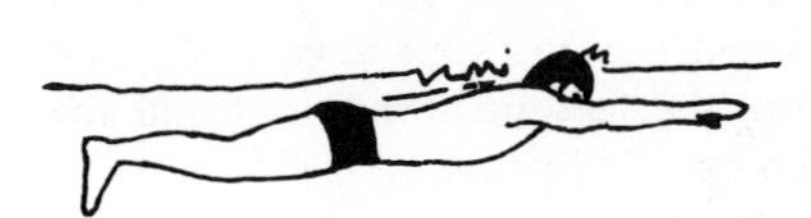

Post-thrust (43 lb)

Fig. 167. Resistance encountered by a breaststroke swimmer at selected instants in his stroke. (*Based on data in Kent and Atha, "Selected Critical Transient Body Positions in Breaststroke and Their Influence upon Water Resistance."*)

Leg Action

For many years the so-called wedge kick was believed to be the optimum breaststroke leg action, and in some quarters this belief persists to the present day. In this type of kick, the recovery phase of the leg action began with the legs together and extended. The legs were then drawn up with the heels held together and the knees moving outward away from each other. At the completion of the recovery phase, the ankles were dorsi-flexed, the heels were together, the knees were spread well apart, and the legs as a whole bounded a near-horizontal diamond-shaped area. The propulsive or driving phase began with the feet being thrust outward and backward to form a wide V or wedge between the legs. The final movement consisted of a vigorous slamming together of the legs, which, so it was thought, produced a forceful backward expulsion of the wedge of water lying between them, and the propulsion that the swimmer experienced.

In relatively recent years the wedge kick has been replaced by a kick (or kicks) in which the width of the leg action is less than in the wedge kick—and thus produces less resistance—and in which the emphasis is on a forceful backward push (primarily with the soles of the feet) rather than on a slamming together of the legs. [*Note:* The reason for the apparent uncertainty as to just how many different kicks have come to supplant the wedge kick lies in the fact that the terms used to describe the different leg actions vary considerably from one authority to the next. Thus one person will refer to an action as if it were simply a variant of another, while a second will give each action a different name. Then, as if this isn't enough in itself, such people only seldom take the care to accurately and completely define the action(s) to which they are referring. Inevitably, the result is confusion—a confusion, it might be observed, that is thoroughly deserved by those who create it!]

For the purposes of further discussion two kicks will be defined here. These will be referred to as the *frog kick* and the *whip kick*, respectively.

Frog kick.

> The recovery starts from the streamlined glide position. The legs are drawn up with the knees dropped slightly, turned outward, and separated more than the feet. As the legs are drawn up, the feet are turned out and the ankles dorsiflexed in preparation for the propulsive action.
>
> In one continuous propulsive movement the knees are rotated inward as the soles of the feet thrust outward, backward and together. As the legs are brought together and extended the feet are also extended.[43]

Whip kick.

> The whip kick (Fig. 168) is a kick in which relaxed knees drop toward the bottom and separate slightly as the heels are drawn toward the buttocks. The ankles are dorsiflexed and the feet turned outward to grip the water as the legs are whipped out, back, and together. As the legs are brought together and extended the feet are also extended.
>
> In this kick the knees separate but remain closer together than the feet during the recovery and there is little or no rotation of the femur.

[43]Mary E. Over, "A Comparison of the Force and Resistance of the Frog Kick and Whip Kick Used in Swimming the Orthodox Breast Stroke" (M.A. thesis, Long Beach State College, 1963), p. 5.

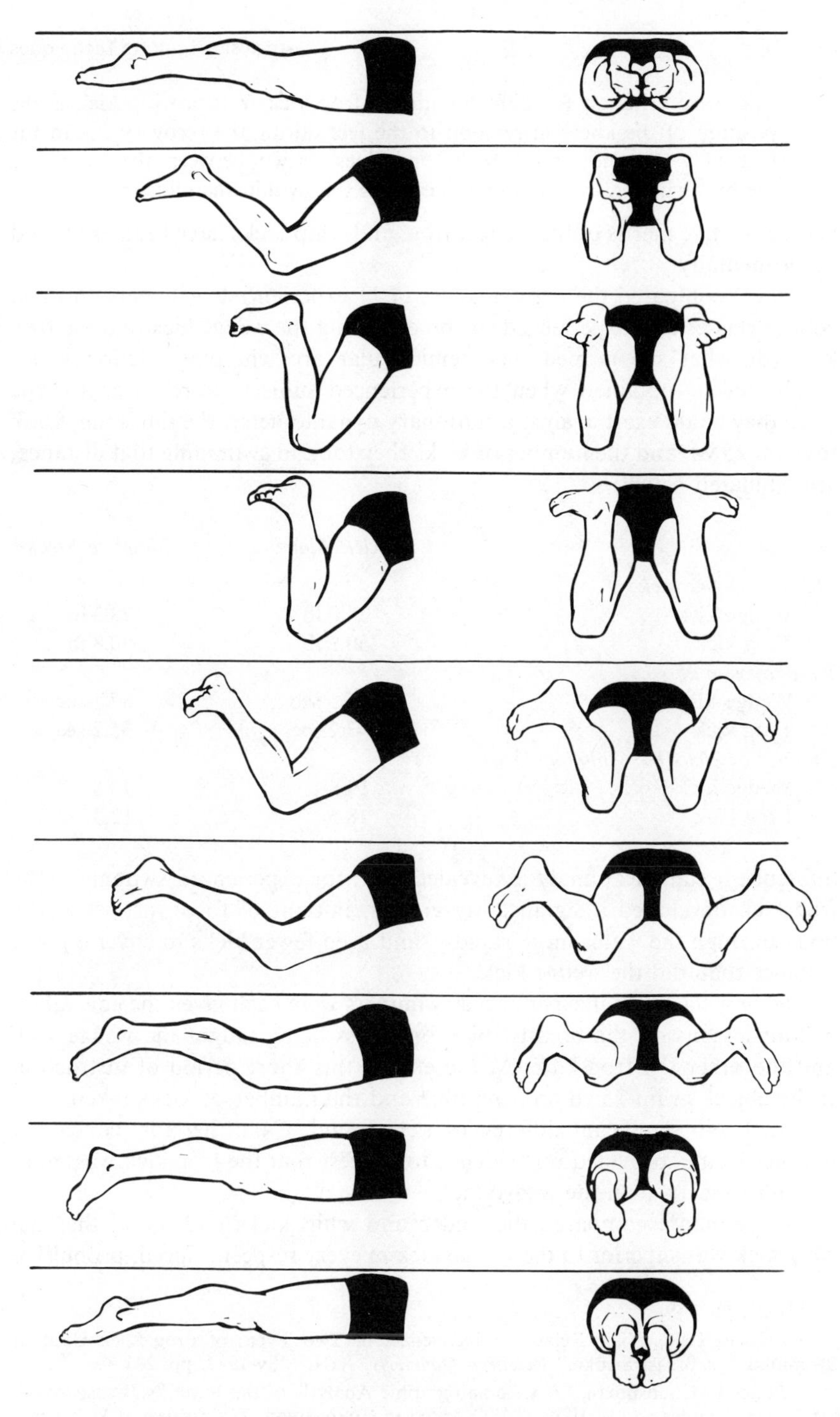

Fig. 168. Breaststroke whip kick. (*Reproduced with permission from Counsilman,* The Science of Swimming.)

> The basic criterion for differentiating a frog kick from a whip kick is the position of the knees in relation to the feet during the recovery. . . . In the frog kick the knees separate beyond lines drawn between the heels and the hips while in the whip kick the knees stay within such lines.[44]

The relative merits of the wedge, frog, and whip kicks have been examined experimentally.

Cake[45] compared the performances of 11 experienced swimmers and two college classes of inexperienced swimmers, using the wedge kick and the frog kick (or what she termed the "semicircular arc whipping" action kick). The results obtained when the experienced subjects were tested for the force they could exert against a stationary dynamometer, the times they took to swim 25 yd, and the number of kicks they took in swimming that distance, are tabulated below:

	Kick Alone	*Complete Stroke*
Mean force per kick		
Wedge kick	28.0 lb	29.3 lb
Frog kick	30.9 lb	30.8 lb
Time to swim 25 yd		
Wedge kick	46.5 sec	37.1 sec
Frog kick	43.2 sec	35.2 sec
Number of kicks in swimming 25 yd		
Wedge kick	19.8	13.8
Frog kick	18.5	12.3

Subsequent statistical analysis revealed that, for experienced swimmers, the frog kick developed a significantly greater amount of force, propelled the body through the water more rapidly, and used fewer kicks to cover a given distance than did the wedge kick.

The two classes of inexperienced swimmers were each given the equivalent of four lessons on the breaststroke, one class being taught the wedge kick and the other the frog kick. At the end of this short period of instruction each subject swam 25 yd and the time and the number of kicks taken were recorded. No significant differences were found. Cake therefore concluded that her results provided no evidence to suggest that the frog kick was more difficult to learn than the wedge kick.

Counsilman[46] compared the wedge and whip kicks and found that the whip kick was superior to the wedge kick in every respect—speed, propulsive

[44] *Ibid.*, pp. 7–9.

[45] Frances Cake, "The Relative Effectiveness of Two Types of Frog Kick Used in Swimming the Breast Stroke," *Research Quarterly*, XIII, May 1942, pp. 201–4.

[46] James E. Counsilman, "A Cinematographic Analysis of the Butterfly Breaststroke" [M.S. thesis, University of Illinois, 1948 (cited in Counsilman, *The Science of Swimming*, pp. 117–18)].

force, and economy of movement. In addition it could be used at a faster tempo than the wedge kick.

Deciding that the limitations of the wedge kick had already been adequately demonstrated, Over[47] confined her attention to comparing the frog and the whip kicks. With an experienced swimmer as the subject she determined the propulsive force he could exert against a line being unreeled at approximately 2.36 fps, the resistive force he encountered when being towed at 3.65 fps, and the "free velocity" he could attain when using the frog and whip kick leg actions. Among her conclusions were the following:

1. When speeds are comparable, the number of whip kicks taken to cover a given distance will be greater than the number of frog kicks necessary to cover the same distance. This, she said, suggests that the per kick efficiency of the frog kick is greater than that of the whip kick.
2. The frog kick is more powerful than the whip kick when executed against resistance.
3. When measured at a speed similar to the whole stroke speed, use of the frog kick incurred more resistance to the swimmer's forward motion than did use of the whip kick.

Several other questions relating to the leg action used in the breaststroke have been the subject of study. For example, Russian researchers Belokovsky and Ivanchenko[48] have recently reported the results of a series of studies concerned with the amplitude and force of the leg actions employed in breaststroke swimming. In the first of these studies, they determined the amplitude (or range) of the hip and knee joint movements of "top-level swimmers" and concluded that

> . . . the average angle of flexion in the hip joint [at the end of the recovery or preparatory phase of the leg cycle] was 137.4 ± 8.5°. The angle . . . for male swimmers ranged from 130 to 148° and that of female swimmers from 124 to 131°. These can be compared to the swimming technique of 1950's–1960's in which this angle was only 107.0 ± 15.6°. While the angle of flexion for the knee joint of male swimmers reached a right angle, that of female swimmers was 100–124°. In the last 5 years angular changes in the hip joints have tended to decrease by one-half and at present these angles are 34.0 ± 8.5°. Working amplitudes of female swimmers are 44–48° and those of male swimmers are less, 24–32°. At present, the mean knee joint angle in the preparatory phase is 40.1 ± 5.1° compared to 30.2 ± 2.8° in 1950–1960.[49]

[47]Over, "A Comparison of the Force and Resistance of the Frog Kick and the Whip Kick Used in Swimming the Orthodox Breaststroke."

[48]V. Belokovsky and E. Ivanchenko, "A Hydrokinetic Apparatus for the Study and Improvement of Leg Movements in the Breaststroke," in *Swimming II*, ed. by J. P. Clarys and L. Lewillie (Baltimore: University Park Press, 1975), pp. 64–69.

[49]*Ibid.*, p. 67.

They next investigated the forces exerted by the swimmers as they executed simulated breaststroke kicks in which the initial flexion at the hips and knees was made to vary from 90°–180° and 40°–90°, respectively. Then, having arrived at what appeared to be the best combination (140° and 50°–60° of hip and knee flexion, respectively), they trained a group of swimmers for 1½ months using a device that limited their hip and knee flexion, at the end of the preparatory phase, to these amounts. This procedure resulted in significant improvements in their "swimming time over competitive distances"—presumably, as compared to other swimmers who underwent similar training without the aid of the special device. Finally, recognizing that the amplitudes of swimming movements are inversely related to the stroke rate, Belokovsky and Ivanchenko determined the range of stroke rates that would permit the use of the recommended pattern of leg movements and found that "the optimal rate is about 65 cycles per min [1.08 cycles/sec]. Exceeding the optimal rate by 10 cycles per min [0.17 cycles/sec] results in substantial changes in swimming technique and a decrease in the swimming speed."[50]

Arm Action

The arm action during the pull phase in the breaststroke is somewhat similar to that used in the butterfly stroke except that the backward action of the hands relative to the shoulders, ends at or near the line of the shoulders. At this point the breaststroke swimmer bends his arms and brings them in toward the midline of his body preparatory to thrusting them forward again and thence into the next arm cycle.

The butterfly swimmer begins to bend his arms earlier than does his breaststroke counterpart and then, instead of beginning the recovery of his arms, pushes backward beneath his body until his arms are almost fully extended and his hands are close to his thighs. Only then, with the propulsive possibilities apparently exhausted, does he begin the recovery of his arms.

Breathing

The breathing action in the breaststroke is identical to that already described with reference to the butterfly stroke (pp. 360–61). The head is lifted and the breath taken at the end of the pull phase when the arms are approximately level with the shoulders and when the latter are relatively high in the water.

Kinnear[51] has pointed out that this taking of the breath late in the pull phase of the arm stroke "conforms to the Russian idea of performing all breathing movements outside a propulsive phase of the arm action in all

[50] *Ibid.*, p. 69.

[51] A. D. Kinnear, "Breaststroke Today," *Swimming Technique*, IV, January 1968, p. 112.

strokes (in relation to the fixing of the rib cage for generating maximum power)." He also refers to the question of minimizing the number of breaths taken during a race—a question already mentioned several times in this chapter:

> All swimmers favor breathing every arm stroke although some breath holding can be seen occasionally in the shorter distance races over 100 metres—but in no set pattern. This, I must admit, does surprise me because breath holding with no need to raise the head must result in a more stable stroke—it is done in butterfly and sprint crawl; why not in breaststroke?

STARTS

When the gun is fired, the swimmer endeavors to get away from the block quickly and with as much forward speed as he can reasonably muster. Unfortunately these two objectives (quickness off the block and maximum forward speed) are somewhat incompatible, for if the man leaves the block as quickly as he possibly can, the horizontal impulse developed is such that his forward speed is less than it could be. Conversely, if he takes the time necessary to develop a maximum horizontal impulse (and thus maximum horizontal speed), he will leave the block rather later than he might otherwise. The swimmer's task therefore is to arrive at that blending of quickness off the block and forward speed that affords him the best results overall.

Front Crawl, Butterfly, and Breaststroke

Apart from relatively minor variations in the angle at which the body enters the water, the starting technique used up to the point of entry is the same for front crawl, butterfly, and breaststroke events.

At the command "Take your marks" the swimmer moves from his erect-standing preparatory position atop the block and assumes his starting position. In this position he generally has his feet 6–12 in. apart, his toes curled over the forward edge of the block, his knees bent slightly, his hips well flexed, and his head, neck, and trunk inclined in a forward and downward direction. While several alternative arm positions may be used, the majority of good swimmers place their arms in one or the other of the following positions:

1. extended near-vertically downwards with the hands gripping the front edge of the starting platform—the position used in the so-called *grab start*,
2. hanging vertically downward from the shoulders, or slightly forward of a vertical through the shoulders,
3. extending back in line with the upper trunk so that the hands are level with or slightly above the hips.

The arm movements that follow the firing of the gun depend in part on the arm position the swimmer adopts in the "set" position. Those who use the grab start generally retain their grip on the block for half a second or more after the gun has been fired and then swing their arms vigorously forward and upward. Those who adopt a position with the arms hanging from the shoulders normally use one of two alternatives—either they swing their arms straight back and then forward again or they swing them in a wide circular action, forward, upward and outward, backward, downward, and then forward again. Those who adopt a "set" position with the arms back generally thrust them upward and backward a little and then swing them vigorously downward and forward.

Which of these alternative positions and arm actions is the best—a question which has been addressed in a number of studies on starting techniques—depends on the extent to which they meet a swimmer's need for quickness off the block and a large forward speed at takeoff.

Bowers and Cavanagh[52] conducted a study in which six female competitive swimmers performed four starts using the grab-start technique and a further four starts using the "conventional (circular armswing)" technique and found that:

1. The mean reaction time (gun to first movement), the mean block time (gun to takeoff), and the mean time to cover 10 yd (gun to 10 yd), were significantly less (0.15, 0.17, and 0.18 sec, respectively) when the grab start was used.
2. The horizontal and vertical components of velocity of the center of gravity at takeoff were not significantly different between the two starting styles. This, they suggested, was an indication that any additional propulsive force obtained by the use of the grab start is only sufficient to account for the losses "due to the absence of the armswing."

On the basis of these findings, they concluded that "The major reason for the superiority of the grab start over the conventional circular armswing start would appear to be that the swimmer is able to leave the block more quickly yet without a decrement in velocity at takeoff."

To gain further insights into the mechanics of the grab start, Cavanagh, Palmgren, and Kerr[53] developed a measurement system to record the forces exerted on the block by the swimmer's hands. They then conducted a pilot

[52]J. E. Bowers and P. R. Cavanagh, "A Biomechanical Comparison of the Grab and Conventional Sprint Starts in Competitive Swimming," in *Swimming II*, ed. by J. P. Clarys and L. Lewillie (Baltimore: University Park Press, 1975), pp. 225–32.

[53]P. R. Cavanagh, J. V. Palmgren, and B. A. Kerr, "A Device to Measure Forces at the Hands during the Grab Start," in *Swimming II*, ed. by J. P. Clarys and L. Lewillie (Baltimore: University Park Press, 1975), pp. 43–50.

(or preliminary) study to determine these forces in the case of an experienced competitive swimmer and found that the hands exerted an upward and forward force on the block throughout the start.

This finding is of particular interest because it suggests an explanation for the results obtained in the study of Bowers and Cavanagh. The downward and backward reaction that the starting block exerts on the hands of the swimmer (in response to the forces that the hands exert on the block) tends to accelerate the swimmer in that direction and to rotate him forward about an axis through his ankles or feet. Further, because the swimmer's body is in a very compact position at this time—and, thus, has a small moment of inertia—the angular acceleration it experiences is relatively large. The swimmer's body is thus rotated rapidly forward into a position from which his legs can drive him vigorously forward and out over the water. The forces exerted via the swimmer's hands and the rapid forward rotation that they cause would seem to be a logical explanation for the finding that the mean block time for the grab start was significantly less than that for the conventional start.

The role of the arms as a "brace" against which the legs can be tensed before the release of the hands—a role suggested by Cavanagh et al.—might just as readily explain the lack of a significant difference in the takeoff velocities for the grab and conventional starts. For, if the pre-tensing of the extensor muscles of the legs leads to an increase in the forces exerted by these muscles by an amount equal to the expected decrease in the forces derived from the swing of the arms, it is only logical that the takeoff velocity for the grab start should not be significantly different from that for the conventional start.

Maglischo and Maglischo[54] compared the results obtained when three different techniques—arms back, arms down with straight backswing, and arms down with circular backswing—were used. Following tests in which 10 members of a college swimming team executed 10 starts with each type of arm action, they found that:

1. The speed at which the first 15 ft were traveled was significantly faster when using the circular-backswing start than when using the straight-backswing start.

2. There was no significant difference between the arms-back and straight-backswing methods.

3. There was no significant difference between the arms-back and circular-backswing methods, although a trend was noted in favor of the circular-backswing method.

[54]Cheryl W. Maglischo and Ernest Maglischo, "Comparison of Three Racing Starts Used in Competitive Swimming," *Research Quarterly*, XXXIX, October 1968.

These results are summarized below, with the methods ranked in order from most to least effective and with those that were found to be not significantly different linked by braces:

Circular backswing }
{Arms back }
{Straight backswing

As his arms begin to move forward toward the end of their swing, the swimmer thrusts forcefully downward and backward against the block. The reaction to this thrust, which continues as the swimmer's body moves forward and out over the water, serves not only to propel the swimmer into the air but also to impart to him the forward rotation necessary to ultimately bring him into position for entry.

Angle of Takeoff. In studying motion-picture films of the starts of a large number of highly skilled college swimmers, Heusner[55] noted that there was little uniformity in the angle of takeoff from the starting block—the angle varied from 5°–22° without any apparent relationship to body build, stroke mechanics, or competitive success. He therefore set out to determine the optimum angle of takeoff in a given set of circumstances. To do this, he derived a mathematical expression for the time required to dive, glide, and swim the first length of a race; validated it by comparing computed theoretical times with those actually recorded in practice; and then noted the results obtained when various angles of release were used in the computation. His results indicated that the optimum angle of takeoff for the average competitor under normal conditions was 13°. In addition, he observed that, although a deviation of less than 9° from this optimum angle will not affect a swimmer's time by an amount measurable with a stopwatch calibrated in tenths of a second, the fact that several men often finish a race within 0.1 sec of each other suggests that any deviation from the optimum angle could be costly. Finally, in terms of the time taken to swim 25 yd, he found that if all else is equal, a tall man can swim faster than a short one, a light man can swim faster than a heavy one, a man with good vertical jumping ability can swim faster than one with less vertical jumping ability, and a man with good gliding ability (distance glided in 10 sec) can swim faster than one with poor gliding ability.

Once the swimmer enters the water—at an angle of approximately 15° in crawl and butterfly events and approximately 20° in breaststroke events[56]—he assumes the most streamlined position possible until his forward speed

[55]William W. Heusner, "Theoretical Specifications for the Racing Dive: Optimum Angle of Take-off," *Research Quarterly*, XXX, March 1959, pp. 25–37.

[56]Counsilman, *The Science of Swimming*, p. 140.

slows to his normal swimming speed. At this point he begins stroking and kicking.

Some other aspects of starting have been investigated by researchers. Tuttle, Morehouse, and Armbruster[57] compared the starting times (gun until feet leave block) recorded by 18 well-trained swimmers starting from a surface placed flat on the pool deck and from the same surface inclined at 20° to the pool deck to form a "starting block." The results "showed quite conclusively that starting blocks are a disadvantage to swimmers in leaving the mark at the start of the race, as far as starting time is concerned." Elliott and Sinclair[58] examined the effects of inclining the top surface of elevated starting blocks. They found that below 10° (the upper limit imposed by F.I.N.A., the international swimming federation, in 1968) "there is no advantage to be gained by building a sloping starting block from the point of view of eliciting the greatest horizontal component of velocity. The regulations may just as well require a horizontal starting block, which is associated with ease of building and also stability of balance."

Back Crawl

In the orthodox back crawl start the swimmer starts in the water with his hands on the grip provided for the purpose and his feet on the end wall of the pool. On the command "Take your marks" he flexes his arms and pulls himself upward and toward the starting block, getting as much of his body above the water as he can. Then, once the gun has been fired, he releases his grip, drops his head back, swings his arms sideways to a position overhead, and drives vigorously against the wall with his feet. These actions result in him being projected out above the surface of the water and ultimately into a streamlined gliding position.

The rules governing the initial position adopted by the swimmer vary from one administrative body to another. The international rules (those laid down by F.I.N.A. and used in A.A.U. meets in the United States) require that the swimmer's feet be under the surface of the water and that he not be "standing in or on the gutter or curling his toes over the lip of the gutter."[59] The N.C.A.A. rules require only that the swimmer not remove himself completely from the water as he assumes his starting position. They place no restriction on the use of the gutter for the purposes of obtaining a firm grip with the feet.

[57]W. W. Tuttle, Lawrence E. Morehouse, and David Armbruster, "Two Studies in Swimming Starts," *Research Quarterly*, X, March 1939, pp. 89–92.

[58]Geoffrey M. Elliott and Helen Sinclair, "The Influence of Block Angle on Swimming Sprint Starts," *First International Symposium on Biomechanics in Swimming, Waterpolo and Diving Proceedings*, ed. by L. Lewillie and J. P. Clarys, Université Libre de Bruxelles Laboratoire de L'effort, 1971.

[59]*Amateur Athletic Union Official Swimming Handbook*, New York: Amateur Athletic Union of the United States, 1973, p. 11.

These differences in the rules and the greater range of starting positions that the N.C.A.A. rules permit have stimulated a good deal of discussion in recent years. Triggering much of this discussion has been a start developed and used by 1964 Olympic silver medalist, Gary Dilley. In this start Dilley stood with his feet in the gutter, his legs straight, his trunk inclined forward over the top of the starting block, and his hands holding the starting grip. At the firing of the gun he lowered his hips slightly, threw his head back, pushed hard with both feet, and flung his arms back and over his head.

In an attempt to determine the merits of this technique Rea and Soth[60] conducted a study in which their subject (Dilley) did four starts using the conventional start and four starts using the modified (or Dilley) start. On averaging the time taken to reach a mark 20 ft from the start they obtained the following figures:

Conventional start	2.695 sec
Modified start	2.519 sec
Advantage in favor of modified start	0.176 sec

Since this study involved only one subject, however, and one who was admitted to be less than expert with the conventional start—"He has never been noted for an above average backstroke start and, therefore, he wanted to see if this new modified start could help him"—the significance of the findings is very much open to question.

Fortunately a much more comprehensive study has since been reported by Stratten[61], and the results shed a good deal of light on the subject of back crawl starting techniques. In this study, 13 subjects (5 highly skilled back crawl swimmers, 5 competitive swimmers in other events, and 3 noncompetitive swimmers) were tested using each of three types of start—the F.I.N.A. start, the orthodox N.C.A.A. start, and the Dilley start. Each swimmer performed each start 45 times over a period of several days and the following times were recorded:

1. the time from the starting signal until the feet left the wall,
2. the time from the feet leaving the wall until the swimmer reached a mark 20 ft from the start,
3. the total time from the starting signal until the swimmer reached the 20-ft mark.

[60]William M. Rea and Scott Soth, "Revolutionary Backstroke Start," *Swimming Technique*, III, January 1967, pp. 94–95.

[61]Gaye Stratten, "A Comparison of Three Backstroke Starts," *Swimming Technique*, VII, July 1970, pp. 55–60.

The results obtained are listed below in order (shortest times first, longest last). Where the statistical analysis revealed that the difference between starts was not significant, the starts concerned are linked with a brace.

Time from gun until toes left wall:

F.I.N.A. }
N.C.A.A. }
Dilley

Time from toes leaving wall until swimmer reached 20-ft mark:

Dilley
N.C.A.A.
F.I.N.A.

Time from gun until swimmer reached 20-ft mark:

N.C.A.A.
F.I.N.A. }
Dilley }

The results for the first two times are consistent with what might have been expected. First, the start in which the swimmer has the greatest distance to travel before he leaves the block (the Dilley start) requires more time between the gun and the toes leaving the wall than do the other two starts. Second, since this increase in time almost certainly permits the swimmer to be traveling faster when he eventually does leave the block—and since, too, he travels a greater proportion of the distance through the air—the time between the feet leaving the wall and the swimmer reaching the 20-ft mark is less with the Dilley start than with either of the other two. Finally, the higher initial starting position and the greater distance traveled through the air probably account for the significant advantage of the N.C.A.A. start over the F.I.N.A. start in terms of the time taken from the feet leaving the wall until the swimmer reaches the 20-ft mark.

The results for the total time taken to reach 20 ft reflect the relative importance of the differences already noted—the advantage that the N.C.A.A. start has over the Dilley start in terms of quickness off the mark offsetting the advantage that the Dilley start has thereafter.

TURNS

Front Crawl Flip Turn

The turn begins with a strong pulling motion that brings both arms alongside the swimmer's body. To bring the arms to this position (a position that facilitates the rotation to follow) the swimmer either stops one arm as it reaches the end of its pull and pulls the other one through to join it, or stops one arm at entry and allows the other to catch up with it before executing a two-handed pull back to the hips. In either case, the pulling action is accompanied by a flexion of the neck and spine, which drives the head and shoulders forward and downward below the surface of the water, and by a bringing together of both legs. The increased resistance experienced by the head and shoulders as they move out of alignment with the rest of the body, together with the moments evoked by a dolphinlike kick of both legs and a pressing downward and forward with the hands, cause the swimmer to somersault forward (Fig. 169). With almost half the somersault completed and his tucked legs passing overhead, the swimmer begins to twist about his long axis. This twist (perhaps facilitated by his being in a position of marked hip flexion at the time it is initiated) carries the swimmer's feet around so that by the time they strike the end wall of the pool they are lying roughly parallel to the surface and some 6–18 in. below it. From this position the swimmer drives forcefully away from the wall and, completing the twist necessary to carry him into a prone position, assumes an extended, streamlined position. When the speed of this gliding action has been slowed to his normal swimming speed, he settles once more into his normal cycle of stroking and kicking actions.

There are two major variants of the front crawl flip turn—the *pike turn*, in which the body is flexed at the hips and the legs are kept extended during the somersault, and the *tuck turn*, in which the somersault is performed with the body flexed at hips and knees. To determine which of these alternative techniques afforded the faster turn, Ward[62] conducted a study in which 14 members of a college life-saving class were matched and then divided into two groups. Each group was given instruction (10 15-min sessions) in one of the two techniques. Each subject then performed 10 trials using the technique in which he had received instruction. These trials were recorded on film, and the time from the head reaching a given vertical plane until

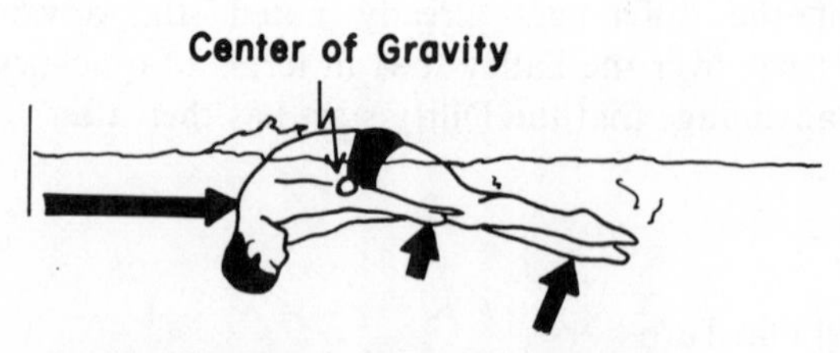

Fig. 169. Forces producing somersaulting rotation in the front crawl flip turn

[62]Thomas A. Ward, "A Cinematographical Comparison of Two Turns," *Swimming Technique*, XIII, Spring 1976, pp. 4–6, 9.

the feet contacted the wall (the "in time") and the time from wall contact until the head again reached the vertical plane (the "out time") were determined. The subsequent statistical analysis revealed that the tuck turn was significantly faster than the pike turn with respect to "in time," "out time," and total time ("in time" plus "out time"). Ward concluded, therefore, that "the tuck [turn] is the superior turn for most beginners."

Butterfly and Breaststroke Turns

Whereas the flip turn involves rotation about the swimmer's transverse and longitudinal axes, the turn most commonly used in butterfly and breaststroke events involves rotation that is primarily about the swimmer's frontal axis.

To execute an orthodox butterfly or breaststroke turn, the swimmer touches the end wall with both hands, bends his arms a little as he reduces his forward speed to zero, and simultaneously begins to draw his knees up underneath him. This rotation of his lower body toward the wall continues as he releases one hand and turns his head and trunk in the direction of the turn. These actions place his body in a near-vertical plane as the feet continue to move toward the wall. The rotation of the swimmer about his frontal axis—most obvious at this time—is accelerated as the remaining hand pushes off from the wall. Then, once his feet have been placed against the wall, the forceful extension of his hip, knee, and ankle joints begins. The reaction to the forces thus exerted drives the swimmer away from the wall and eventually into the streamlined, prone-gliding position from which he resumes his normal swimming action.

Back Crawl Flip Turn

The flip turn used in backstroke events also involves rotation about the swimmer's frontal axis.

The turn begins with the swimmer reaching back to place his hand in an appropriate position for the movements to follow—palm flat against the end wall and fingers horizontal and pointing inward toward the midline of the body. Once this has been accomplished, the swimmer contracts the muscles he would normally use if he were going to sweep his arm laterally to a position by his side. Since his hand is fixed against the wall, however, the contraction of these muscles has the effect of rotating the body around toward the arm. To decrease the moment of inertia of his body relative to the axis of rotation, and to thereby increase the speed of his turn, the swimmer flexes his knees and draws them up toward his chest.

The turn continues with the swimmer's body in this position until his feet come into contact with the wall. Then, with both hands moving to a position overhead, he begins the extension of his hip, knee, and ankle joints that will drive him away from the wall and into his glide.

Chapter 15

TRACK AND FIELD: RUNNING

Basic Considerations

In track events an athlete's objective is simply to cover a given distance (either on the flat or over obstacles) in the least possible time. The speed at which the athlete runs (and thus, the time it will take him to cover a given distance) is equal to the product of two factors:

1. the distance he covers with each stride he takes, his *stride length*, and
2. the number of strides he takes in a given time, his *stride frequency* (also referred to as *stride cadence* or *rate of striding*).

Thus, a distance runner who has a 6-ft stride and takes three strides per second runs at a speed of 18 fps:

$$\begin{aligned}\text{Speed} &= \text{stride length} \times \text{stride frequency}\\ &= 6 \text{ ft} \times 3 \text{ per second}\\ &= 18 \text{ fps}\end{aligned}$$

Now if the distance runner were somehow able to increase his stride frequency to four strides per second while maintaining the same stride length as before, his speed would be markedly increased:

$$\begin{aligned} & \qquad \textit{Stride Length} \quad \textit{Stride Frequency} \\ \text{Original speed} &= 6 \text{ ft} \times 3 \text{ per second} = 18 \text{ fps}\\ \text{New speed} &= 6 \text{ ft} \times 4 \text{ per second} = 24 \text{ fps}\end{aligned}$$

However, if this increase in stride frequency were accompanied by a decrease in stride length to $4\frac{1}{2}$ ft, the effort he made to bring this change about would have been to no avail:

	Stride Length		*Stride Frequency*	
Original speed =	6 ft	×	3 per second	= 18 fps
New speed =	$4\frac{1}{2}$ ft	×	4 per second	= 18 fps

In other words, the increase in stride frequency would be matched by a comparable decrease in stride length, and his running speed would be unaltered. From all this, it is apparent that if a runner is to improve his speed, he must bring about an increase in one parameter without causing the other to be reduced a comparable (or, worse yet, a more than comparable) amount. (*Note:* An increase in one parameter accompanied by a decrease in the other may result in an improvement in speed if the decrease is less than comparable to the increase; e.g.,

	Stride Length		*Stride Frequency*	
Original speed =	6 ft	×	3 per second	= 18 fps
New speed =	5 ft	×	4 per second	= 20 fps)

Since running speed is completely dependent on the magnitudes of the stride length and the stride frequency, it is important to consider the factors that determine these magnitudes.

Stride Length

The length of each stride taken by a runner may be considered as the sum of three separate distances:

1. the *takeoff distance*—the horizontal distance that his center of gravity is forward of the toe of his "takeoff" foot at the instant the latter leaves the ground (*A* in Fig. 170),
2. the *flight distance*—the horizontal distance that his center of gravity travels while he is in the air (*B* in Fig. 170), and
3. the *landing distance*—the horizontal distance that the toe of his leading foot is forward of his center of gravity at the instant he lands (*C* in Fig. 170).

The first of these three contributions depends on the athlete's physical characteristics (notably, the length of his legs and the range of movement he has at the hip joints) and on the position of his body. In the latter respect, the extent to which he extends his leg before his foot leaves the ground, and the angle that his leg makes with the horizontal at this time, are of particular importance. Of these factors, the length of his legs is the one over which the athlete would seem to have least control. It has been suggested by Slocum and

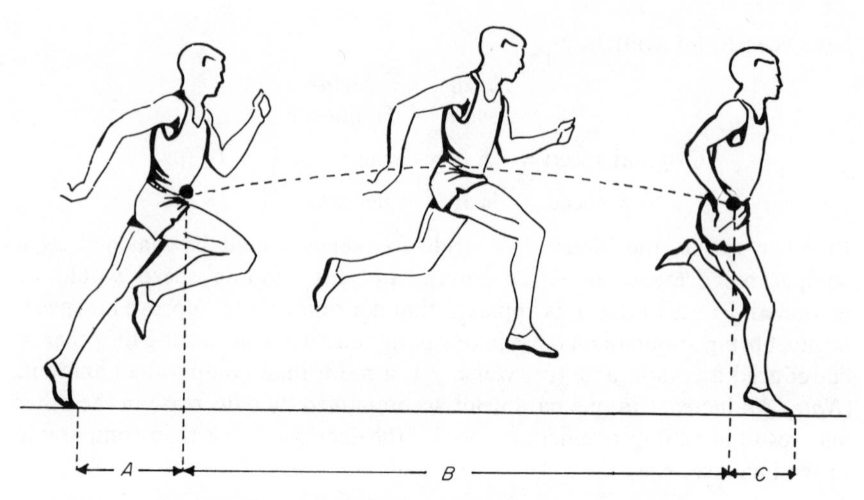

Fig. 170. Contributions to the total length of a runner's stride

Bowerman,[1] however, that the length of the leg may effectively be increased if, instead of the leg pivoting about the hip joint, "the leg plus the pelvis and the low back" form "the lower-extremity lever system" and the lower back "joint complex becomes the pivotal point." The angle that the leg makes with the horizontal at the instant the foot breaks contact with the ground is subject to considerable variation. For example, in the case of the sprinter in Fig. 171, the angle varies from approximately 30° as he leaves the blocks to approximately 60° as he approaches full stride. Correspondingly, the horizontal distance from toe to center of gravity decreases from 36 to 16 in.

During that part of the running stride in which he is airborne, the horizontal distance that the athlete travels is determined by the factors that govern the flight of all such projectiles, viz., the speed, angle, and height of release and the air resistance encountered in flight. By far the most important of these is the speed of release, a quantity primarily determined by the ground-reaction forces exerted on the athlete. These in turn are a result of the forces, mainly from the extension of hip, knee, and ankle joints, that the runner exerts against the ground. While the influence of air resistance on running speed is certainly not confined to the airborne phase of the stride, it is in causing variations in the horizontal distance that the runner travels during this phase that air resistance probably has its greatest effect.

The horizontal distance from the toe of the leading foot to the line of gravity at the instant the athlete lands is invariably the smallest of the three contributions to the total length of the stride. Its magnitude—rarely more than a few inches in good running—is limited by the need to ensure that the ground-reaction forces evoked as the foot lands are as favorable as possible.

[1]Donald B. Slocum and William Bowerman, "The Biomechanics of Running," *Clinical Orthopaedics*, XXIII, 1962, p. 41.

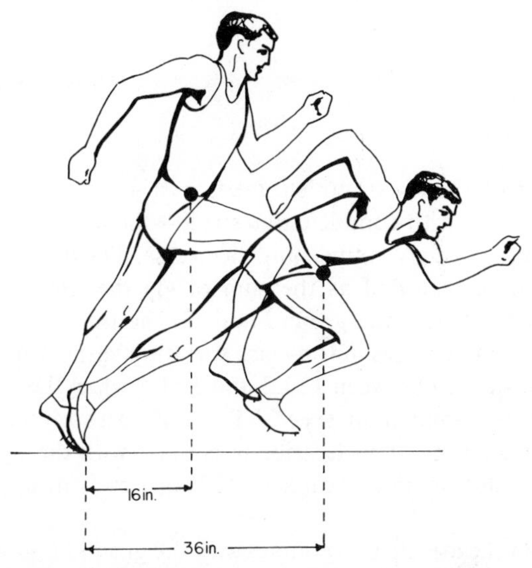

Fig. 171. The distance that a runner's center of gravity is forward of his foot at the instant the foot leaves the ground varies with the inclination of his body at that time.

Thus, while swinging the foreleg forward just before the foot lands might seem a logical way for a runner to increase his stride length, the forward motion of the foot as it hits the ground evokes a backward reaction (a kind of "propping" or "braking" reaction) that reduces the runner's forward speed. In other words, the small gain in stride length is likely to be achieved only at the expense of a more-than-comparable reduction in the length and/or frequency of the strides that follow. (*Note:* This swinging forward of the lower leg, or "overstriding," is precisely the technique that runners use to slow down once they have passed the finish line—Fig. 172).

Fig. 172. Slowing down at the finish of a race (*Photograph courtesy of Mike Conway*)

Stride Frequency

The number of strides an athlete takes in a given time is determined by how long it takes him to complete one stride—the longer this takes, the less strides he can take in a given time, and vice versa. The time taken to complete one stride may be regarded as the sum of (1) the time during which the athlete is in contact with the ground and (2) the time he spends in the air. The ratio of these two times in top-class sprinting varies from approximately 2 : 1 during the start to between 1 : 1.3 and 1 : 1.5 when the athlete is running at maximum or near-maximum speed.[2] Thus, while a sprinter spends approximately 67% of the time of each stride in contact with the ground during his first few strides, this figure decreases to 40% or less as he approaches his top speed.

The time that the athlete is in contact with the ground is governed primarily by the speed with which the muscles of his supporting leg can drive his body forward and then forward and upward into the next airborne phase. While the factors that limit this speed are still anything but clearly understood, a study by Slater-Hammel[3] does cast some light on the subject. In an effort to determine whether stride frequency in sprinting was limited by the ability of the athlete's nervous and muscular systems to produce rapid contraction and relaxation of the appropriate leg muscles, Slater-Hammel compared the rate of "striding" in sprinting with that attained when pedaling on a stationary bicycle. Because the rates in cycling (5.6–7.1/sec) were far higher than those in sprinting (3.10–4.85/sec)—in fact far higher than "any possible rate in sprinting"—Slater-Hammel concluded that the rate at which the legs were moved in sprinting was not limited by a neuromuscular mechanism. Instead, he suggested, "The present indications are that the rate of leg movement in sprinting is determined by the load, the weight the muscles must move."

Like other parameters associated with the motion of a projectile, the time the athlete spends in the air is determined by his velocity and the height of his center of gravity at "takeoff" and by the air resistance he meets in flight.

Studies on Stride Length and Stride Frequency

The stride length and stride frequency of sprinters of varying abilities have been the subject of a number of research papers. Notable among these are two in which the stride characteristics of some of the world's best male and female sprinters were investigated.

[2]Fred Housden, "Mechanical Analysis of the Running Movement," in *Run, Run, Run*, ed. by Fred Wilt (Los Altos, Calif.: Track & Field News Inc., 1964), pp. 240–41.

[3]Arthur Slater-Hammel, "Possible Neuromuscular Mechanism as Limiting Factor for Rate of Leg Movement in Sprinting," *Research Quarterly*, XII, December 1941, pp. 745–56.

In the first of these two papers, Hoffman[4] examined the performances of 56 male sprinters—including Figuerola (Cuba); Hary (Germany); Hayes, Metcalf, Murchison, Norton, and Owens (U.S.A.); Radford (England); and Sime (U.S.A.)—with best times for 100 m ranging from 10.0 to 11.4 sec. This study revealed that:

1. A very close relationship existed between an athlete's standing height and his average stride length during a 100-m race. A similarly close relationship existed between the athlete's leg length—measured from the greater trochanter on his femur to the sole of the foot—and his average stride length. (*Note:* The average stride length was computed by dividing the distance from the start line to the end of the stride immediately before the finish line was reached, by the total number of strides taken. Thus, if the athlete's foot landed 1 m from the finish line at the end of his penultimate stride in the race, and he had taken 50 strides up to that point, his average stride length was $\frac{99}{50} = 1\frac{49}{50} = 1.98$ m.) On the average, the average stride length was equal to 1.14 times the athlete's height or 2.11 times his leg length.

2. A very close relationship existed between each of these same two measurements (height and leg length) and the average stride frequency, although in this case the relationships were inverse ones. In other words, the average stride frequency decreased as the height and leg length of the athletes increased.

3. The maximum stride length (defined by Hoffman as the average length of four strides taken between 50 and 60 m in a 100-m race) could be fairly accurately estimated by adding 18 cm (7 in.) to the average stride length. The largest value of the maximum stride length actually measured was the 237 cm (7 ft $9\frac{1}{4}$ in.) of Schmidt (Poland), Olympic triple jump champion in 1960 and 1964, and Archipezuk (U.S.S.R.). However, estimated values of the maximum stride length for several of the sprinters for whom it was not possible to obtain actual measurements exceeded this figure. The highest estimate was the 246 cm (8 ft 1 in.) for Sime and Collymore (both U.S.A.).

4. On the average, the maximum stride length was equal to 1.24 times the height of the athlete. If only the best 12 sprinters (100-m times 10.7 sec or better) were considered, this value increased to 1.265 times the athlete's height. (*Note:* From data gathered on 20 male students at the University of Helsinki and on 12 of the "best male runners at Stanford University," Rompotti[5] concluded that "the normal full speed running stride length is

[4]Karol Hoffman, "The Relationship between the Length and Frequency of Stride, Stature and Leg Length," *Sport* [Belgium], VIII, No. 3, July 1965 (French).

[5]Kalevi Rompotti, "A Study of Stride Length in Running," in *International Track and Field Digest*, ed. Don Canham and Phil Diamond (Ann Arbor, Mich.: Champions on Film, 1957), pp. 249–56.

1.17 × Height ± 4 in." Considering the obvious differences in the abilities of their respective samples, the results of Hoffman and Rompotti seem to be in very good agreement.)

In the second study (also conducted by Hoffman[6]) the subjects were 23 female sprinters with best 100-m times between 11.0 and 12.4 sec. As in the study of male sprinters, this group contained many outstanding performers including Heine (Germany); Hyman (England); Kirzenstein and Klobukowska (Poland); Leone (Italy); and McGuire, Rudolph, and Tyus (U.S.A.). In general, the results obtained were very similar to those obtained in the earlier study (e.g., on the average, the average stride length was equal to 1.15 times the athlete's height and 2.16 times the leg length). Probably the most important (and perhaps surprising) finding was that "the best female sprinters when compared with male sprinters of the same class, body height, length of leg and length of stride, run . . . about 1 second slower over 100 meters because of markedly lower frequencies of stride."

Summary

The relationships between the time taken to run a given distance and the factors that determine that time are summarized in Fig. 173.

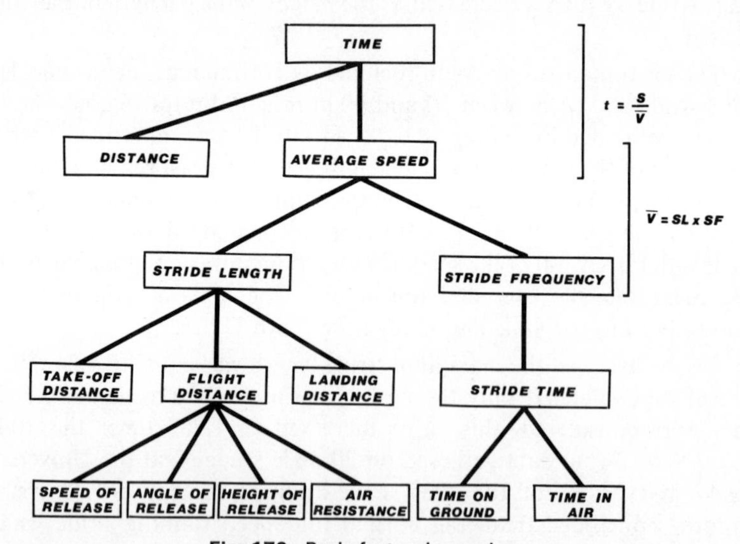

Fig. 173. Basic factors in running

[6]Karol Hoffman, "The Length and Frequency of Stride of the World's Leading Female Sprinters," *Treatises, Texts and Documents WSWF in Poznan Series*, Treatise No. 17, 1967 (Polish).

Techniques

Sprint Start

Of all the sports techniques that have been subjected to biomechanical analysis, few have been more thoroughly examined than the sprint start.

At the starter's command "On your marks" the athlete moves forward and adopts a position with his hands shoulder width apart and just behind the starting line, his feet in contact with both the ground and his starting blocks, and the knee of his back leg resting on the ground [Fig. 174(a)]. On the command "Set" he lifts the knee of his back leg off the ground, thereby elevating his hips and shifting his center of gravity forward [Fig. 174(b)]. Finally, when the gun is fired, he lifts his hands from the track, swings his arms vigorously (one forward, one backward), and with a forceful extension of both legs drives his body forward away from the blocks and into his first running strides [Fig. 174(c)–(e)].

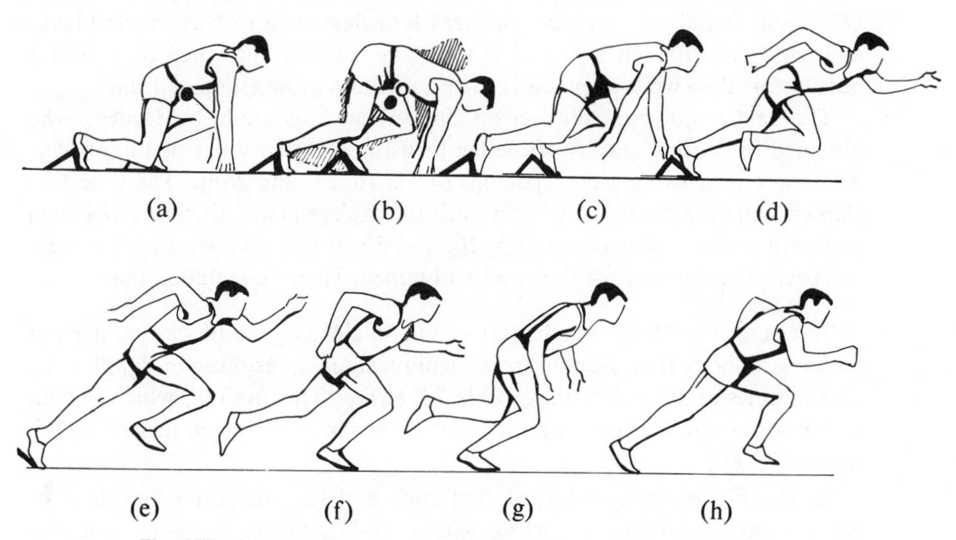

Fig. 174. An example of good technique in the crouch sprint start

There are three main types of crouch start—the bunch or bullet, the medium, and the elongated. The principal difference among these three types lies in the longitudinal distance between the feet (i.e., in the distance from the toes of one foot to the toes of the other, as measured in the running direction). In the bunch start the toes of the back foot are placed approximately level with the heel of the front foot. The toe-to-toe distance is therefore of the order of 10–12 in. In the medium start the knee of the back leg is placed so that it is opposite a point in the front half of the front foot, when the athlete is in the "On your marks" position. Such a placement yields a toe-to-toe distance of somewhere between 16 and 21 in. Finally, in the elongated start the knee of the back leg is placed level with or slightly behind the heel of the front foot, in the "On your marks" position. The resulting toe-to-toe distances are of the order of 24–28 in.

A number of attempts have been made to determine which of these three types of start yields the best results in terms of sprinting performances.

Among the first were those of Kistler[7] and Dickinson.[8] With 30 trained sprinters as subjects, Kistler measured the maximum force exerted at right angles to the face of each starting block—the face of the front block was inclined at an angle of 45° and the face of the back one at a little less than 90°—and found that the sum of these two maximum forces increased as the distance between the blocks increased. In other words, starts from an elongated position were more forceful than those from either bunch or medium positions. Seemingly in contradiction to the findings of Kistler, Dickinson found that trained sprinters took less time to leave their blocks and less time to reach a line $7\frac{1}{2}$ ft from the starting line when they used a bunch start than when they used either a medium or an elongated start.

The next major contribution on this subject was made by Henry[9] who obtained force-time graphs of the leg thrust during the starts of 18 sprinters who made runs using block spacings of 11, 16, 21, and 26 in. The time that elapsed from the firing of the gun until the athlete made his first movement and until he reached markers at 5, 10, and 50 yd from the starting line were recorded. On the basis of the results obtained, Henry concluded that:

1. Use of the 11-in. bunch start results in clearing the blocks sooner but with less velocity than secured from medium stances, resulting in significantly slower times at 10 and 50 yd—Table 22. (*Note:* This finding, which is consistent with those of Kistler and Dickinson, has been discussed in some detail on pp. 76–77.)

2. The highest proportion of best runs and the smallest proportion of poorest runs result from starting with a 16-in. stance. A 21-in. stance is nearly as good.

3. An elongated stance of 26 in. results in greater velocity leaving the blocks but the advantage is lost within the first 10 yd.

[7]J. W. Kistler, "A Study of the Distribution of the Force Exerted Upon the Blocks in Starting from Various Starting Positions," *Research Quarterly*, V (Supplement), March 1934, pp. 27–32.

[8]A. D. Dickinson, "The Effect of Foot Spacing on the Starting Time and Speed in Sprinting and the Relation of Physical Measurements to Foot Spacing," *Research Quarterly*, V (Supplement), March 1934, pp. 12–19.

[9]Franklin M. Henry, "Force-Time Characteristics of the Sprint Start," *Research Quarterly*, XXIII, October 1952, pp. 301–18.

Table 23. Time and Velocity Characteristics of Crouch Starts Using Various Block Spacings (Mean Values)*

	Block Spacing			
	11 in.	*16 in.*	*21 in.*	*26 in.*
Time on Blocks (i.e., time from gun to front foot leaving)	0.345 sec	0.374 sec	0.397 sec	0.426 sec
Block Velocity (i.e., horizontal velocity of athlete as he leaves blocks)	6.63 fps	7.41 fps	7.50 fps	7.62 fps
Time to 10 yd	2.070 sec	2.054 sec	2.041 sec	2.049 sec
Time to 50 yd	6.561 sec	6.479 sec	6.497 sec	6.540 sec

*Adapted from data in Henry, "Force-Time Characteristics of the Sprint Start," p. 306.

Henry's conclusion favoring the use of the medium starts has subsequently been supported by the results of studies by Sigerseth and Grinaker[10] and Hogberg.[11]

Menely and Rosemier[12] examined the effect of using four different starting positions on the time taken to reach 10 yd and 30 yd and found that use of a "hyperextended" starting position (medium block spacings but with "the front foot . . . placed as close to the starting line as feasible") resulted in significantly shorter times to both distances than those recorded when any of the three orthodox types of start was used. No other significant differences were found. (*Note:* In this study, no starting gun was used. Each subject began to run when he was ready and times were measured from the instant the left hand broke contact with the track.)

Numerous other aspects of crouch starting have been investigated: Bresnahan[13] studied the movements of trained sprinters as they left their marks and found that the order in which a right-handed sprinter broke his contacts with the track was left hand ($\bar{X} = 0.172$ sec), right hand ($\bar{X} = 0.219$ sec), right foot ($\bar{X} = 0.286$ sec), and left foot ($\bar{X} = 0.443$ sec). He also expressed the view that this sequence is "the only correct form" and all others are "to the detriment of a sprinter in starting."

[10]Peter O. Sigerseth and Vernon F. Grinaker, "Effect of Foot Spacing on Velocity in Sprints," *Research Quarterly*, XXXIII, December 1962, pp. 599–606.

[11]Paul Hogberg, "The Effect of the Starting Position on the Straight Speed Short Distance Races," *Svensk Idrott*, XXV, No. 20 (cited in *Index and Abstracts of Foreign Physical Education Literature*, Indianapolis, Ind., Phi Epsilon Kappa Fraternity, IX, 1964).

[12]Ronald C. Menely and Robert A. Rosemier, "Effectiveness of Four Track Starting Positions on Acceleration," *Research Quarterly*, XXXIX, March 1968, pp. 161–65.

[13]George T. Bresnahan, "A Study of the Movement Pattern in Starting the Race from the Crouch Position," *Research Quarterly*, V (Supplement), March 1934, pp. 5–11.

Payne and Blader[14] studied the forces exerted against the starting blocks in over 150 starts made by 17 international-class sprinters and found:

1. The pattern of forces evoked was characteristic of the particular athlete.

2. In general, both rear and front feet started to exert forces on the blocks at the same instant, rarely being separated by more than 0.01 sec.

3. A strong rear leg action was characteristic of the better starts.

4. The resultant force evoked by the athlete while on the block acted first in front of his center of gravity and then behind.

Baumann[15] determined selected characteristics of the sprint starts of 30 experienced sprinters grouped on the basis of recent performances over 100 m—Group 1 (10.2–10.6 sec), Group 2 (10.9–11.4 sec), and Group 3 (11.6–12.4 sec)—and found that:

1. The horizontal displacement-time, velocity-time, and acceleration-time curves for the start and first few strides "are characteristic for the individual subject like a fingerprint." (*Note:* Payne and Blader[16] reported a similar finding concerning the pattern of forces exerted against the blocks.)

2. There was a significant difference between the reaction time of the rear foot and that of the front foot—the mean differences were 0.016 sec, 0.019 sec, and 0.027 sec for Groups 1, 2, and 3, respectively. (*Note:* These results are in general agreement with those or Henry[17] and Payne and Blader[18] except that the latter reported a difference in reaction times substantially less than that reported by Baumann for a comparable group—the top sprinters in Britain and Germany, respectively.)

3. The good sprinters had a greater proportion of their body weight supported on the hands when they were in the "Set" position than did the sprinters of lesser ability. The proportions for each group (expressed as percentages of body weight) were: Group 1, 73–82%; Group 2, 62–75%; and Group 3, 52–67%.

[14]A. H. Payne and F. B. Blader, "A Preliminary Investigation into the Mechanics of the Sprint Start," *Bulletin of Physical Education*, VIII, April 1970, pp. 21–30.

[15]W. Baumann, "Kinematic and Dynamic Characteristics of the Sprint Start," in *Biomechanics V-B*, ed. by Paavo V. Komi (Baltimore: University Park Press, 1976).

[16]Payne and Blader, "A Preliminary Investigation into the Mechanics of the Sprint Start," p. 27.

[17]Henry, "Force-Time Characteristics of the Sprint Start," p. 306.

[18]Payne and Blader, "A Preliminary Investigation into the Mechanics of the Sprint Start," p. 27.

Jackson and Cooper[19] used 12 male subjects with no specialized training in sprinting to investigate the effects of varying the distance between the hands, and the angle of the knee joint of the back leg, in the "Set" position. Their results supported the use of a narrow hand position (8 in. between the thumbs as opposed to 20 in.) and indicated that use of a back knee angle of 180° resulted in significantly slower times to 10 yd and to 30 yd than did back knee angles of 90° and 135°. There was no significant difference in the times to either distance when trials with angles of 90° and 135° were compared.

The optimum time for holding a runner in the "Set" position before firing the gun,[20] the effects of variations in the height of the hips when the runner is in the "Set" position,[21] and the effect of using starting blocks instead of holes[22] were also studied by those who pioneered research into the crouch start. Regrettably, however, each of these factors was considered in terms of the time that elapsed between the firing of the gun and the removal of the back foot from its starting block—a time that appears to have very little to do with overall sprinting performance. Thus, although the results of these studies have quite frequently been referred to in the literature, their practical significance is at least open to question.

The time taken by a sprinter to reach top speed has been commented on by a number of writers. Bunn,[23] for example, has stated that full speed is gained in about 10 yd and that even the slowest runners are "fully accelerated" in 15 yd. These figures are somewhat smaller than those found by others. Hill,[24] for instance, states that a sprinter reaches his top speed between 30 and 50 yd, while Henry[25] implies that an even greater distance might be involved when he states that top speed is attained approximately 6 sec after the gun is fired.

Although making no specific reference to the point at which top speed is reached, Henry and Trafton[26] have given a clear indication that a sprinter's

[19]Andrew S. Jackson and John M. Cooper, "Effect of Hand Spacing and Rear Knee Angle on the Sprinter's Start," *Research Quarterly*, XLI, October 1970, pp. 378–82.

[20]George A. Walker and Thomas C. Hayden, "The Optimum Time for Holding a Sprinter between the 'Set' and the Stimulus (Gun Shot)," *Research Quarterly*, IV, May 1933, pp. 124–30.

[21]Ray A. White, "The Effect of Hip Elevation on Starting Time in the Sprint," *Research Quarterly*, VI (Supplement), October 1935, pp. 128–33.

[22]Thomas C. Hayden and George A. Walker, "A Comparison of the Starting Time of Runners Using Holes in the Track and Starting Blocks," *Research Quarterly*, IV, May 1933, pp. 117–23.

[23]John W. Bunn, *Scientific Principles of Coaching* (Englewood Cliffs, N.J.: Prentice-Hall, Inc., 1962), pp. 109–10.

[24]A. V. Hill, *Muscular Movement in Man* (New York: McGraw-Hill Book Co., 1927), p. 51.

[25]Franklin M. Henry, "Research on Sprint Running," *Athletic Journal*, XXXII, February 1952, p. 32.

[26]Franklin M. Henry and Irving R. Trafton, "The Velocity Curve of Sprint Running," *Research Quarterly*, XXII, December 1951, p. 412.

gain in speed after the first 15–20 yd is relatively small. Their experimental testing of a theoretical sprint velocity curve revealed that "90 percent of the maximum velocity is reached by 15 yards and 95 percent by 22 yards."

Despite a considerable volume of research into the biomechanics of the crouch start, many basic questions remain unanswered. For example: Do the results obtained from studies of male sprinters also hold true in the case of female sprinters, or do the acknowledged differences in strength, anatomical structure, etc., cause the optimum block spacings and body position for female sprinters to be different from those of their male counterparts? What is the optimum angle for each starting block? What is the optimum design for starting blocks? (Should the faces of the blocks be curved or flat? What materials should be used on the face of the blocks? Should the tips of the spikes or the soles of the shoes bear on the face of the blocks? Should the block be large enough to support the heel as well as the forward part of the foot? Or do all of these factors have so little effect on the final outcome that they can be considered irrelevant?) These, and many other questions, await the attention of those interested enough to seek the answers.

Sprinting

The basic sprinting action is of considerable importance not only in track and field but in many other sports as well. Although success in sprinting obviously depends on an athlete's ability to combine the actions of his legs, arms, trunk, etc., into a smoothly coordinated whole, for the purpose of the analysis that follows, the position and movements of each body part are considered separately.

Legs. The action of the legs in running is cyclic. Each foot in turn lands on the ground, passes beneath and behind the body, and then leaves the ground to move forward again ready for the next landing. This cycle can be conveniently subdivided into:

1. a *supporting phase* that begins when the foot lands and ends when the athlete's center of gravity passes forward of it,
2. a *driving phase* that begins as the supporting phase ends and ends as the foot leaves the ground, and
3. a *recovery phase* during which the foot is off the ground and is being brought forward preparatory to the next landing.

Supporting phase. The function of the supporting phase is to arrest the athlete's downward motion—a downward motion imparted to him by gravity during the time he is airborne—and to allow him to move into position to drive his body forward and upward into the next stride with the minimum loss of momentum.

To ensure that the forces involved in reducing his downward motion to zero are within limits that can readily be tolerated, the athlete instinctively sees to it that the distances through which they act are appropriately large (work-energy relationship, p. 103). Thus when his foot makes contact with the track, the flexion of his hip, knee, and ankle joints is increased to cushion the shock of the impact. In the course of this process (and despite the views occasionally expressed by coaches that this does not or should not occur) the heel of the foot is lowered to the track.

The manner in which a sprinter's foot makes contact with the track has been studied by Nett. Following an analysis of top-class athletes, he reported that:

> In the 100 meter and 200 meter runs, the ground is contacted first on the outside edge of the sole, high on the ball (joints of the little toe). . . . In the 400 meter run, which is run at a somewhat slower pace, the contact point lies a bit further back toward the heel; the foot plant is now somewhat flatter. . . . In the further course of the motion . . . even in the case of sprinters . . . the heel contacts the ground.[27]

Whether the athlete's forward momentum is reduced during the supporting phase depends on the nature of the horizontal forces his foot exerts on the ground or, more precisely, on the equal and opposite forces the ground exerts on his foot during this time.

The magnitude and direction of the horizontal force that the foot exerts as it lands are governed by the velocity of the foot *relative to the ground* at that instant. If the foot is traveling forward at the instant it strikes the ground, it will tend to continue to do so (Newton's first law) and will thus exert a forward horizontal force against the ground. In reaction the ground will exert a backward horizontal force that will retard the athlete's forward motion. If the athlete's foot is traveling neither forward nor backward at the instant it strikes the ground, the ground reaction is entirely vertical and the athlete's horizontal motion is unaffected. Finally, if the athlete's foot is moving backward at the instant it lands, a forward horizontal reaction is evoked and the athlete's forward momentum is increased.

Now it should be apparent from the foregoing discussion that for the second function of a supporting phase to be effectively served, the athlete's foot must not be moving forward at the instant it lands. To understand how a sprinter prevents this from occurring, or at least endeavors to prevent it from occurring, it is necessary to examine in some detail those factors that determine the velocity of the foot at the instant it strikes the track. When the athlete is airborne prior to his foot touching down, his center of gravity is moving forward with a horizontal velocity determined at the moment he

[27]Toni Nett, "Foot Plant in Running," *Track Technique*, No. 15, March 1964, pp. 462–63.

left the ground (ignoring the effects of air resistance). Those parts of his body that are not moving forward or backward relative to the center of gravity have this same horizontal velocity. The other parts of his body have horizontal velocities larger or smaller than that of the center of gravity depending on the direction (forward or backward) in which they move. Thus, for example, if an athlete's center of gravity is moving forward at 30 fps and he has one foot moving backward with a horizontal velocity of 10 fps relative to his center of gravity and the other moving forward with a horizontal velocity of the same magnitude relative to the center of gravity, the actual horizontal velocities of the feet are, respectively, 20 and 40 fps. In other words, the horizontal velocity of a body part is equal to the horizontal velocity of the center of gravity of the body, plus the horizontal velocity of the part relative to the center of gravity.

It is obvious therefore that the only way in which the athlete can ensure that his foot is not moving forward at the instant it strikes the track (and thus that there are no retarding horizontal forces evoked at this instant) is to have it moving backward relative to his center of gravity with a horizontal velocity at least equal to that at which his center of gravity is moving forward. For example, if his center of gravity is moving forward at 30 fps, his foot must be moving backward relative to the center of gravity at no less than 30 fps to achieve the desired result.

Although it must be stressed that the horizontal velocity of the foot is the sole determinant of whether there is a braking or retarding effect when the foot lands, it has frequently been observed that such an effect is produced if the foot is more than a few inches forward of a vertical line through the athlete's center of gravity at this time. This would seem to imply, therefore, that there is little likelihood of the foot acquiring the necessary backward velocity relative to the center of gravity until it is almost directly below this point (that is, of course, assuming that the athlete does not employ a deliberate "pawing" action as he brings the foot down to the track).

With respect to the placement of the foot relative to the center of gravity, Deshon and Nelson[28] found a significant positive correlation between (1) the angle the leg made with the ground at the instant the foot landed and (2) the speed of running. They concluded, therefore, that "efficient running is characterized by . . . placement of the foot as closely as possible beneath the center of gravity of the runner."

There is also some evidence to suggest that even when the foot is placed below or almost below the center of gravity, its backward velocity relative to that point is still insufficient to completely eliminate any retarding effect.[29,30]

[28]Deane E. Deshon and Richard C. Nelson, "A Cinematographical Analysis of Sprint Running," *Research Quarterly*, XXXV, December 1964, pp. 453–54.

[29]A. H. Payne, W. J. Slater, and T. Telford, "The Use of a Force Platform in the Study of Athletic Activities," *Ergonomics*, XI, March 1968, pp. 123–43.

[30]Akira Tsujino, "The Kick in Sprint Running," *Kobe Journal of Medical Science*, XII, January 1966, pp. 1–26.

Driving phase. The athlete's task during the driving phase is to drive or thrust downward and backward against the ground. This drive, brought about by the forceful extension of his hip, knee, and ankle joints, causes his body to be projected forward and upward into the next stride.

The athlete's velocity as his foot leaves the ground is a function of the work done by the extensor muscles of his hip, knee, and ankle joints during the driving phase (work-energy relationship). Since the distance through which each of these muscle groups exerts force is thus important in determining the body's velocity at "takeoff" (and consequently in determining the athlete's stride length), it is obviously desirable that they should be contracted through the greatest range possible. Failure to obtain complete extension of hip, knee, and ankle joints is probably one of the most common faults to be found in the techniques of sprinters. (*Note:* While cases of incomplete leg extension during the driving phase are very common, cases in which the contours of the leg muscles give the impression that the leg joints are not completely extended even though the limits of the joint ranges have been reached are almost equally common. Coaches must therefore exercise considerable care if they are to accurately determine whether or not an athlete is obtaining complete extension during the driving phase.)

Recovery phase. During the recovery phase the athlete's foot is brought forward from behind his body to that point at which it next makes contact with the track.

As soon as the foot breaks contact with the track, the thigh of the same leg begins its forward rotation about an axis through the hip joint. This deliberate action (possibly supplemented by an involuntary physiological reaction to the stretching that the flexor muscles of the leg experience during the driving phase) results in the leg bending sharply at the knee and the foot being lifted to a position close to the buttocks. Once seen as an unnecessary and therefore wasteful motion, this high kickup of the foot is now recognized as an inevitable consequence of other important actions. In addition, it is seen to be of value in its own right, for it reduces the moment of inertia of the whole limb to a minimum (relative to a transverse axis through the hip joint) and thus enables it to be rotated forward about the hip joint rather more quickly than would otherwise be the case.

When the athlete's thigh reaches a horizontal or near-horizontal position, his lower leg swings forward about an axis through his knee and the whole limb begins its descent to the track.

The height to which the athlete brings his knee during the forward and upward swing of the thigh has been remarked upon by a number of researchers. Fenn[31] observed that proficient runners tend to raise the knee

[31]Wallace O. Fenn, "Work Against Gravity and Work due to Velocity Changes in Running," *American Journal of Physiology*, XCIII, 1930, pp. 433–62.

high as the free-swinging leg is thrust forward; Deshon and Nelson[32] concluded that a high knee lift was one of a number of factors characteristic of efficient running; and Sinning and Forsyth[33] found a "more acute angulation between the trunk and the thigh as running velocity increased."

Arms. Throughout the various phases of an athlete's leg action, his hips are rotated back and forth in a roughly horizontal plane. When the left knee is brought forward and upward in the recovery phase of the left leg cycle, the hips (viewed from above) rotate in a clockwise direction. The limit of this clockwise rotation is reached when the knee reaches its highest point in front of the body. Then, as the left foot is lowered toward the track and the right leg begins its forward and upward movement, the hips begin to rotate in a counterclockwise direction. The limit of this counterclockwise hip rotation is reached as the right knee reaches its highest point in front of the body. At this point the cyle is complete.

These rotary actions of the hips evoke contrary reactions in the athlete's upper body, for, as the athlete's left knee is swung forward and upward, his right arm is swung forward and upward and his left arm backward and upward, to balance this leg action. Then, as the left foot is lowered and the right leg starts to move forward, the actions of the arms are reversed.

Although the shoulders might also be rotated to balance the hip action, such rotation would of necessity be relatively slow. Thus, to avoid the complications that this slowness might introduce, good sprinters use an arm action of such range and vigor that there is no need for a contribution from the shoulders to achieve the required equality between hip action and upper body reaction.

In this arm action, the arms are flexed to about a right angle at the elbow and swung backward, forward, and slightly inward about an axis through the shoulders. At the forward limit of the swing the hands (generally held in a lightly clenched fashion) are at about shoulder height and at the backward limit are level with, or slightly behind, the hip.

Trunk. During the support and driving phases, the athlete exerts vertical and horizontal forces against the ground. The equal and opposite reactions that these evoke tend to accelerate the athlete in the direction in which they act and, if they do not act through his center of gravity, to angularly accelerate him. Thus at the instant depicted in Fig. 175, the athlete is acted upon by a vertical reaction R_V, which tends to accelerate him upward and rotate him forward about his transverse axis, and a horizontal reaction R_H, which tends to accelerate him forward and rotate him backward about the same axis. In addition, the athlete is acted upon by an air resistance force A, which in

[32]Deshon and Nelson, "A Cinematographical Analysis of Sprint Running," pp. 451–55.

[33]Wayne E. Sinning and Harry L. Forsyth, "Lower-Limb Actions While Running at Different Velocities," *Medicine and Science in Sports*, II, No. 1, Spring 1970, p. 31.

general opposes his motion and tends to rotate him backward. The moment tending to rotate the athlete backward is equal to

$$R_H y_H + A y_A$$

while that tending to rotate him in the opposite direction is equal to

$$R_V x$$

where y_H, y_A, and x are the lengths of the respective moment arms. What happens to the athlete as a result of these opposing tendencies obviously depends on which of the two is the greater. And it is here that the inclination of the sprinter's trunk comes into the reckoning, for it very largely determines the position of the athlete's center of gravity and thus, too, the lengths of the moment arms. For example, if the athlete in Fig. 175 were to increase the forward inclination of his trunk, his center of gravity would move forward and downward, decreasing y_H and increasing x. (*Note:* The magnitude of y_A would almost certainly change also, although the exact nature of this change would be difficult to predict.)

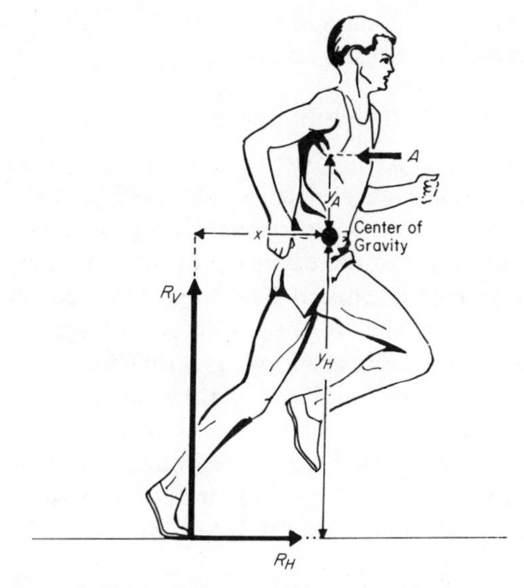

Fig. 175. The optimum inclination of the trunk is determined by the moments of the eccentric forces that act on the runner.

By making appropriate adjustments in the inclination of his trunk and thus modifying the moments involved, the good sprinter controls the rotation of his body about its transverse axis. When he drives downward and backward against his starting blocks, the horizontal component of the ground-reaction force is very large. Therefore, to prevent the backward-rotating effect of this force becoming overwhelmingly dominant, the sprinter leans well forward, keeping the moment arm of the horizontal reaction small and that of the vertical reaction large. In succeeding steps the athlete's progressively greater forward speed makes it increasingly difficult for him to exert horizontal forces of the same magnitude as he did at the outset. Thus to prevent the forward-rotating tendency of the vertical reaction becoming dominant and perhaps causing him to stumble forward, the athlete steadily raises his trunk as the horizontal forces decrease in magnitude. By the time he has reached top speed, the horizontal forces he exerts against the ground have been reduced to the point where their overall accelerating effect is just sufficient to balance the retarding effect of air resistance.[34] The backward-rotating tendencies of these two forces have been similarly reduced and the need for a pronounced forward lean of the trunk no longer exists. There is still, however, a need to combat the small backward-rotating tendencies of air resistance and the horizontal reaction. If this is not done, the athlete's body will eventually rotate into a position from which he cannot apply the horizontal forces against the ground necessary to maintain top speed. (Under such circumstances, commonly seen in the concluding stages of 440-yd races, the athlete may lose speed to such an extent that he appears to be running on the spot.) For these reasons most good sprinters retain a slight forward lean of the trunk even when running at top speed.

Middle- and Long-Distance Running

As the length of the race increases beyond 440 yd—normally regarded as the longest sprint event—the athlete's stride length and stride frequency are both substantially reduced; so too are the range and vigor of most of his actions. The forcefulness of the extension of his hip, knee, and ankle joints during the driving phase is reduced. The extent to which his foot rises toward his buttocks and the height his knee is raised in front during the recovery phase are both reduced. His arms swing through a lesser range than they would if he were sprinting, and part of their function of balancing the leg action may be assumed by the shoulders rotating in opposition to the hips. Finally, with reductions in both the air resistance and the horizontal ground reaction, the forward inclination of his trunk when running at a constant speed is generally less than it would be if he were sprinting.

Hurdling

Expressed simply, hurdling is a specialized form of running in which, for most of a race, 1 stride in 4 (in the 120-yd high hurdles) and 1 stride in somewhere between 12 and 18 (in the 440-yd intermediate hurdles) is exaggerated to allow the athlete to safely negotiate the hurdles.

[34]If the effects of air resistance were greater than those of the horizontal reaction, the athlete would lose speed. Conversely, if the effects of air resistance were less than those of the horizontal reaction, he would gain speed. Thus, for the athlete to run "at top speed" these contrary effects must be equal in magnitude.

High Hurdles

In the 120-yd high hurdles, the distance from the start line to the first hurdle is 15 yd; the distance between the 3-ft 6-in. hurdles is 10 yd; and the distance from the last hurdle to the finish line is 15 yd.

Approach. While essentially the same as that of a sprinter, the starting technique employed by a high hurdler differs in two important respects:

1. The foot that he has forward in his starting position must allow him to take the appropriate number of strides to the first hurdle and to arrive for his takeoff on the correct foot. For example, if he takes eight strides to the first hurdle (as do most good high hurdlers) and takes off from his left foot, this foot must be the forward one when he is in his starting position.
2. The need for the athlete to be in a suitable position for takeoff at the first hurdle requires that his trunk be brought to an upright, or near-upright, position earlier than if he were competing in a flat sprint event.

Takeoff. The first part of the high hurdler's action at takeoff is almost identical to that of a normal sprinting stride. Once his takeoff foot has landed at the end of the last stride before the hurdle, his other foot (the lead foot) is brought up high under his buttocks [Fig. 176(a)]. Then with the lead leg well flexed—to reduce its moment of inertia and facilitate its rotation about an axis through the corresponding hip joint—the lead knee is swung forward and upward [Fig. 176(b)]. When the thigh of the leading leg nears the limit of its forward and upward movement, its momentum is transferred to the lower leg. This transfer of momentum, probably supplemented by a contraction of the muscles that extend the knee joint, ultimately brings the leading leg into a near-straight position [Fig. 176(c) and (d)]. These actions of the leading leg tend to shift the hurdler's center of gravity in a forward and upward direction and to rotate his body backward. However, partly in reaction to the movements of the lead leg and partly as a result of additional muscular forces, the hurdler's trunk is brought forward and downward at the same time as the lead leg is being swung forward and upward [Fig. 176(a)–(d)]. This action of the trunk serves as a counter to the lifting and backward-rotation tendencies of the lead leg.

While the lead leg and trunk are being brought toward each other, the takeoff leg first supports the athlete and then drives him forward and upward toward the hurdle. As with the first part of the lead leg action, these actions of the takeoff leg are very like those used at the equivalent stages in sprinting.

The distance at which the hurdler takes off in front of the hurdle (i.e., the distance from the toe of his takeoff foot to the line of the hurdle) depends on his height, his leg length, his speed, and his technique. Of these factors, the last two—the only two over which the hurdler has any control—are of greatest practical importance. For a given takeoff distance, the greater the

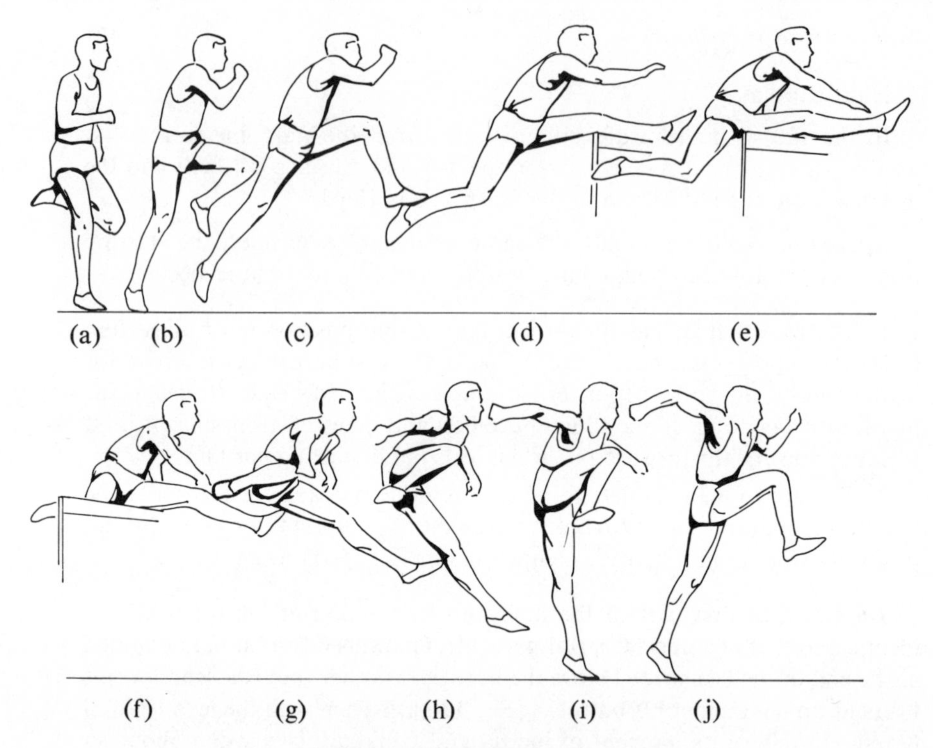

Fig. 176. Technique in the high hurdles

hurdler's horizontal velocity, the less the time he has in which to get his lead foot high enough to clear the hurdle. Thus, to allow himself sufficient time for this purpose (and to avoid disaster!), a good hurdler instinctively modifies his takeoff distance in keeping with changes in his horizontal velocity (i.e., he increases his takeoff distance as he increases his horizontal velocity and vice versa). The normal pattern of changes in takeoff distances during a high hurdles event is thus one in which the distances increase steadily over the first few hurdles (as the hurdler gradually attains his top speed) and decrease over the final few hurdles (as fatigue begins to reduce his speed). The hurdler's technique, and particularly the speed with which he can bring his lead foot up to the height of the hurdle, is also important in determining the takeoff distance. For, if all else is equal, the faster the hurdler's lead leg action, the closer to the hurdle he can afford to be at the instant of takeoff. While takeoff distances clearly vary from one athlete to another, and from one hurdle to another for the same athlete, a distance of 7 ft is generally regarded as average for most athletes. (*Note:* Takeoff distances reported by Doherty[35] for six top-class hurdlers ranged from 5 ft 9 in. to 8 ft and averaged 6 ft $10\frac{1}{2}$ in.)

Flight. Once the hurdler leaves the ground, the only forces acting on him (assuming he doesn't hit the hurdle!) are gravity and air resistance. The

[35] J. Kenneth Doherty, *Modern Track and Field* (Englewood Cliffs, N.J.: Prentice-Hall, Inc., 1963), pp. 135–36.

former has no effect on his forward velocity but acts simply to bring him back to the ground. The time taken for this to be effected, and thus the time before the hurdler can next drive against the ground, is governed mainly by his vertical velocity at takeoff—the greater his vertical velocity at this instant, the greater the time during which he is airborne. And, since the height that the hurdler's center of gravity rises in flight is directly related to his vertical velocity at takeoff, this means that the hurdler should skim across the hurdle with his center of gravity as low as safety and a balanced landing will permit if he is to get back to the ground and start driving again as quickly as possible. The air resistance that he encounters while he is airborne acts to reduce his forward velocity. Although the magnitude of this force is probably fairly small—especially where the hurdler's forward "dip" is pronounced and his frontal area is therefore very small—the longer it acts, the greater is the reduction it effects in the hurdler's forward velocity. This is a further reason for him to seek to get back on the ground as soon as he can.

During the initial part of the flight the hurdler's lead leg and trunk continue to move toward each other [Fig. 176(c)–(e)]. These actions bring his center of gravity close to the lower limits of his body and thus reduce the height to which the athlete must raise his center of gravity in order to clear the hurdle. This effect is further enhanced by the forward and downward motion of the leading arm [Fig. 176(c)–(e)], and with some hurdlers by a dropping forward of the head so that the face is roughly parallel with the track.

Once the knee of the leading leg has crossed the hurdle, the motions of the trunk and lead leg are reversed, with the downward and backward action of the lead leg producing an upward and backward reaction of the trunk.

Although it may look very different, the action of the trailing leg during the flight phase is in reality very similar to the action during the recovery phase in normal sprinting. The one major difference is that as the trailing leg is brought forward, the thigh is lifted outward so that instead of passing vertically beneath the body (as in sprinting) it passes horizontally out to the side. This action permits the athlete to keep much lower than would otherwise be possible as he passes over the hurdle.

The upper-body reaction to the movements of the hurdler's trail leg generally takes the form of a contrary motion of the athlete's leading arm [Fig. 176(d)–(j)]. However, if the athlete has an insufficient forward "dip" of his trunk, the moment of inertia of the leading arm may be so small that the arm must sweep back very fast indeed if it is to provide the needed reaction. And, since a fast-sweeping arm action like this tends to drag the athlete's shoulders and trunk away from their straight-to-the-front alignment and thus to interfere with the maintenance of balance and forward speed, such a situation is clearly to be avoided.

Landing. The hurdler lands with his body nearly-erect [Fig. 176(i)] at a distance of approximately 4½ ft from the hurdle—landing distances reported by Doherty[36] ranged from 4–5 ft and averaged 4 ft 6½ in.—and then drives vigorously forward into his next running stride [Fig. 176(j)]. This forward drive is greatly facilitated by a strong forward and upward action of the trail leg, for such an action not only puts the trail leg in an ideal position to swing forward into the next stride [Fig. 176(i) and (j)] but it also tends to arrest the backward rotation of the trunk evoked by the movement of the lead leg—a rotation that, if unrestrained, would place the hurdler in such a position that he could not drive off effectively into his next stride. (*Note:* Failure to obtain a high, forward, and upward recovery of the trail leg is probably the most common cause of a shortened first stride and a consequent difficulty in covering the distance between hurdles in the required three strides.)

The speed of the leading foot relative to the ground determines whether or not the body experiences a momentary braking (or deceleration) at the instant of landing—if the foot is not moving backward at a speed at least equal to that at which the body is moving forward, braking must inevitably occur. Since this situation most frequently occurs when the athlete's foot lands well forward of his line of gravity, a placement of the foot below the center of gravity is generally regarded as a characteristic of good hurdling technique.

Between Hurdles. The distance from the point of landing to the point of takeoff for the next hurdle is negotiated in three running strides. Of these, the first is invariably the shortest and the second is usually the longest. The third, shortened by a few inches in preparation for takeoff, is usually somewhere between the other two. A striding pattern suggested by Mitchell[37] illustrates these differences:

Landing	4 ft 6 in. from the hurdle
Length of stride 1	5 ft 6 in.
Length of stride 2	6 ft 9 in.
Length of stride 3	6 ft 3 in.
Takeoff	7 ft 0 in. from the hurdle.

Intermediate Hurdles

In the 440-yd intermediate hurdles, the distance from the start line to the first hurdle is 49¼ yd, the distance between the 3 ft hurdles is 38¼ yd, and the distance from the last hurdle to the finish line is 46½ yd.

Just as the actions of middle- and long-distance runners might be regarded as less pronounced versions of the actions employed by sprinters, so too might those of intermediate hurdlers be regarded vis-à-vis their high hurdling counterparts. For with the height of the hurdle reduced by 6 in., the athlete

[36] *Ibid.*, pp. 135–36.

[37] Les Mitchell, "Some Observations on the High Hurdles," *Track Technique*, No. 37, September 1969, p. 1187.

can maintain his speed over the obstacles without resorting to the same vigorous, exaggerated action that the high hurdler must use to achieve the same end.

The greatest technique problem facing the intermediate hurdler (aside from his actual hurdling action) is unquestionably that of how many strides to take between hurdles. And in this there are a number of important considerations. First, his choice of the number of strides to use must be reasonably consistent with the length of his normal running stride, for any marked understriding or overstriding will undoubtedly reduce his speed. (*Note:* Taking typical values for takeoff and landing distances and for the first stride after clearing the hurdle, Le Masurier[38] computed the average stride length necessary to cover the distance between hurdles in a given number of strides:

Number of Strides between Hurdles	*Average Stride Length*
13	8 ft 2 in.
14	7 ft 7 in.
15	7 ft 0 in.
16	6 ft 6 in.
17	6 ft 1 in.

He contends, though, "that for a hurdler to be able to flow economically between hurdles his natural stride must exceed the above figures by several inches.") Second, unless he can hurdle reasonably well from either foot, it will be necessary for him to use an odd number of strides between hurdles. Third, due allowance must be made for the fact that as fatigue sets in, his stride length will tend to decrease.

While these often conflicting requirements have been met in a variety of ways over the years, a general trend toward (1) the use of fewer strides between the hurdles and (2) the use of an even number of strides between hurdles at some point in the race is apparent. For example, whereas 17 strides between hurdles, or 15 strides between the first few hurdles and 17 strides between the remainder, were common patterns in the early days of the event, the medalists in the 1968 Olympic Games used the following stride patterns:

First:	David Hemery (Great Britain)	13 strides to the sixth hurdle, 15 strides thereafter	48.1 sec
Second:	Gerhard Hennige (West Germany)	13 strides to the sixth hurdle, 15 strides thereafter	49.0 sec
Third:	John Sherwood (Great Britain)	13 strides to the sixth hurdle, 14 strides to the eighth hurdle, 15 strides thereafter	49.0 sec

[38]John Le Masurier, "Some Factors of Performance in the 400 Metres Hurdles," *Athletics Weekly*, XXIII, September 13, 1969, p. 14.

Edwin Moses (U.S.A.), winner of the 400-m hurdles at the 1976 Olympic Games, used 13 strides between the hurdles throughout that race and, as the following excerpts from a recent interview indicate, has used 12 strides between hurdles in practice:

> Q: "You said once that 15 strides actually slowed you down. When did you start working on going 13s all the way?"
>
> A: "I've never run 15s. At Florida [my first serious meet in the IH] I ran 13s until the 6th hurdle, ran 14s for two because I wasn't sure of the pattern and then 13s to the finish. I knew even then I could do 13s all the way, but I didn't do it because the race was new. So after that, I worked on 13s all the way. I just did everything in my workouts with 13 strides."
>
> Q: "Is 12 strides one of the keys to running that fast [45–46 sec]?"
>
> A: "Yes and I've worked on that in practice. I'm doing it comfortably, more so than before the Games when I was concentrating on 13s. I didn't want to work hard on 12s then because I didn't want to upset the pattern I had."[39]

Steeplechase

In the 3000-m steeplechase, an athlete must negotiate a solidly constructed 3-ft hurdle a total of 28 times, and a water jump (consisting of a 3-ft hurdle fixed in place in front of a 12-ft square of water) a total of 7 times.

The orthodox technique for clearing the hurdles is essentially the same as that used by athletes in the intermediate hurdles, i.e., a simplified form of the high hurdling technique previously described. As in the intermediate hurdles, the ability to hurdle off either foot—although relatively rare among steeplechasers—is a distinct advantage, for it reduces the extent of the adjustment that the athlete may be called on to make to get into an appropriate position for takeoff. (It should be noted here that, unlike high and intermediate hurdlers, steeplechasers do not normally strive to cover the distances between obstacles in a set number of strides.)

The technique used in negotiating the water jump is shown in Fig. 177. As the athlete approaches the barrier, he adjusts his speed and his striding—the latter with the aid, perhaps, of a checkmark—so that he will arrive at the optimum point for takeoff with sufficient speed to allow him to complete the jump efficiently. Once his takeoff foot is grounded, he brings his leading leg forward, with knee well bent, in preparation for the spring onto the 5-in. wide rail (Fig. 177). With his eyes focused on the rail to minimize the risk of a faulty foot placement, he leaps into the air and places the arch of his foot against the near upper edge of the rail (Fig. 177). Then follows a period of support during which the athlete maintains a compact body position as he

[39]Jon Hendershott, "T & FN Interview: Edwin Moses," *Track and Field News*, XXIX, September 1976, pp. 33–34.

rotates forward over and beyond the line of the rail. This compact position with the supporting leg well bent and the trunk hunched forward over it minimizes the athlete's moment of inertia and thereby enables him to rotate quickly forward into position for the next stage in the sequence. Once his center of gravity has passed some distance forward of the rail, and the sole of his foot has rotated so that his spikes are digging into its forward vertical face, the athlete drives backward and downward against the rail by vigorously extending the hip, knee, and ankle joints of his supporting leg. The reaction to this leg drive projects the athlete out over the water to a landing on the opposite foot some 12–18 in. from the water's edge—a "long jump" of some $10\frac{1}{2}$–11 ft. (*Note:* While a few top-class athletes have made a practice of jumping completely over the water to land on the track beyond, the enormous amount of energy that the athlete expends in performing such a feat raises grave doubts as to its efficiency.) The whole action is completed by the athlete's stepping forward out of the water and onto his free leg.

Fig. 177. An example of the techniques used in clearing the water jump in the 3000-m steeplechase

Chapter 16

TRACK AND FIELD: JUMPING

The standard jumping events in track and field are the long jump, the triple jump (formerly known as the hop, step, and jump), the high jump, and the pole vault.

In each of these events, the athlete's objective is to obtain a maximum displacement of his center of gravity in a given direction—in the long and triple jumps, in a horizontal direction; in the high jump and the pole vault, in a vertical direction—and then, in keeping with the rules governing the event, to extract as much credit as he can for having achieved this displacement. In the horizontal jumps this latter means that the athlete endeavors to get his feet as far forward as he can without falling back on landing. In the jumps for height it means that the athlete strives to have his body pass over a bar set close to, or perhaps even above (p. 124), the maximum height that his center of gravity attains.

LONG JUMP

Basic Considerations

The distance with which an athlete is credited in the long jump may be considered to be the sum of three lesser distances:

1. The horizontal distance between the front edge of the takeoff board and the athlete's center of gravity at the instant of takeoff (the takeoff distance, L_1 in Fig. 178).

2. The horizontal distance that his center of gravity travels while he is airborne (the flight distance, L_2 in Fig. 178).

3. The horizontal distance between his center of gravity at the instant his heels hit the sand and the marks in the sand from which the distance of the jump is ultimately measured (the landing distance, L_3 in Fig. 178).

Fig. 178. Contributions to the length of a hang-style long jump

The contribution that each of these distances makes to the total distance jumped is indicated by the data in Table 24.

*Table 24. Relative Contributions to Distance in the Long Jump**

	Distance	*Percentage of Total*
Takeoff distance	9½ in	3½
Flight distance	19 ft 7½ in	88½
Landing distance	1 ft 9 in	8
Total distance	22 ft 2 in	

*Based on unpublished data from 25 trials by Swiss and West German long jump specialists and decathletes (*Data courtesy Benno M. Nigg, Eidg. Technische Hochschule, Zürich, Switzerland*).

The takeoff distance is a function of the accuracy with which the athlete places his foot on the takeoff board, his physique, and his body position at the instant of takeoff.

An athlete's flight distance is governed by the same four variables that determine the motion of all such projectiles—his speed, angle, and height of takeoff and the air resistance he encounters in flight.

The athlete's speed at the instant of takeoff—by far the most important of these variables—depends on the speed he develops in his run to the board and on the losses in speed associated with the adjustments he makes in preparation for takeoff. The ideal combination therefore is one of maximum (controlled) speed in the run-up together with a minimum loss of speed in preparing for takeoff.

A combination of the horizontal speed developed in the run-up and the vertical speed (or lift) acquired at takeoff determines the athlete's angle of takeoff.:

$$\text{Angle of takeoff} = \arctan\left(\frac{\text{vertical speed at instant of takeoff}}{\text{horizontal speed at instant of takeoff}}\right)$$

The lift that the athlete develops at takeoff is very much influenced by the speed of his run-up. The faster his run, the less time his foot spends on the ground at takeoff and the less vertical speed he is capable of developing. Thus, because the horizontal speeds attained by the end of the run-up are so great (equivalent to the average speeds for a 9.55–10.52-sec 100-m) and the times of takeoff are so small (0.11–0.13 sec),[1] the angles of takeoff used by top-class jumpers are rather less than the near-45° angles that might otherwise be expected (Table 25).

The height of takeoff (i.e., the difference between the height of the athlete's center of gravity at the instant of takeoff and at the instant he touches down in the pit) depends on his physique, and his body position at both instants. Of these, it is only the latter over which he can exercise any real measure of control—in the first instance by driving his arms, leading leg, head, and trunk as high as he reasonably can and in the second by delaying his landing for as long as feasible.

Under "normal" circumstances the effects of air resistance are probably so small that they may be disregarded.

The landing distance depends on the athlete's body position as he touches down in the pit and on the actions he employs to avoid falling backward and reducing the measured length of his jump. The principal factors influencing the body position at touchdown are the initial body position (i.e., its position as the athlete leaves the ground), the rotation imparted to the body during the takeoff, and the movements made in the air to minimize the effects of this rotation and to position the body for landing. Whether the athlete sits back in the sand or rotates forward over his feet depends primarily on the magnitude and direction of the ground-reaction forces exerted on him during the landing. If his feet land beneath or just slightly forward of his center of gravity, the athlete is subjected to a near-horizontal ground-reaction force that serves to dramatically increase his forward angular momentum. This results in the athlete rotating quickly forward over his feet, often sprawling headlong in an effort to regain control. If he lands with his legs fully extended and in a near-horizontal position, the ground-reaction force to which he is subjected passes in front of his center of gravity, slowing his forward motion and reversing the direction of his angular momentum to such an extent that he inevitably sits back. The athlete's task, therefore, is to strike the best possible compromise between these two extremes—to evoke a ground-reaction force

[1] Vladimir Popov, *Training for the Long Jump*, trans. by Masami Okamoto (Tokyo: Baseball Magazine Co., 1969), p. 27.

*Table 25. Speeds and Angles of Takeoff for Top-Class Long Jumpers**

Athlete	*Distance of Jump Analyzed*	*Speed of Takeoff* (fps)	*Angle of Takeoff*	*Optimum Angle of Takeoff for Given Speed*†
G. Bell (U.S.A.)	25 ft 5¾ in.	29.85	21°50′	43°33′
R. Boston (U.S.A.)	27 ft 2 in.	31.29	19°50′	43°39′
J. Owens (U.S.A.)	26 ft 8¼ in.	30.18	22°	43°35′
I. Roberson (U.S.A.)	26 ft 7¼ in.	30.56	19°10′	43°37′
E. Shelby (U.S.A.)	24 ft 7¼ in.	28.86	21°10′	43°27′
	25 ft 0 in.	29.52	20°50′	43°31′
	26 ft ½ in.	30.50	20°	43°37′
I. Ter-Ovanesian (U.S.S.R.)	24 ft 6¾ in.	28.86	18°50′	43°27′
	24 ft 11¼ in.	28.86	18°20′	43°27′
	25 ft 7½ in.	29.19	24°20′	43°28′
	26 ft 10½ in.	30.34	21°10′	43°35′

*Based upon data reported in Popov, *Training for the Long Jump*, p. 38.
†The values in this column are those that yield a maximum horizontal displacement when the athlete's center of gravity at takeoff is 1.5 ft above his center of gravity on landing. (The effects of air resistance have been ignored in these computations.)

that allows him to get the best possible forward extension of the legs on landing (i.e., the best L_3) and yet still have his center of gravity pass safely forward over his feet.

Summary

The relationships between the distance with which a long jumper is credited and the factors that determine that distance are summarized in Fig. 179.

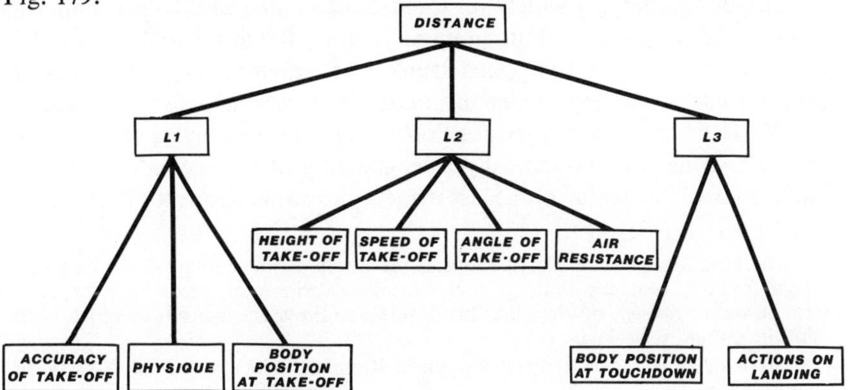

Fig. 179. Basic factors in long jumping

Techniques

For the purposes of analysis the long jump may be considered to consist of four consecutive parts: the run-up, the takeoff,[2] the flight, and the landing.

Run-Up

The purpose of the run-up is to get the athlete to the optimum position for takeoff with as much speed as he can control during that part of the jump.

The length of run-up that an athlete should use depends on the percentage of his top sprinting speed that he is capable of controlling at takeoff and on his ability to maintain a consistent pattern of striding from one jump to the next. Considering only the first of these factors, the findings of Henry[3] (p. 393) suggest that if a jumper is capable of controlling 100% of his maximum sprinting speed he should use a run-up equal in length to the distance he can sprint in 6 sec—some 150–180 ft, depending on his speed. However, if he can handle "only" 95% of his top speed, Henry's findings suggest that a run-up as short as 66 ft might be sufficient. (*Note:* Since most long jumpers do not use a crouch start and do not run with maximum effort from the very beginning of the run-up—as did the subjects in Henry's study—and since allowance should be made for an additional 3 to 4 strides during which the athlete prepares himself both physically and mentally for the takeoff, these figures are some 20–30 ft less than the theoretical lower limits for the respective cases.) As the length of the run-up (and thus the number of strides taken) increases, the scope for errors also increases. For this reason, the long jumper must weigh carefully the gains he may expect from a long run, and the extra speed it affords him, against the losses he may incur as a result of the increased opportunity for errors in striding. In practice, most top-class long jumpers use run-ups between 130 and 150 ft (or 17 to 23 running strides) in reaching a compromise between these conflicting requirements of speed and accuracy.

The transition from the run-up to the takeoff (variously referred to as the "coast" or "gather") is widely regarded as being one of the most important parts in the technique of long jumping. During the last 3 to 4 strides of his approach the athlete brings his trunk into an upright (or near-upright) position and lowers his center of gravity in preparation for the takeoff to follow. These changes in body position produce accompanying changes in the athlete's stride length—generally a lengthening of the second-to-last stride and a relative shortening of the last stride—and stride frequency (Table 26).

[2]The term *takeoff* is here used to mean the whole period during which the athlete's takeoff foot is in contact with the ground immediately prior to his becoming airborne. The term *instant of takeoff*, on the other hand, refers to the exact moment at which contact with the ground is broken.

[3]Franklin M. Henry, "Research on Sprint Running," *Athletic Journal*, XXXII, February 1952, p. 32.

*Table 26. Stride Lengths and Stride Frequencies in the Last Six Strides of the Run-Up**

		Strides					
		Sixth Last	*Fifth Last*	*Fourth Last*	*Third Last*	*Second Last*	*Last*
V. Popov (U.S.S.R.)	24 ft $11\frac{1}{4}$ in.						
Stride length		7 ft 7 in.	7 ft $5\frac{1}{2}$ in.	7 ft $7\frac{3}{4}$ in.	7 ft $4\frac{1}{2}$ in.	7 ft 11 in.	7 ft $1\frac{3}{4}$ in.
Stride frequency (strides per second)		4.05	4.05	4.1	4.27	4.3	4.7
E. Shelby (U.S.A.)	25 ft 8 in.						
Stride length		7 ft 0 in.	7 ft $1\frac{1}{2}$ in.	7 ft $5\frac{1}{2}$ in.	7 ft $\frac{3}{4}$ in.	8 ft 6 in.	6 ft $10\frac{1}{4}$ in.
Stride frequency (strides per second)		4.57	4.4	4.4	4.57	4.1	5.0
I. Ter-Ovanesian (U.S.S.R.)	26 ft $10\frac{1}{2}$ in.						
Stride length		7 ft $8\frac{3}{4}$ in.	7 ft $9\frac{3}{4}$ in.	7 ft $8\frac{1}{2}$ in.	7 ft $5\frac{1}{2}$ in.	8 ft 2 in.	6 ft $9\frac{1}{4}$ in.
Stride frequency (strides per second)		4.2	4.15	4.25	4.4	4.3	5.3

*Adapted from data presented in Popov, *Training for the Long Jump*, pp. 15, 23.

Takeoff

The purpose of the takeoff is to obtain vertical velocity (or lift) while retaining as much horizontal velocity as possible. As the athlete's foot lands (generally heel first) at the end of the last stride of the run-up, the hip, knee, and ankle joints flex a little to cushion the shock of the impact and to position the leg for the vigorous extension to follow moments later. Then, with the center of gravity moving forward, over, and beyond the takeoff foot, the arms and free (or lead) leg are driven vigorously upward in concert with a forceful extension of the hip, knee, and ankle joints of the takeoff leg. The reaction to this extension drives the athlete into the air and, if it acts eccentrically, as is usually the case, causes the athlete to rotate about a transverse axis through his center of gravity. While it is possible to obtain either a backward or forward rotation (or, indeed, no rotation at all) at this time, in good jumping the rotation imparted is almost invariably in a forward direction.

Flight

Once the athlete is in the air, his sole objective should be to assume the optimum body position for landing and, except for the forward rotation that he will almost certainly have acquired at takeoff, this would be a relatively easy task. The forward rotation, however, tends to bring his feet beneath his center of gravity at the very time (i.e., the instant of landing) when he wants them to be well forward of this point. Thus, the athlete's principal problem, in seeking to achieve his objective, is minimizing the undesirable effects of his forward rotation.

There are three in-the-air techniques in common use—the sail, the hang, and the hitch-kick.

In the sail technique (Fig. 180) the athlete brings both legs together shortly after takeoff and continues the remainder of the flight in a sitting position, either with both legs fully extended or with the knees bent at about a right angle. While the virtue of this technique lies in its simplicity—it is the technique that most people would use naturally—it has the great weakness of placing the athlete's mass close to his transverse axis and thus of facilitating the forward rotation that he needs to inhibit to obtain a good landing position.

In the hang technique (Fig. 178) the athlete reaches forward with his leading leg and then sweeps it downward and backward until he has both legs together and somewhat behind the line of his body. This sweeping movement of the leading leg (and the downward and backward swing of the arms that usually accompanies it) produces a contrary reaction in the athlete's upper body. It also has the effect of extending his body, thereby increasing his moment of inertia about his transverse axis and lessening the rate at which he rotates forward. The circular swing of the arms continues until both are high

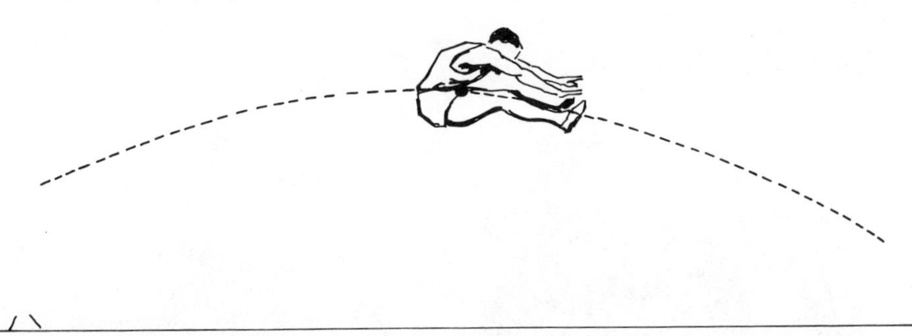

Fig. 180. The in-the-air position adopted in the sail technique

overhead, at about which time the athlete bends his knees and begins the forward movement of his legs in preparation for landing. It is important that the legs be well bent as they are brought forward under the body, for otherwise the forward and downward reaction of the trunk, which their movement evokes, might well be large enough to bring the trunk into a position where it limits the height to which the legs can be raised. Some athletes, generally those with good hip and trunk flexibility, keep their legs fairly straight and swing them wide on each side of the body as they bring them through. This serves the same purpose as bending them (i.e., it keeps their moment of inertia about a transverse axis relatively small) and may also assist in overcoming the problem of the trunk restricting the lifting of the legs by having them outside the trunk rather than under it.

There are two main variants of the hitch-kick or "running-in-the-air" technique, named according to the number of in-the-air strides involved. Unfortunately, to the general confusion of those interested in the event, there is some lack of agreement as to when the counting of strides should begin. Some coaches and athletes count the "stride" from the takeoff foot to the lead foot, which begins the instant the athlete becomes airborne, as the first stride and describe the two variants as the $2\frac{1}{2}$ and the $3\frac{1}{2}$ hitch-kick. Others commence the count one stride later. To them these same two variants are the $1\frac{1}{2}$ and the $2\frac{1}{2}$ hitch-kick. The former of these counting systems is used here.

The first part of the action in the hitch-kick (Fig. 181) is similar to that in the hang in that once the athlete has become airborne, he extends his leading leg forward and then sweeps it downward and backward beneath him. Coordinated with this movement is a pulling through of the takeoff leg (knee well bent, heel passing close to the buttocks) and a downward and backward swinging of the arm on the side opposite the lead leg.

Because of the difference between the moments of inertia of the legs, the angular momentum of the lead leg as it swings downward and backward far exceeds that of the takeoff leg that is simultaneously moving in the opposite direction. Therefore, to provide the necessary "balance" between the angular momentum of the body parts moving in one direction (the action) and the

(a) (b)

(e) (f)

Fig. 181. Technique in the execution of a $2\frac{1}{2}$-step hitch-kick

angular momentum of the parts moving in the opposite direction (the reaction), the athlete's trunk rotates backward as his lead leg and arm swing downward and backward. And this, of course, is the whole purpose of the exercise—to move the athlete's trunk into a position from which an optimum landing position can later be obtained.

At the end of this in-the-air stride the athlete's legs are in a position that is essentially the reverse of that at takeoff. Then, in a $2\frac{1}{2}$ hitch-kick, the leg that is to the rear at this point is brought forward to join the other in preparation for landing. To minimize the undesirable forward rotation of the trunk, in reaction to this movement, this rear leg should be brought forward with

(c) (d) (g) (h) (i)

the knee fully flexed. In a $3\frac{1}{2}$ hitch-kick a further full stride is taken before both feet are brought together for landing. In other words, once the athlete has reached that point at which the positions of his legs at takeoff have been reversed, the takeoff leg (the leg in front at this time) is swept downward and backward while the opposite leg is brought forward close under the buttocks. At the completion of this second reversal of the leg positions, the rear leg is brought forward for the final half stride to complete the sequence.

The question of which in-the-air technique a given jumper should use has been investigated by Khadem and Huyck.[4] They concluded that "Because of the time involved in the execution of the various in-flight techniques, it seems that the optimum technique for the individual is determined by his own abilities." Their suggested optimum techniques for performers of various abilities are a follows:

Under 6 m (19 ft 8 in.)	sail
6.0–6.5 m (19 ft 8 in.–21 ft 4 in.)	sail or hang
6.5–7.0 m (21 ft 4 in.–23 ft)	hang
7.0–7.5 m (23 ft–24 ft 7 in.)	$2\frac{1}{2}$ hitch-kick
Over 7.5 m (24 ft 7 in.)	$3\frac{1}{2}$ hitch-kick

Throughout any discussion of the flight phase in long jumping it is important to recognize that whatever the athlete does while he is in the air can have no effect whatsoever on the path followed by his center of gravity. This latter is determined at the instant of takeoff and can only be altered by the introduction of some force *external* to the jumper. And, ignoring air resistance, the conditions that normally apply in long jumping do not allow for such an external force. Certainly the assertion, occasionally found in coaching texts, that use of the hitch-kick aids in propelling the jumper forward through the air is completely without foundation.

Landing

Two sets of factors must be taken into account in deciding the optimum body position at the instant the athlete touches down in the pit:

1. those that influence the distance between the takeoff board and the athlete's heel marks in the pit,
2. those that determine whether the athlete passes forward over his feet or falls back in the sand.

With each of these, the most important single feature of the body position is the inclination of the athlete's trunk.

[4]Achmed El Khadem and Bill Huyck, "Long Jump Technique Analysis," *Track Technique*, No. 24, June 1966, p. 758.

If the athlete deliberately leans well forward during the final moments of the flight, his legs are lifted in reaction to this movement and his touchdown is slightly delayed. This increase in his time of flight allows him to be carried farther along his parabolic flight path than would otherwise be the case. On the other side of the ledger, the forward inclination of the athlete's trunk reduces his landing distance (assuming he doesn't fall backward) by moving his center of gravity nearer to his feet than it would be if he were to retain a more upright position. If he assumes a position in which his trunk is erect, or inclined slightly backward, these various effects are reversed—his time of flight is decreased while his landing distance is increased.

If the athlete leans well forward just before landing, the angle at which his legs are inclined to the horizontal at touchdown is less than it would be if he maintained a more upright body position. Under such circumstances, the magnitude and direction of the reaction force evoked from the ground are generally such as to make it considerably more difficult for the athlete to avoid sitting back.

It is clear therefore that the athlete must strike a compromise between (1) obtaining the maximum distance between the takeoff board and the marks of his heels in the sand and (2) preserving the ability to rotate forward over his feet.

McIntosh and Hayley[5] undertook a cinematographical analysis to shed some light on this question and found that when their subject used a "jack-knife" landing position (i.e., trunk inclined forward with arms extended toward the feet), he touched down in the pit 15.7 in. behind the point at which a continuation of the trajectory of his center of gravity would have reached the ground. When he used an "extended" position (i.e., trunk inclined backward slightly, hands beside hips), this distance was reduced to 3.3 in. They concluded therefore that for their jumper and the two trials investigated, the "extended position appeared to have an advantage of about 12 inches over the jackknife position." It is perhaps worth mentioning here that, contrary to the expectations of the investigators, the jumper's feet came down in the pit nearer to the takeoff board than did the trajectory of the center of gravity.

Once the athlete's heels have cut the sand, he flexes his knees to cushion the shock of the impact and, as he begins to rotate forward, he thrusts his head and shoulders over (or between) his knees to facilitate his forward rotation. If at the instant of landing the athlete's arms have been level with, or behind, his trunk, these actions are supplemented by a forward and upward swing of the arms—an action that evokes a contrary angular reaction in the rest of the body and thus assists in rotating the athlete forward over his feet.

[5]Peter C. McIntosh and H. W. B. Hayley, "An Investigation into the Running Long Jump," *Journal of Physical Education*, XLIV, November 1952, pp. 105–8.

TRIPLE JUMP

Basic Considerations

The distance with which an athlete is credited in the triple jump may be broken down into a series of consecutive parts in much the same way as has already been described in the long jump (pp. 408–9). However, whereas distance in the long jump may be considered simply as the aggregate of one takeoff, one flight, and one landing distance, in the triple jump the distance achieved is equal to the sum of three different takeoff distances, three different flight distances, and three different landing distances—not to mention the lengths of the jumper's feet, which, strictly speaking, should also be taken into account.

While the basic factors influencing each of these distances in each phase of the triple jump are essentially the same as those that apply in the case of the long jump, in the triple jump the takeoff and landing for each of the first two phases (the hop and the step) must be modified somewhat to allow for the phase or phases that follow. For example, a triple jumper who obtained the maximum (takeoff + flight + landing) distance of which he was capable in the hop phase would not produce his best effort simply because the distances he obtained in the succeeding two phases would be very much below what he could otherwise achieve. In other words, the distance he might gain with a maximum effort in the hop would be more than lost in the step and jump phases.

The optimum distribution of effort, or distance, throughout the three phases has been the subject of much discussion. Probably the two most important contributions to this discussion have been those of Tan[6] and Nett.[7] On the basis of a study of the top performers over several decades, Tan advocated a 10: 7: 10 ratio for beginners learning the event and a 10: 8: 9 ratio for top performers. Nett analyzed a total of 38 competitive jumps by Josef Schmidt (Poland), twice the Olympic Champion in this event, and Vilhjalmar Einarsson (Iceland) and concluded that the most economical ratio for the "flat" technique (low hop and step, high jump), of which these two jumpers were leading exponents, was 7: 6: 7.

Techniques

The run-up serves the same function and is performed in essentially the same manner as in the long jump. The only difference of any real consequence

[6]Pat Tan Eng Yoon, "The Triple Jump," in *International Track and Field Coaching Encyclopedia*, ed. by Fred Wilt and Tom Ecker (West Nyack, N.Y.: Parker Publishing Co., Inc., 1970), pp. 206–7.

[7]Toni Nett, "Practical Ratios for Triple Jumpers," *Track Technique*, No. 6, December 1961, p. 191.

occurs in the final three to four strides before takeoff where, because the emphasis on gaining height is less than in the long jump, the adjustments in stride length and frequency are generally less pronounced.

An example of the sequence of movements performed by a top-class triple jumper, from the instant of takeoff from the board until the final touchdown in the pit, is shown in Fig. 182. The following points should be noted:

(a)[8] The athlete's body position at the instant of takeoff is very similar to that of a long jumper—an essentially erect position of the trunk; a full extension of the hip, knee, and ankle joints of the takeoff leg, reflecting a forceful downward and backward drive against the board; and a high position of the leading knee and of both elbows, reflecting the vigorous action of the free limbs and their contribution to the forces exerted via the athlete's takeoff leg. The athlete's attention is focused more directly forward than it would be at this point in a long jump. Because an angle of takeoff as great as that used in a long jump would produce a high, long hop and forces at touchdown that the jumper would be unable to control sufficiently to make an effective takeoff into the step phase, the angle of takeoff here is somewhat less than that normally used in long jumping.

(b) The left leg has been lowered and is moving backward while the flexed right leg is being brought forward in preparation for the next landing. Because of the difference in the moments of inertia of these two limbs, some slight backward rotation is imparted to the athlete's trunk as a result of their motion (cf. hitch-kick, p. 415). The athlete's trunk is in a near-erect position and remains so throughout the hop—a markedly forward or backward lean is generally to be avoided, as this impairs the athlete's ability to land and drive upward and forward into the next phase.

(c) The athlete's arms are now moving outward and backward in unison. With both arms moving like this in roughly the same plane, the action of one serves to balance the contrary action of the other, and the remainder of the body is unaffected except for its slight forward movement to compensate for (or "balance") the weight that has been shifted backward.

(d) The arms have reached their backward limit and are about to be swung forward. The right foot, having first been reached well forward, is now being swept forcefully downward and backward to ensure a so-called "active landing." The purpose of this action is to try to ensure that the backward speed of the foot relative to the athlete's center of gravity exceeds the forward speed of his body and that, as a result, no retarding horizontal forces are evoked as the foot lands (cf. sprinting, p. 395). If this backward speed of the foot exceeds the forward speed of the body, the ground-reaction force evoked serves to drive the athlete forward into the next phase.

[8]The letters in parentheses refer to the corresponding positions in Fig. 182.

Fig. 182. Technique in the triple jump

(c) (d)

(g) (h)

(k) (l)

Fig. 182. *Continued*

(m)

(n)

The other leg is left well behind during these latter stages of the hop so that once the athlete touches down, the amplitude of its swing forward and then upward can be as great as possible. In this way it can acquire a large amount of momentum that can later be transferred to the athlete as a whole.

(e) The landing to complete the hop has just been effected. The hip, knee, and ankle joints are about to flex—to cushion the shock of the impact and to place these joints in the optimum position for the leg drive preceding the takeoff into the step.

The forward swing of the arms and left leg is well advanced at this point. Since the time the athlete spends on the ground is small—approximately

0.16 sec according to a study of 12 world-class performers[9]—this forward movement of the arms and free leg must start early (almost certainly before touchdown) and be fast if it is to be completed in time for takeoff.

The right shoulder is slightly lower than the left as the athlete leans to his right to bring his center of gravity over his supporting foot. This lateral shifting of the trunk to one side at the end of the hop and to the other at the end of the step can readily be observed by watching a triple jumper from in front or behind.

(f) Another strong takeoff position, very similar to that shown in (a).

(g) The knee lift evident in the previous position is increased, the trunk is held erect, and the arms are used for balance.

(h) The right leg has been flexed and brought forward, and in reaction to this movement the athlete's trunk has acquired some forward lean. The arms are beginning to sweep outward and backward for the second time.

(i), (j), and (k) The landing at the conclusion of the step and the takeoff into the jump are almost identical to the corresponding actions at the end of the hop phase [cf., (d), (e), and (f)].

Because the athlete's horizontal velocity is decreasing, the time he spends on the ground at each successive takeoff gets progressively longer—e.g., Kreer[10] reported successive takeoff times of 0.133 sec, 0.155 sec, and 0.180 sec for two world record jumps by V. Saneev (U.S.S.R.). Thus many jumpers who feel they have too little time to complete the lengthy double-arm swing as they go into the step find they can do so comfortably and to their advantage as they take off for the jump.

(l) and (m) Except that the athlete has much less speed with which to work, the techniques employed in the jump phase are essentially the same as those used in the long jump. This relative lack of speed limits his time in the air and the movements he can perform. The sail and hang techniques are those most widely used in this final phase. (*Note:* Since the distances achieved in this final jump range between 17 and 20 ft for top performers, the use of these techniques is in accord with the findings of Khadem and Huyck[11] mentioned earlier, p. 418).

(n) Approaching the landing, the athlete is in an excellent position with his legs fully extended and his feet well forward of his center of gravity.

[9]Geoffrey H. G. Dyson, *The Mechanics of Athletics* (London: University of London Press Ltd., 1967), p. 161.

[10]V. Kreer, "The World Record of Victor Saneev," *Track and Field*, No. 11, 1973. Translated by Michael Yessis and reported in *Yessis Review of Soviet Physical Education and Sports*, IX, June 1974, p. 39.

[11]El Khadem and Huyck, "Long Jump Technique Analysis," p. 758.

HIGH JUMP

Basic Considerations

In high jumping the height that an athlete clears may be regarded as the sum of three separate heights:

1. the height of the athlete's center of gravity at the instant of takeoff (H_1 in Fig. 183),
2. the height that he raises his center of gravity during the flight (H_2 in Fig. 183), and
3. the difference between the maximum height reached by his center of gravity and the height of the crossbar (H_3 in Fig. 183).

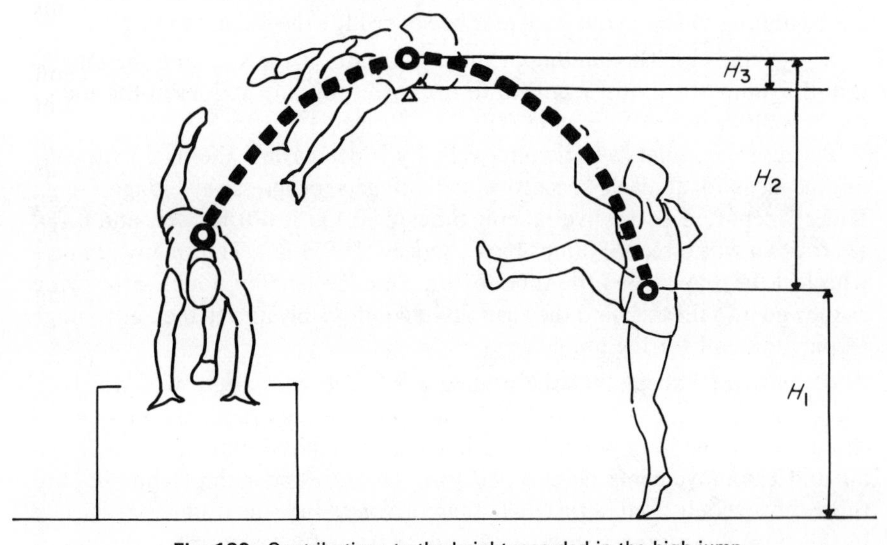

Fig. 183. Contributions to the height recorded in the high jump

Values obtained by the author[12] for a jump of 7 ft by then-world-record holder Pat Matzdorf (U.S.A.) give some indication of the relative importance of these heights to the total performance (Table 27).

The height of the athlete's center of gravity at the instant of takeoff depends on his physique and on his body position at that instant.

Since tall, long-legged athletes have higher centers of gravity at takeoff than do shorter ones, such people have a distinct advantage in the high

[12]James G. Hay, "A Kinematic Analysis of the High Jump," *Track Technique*, No. 53, September 1973, p. 1697.

Table 27. Relative Contributions to Height in the High Jump

	Height	*Percentage of Height of Bar*
Height of CG at takeoff	4 ft 8.7 in.	67.5
Height CG is lifted	2 ft 6.8 in.	36.6
Difference between maximum CG height and height cleared	−3.4 in.	−4.1

jump. This is borne out by the results obtained by Tanner[13] in his study of the physiques of Olympic athletes. After studying 137 athletes, including 10 high jumpers, he concluded that "The high jumpers are tall men; the shortest of our ten is . . . 6 ft $\frac{1}{2}$ in. They have the longest legs relative to the trunk of all the athletes (with the possible exception of the hammer throwers)."

Although in practice the style of jumping employed may make it impossible to obtain (or even undesirable to attempt), the optimum body position in terms of the height of the center of gravity at takeoff is one with the trunk erect, both arms high, lead leg extended and high, and jumping leg fully extended and vertical.

The height that the athlete's center of gravity rises in flight is governed by his vertical velocity at takeoff. This in turn is governed by his vertical velocity at the instant his jumping foot touches down and by the vertical impulse transmitted via this foot to his body during the takeoff (impulse-momentum relationship, p. 76).

The athlete's vertical velocity at touchdown depends primarily on his actions during the last one to two strides of his run-up. If at the end of his penultimate stride the athlete has sunk low over his supporting leg and then taken a low, fast step onto his takeoff foot, his center of gravity is likely to have little or no downward vertical velocity at the instant this foot touches down. On the other hand, if by failing to sink low at the end of his penultimate step he makes his last step like those that have preceded it, the athlete's downward vertical velocity at touchdown is likely to be relatively large. And, since he must first arrest this downward motion before he can begin to drive his body upward, this large downward velocity acts to his detriment. In fact, although it has yet to be convincingly demonstrated in practice, the ideal would be to have the athlete's center of gravity moving upward (and forward, of course) at the instant his takeoff foot contacted the ground. [*Note:* An

[13] J. M. Tanner, *The Physique of the Olympic Athlete* (London: George Allen and Unwin Ltd., 1964), p. 105.

upward vertical velocity at touchdown necessarily implies a reduction in the vertical distance through which the athlete can exert forces once his takeoff foot is grounded. This, in turn, can lead to a reduction in the vertical impulse exerted during the takeoff itself. A zero or upward vertical velocity at touchdown is only desirable, therefore, if it does not lead to a corresponding (or more than corresponding) reduction in this vertical impulse.]

The magnitude of the vertical impulse that the athlete exerts against the ground and that the ground in reaction exerts against him, is by definition, dependent on the magnitude of the forces involved and the length of time during which they act.

The forces involved are those resulting from the swing of the athlete's arms and leading leg and from the extension of the hip, knee, and ankle joints of his jumping leg. The magnitude of these forces depends on such factors as the speed with which the free limbs are moved, the strength of the muscles of his jumping leg, and the manner in which the various movements are coordinated or timed.

One other very important factor influences the magnitude of the vertical forces. It is the amount of rotation (or angular momentum) that the athlete must acquire at takeoff so that he can assume the required position by the time he reaches the peak of his jump. With a style like the straddle (Fig. 184), which requires that the athlete obtain a considerable amount of rotation at takeoff, the vertical forces that he can exert are somewhat smaller than if he were to use a style requiring less rotation. In other words, the athlete acquires his layout position at the expense of some vertical lift. And what he must try to do, of course, is to see that whatever he may gain by being in a good layout position at the peak of the jump is not more than offset by the losses he incurs at takeoff.

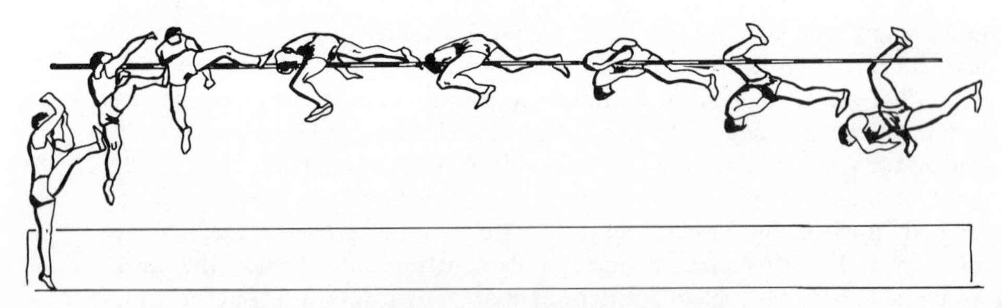

Fig. 184. Technique in the straddle style

The time during which the athlete's foot is in contact with the ground at takeoff, and thus the time during which it is possible for him to exert vertical forces, has been studied by a number of investigators. These studies have resulted in a number of important conclusions:

1. The duration (or time) of takeoff is a function of the style of jumping employed. Athletes who use the straddle style popularized by Russian athletes in the late 1950s and early 1960s (Fig. 184) generally have takeoff times in the 0.17–0.23 sec range,[14,15,16] while those who use the flop style popularized by Fosbury in the late 1960s (Fig. 68) tend to have takeoff times in the 0.12–0.17 sec range.[17,18,19]

2. The time of takeoff is a function of the action of the free limbs and, in particular, of the leading leg. Irrespective of the style employed, athletes who use a straight lead leg action generally have longer times of takeoff than do those who use a bent lead leg.[20,21] There is also some evidence that those athletes who use a double-arm action at takeoff have longer takeoff times than those who use an alternate arm action.[22]

3. Given the style and the specific actions of the free limbs, there appears to be an optimum time of takeoff for each athlete. Further, within limits that are specific to the athlete concerned, it appears that the shorter the time of takeoff the greater the vertical lift (H_2) the athlete obtains. At first glance this finding seems to be in direct conflict with the notion that the greater the vertical impulse at takeoff, the greater the vertical velocity at takeoff and therefore the greater the height attained. (*Note:* Impulse = force × time.) However, if the athlete is somehow able to increase the magnitude of the vertical forces he exerts by decreasing his takeoff time, and if these increases in force are effectively greater than the decreases in time, the conflict is dispelled. Under such circumstances the vertical impulse increases as the height of the jump increases, even though the takeoff time is meanwhile decreasing. How an athlete achieves these increases in vertical force while he simultaneously decreases the time of takeoff has yet to be completely explained. It would appear, though, that by shortening the time between (1) the forced eccentric contraction of the muscles of the jumping leg during the first part of the takeoff and (2) the explosive concentric contraction of these same muscles during the final part of the takeoff, the forces these muscles exert can be increased.

[14]Vladimir M. Dyatchkov, "The High Jump," *Track Technique*, No. 34, December 1968, p. 1070.

[15]Nikolay Ozolin, "The High Jump Takeoff Mechanism," *Track Technique*, No. 52, p. 1671.

[16]Benno M. Nigg, *Sprung, Springen, Sprünge* (Zurich: Juris Verlag Zuerich, 1974), pp. 75–104.

[17]Ozolin, "The High Jump Takeoff Mechanism," p. 1671.

[18]Nigg, *Sprung, Springen, Sprünge*, pp. 75–104.

[19]Jesus Dapena, "The Mechanics of the Fosbury Flop—Part 1: Translation," submitted to *Medicine and Science in Sports*.

[20]Dyatchkov, "The High Jump," p. 1070.

[21]Nigg, *Sprung, Springen, Sprünge*, p. 92.

[22]Jesus Dapena, personal communication, December 16, 1973.

The difference between the maximum height reached by the athlete's center of gravity and the height he clears (often referred to as the efficiency of his bar clearance) depends on the athlete's body position at the peak of the jump and on his body movements as he crosses the bar.

His body position depends primarily on the style of jumping he uses—and here the athlete has a greater choice of styles than exists in any other track or field event. For, in the high jump, there are no less than six named styles (scissors, modified scissors or back layout, eastern cutoff, western roll, straddle, and Fosbury flop) and several unnamed ones from which to choose.

The simplest of these, the first to evolve historically, and the one most widely used by beginners, is the scissors. In this style the athlete takes off from the foot farther from the bar and rises to cross the bar with his trunk erect and his legs in a near-horizontal position (Fig. 185). In this body position the distance between the athlete's center of gravity and the greatest height he clears is generally something like 10–12 in.

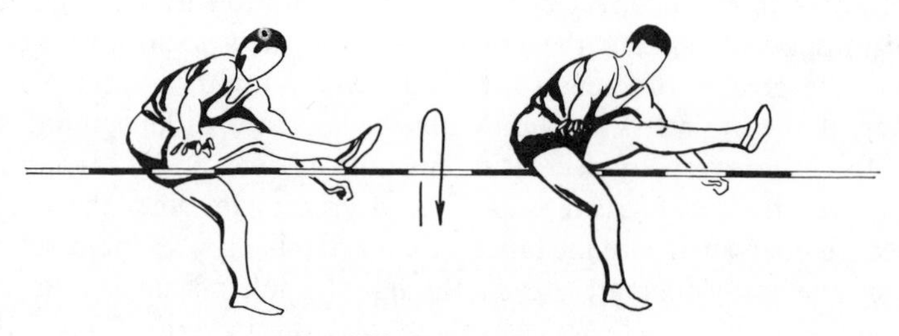

Fig. 185. The scissors style

If the athlete rotates backward or sideways as he rises to the bar, he can arrive in a clearance position in which this distance of 10–12 in. is substantially reduced—and this is exactly what some of the early high jumpers did. Some rotated backward to arrive in a position in which they were stretched out on their backs as they passed over the bar. This modified scissors or back layout style was especially hazardous because the rotation that carried the athlete into a horizontal position at bar level also tended to carry him into an inverted landing position and this, coupled with the meager pits in use at the time, discouraged most jumpers from using the style. (*Note:* With the advent of foam rubber and air-filled pits over the last decade or so there have been some signs of a reawakening of interest in this style of jumping.) The jumpers who rotated sideways as they rose to the bar evolved a style that later became known as the eastern cutoff, owing to the popularity it enjoyed among jumpers in the Eastern United States (Fig. 186). Although in theory this style could allow the jumper to reduce the distance between his

Fig. 186. The eastern cutoff style

center of gravity and the bar to a mere inch or two (perhaps even to zero), in practice it made such heavy demands on trunk and hip flexibility and on gymnastic ability that its exponents rarely achieved differences much less than 6–8 in.

The next style to evolve historically was the western roll—so named because a jumper from California developed it and popularized it among athletes on the West Coast of the United States. In this style the athlete takes off from the foot nearer the bar and, with something closely akin to a hopping motion, rises to cross the bar lying on his side with the knee of his takeoff leg tucked in against his chest (Fig. 187). At its best the western roll probably allows the athlete to reduce the difference between the peak height of his center of gravity and the bar to approximately 6 in.

Fig. 187. The western roll style

In the straddle style the athlete takes off from the foot nearer the bar and rises to a clearance position in which he "lies" face down along its length. If his body is stretched out, his center of gravity may be as little as 4 in. above the bar. On the other hand, if his body is wrapped around the bar (in a so-called drape- or dive-straddle), this distance may be reduced by as much as 2-4 in.

The latest of this long line of styles, the Fosbury flop, incorporates a takeoff from the foot farther from the bar and a clearance position in which the athlete is arched backward over the bar (Fig. 68). Such a position affords the possibility of the center of gravity being outside the body and perhaps even passing through or below the bar while the jumper passes over it (p. 124).

The clearance action of Dick Fosbury (U.S.A.), the originator of this style, has been described by Kerssenbrock[23] as "a model of perfect economy." In support of this contention he points out that "The center of gravity in the normal attitude of the body (erect) is found about $\frac{3}{4}$ in. before the sacral vertebra in the direction of the stomach wall, i.e., nearer to the back-side than to the stomach side" and that as a result it is easier for a jumper using the Fosbury flop to get his center of gravity outside his body than it is for a jumper using the straddle style.

In the course of these successive developments in high-jump styles, little attention has been given to whether the new style of the moment is indeed the best possible, or ultimate, style. If the height that an athlete clears in the high jump is considered to be the sum of H_1, H_2, and H_3, it can be argued that the optimum technique is one that maximizes the first two of these and optimizes the third—or, if this is not possible, the one that provides the best possible combination of the three. After considering the factors that influence the magnitude of these three contributing heights, the author[24] proposed an ultimate high-jump style (Fig. 188) that has the following characteristics: a frontal 90° approach to the bar so that the best possible clearance position, a front pike, can be adopted with the least difficulty; a fast approach run and a bent lead leg action to maximize the vertical impulse at takeoff and thus the height that the center of gravity can be raised; a takeoff position with the arms fully extended overhead, the trunk erect and stretched, and the takeoff leg fully extended at knee and ankle joints to maximize the height of the center of gravity at takeoff; and a front pike or jackknife clearance action to optimize bar clearance.

In a number of the styles described above, the athlete performs certain movements in the air so that the various parts of his body can pass over and around the bar without dislodging it. Because his angular momentum is con-

[23]Klement Kerssenbrock, "Analyzing the Fosbury Flop," *Track Technique*, No. 41, September 1970, p. 1292.

[24]James G. Hay, "The Hay Technique—Ultimate in High Jump Style?" *Athletic Journal*, LIII, March 1973, p. 46–48, 113–15.

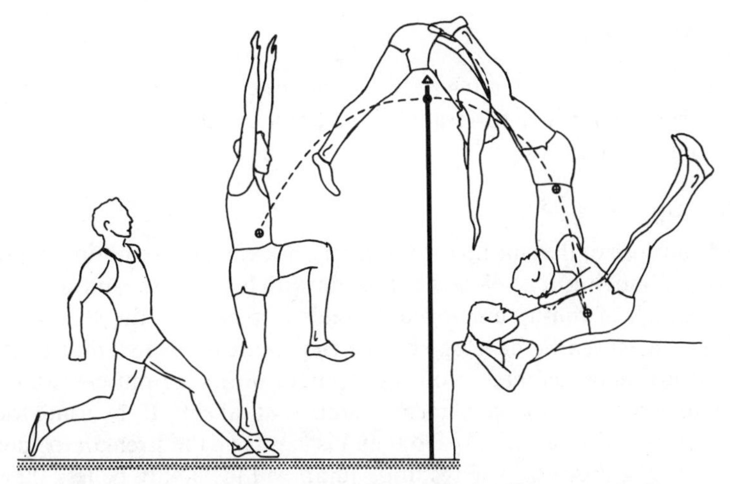

Fig. 188. A frontal approach and a front-piked position over the bar may be features of the ultimate in high-jumping techniques.

served during this time, each of these movements or actions that the athlete initiates must be accompanied by an equal and opposite reaction in some other part of his body.

Summary

The relationships between the height that an athlete clears and the factors that determine that height are summarized in Fig. 189.

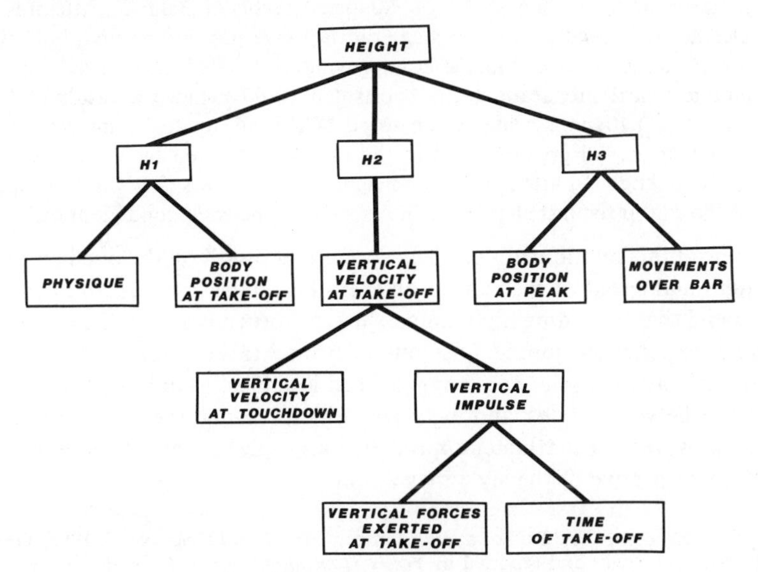

Fig. 189. Basic factors in high jumping

Techniques

For the purposes of analysis the techniques in high jumping are considered here under four subheadings—the run-up, the takeoff, the bar clearance, and the landing.

Run-Up

The purpose of the run-up is to bring the athlete into the optimum position for takeoff, moving at a velocity consistent with his strength and skill.

The length of run-up employed depends primarily on the athlete's ability to utilize the speed he develops by the time he reaches his point of takeoff. This in turn depends mainly on the strength of his leg muscles and on his ability to coordinate the movements required at takeoff. If, by using too long a run-up, he develops more speed than his legs have the strength to "control" at takeoff, the height of the resulting jump will inevitably be less than he is capable of producing. Similarly, if he arrives at the takeoff traveling so fast that he has insufficient time to complete the required sequence of movements, his performance again suffers.

Chistyakov[25] has reported some interesting figures to show how the speed that an athlete can utilize at takeoff increases as his strength and skill are developed:

> . . . with growth of physical qualities and technical mastery, the speed of the run increases. For example, Master of Sports, G. Kutganin, in the spring of 1964, having a best result of 2.02 meters [6 ft $7\frac{1}{2}$ in.], made successful jumps running with a speed of 7.0 meters/second [23.0 fps]. Attempts to increase the speed at a given stage did not give positive results—but, after a year, in the spring of 1965, after improving his results in specialized physical and technical preparation, the sportsman could achieve a result of 2.10 meters [6 ft $10\frac{3}{4}$ in.], bringing the speed of the run up to 7.3 meters/second [24.0 fps]. A still greater speed in the run was achieved by V. Brumel and A. Kharskogo. In attempts at the height of 2.15 meters [7 ft $\frac{1}{2}$ in.] the speed of the run in the last step by them reached 7.5 meters/second [24.6 fps].

In view of this relationship between strength and skill on the one hand and the length and speed of the run-up on the other, most teachers and coaches recommend that beginning high jumpers use a short run of some 5 to 7 strides and more experienced jumpers use one of from 7 to 11 strides.

The angle at which the bar is approached is subject to considerable variation both between styles and between athletes using the same style. In general, however, most athletes approach at an angle somewhere between 20° and 40°, irrespective of the style they employ.

[25] Urii Chistyakov, "The Run of a High Jumper," *Track and Field*, No. 8, 1966. Translated by Michael Yessis and reported in *Yessis Translation Review*, I, December 1966, p. 108.

Although a straight-line approach to the bar is generally used with most other styles, exponents of the eastern cutoff and the Fosbury flop (including Fosbury himself) commonly use a curved approach. In the Fosbury flop, the athlete who uses a curved approach acquires angular momentum about an axis parallel to the bar as he straightens up from the inward-leaning position required to negotiate the curve and into the vertical position from which he leaves the ground. An athlete who uses a straight approach generally acquires the same angular momentum by leaning toward the bar during the takeoff. As in other jumping styles (e.g., the western roll and the straddle), this leaning toward the bar has a detrimental effect on the athlete's ability to generate lift.[26,27]

During the last few strides of the run-up the athlete adjusts his body position in preparation for the takeoff. These adjustments take the following form:

1. The trunk is brought from a position in which it is inclined forward to one in which it is inclined backward. With top-class jumpers, the greater part of this change in the inclination of the trunk occurs as a result of a forceful forward movement of the hips at the beginning of the last stride of the run-up.

With the trunk inclined backward at the end of the last stride, the athlete's center of gravity has greater horizontal and vertical distances to travel, from the instant of touchdown to the instant of takeoff, than it would if the trunk were erect. The athlete is therefore able to take more time to coordinate his takeoff movements or to take the same time and approach the takeoff at a slightly higher speed.

2. The center of gravity is gradually lowered by the athlete increasing the flexion of the knee of his supporting leg as his body passes forward over the grounded foot at the end of each stride.

There is some evidence to suggest that the amount the center of gravity is lowered is related to the height that the athlete jumps. For example, Dyatchkov[28] has reported that when Kashkarov jumped 6 ft 7 in. in 1955, "the angle of bend of the knee joint of the lead leg amounted to 90 degrees" as his center of gravity passed forward over the grounded foot, while 1 yr later when the same jumper improved to clear 6 ft 10½ in., this angle was only 86°. A similar decrease from 99° to 88° was also noted when Stepanov improved from 6 ft 7½ in. to 6 ft 11¾ in. Nigg[29] analyzed the performances of Dwight Stones (U.S.A.) during a competition in which he set a new world record of 7 ft 6½ in. and obtained similar results (Fig. 190).

[26]Dapena, "The Mechanics of the Fosbury Flop—Part 1: Translation."

[27]Jesus Dapena, "The Mechanics of the Fosbury Flop—Part 2: Rotation," submitted to *Medicine and Science in Sports.*

[28]Dyatchkov, "The High Jump," p. 1059.

[29]Nigg, *Sprung, Springen, Sprünge*, p. 90.

3. The lowering of the center of gravity—variously referred to as the "squat," "sink," or "gather"—is accompanied by a decrease in the degree to which the knee of the supporting leg is extended as the athlete "takes off" into the next stride. This incomplete extension of the leg assists in keeping the athlete's center of gravity low.

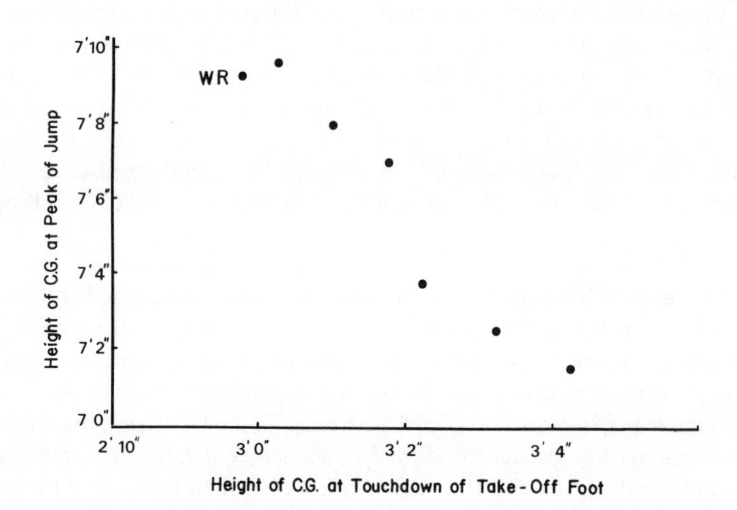

Fig. 190. The height of the center of gravity at the touchdown of the takeoff foot and at the peak of the jump for Dwight Stones, U.S.A. WR = World Record (*Adapted from data in Nigg,* Sprung, Springen, Sprünge.)

The data in Fig. 191 illustrate the extent of these various adjustments in body position in the case of some world-class Russian high jumpers. All three of these men were exponents of the straddle style.

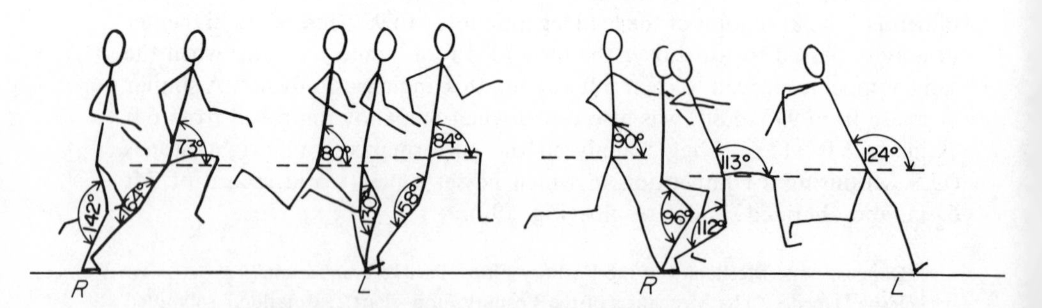

Fig. 191. Adjustments made during the last three strides of the run-up (*Mean values for Russian jumpers, Kashkarov, Sitkin, and Stepanov, based on data in Dyatchkov, "The High Jump."*)

As in the case of the long jump (p. 413), the changes in body position during the final few strides before takeoff produce concomitant changes in the length of the strides involved. Most important of these is a gradual increase in length up until the second to last stride (almost invariably the longest of all the strides in the run-up) and a decrease in length of the final stride. The lengths of the last four strides in a series of high jumps by Valeriy Brumel (U.S.S.R.) have been reported by Chistyakov[30] and are presented here (Table 28) to give an indication of the magnitude of the changes involved in the case of a jumper using the straddle style.

*Table 28. Lengths of the Last Four Run-Up Strides of Valeriy Brumel (U.S.S.R)**

Length of Stride				Height of Jump
Fourth Last	*Third Last*	*Second Last*	*Last*	
6 ft 8 in.	7 ft ¼ in.	7 ft 6½ in.	6 ft 5½ in.	6 ft 11½ in.
6 ft 8 in.	7 ft 3½ in.	7 ft 6½ in.	6 ft 2½ in.	6 ft 11½ in.
6 ft 10¾ in.	7 ft 3 in.	7 ft 6½ in.	6 ft 2¾ in.	7 ft ½ in.
6 ft 11 in.	7 ft 1 in.	7 ft 5½ in.	6 ft 4½ in.	7 ft 2¼ in.
6 ft 10 in.	7 ft ¼ in.	7 ft 7½ in.	6 ft 0 in.	7 ft 2¼ in.

*Adapted from data in Chistyakov, "The Run of a High Jumper."

Takeoff

The takeoff begins with the grounding of the heel of the jumping foot at the end of the last stride of the run-up. At this instant the athlete's body is generally inclined backward, with the foot of his leading leg close to the ground and well behind his body and (depending on whether he is using a double-arm swing or the natural one-forward-one-backward arm action) both arms behind his body or one forward and one backward "balancing" the respective legs.

Once the sole of the jumping foot has been grounded, the knee of the jumping leg flexes to reduce the effect of the impact and to put the leg into the optimum position for the forceful extension to follow. According to Dyatchkov,[31] a theoretical consideration of the structure of the leg suggests that the knee should be flexed to an angle of 125° if the best results are to be obtained. However, he has found that in practice most high jumpers do not achieve such a pronounced flexion of the knee—"Among most of the high jumpers the maximum angle of the flexion in the jumping leg varies between 135°–140°."

[30]Chistyakov, "The Run of a High Jumper," p. 108.

[31]Dyatchkov, "The High Jump," p. 1069.

During these initial movements of the takeoff, the lead leg, which has been deliberately left well behind the body during the last stride of the run, begins to swing forward and upward in unison with the arm or arms that accompany it. This forward and upward swing of the lead leg and arm(s) serves three principal functions:

1. It increases the magnitude of the vertical forces exerted against the ground, the vertical forces that the ground exerts on the athlete in reaction, and thus the athlete's vertical velocity at takeoff.

2. It imparts angular momentum to the athlete's body. As the swing of the lead leg and arm(s) slows down, the angular momentum that these limbs possess is transferred to the body as a whole.

3. It increases the height of the athlete's center of gravity at the instant of takeoff.

There are two main types of lead leg swing—the *bent leg* and the *straight leg*.

The bent-leg swing is used by beginners, who tend to use it naturally, and by most trained exponents of the dive straddle and the Fosbury flop. In using the dive-straddle style, an athlete must acquire a certain amount of forward rotation at takeoff in order to get his body into the desired layout position by the time he reaches the peak of his jump. A vigorous straight-leg action imparts a large amount of backward angular momentum to the athlete's body and makes it difficult for him to acquire angular momentum in the opposite direction. For this reason, athletes who use the dive straddle either reduce the vigor of their straight-leg action (compared with what they would use with a straight straddle style) or use a bent-leg action. If a bent-leg action is used, as it usually is, the lead leg makes a contribution to the vertical velocity at takeoff and to the height of the takeoff—although probably not as much as a straight lead leg would in either case—and, at the same time, keeps the backward angular momentum that must be overcome within acceptable limits. If an exponent of the Fosbury flop style were to use a straight lead leg swing, he might well increase both his vertical velocity and height at takeoff (relative to what he could achieve with a bent-leg swing) but he could expect to encounter similar difficulties to those of his dive-straddle counterpart in obtaining the angular momentum he requires about his transverse axis. In addition, because of the large moment of inertia he would have about his long axis, it would be very difficult for him to achieve sufficient rotation to get around onto his back by the time he reached the peak of his jump.

With the straight-leg technique the athlete brings the leading leg forward with a bend at the knee just sufficient to allow the foot to clear the ground. Then, once the foot has passed forward of the supporting (jumping) leg, the knee is extended and the lead leg continues forward and upward as one unit.

The straight-leg swing of the lead leg is generally regarded as being better than the bent-leg action for those styles of jumping that would allow the use of either. Dyatchkov,[32] for example, has computed the upward force attributable to the straight-leg swing of Kashkarov (258 lb), Degtyarev (240 lb), and Stepanov (236 lb) and shown that it far exceeds that obtained by Nilsson (106 lb), who used a bent-leg swing. He also points out that the computed upward forces for the first three athletes far exceeded their body weights (194, 168, and 159 lb, respectively). This, he concluded, "speaks for the fact that the jumper moves upward by the swing of the leg alone."

Once the lead leg and the upper arm(s) reach the horizontal or near-horizontal, the athlete drives down vigorously against the ground by extending the hip, knee, and ankle joints of his jumping leg. The reaction to this leg drive propels the athlete into the air. In addition, if it acts eccentrically, as is usually the case, it also contributes to the angular momentum that the athlete acquires at takeoff.

Bar Clearance

Regardless of the style used, most of what takes place while the athlete is in the air may be directly attributed to the nature of his takeoff (i.e., to his center-of-gravity height, his velocity, and his angular momentum at that instant). Nonetheless, in some styles (notably the eastern cutoff, the straddle, and the Fosbury flop) the athlete also makes use of a judicious interplay between the action of one part of his body and the reaction it evokes in some other part to effect a clearance of the bar.

In using the straddle style, an athlete's greatest problem once he has reached the peak of his jump is to avoid knocking the bar off with his trailing leg. One method used to overcome this problem is to twist the trunk away from the bar. The reaction to this movement is a contrary rotation of the legs that tends to lift the trailing leg clear (Fig. 192, taken from a photograph of Valeriy Brumel, shows this action to perfection). It might be noted here that the fairly common practice of instructing a straddle-style jumper to swing his outside arm under the bar in order to better rotate his trail leg clear is unlikely to be successful, for the reaction to such a movement—a lowering of the trailing leg on the takeoff side of the bar—is the exact reverse of what is required.

In another method designed to facilitate the clearance of the trailing leg, the athlete drives his head and trunk and his lead leg down toward each other on the far side of the bar (Fig. 193). Since the movements he initiates in the air cannot alter the path followed by his center of gravity, these actions inevitably result in his hips being elevated—i.e., the downward movements of the head, trunk, and lead leg are "balanced" by the upward movement of the

[32] *Ibid.*, pp. 1070–71.

hips, and the center of gravity continues undisturbed on its predetermined path. This elevation of the hips makes it easier for the athlete to lift his trailing leg over the bar.

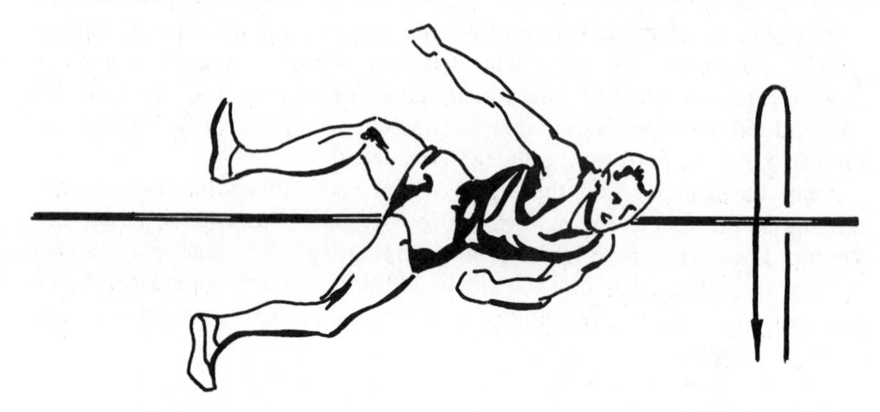

Fig. 192. Clearing the trailing leg in the straddle style—Valeriy Brumel (U.S.S.R.)

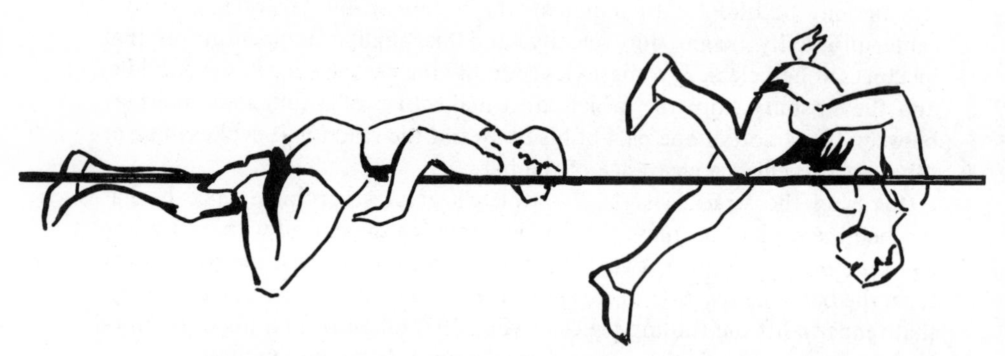

Fig. 193. An alternative method of clearing the trailing leg in the straddle style—Yuriy Stepanov (U.S.S.R.)

The procedure used to get the legs over the bar in the Fosbury flop makes use of a similar interplay between body parts. In this style the athlete quickly extends his knees as soon as his legs reach the point at which they are in imminent danger of dislodging the bar. This motion, which lifts the athlete's legs and feet clear of the bar, is accompanied by a contrary reaction in the athlete's upper body. Since one of the side effects of this combined action and reaction is a lowering of the athlete's hips relative to the rest of his body, it is important that the extension of the knees not be initiated before the hips have themselves crossed the bar. If it is initiated too soon, the athlete lessens the danger of dislodging the bar with one part of his body merely to increase it with another.

Landing

The landing is generally made on either the foot of the leading leg (scissors), the foot of the jumping leg (eastern cutoff), the hands and the foot of the jumping leg (western roll), the side or back (straddle), or the back (Fosbury flop). The part of the body on which the landing is made in the back layout style is not easily predicted!

POLE VAULT

Basic Considerations

For the purposes of analysis, the height that a pole-vaulter clears may be regarded as the sum of four separate parts:

1. the height of his center of gravity at the instant of takeoff (H_1 in Fig. 194),
2. the height that his center of gravity is raised while he is on the pole (H_2 in Fig. 194),
3. the height that his center of gravity is raised while he is airborne, i.e., once he has released the pole (H_3 in Fig. 194), and
4. the difference between the maximum height reached by his center of gravity and the height of the crossbar (H_4 in Fig. 194).

The factors that influence the magnitude of H_1, H_3, and H_4 are identical to those that determine the magnitude of the corresponding heights in high jumping—physique and body position (H_1), vertical velocity at release (H_3), and body position and movements initiated in the air (H_4).

The factors that determine the magnitude of H_2 are perhaps best considered in terms of mechanical energy changes. At the instant of takeoff the vaulter has a large amount of kinetic energy (equivalent to speeds of 25–30 fps) and a relatively small amount of potential energy (equivalent to a center of gravity height of $3\frac{1}{2}$ to 4 ft). In addition, if the vaulter is using a flexible pole, it is likely that he has already "stored" some energy in the pole by bending it. At the instant he releases the pole, the vaulter has a large amount of potential energy (equivalent to a height of up to 19 ft) and a relatively small amount of kinetic energy. The difference between the total mechanical energy that he possesses as he releases the pole and that at takeoff is equal to the algebraic sum of the work he does during the ascent and the losses in mechanical energy to other energy forms (heat, sound, etc.) incurred en route. (*Note:* Whereas the total mechanical energy of an athlete who is airborne remains constant throughout his flight, the vaulter's total mechanical energy is subject to change, since he still retains contact with the ground via the pole.)

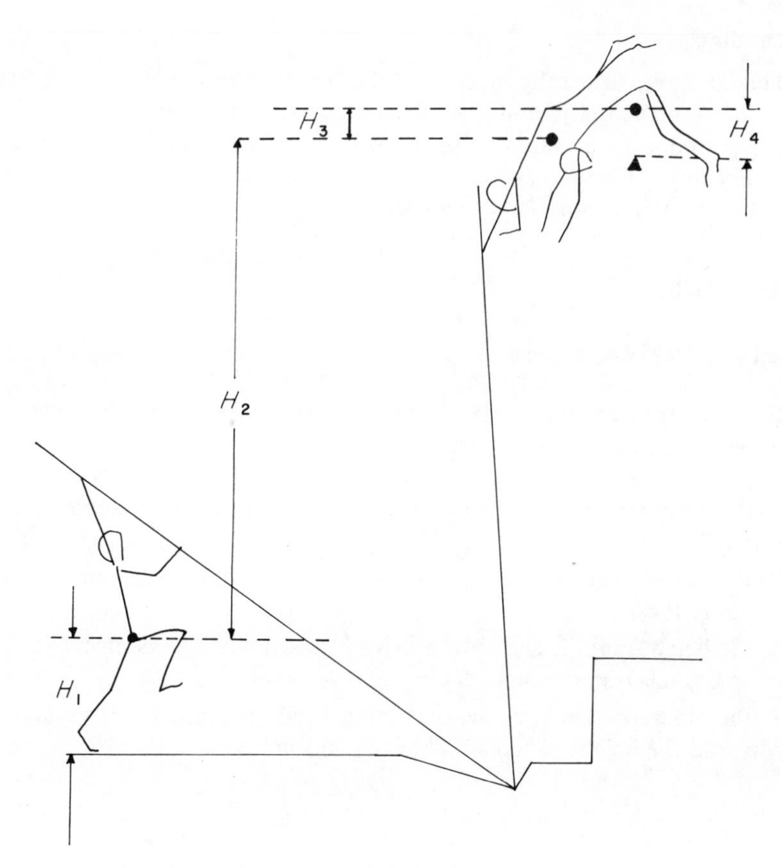

Fig. 194. Contributions to the height recorded in the pole vault

These various energy changes may be summarized in the following equation:[33]

$$\begin{matrix}\text{Potential} \\ \text{energy at} \\ \text{release}\end{matrix} + \begin{matrix}\text{kinetic} \\ \text{energy at} \\ \text{release}\end{matrix} = \begin{matrix}\text{kinetic} \\ \text{energy at} \\ \text{takeoff}\end{matrix} + \begin{matrix}\text{potential} \\ \text{energy at} \\ \text{takeoff}\end{matrix} + \begin{matrix}\text{strain} \\ \text{energy at} \\ \text{takeoff}\end{matrix} + \begin{matrix}\text{work done} \\ \text{during} \\ \text{ascent}\end{matrix} - \begin{matrix}\text{mechanical} \\ \text{energy} \\ \text{losses}\end{matrix}$$

This equation can be rearranged to yield the following expression for ΔPE, the difference between the potential energies at release and takeoff:

$$\Delta\text{PE} = \begin{matrix}\text{kinetic} \\ \text{energy at} \\ \text{takeoff}\end{matrix} + \begin{matrix}\text{strain} \\ \text{energy at} \\ \text{takeoff}\end{matrix} + \begin{matrix}\text{work done} \\ \text{during} \\ \text{ascent}\end{matrix} - \begin{matrix}\text{mechanical} \\ \text{energy} \\ \text{losses}\end{matrix} - \begin{matrix}\text{kinetic} \\ \text{energy at} \\ \text{release}\end{matrix}$$

Since this difference is equal to the product of the vaulter's weight (a constant) and H_2, an examination of these five terms should reveal the basic factors upon which the magnitude of H_2 depends.

[33]Since it can readily be demonstrated that the kinetic and potential energies possessed by the pole at takeoff and at release are very small, these have been disregarded.

Kinetic Energy at Takeoff

The kinetic energy that the vaulter possesses as he leaves the ground derives from the work he does to build up speed during his run-up and from the additional work he does at takeoff to add a vertical component to his motion.

Strain Energy at Takeoff

The amount of energy "stored" in the pole at the instant of takeoff is a function of the material of which the pole is made and of the forces that are exerted upon it.

Poles of fiberglass, bamboo, or any similarly flexible material can more readily be deformed than the aluminum and steel poles in widespread use 15–20 yr ago. Thus it is more likely that a vaulter can usefully "store" energy in a modern fiberglass pole than in one of the earlier metal poles. This, of course, is one of the advantages that present-day vaulters have over their predecessors.

The forces that the vaulter exerts on the pole at takeoff are transmitted to the pole via his hands. To examine how these forces may be used to deform the pole, and thus to invest in it a certain amount of strain energy, it is desirable to resolve the forces into components acting perpendicular and parallel to the long axis of the pole.

If the components acting perpendicular to the long axis of the pole are both directed forward and upward, their combined effect is to rotate the pole upward about a transverse axis through its base. This is the situation that normally prevails when a skilled vaulter uses a metal pole [Fig. 195(a)]. However, if the two perpendicular components act in opposite directions—the one applied by the lower hand in a forward and upward direction and the other applied by the upper hand in a backward and downward direction—the results are rather different [Fig. 195(b)]. Then if the component of the force applied by the lower hand is greater than that exerted by the upper hand, the pole rotates toward the vertical (as a result of the dominance of the force exerted via the lower hand) and bends upward (as a result of the "coupling" effect produced as the hands exert opposing forces against a pole that is itself fixed at one end). The extent to which the pole is bent in this process is governed by the magnitude of the perpendicular forces exerted and by the distance between their lines of action, i.e., the distance between the vaulter's hands.

The components acting parallel to the long axis of the pole may be regarded as eccentric forces that, like all eccentric forces, tend to both translate and rotate the body upon which they act. In this case, the tendency to translate the body is thwarted by the contrary forces exerted by the back of the box on the butt of the pole, and the pole merely becomes more firmly "fixed" in the box than before. With this "fixing" of the butt end of the pole,

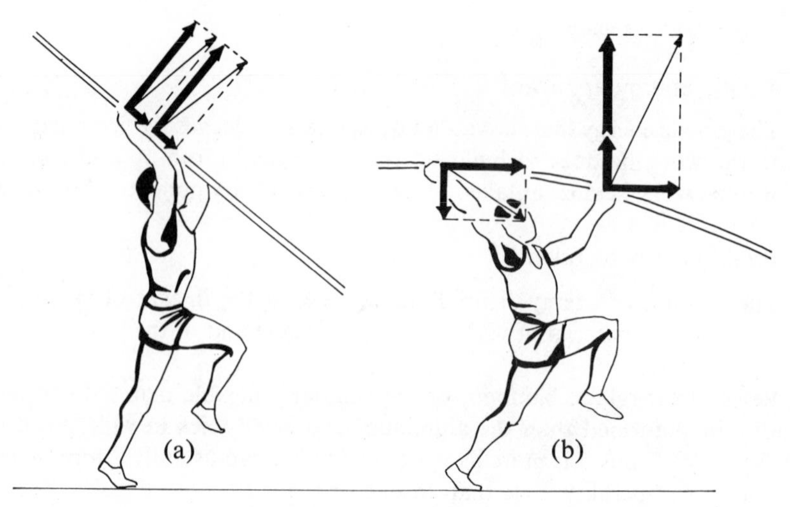

Fig. 195. Forces exerted on the pole via the hands (a) using a metal pole, (b) using a flexible fiberglass pole

the only way in which the eccentric parallel forces can produce a rotation of the body on which they act is by causing the pole to bend—and this is exactly what happens. The amount that the pole is bent in this way is determined primarily by the magnitudes of the parallel forces involved, and these, in turn, are governed by the athlete's actions at takeoff. If the athlete drives upward and forward across the line of the pole—as he would if he were using a metal pole—the magnitudes of the parallel forces are relatively small and their tendency to bend the pole is minimal [Fig. 195(a)]. On the other hand, if he drives forward into the pole—the common practice when using a fiberglass pole—the magnitudes of the parallel forces and the resulting bending of the pole are both correspondingly greater.

The relationship between the angle of takeoff and the amount of pole bend obtained and differences between the angles of takeoff of vaulters using fiberglass poles and those using other poles have been reported in a number of studies. A study conducted by the author[34] revealed a significant relationship indicating that the smaller the angle of takeoff, the greater the magnitude of the pole bend obtained by a vaulter. Lindner[35] reported the angles of takeoff of four top-class vaulters who used fiberglass poles—Hansen (25.5°), Pennel (22°), Preussger (22°), and Reinhardt (14.5°)—and Ganslen,[36] in referring to these figures obtained by Lindner, noted that "In earlier studies some thirty years ago, of the bamboo pole, the takeoff angle was found to average 25–30 degrees for vaults between 12½–14 feet."

[34]James G. Hay, "Pole Vaulting: A Mechanical Analysis of Factors Influencing Pole-Bend," *Research Quarterly*, XXXVIII, March 1967, pp. 34–40.

[35]Erich Lindner, *Sprung und Wurf* (Schorndorf bei Stuttgart: Verlag Karl Hofmann, 1967), p. 90.

[36]Richard V. Ganslen, *Mechanics of the Pole Vault*, 7th ed. (Denton, Tex., 1970), p. 41.

Work Done during the Ascent

Once the vaulter leaves the ground, he and the pole form what has frequently been referred to as a *double pendulum.* At the same time as the vaulter is swinging on the pole about an axis that passes transversely through his hands (the man pendulum), he and the pole together are rotating about a transverse axis through the base of the pole (the man-and-pole pendulum).

The vaulter's first objective, upon leaving the ground, is to bring the pole to a position in which he can most effectively do work to lift his body. To achieve this objective he endeavors to bring his center of gravity[37] as close as he can to the axis about which the man-and-pole pendulum is rotating, thereby facilitating the rotation of this pendulum toward the vertical. For a given height at which the vaulter grips the pole, there are basically two ways in which he may bring his center of gravity toward the axis through the end of the pole: (1) He can straighten his arms, lower his leading leg, and assume a fully extended body position. (2) If he is using a flexible pole, he may also exert forces to increase the bend of the pole.

The forces that the vaulter can exert on the pole to increase its bend derive from two principal sources. If he keeps his lower arm firm (or perhaps even actively pushes with this lower arm), the force thus exerted, together with part of the component of his weight applied via his upper hand in the opposite direction, forms a couple that acts to increase the bend of the pole [Fig. 196(a)]. The moment of this couple, and thus the effect it has on the magnitude of the pole bend, depends primarily on the force exerted via the lower hand. The vaulter may also exert force to increase the bend of the pole by vigorously swinging his legs upward. The reaction to the centripetal force that the pole exerts on the vaulter to make this angular motion possible is a force that the vaulter exerts on the pole in a forward and downward direction [Fig. 196(b)].

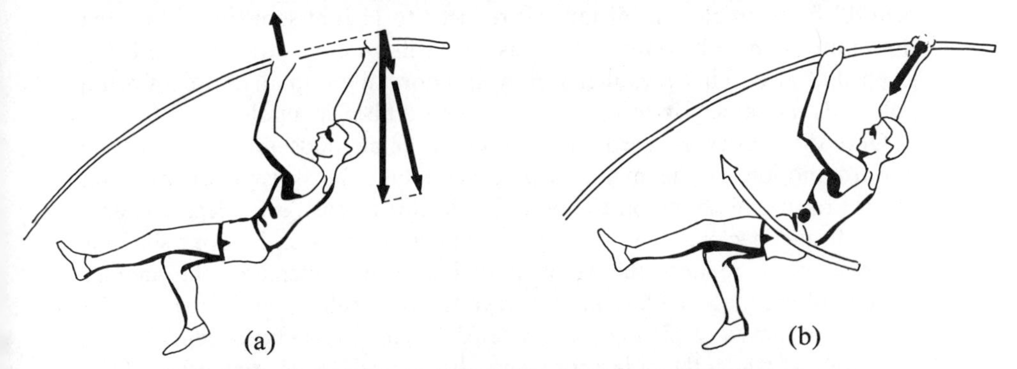

Fig. 196. Components of the force that the vaulter exerts on the pole during the swing due to (a) the action of his left arm and (b) the vigorous upward swing of his legs

[37]Strictly speaking, the center of gravity of the man and pole considered together.

This force, like the parallel forces referred to earlier, acts eccentrically to increase the bend of the pole.

There is a third way in which a vaulter can adjust the moment of inertia of the man-and-pole pendulum during the hang phase—he can alter the position at which he initially grips the pole. If he lowers his grip, he will decrease the moment of inertia; if he raises it, he will produce the reverse effect.

The position on the pole at which a vaulter grips with his upper hand is governed by his ability to bring the pole to the vertical (or near-vertical) position required to effectively complete the vault. If he holds too high (i.e., too far from the butt end of the pole), the moment opposing the motion of the man-and-pole toward the vertical (i.e., the weight of the man-and-pole times the horizontal distance from their combined center of gravity to the base of the pole) is such that the vaulter is unable to bring the pole to the vertical. As a direct consequence, the quality of his performance suffers. If he holds too low, his performance is also likely to suffer, for then the pole tends to come to the desired final position before he has time to effectively complete the sequence of movements designed to project his body upward into the air. It should be clear, therefore, that there is an optimum height at which a vaulter should grip in any given case. Furthermore, since the higher the vaulter can grip and still bring the pole to the required final position, the better his overall performance is likely to be, the good vaulter continually strives to increase the height of the grip that he can use effectively. (*Note:* An advantage that fiberglass poles have over metal poles lies in the higher grip heights that they permit. If a vaulter using a metal pole grips the pole with his top hand at a distance of 13 ft from the end of the pole, this distance remains essentially unaltered throughout the vault. On the other hand, a vaulter who grips a fiberglass pole 13 ft from the end of the pole may have this distance markedly reduced during the early stages of the vault due to the pole being bent. If, for example, this distance is reduced to 11 ft at some instant during the vault, the pole behaves at that instant as if it were a straight pole being gripped at 11 ft. Thus a vaulter who would normally grip at 13 ft on a metal pole finds the same grip too low when he uses a fiberglass pole.)

Once the vaulter has used one or more of the methods above to assist the upward motion of the man-and-pole pendulum, he directs his attention toward raising his body on the pole. To do this he swings his legs forward, upward, and backward, flexing at the hips and knees as he does so. This process, which reduces the moment of inertia and increases the angular velocity of the man pendulum, also, regrettably, results in an increase in the moment of inertia of the man-and-pole pendulum and a consequent slowing in the rate at which the pole approaches the vertical. (*Note:* Extension of the vaulter's body immediately after takeoff produces the reverse effect, slowing the vaulter as it speeds the pole.)

Having thus reached an inverted tuck position, the vaulter next exerts forces down the line of the near-vertical pole to raise his center of gravity above the level of his hands. This lifting of the center of gravity—aided, in the case of flexible poles, by a returning to the vaulter of the energy "stored" in the pole—culminates in the vaulter pushing off from the pole and projecting himself into the air.

Mechanical Energy Losses

During the course of a vault, forces acting between the pole and the box and forces within the pole itself result in a conversion of mechanical energy to other nonmechanical forms (e.g., heat, sound, etc.). Thus, if the vaulter did no work at all between the instants of takeoff and release of the pole, the total mechanical energy possessed by the vaulter and the pole at that latter instant would be less than they possessed at the instant of takeoff.

Kinetic Energy at Release

The vaulter's kinetic energy at the instant he releases the pole is determined primarily by his velocity at that time. Here the optimum velocity is one with a large vertical component (to carry the vaulter's center of gravity high into the air and maximize the value of H_3) and a small horizontal component (to ensure the vaulter a safe passage across the bar).

Summary

The relationship between the height that an athlete clears and the factors that determine that height are summarized in Fig. 197.

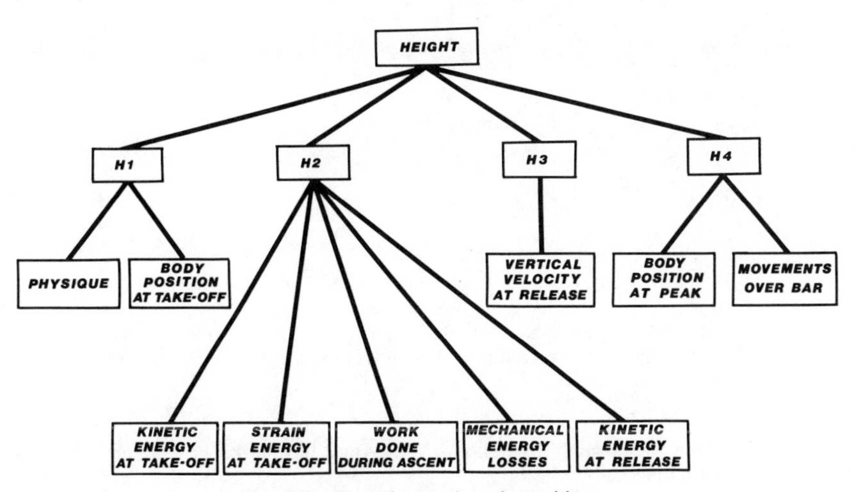

Fig. 197. Basic factors in pole vaulting

Techniques

For the purposes of analysis the act of pole vaulting may be subdivided into the following parts—the carry, the run-up, the plant, the takeoff, the hang, the swing-up and rock-back, the pull-turn-and-push, the clearance, and the landing.

Carry[38]

During the run-up the pole is carried close to the vaulter's right hip with the thumb of his left (lower) hand under the pole supporting its weight and the thumb of his right (upper) hand pressing down to hold the tip of the pole up—Fig. 198. Taking the vaulter's left thumb as the fulcrum, the moment of the downward force exerted via his right hand must be equal to the contrary moment of the weight of the pole, if the pole is to be held steady. The fingers of each hand are wrapped loosely around the pole and supplement the work done by the thumbs.

The height at which the tip of the pole is held as the vaulter runs toward the box is subject to some variation between vaulters. Some use a high carry with the tip of the pole well above head height. This permits the right elbow

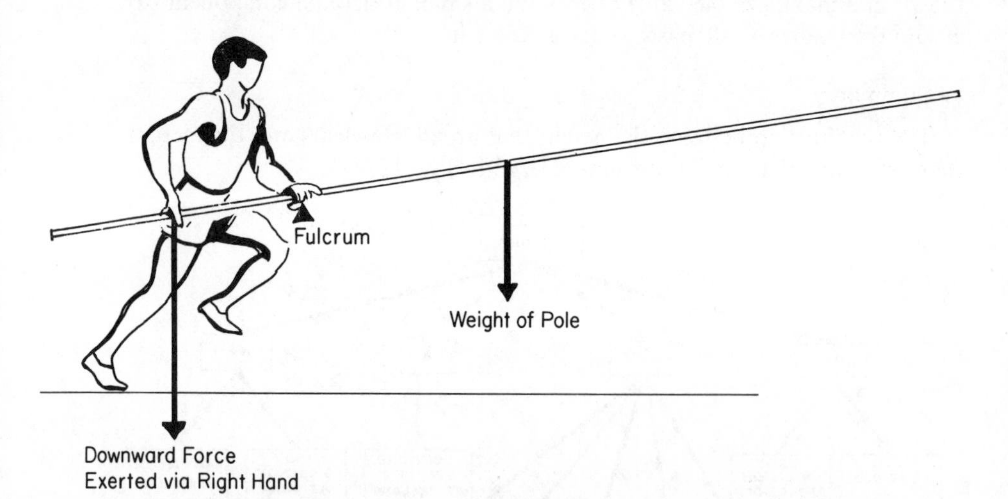

Fig. 198. During the carry, the moment of the downward force exerted by the vaulter's top hand balances the moment of the weight of the pole.

[38]Throughout the ensuing discussion of pole-vault techniques, it is assumed that the vaulter is right-handed.

to be more fully extended than otherwise and makes it easier for the vaulter to apply the required downward force via his right hand. For this reason a high carry is often favored by vaulters using heavy poles and/or high grips, both of which require that a considerable force be exerted via the right hand.

The principal disadvantages in the use of a high carry are the large frontal area presented to the air through which the pole is moved—this can be particularly troublesome on a windy day—and the considerable distance through which the tip of the pole must be lowered during the plant. This latter increases the scope for error beyond that which exists when a lower carry is used. When the tip of the pole is carried at head height (a medium carry), or still lower at approximately hip height (a low or parallel carry), the advantages and disadvantages associated with a high carry tend to become reversed—difficulty in applying the required downward force increases as problems with air resistance and the potential for errors in planting the pole decrease.

The effect of the choice of carry (high, medium, or low) on the speed with which a vaulter can run has been investigated by Ganslen.[39] Following a study in which a number of vaulters were "timed over a set distance with a flying start," it was found that "the men were able to run fastest with the front tip of the pole held at head height (medium carry). The difference in speed of the men was as great as 3 ft per sec over a 50-foot distance."

Run-Up

The length, speed, and accuracy of a pole-vaulter's run-up are subject to precisely the same influences that operate in the case of a long or triple jumper, for athletes in all three events seek the same objective—to obtain the maximum speed that they can effectively use during the takeoff.

Warmerdam,[40] the first man to vault 15 ft and world-record holder for no less than 17 yr, conducted an investigation into various aspects of the run-up. He noted, among other things, a relationship among the length of the run, the average speed attained over the last 50 ft of the run, and the height cleared—the best vaulters using longer runs and attaining higher average speeds than the vaulters of lesser ability.

The remaining parts of the vault—the plant, the takeoff, the hang, the swing-up and rock-back, the pull-turn-and-push, the clearance, and the landing—are probably best considered with reference to a sequence that shows these movements being performed (Fig. 199).

[39]Richard V. Ganslen, *Mechanics of the Pole Vault*, 3rd ed. (St. Louis: John S. Swift and Co. 1957), p. 21.

[40]Cornelius A. Warmerdam, "Factors Associated with the Approach and Take-off in Pole Vaulting" (M.A. thesis, Leland Stanford Junior University, 1941).

Fig. 199. An example of good technique in vaulting with a fiberglass pole

(c)
(d)
(g)
(h)
(k)
(l)
KUUSINEN
SANOMAT
KOFF 3
HAKKAPELIITTA

Fig. 199. *Continued*

(a)[41] The vaulter has approximately two full strides remaining before the point of takeoff is reached. His hips and shoulders are essentially at right angles to the direction in which he is moving (as they must be if he is to avoid impairing his forward speed); the pole is held close to the right hip and almost parallel with the ground; and the right hand is about to initiate the planting of the pole.

There are basically three methods of planting the pole—the underarm, the sidearm, and the overarm plants (Fig. 200). The underarm plant, used by the

[41]The letters in parentheses refer to the corresponding positions in Fig. 199.

majority of metal-pole vaulters, allows the pole to be kept close to the vaulter's body throughout the plant and thus has less tendency than the other two methods to upset his balance in a lateral direction. In addition, a well-timed forward and upward movement of the vaulter's arms can make a contribution to his vertical velocity at takeoff. This is much less likely to happen if the vaulter uses either the sidearm or overarm method. While the underarm plant was favored by the majority of top-class vaulters who used metal poles, the desirability of maintaining a fairly wide separation of the hands when vaulting with a fiberglass pole has led to a widespread preference for the sidearm plant among present-day vaulters. With this type of plant they can

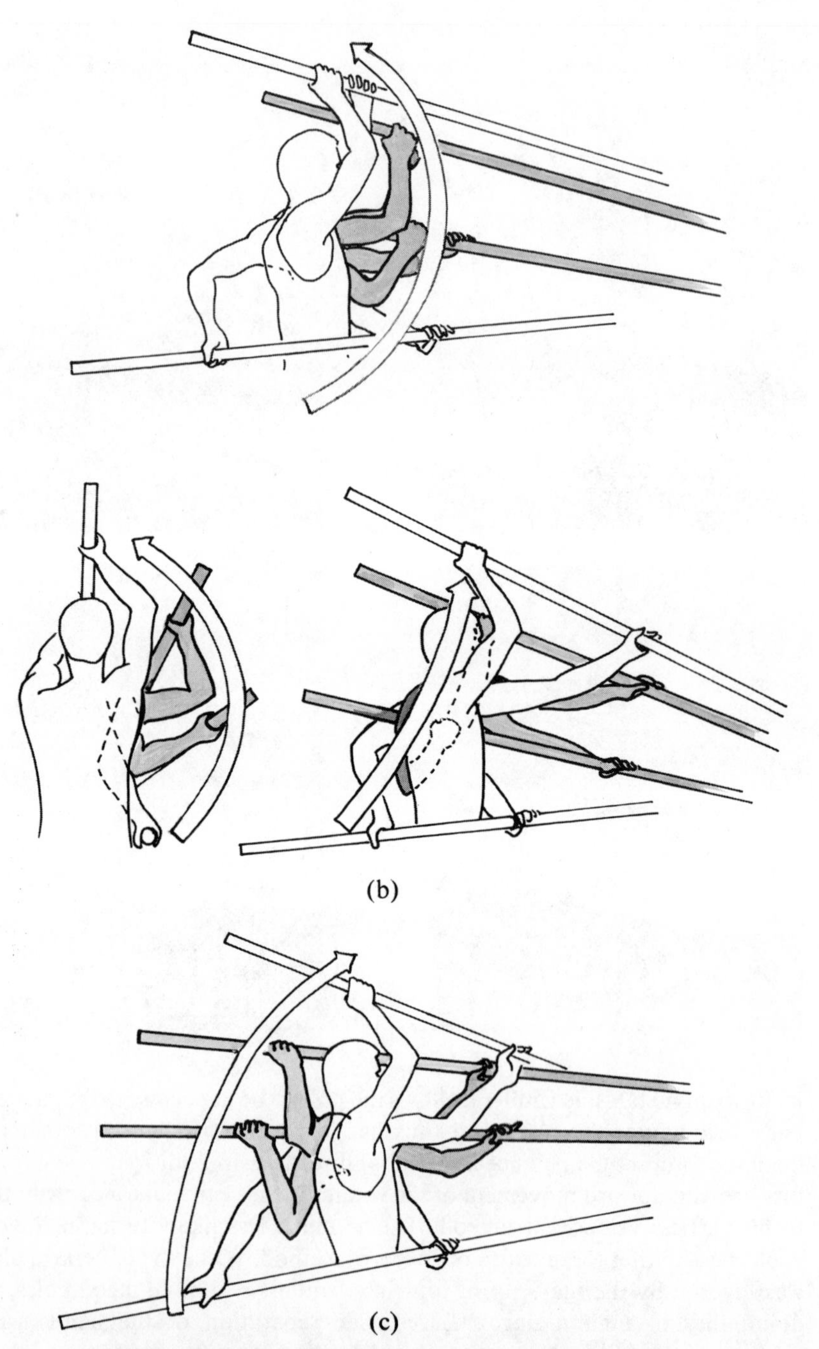

Fig. 200. Techniques used in planting the pole (a) the underarm plant, (b) the sidearm plant and (c) the overarm plant

keep their hands apart without having to twist their shoulders markedly out of alignment to retain contact with the lower hand—something that would be practically inevitable if an orthodox underarm plant were to be used.

(b) (c) The sidearm plant is now well advanced. The vaulter has rotated his right hand under the pole and then raised it to head height. To assist in keeping the pole close to his body during this process he has allowed his shoulders to rotate a little toward the right.

(e) The completion of the last stride before takeoff. The vaulter has his top hand overhead and is stepping in under the pole as it slides the last few inches before striking the back of the box. Well-executed pole plants, regardless of type, result in the vaulter passing through this position.

(f) The takeoff. The vaulter drives vigorously upward with his leading knee and forcefully extends the hip, knee, and ankle joints of his takeoff leg. The takeoff foot is directly below the vaulter's top hand and, although not evident from this figure, on the line of his run-up, which also passes through the butt of the pole.

The position of the vaulter's left (takeoff) foot at this time is critical to the success of the vault. If the foot is too far forward of a perpendicular line through the top hand, the vaulter experiences a sharp jerk at takeoff as he leaps forward against the restraint imposed by his right arm. (*Note:* Many fiberglass-pole vaulters place the takeoff foot forward of the perpendicular line through the vaulter's top hand in order to increase the magnitude of the forces exerted on the pole. For instance, in a survey of the techniques used by 25 champion fiberglass-pole vaulters, Cramer[42] reported that "12 . . . planted the takeoff foot directly below the top hand, 11 planted the takeoff foot between the hands, and two placed the takeoff foot directly below the bottom hand.") If the takeoff foot is placed behind a perpendicular line through the top hand, the vaulter may develop more momentum in his swing than he is able to control later in the vault. In addition, the distant takeoff may result in a reduction in the vertical force he can exert at takeoff and a concomitant difficulty in bringing the pole to the vertical. Serious though these problems are, even greater ones are likely to be experienced if the vaulter places his takeoff foot off-line in a lateral sense, for then he will almost certainly initiate a rotation of his body about the long axis of the pole and thus reduce the height of his vault. This fault, particularly common among beginning vaulters, who tend to "step around the pole" instead of driving straight forward onto it, usually results in the vaulter rotating around the pole to a position parallel (or near-parallel) to the bar. The same type of rotation is often initiated if the vaulter fails to bring the pole directly overhead at takeoff.

[42]John Cramer, "Fiberglass Pole Vaulting by the Champions," *Scholastic Coach*, XXXVII, February 1969, p. 16.

The positions of the vaulter's arms are such as would permit him to exert perpendicular forces to initiate the bending of the pole. On the subject of arm action at takeoff, Cramer's[43] survey revealed that some of the vaulters "held the top arm stiff above the head with no backward give," some "pulled with the shoulder, leaving the arm straight behind the head," some "pulled with the shoulder, slightly flexing the arm behind the head," some "pushed toward the cross-bar," and "one pulled down toward the head." There was an almost equal variety in the reported actions of the lower arm.

The distance between the vaulter's hands is considerably greater than would normally be the case if he were using a metal pole. Ganslen[44] has reported the "hand spreads" used by 19 leading fiberglass-pole vaulters. They range from 8 to 30 in. and average 16.3 in. Cramer[45] obtained similar results —a range from 2 to 30 in. and an average of 17 in.

(g) (h) The so-called hang phase of the vault. The vaulter has extended his arm to cushion the shock as his hands take over their weight-bearing role. This extension, together with the extension and lowering of the right knee, decreases the moment of inertia of the man-and-pole pendulum, thereby facilitating its passage toward the vertical. By increasing his moment of inertia relative to the transverse axis through his hands, it also slows the man pendulum and keeps the vaulter behind the pole.

(i) (j) The vaulter is swinging his legs upward by flexing at the hips. To increase the speed of this action, he decreases the moment of inertia of his legs by bending his knees.

(k) (l) (m) As the pole straightens, the vaulter brings his knees back still farther toward his hands and then, with an extension at the hips and knees, drives his feet high into the air, in the process passing through the so-called *L*, *J*, and *I* positions [Fig. 199(k)–(m), respectively]. Throughout these movements the top arm is kept essentially straight, delaying the pull-up, and the vaulter's center of gravity is kept close to the line of the pole.

(n) (o) The pull-up and turn are delayed until the pole is very nearly vertical and then executed at considerable speed. The rotation of the vaulter's body about its long axis is facilitated if he assumes a position—body extended, legs straight and together—that minimizes his moment of inertia relative to this axis.

The timing of the pull is of some importance, for if it is initiated too soon —as is frequently the case with beginners—the pole is less likely to reach the desired final position, due to the premature increase in the moment of inertia of the man-and-pole pendulum. It is also important that the vaulter's center of gravity remain in line with the pole and not pass forward of this line prior

[43] *Ibid.*, p. 16.

[44] Richard V. Ganslen, *Mechanics of the Pole Vault*, 6th ed. (St. Louis: John Swift and Co., 1965), p. 107.

[45] Cramer, "Fiberglass Pole Vaulting by the Champions," p. 16.

to the pull, for if it is permitted to get away from the pole in this fashion, the reaction to the force that the vaulter exerts on the pole will be an eccentric one and will tend to cause his legs to drop at the very time he most wants to keep them moving upward.

(p) Nearing the end of the push-up, the left arm is almost fully extended and the right arm is about to complete its contribution.

(r) With the push-up completed and the left hand already removed from the pole, the vaulter releases the grip of his right hand and pushes the pole back away from the bar.

(r) (s) (t) The bar clearance. There are three basic methods of clearing the bar—the jackknife, the arch, and the flyaway. A combination of two of these (the arch-flyaway) has been more widely used than any other.

In the jackknife method, the vaulter actively swings both legs downward once they have cleared the bar and then, as his body begins to fall, he extends his hips to lift his trunk and arms clear (Fig. 70). The objective here is to elevate the hips by lowering the legs (cf. methods of clearing the trailing leg in the straddle style of high jumping, pp. 439–40) and perhaps, too, to have the vaulter's body pass over the bar while his center of gravity passes below it. The principal disadvantage with this method lies in the fact that since his body is piked around the bar, the vaulter's timing must be highly precise if he is to avoid dislodging the bar with one of the many body parts in close proximity to it. As a direct consequence of this serious disadvantage, few top-class vaulters of the last 30–40 yr have used this method except as a last resort on otherwise poor vaults.

In the arch method, the vaulter arches his body around the bar and lets the angular momentum he acquired as he left the pole rotate his feet down toward the pit and his trunk and arms up clear of the bar.

With the flyaway method, the vaulter projects himself off the pole at such a speed that he is able to fly upward and across the bar while retaining his body in what is essentially a straight position. His arms, which would otherwise be the lowest parts of his body as he crossed the bar, are flung upward and backward before they endanger the success of the vault.

The arch-flyaway, the method used in the sequence of Fig. 199, involves an arching of the body around the bar, followed, once the chest has passed across the bar, by a flinging upward and backward of the vaulter's arms. This final action, which clears the arms, results in the vaulter's legs coming upward and backward (in reaction) and his chest and abdomen being moved in the opposite direction [Fig. 199(t)]. The latter, however, is unimportant if the movement is timed correctly, for then these parts of the body have already passed safely over the bar. It might be noted here that dislodging the bar with the chest or abdomen, as a result of a premature flinging up of the arms, is a very common fault at this stage of the vault, even among relatively experienced vaulters.

Chapter 17

TRACK AND FIELD: THROWING

The standard throwing events in track and field are the shot put, the discus throw, the javelin throw, and the hammer throw.

In each of these events, the athlete's objective is to obtain as large a displacement of the implement as he can, without infringing the rules governing the recording of a legal throw. The principal rules with which he is concerned are those prescribing the manner in which the implement is to be thrown, the sector in which it must land, the manner in which it is to land (javelin throw), and the forward limits of the area from which he makes his throw.

SHOT PUT

Basic Considerations

Assuming he does not violate any of the rules referred to above, the distance with which a shot-putter is credited is equal to the sum of (1) the horizontal distance that the shot is in front of the inside edge of the stopboard at the instant it is released, and (2) the horizontal distance it travels during the time it is in the air.

The first of these distances (perhaps as much as 1 ft in some cases) is governed by the athlete's body position at the instant of release and by his physique (especially the length of his throwing arm.)

The second is governed by the speed, angle, and height at which the shot is released and by the air resistance encountered during its flight.

The speed of release, unquestionably the most important of these factors (see pp. 40–41), is determined by the magnitude and direction of the forces applied to the shot and by the distance and time over which these forces act (see work-energy and impulse-momentum relationships, pp. 101 and 76, respectively).

The angle of release is also fixed by these same four factors of magnitude, direction, distance, and time. The optimum angle of release is always somewhat less than 45°, because the point at which the shot is released is some distance above the point at which it lands (p. 40). The extent to which the optimum angle differs from 45° depends on the magnitude of the release speed obtained and, to a lesser extent, on the height of release—the less the speed and the greater the height of release, the lower the optimum angle of release (pp. 39–40).

The height of release is governed by the athlete's body position at that instant and by his physique. If all else is equal, a tall, long-armed athlete who attains a position in which his legs, trunk, and throwing arm are fully extended at the instant he releases the shot will achieve greater distances than an athlete who is less generously endowed or who is in some other less effective position.

As already indicated elsewhere (p. 165), the effects of air resistance in shotputting are so small that they can be ignored in all but the most precise analysis.

Summary

The relationships between the distance with which a shot-putter is credited and the factors that determine that distance are summarized in Fig. 201.

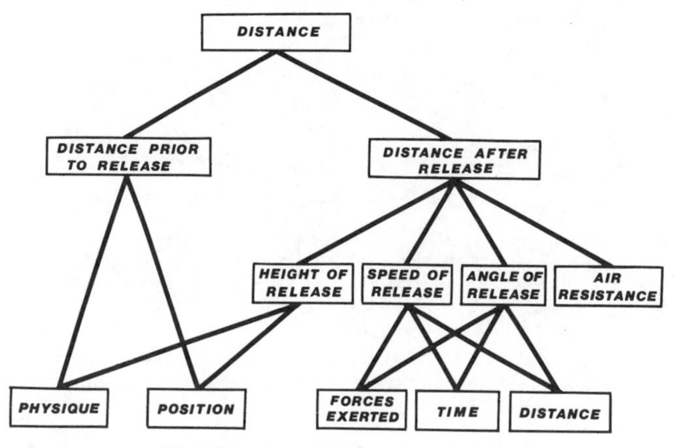

Fig. 201. Basic factors in shot-putting

Techniques

The techniques currently used by shot-putters are the result of a continuing series of developments aimed at increasing the speed with which the shot can be released.

The first major step in this process was the addition, in the very early days of the event, of some form of movement across the circle. By permitting the legs to make two contributions and by increasing both the time and the distance over which forces were exerted on the shot, this enabled athletes to obtain greater release speeds than they had hitherto with standing throws.

Although many variations were tried initially, the preliminary movement across the circle eventually evolved into a side-facing hopping (or gliding) motion (Fig. 202). This became the generally accepted technique and remained so for several decades until in the early 1950s James Fuchs (U.S.A.) developed a modification that substantially increased the contribution made by the muscles responsible for lateral flexion of the trunk and, in addition, permitted an increase in the time and the distance over which the athlete exerted force on the shot. The technique used by Fuchs is shown in Fig. 203.

Fig. 202. Technique in the old side-facing style of shot-putting

In an attempt to further increase the speed of the shot at release (by increasing the distance, time, and number of forces involved) a number of athletes carried Fuch's initial away-from-the-neck position of the shot to its logical conclusion and held it low and at arm's length behind the circle. However, this method yielded no marked success, before the rules were altered to make such a procedure illegal.

Fig. 203. The shot-putting style of James Fuchs (U.S.A.)

The next major development in shot-putting techniques was introduced by Parry O'Brien (U.S.A.), also in the early 1950s (Fig. 204). The O'Brien technique, used by almost every shot-putter of note during the next 20 yr, made provision for a further increase in the distance (and time) through which force could be exerted on the shot and also allowed for an increase in the contribution of force from the muscles of the back.

The most recent and most radical development in shot-putting techniques is the rotational (or discus-style) technique used by world-record holder Aleksandr Barishnikov (U.S.S.R.)—Fig. 205—and by professional Brian Oldfield (U.S.A.). While the advantages and disadvantages of the rotational technique (vis-à-vis that of O'Brien) have been the source of much speculation, there is, as yet, very little research on the subject. However, what research evidence is available has been quite helpful in clarifying the issues involved.

At about the same time as Fuchs was developing his new technique, other athletes were experimenting with weight training as a means of increasing their strength and thereby the magnitude of the forces they could exert against the shot. The successes enjoyed by these athletes and those that followed them have led to the recognition of weight training as an essential part of the training of shot-putters. To what extent recent advances in shot-putting performances are due to the development of improved techniques

Fig. 204. The O'Brien back-facing style of shot-putting

and to what extent they are due to the weight-training regimes followed by present-day champions may never be known. It seems likely, though, that the influence of weight training far exceeds that attributable to improvements in technique.

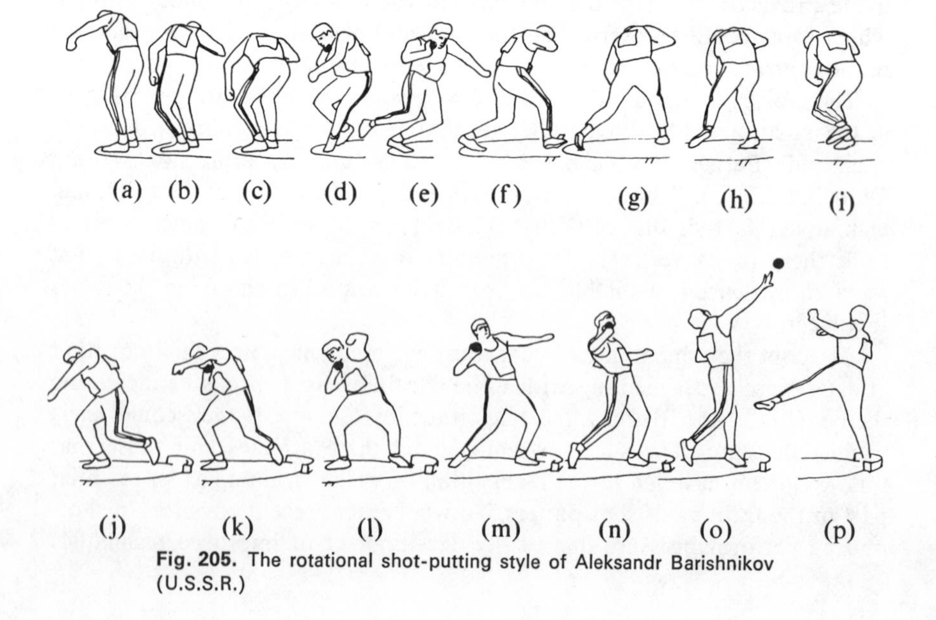

Fig. 205. The rotational shot-putting style of Aleksandr Barishnikov (U.S.S.R.)

The O'Brien, or back-facing, technique is considered in some detail in the following discussion.

Initial Stance

While athletes differ markedly in the routine (or ritual) undertaken in assuming the initial position at the back of the circle, the position finally adopted differs relatively little from one athlete to the next.

In adopting this initial position the athlete places the toes of his right foot, pointing away from the direction of throw, and close to the rear edge of the circle.[1] He then moves his center of gravity over his right foot so that it is supporting his full weight and the weight of the shot. He holds his trunk erect, his left foot on the ground a short distance behind him, and his left arm in a relaxed near-vertical position [Fig. 204(a)]. These positions of the left arm and left foot are used to assist the athlete in maintaining a state of equilibrium—the former by minor adjustments in position and the latter by adjustments in the force exerted against the ground.

The shot is supported on the base of the fingers of the right hand and, in accord with the rules, against the athlete's neck. The manner in which the shot is held influences the forces that can eventually be applied to it. If it is held in the palm of the hand, as tends to be the case with beginners, the forces that can be applied to it as a result of wrist and finger flexion are severely reduced. Thus, although a beginner may initially feel more comfortable holding the shot in the palm of his hand, he will ultimately achieve greater distances supporting it on the base of his fingers.

[1]Throughout this chapter it is assumed that the athlete under discussion throws or puts with his right hand.

Glide

From this preliminary position the athlete initiates a series of movements designed to put him into the optimum position from which to begin his movement across the circle. These generally consist of an upward and backward swing of the left leg accompanied by a lowering forward of the upper body—bringing the athlete into the so-called T position—followed by a flexion of the hip, knee, and ankle joints of the right leg and a downward and forward motion of the left leg [Fig. 204(b) and (c)]. The lowering of the trunk carries the shot to a position outside the circle and thereby increases the distance through which the athlete may exert force on it. It also places the trunk in a position that will later permit the muscles of the back to make a substantial contribution to the release speed of the shot. The flexion of the joints of the supporting leg serves some of these same purposes—it aids in bringing the shot to a low position, thereby increasing the distance over which force may be applied before the shot is released, and it puts the leg in a position to contribute force to accelerate the shot (and the athlete) across the circle.

The downward and forward swing of the left leg, together with the flexion of the right leg, puts the athlete into a low and compact position. As soon as he reaches this position—a light touch of the left foot to the ground or to the right foot is often used as a cue that the position has been reached—the athlete begins his drive across the circle. This consists of a well-coordinated combinaton of three separate movements:

1. A shifting of the athlete's center of gravity beyond the backward limit of the base provided by his right foot. This shift, brought about by the downward and forward swing of the left leg and an accompanying pushing backward of the hips, sets the body in motion across the circle.
2. A vigorous backward swing of the left leg toward the front of the circle.
3. An extension of the knee and ankle joints of the right leg [Fig. 204(d)]. The greater part of this extension is delayed until the previous two movements have placed the athlete's center of gravity in a position that will allow it to be driven toward the front of the circle rather than upward.

As soon as the drive from the right leg has been completed, the right foot is whipped low across the circle to a position near the center and beneath the athlete's center of gravity. Shortly after the right foot has landed—a time of 0.03 sec has been reported[2] for a put made by 1964 Olympic champion and former world-record holder Dallas Long—the left foot is grounded close to the stop board and a little to one side of the direction line [Fig. 204(f)].

[2]Bob Ward, "Analysis of Dallas Long's Shot Putting," *Track Technique*, No. 39, March 1970, p. 1232.

The extent to which the right leg is extended as the athlete drives across the circle varies between athletes. Some use an almost complete extension of both knee and ankle joints, while others confine the extension almost entirely to the knee joint. In the former case the athlete's toes are the last part of his foot to break contact with the ground [Fig. 204(d)], while in the latter the heel has this distinction. Which of these two methods, if either, is generally the better has yet to be resolved. [*Note:* The results of an informal survey conducted by the author suggested that in general, topclass performers prefer to drive from the heel—Matson and Neider (U.S.A.) used both knee and ankle extension and pushed off from the toes, while Feuerbach, Long, O'Brien, and Woods (U.S.A.), Briesenick, Gies, and Hoffman (East Germany), and Komar (Poland) used knee extension almost exclusively and pushed off from the heel.]

The distance that the right foot travels, its orientation as it lands, and the manner in which it is grounded also vary considerably from one athlete to the next.

Tall athletes, who need a greater distance between their feet during the delivery than do those who are shorter, usually restrict the length of the glide to a little less than 3 ft 6 in. and thus place the right foot either on the center line or just behind it in the rear half of the circle. Short athletes—and anything less than 6 ft 1 or 2 in. would generally be regarded as short for this event—usually travel a little more than 3 ft 6 in. during the glide.

The orientation of the athlete's right foot as he lands at the completion of the glide varies from the pointing-to-the-back-of-the-circle position used by Fuchs to the position at right angles to the direction line advocated by Tschiene.[3] However, most top-class athletes—presumably compromising between (1) the difficulty in coordinating the various movements of the delivery inherent in the position used by Fuchs and (2) the dangers of prematurely rotating the hips and trunk to the front inherent in that proposed by Tschiene —use a position roughly midway between the two.

While at least one authority on shot-putting[4] has stated that a flat-footed landing should be made at the end of the glide—mainly on the grounds that the extension of the right ankle can then proceed without the delay necessitated by having to first flex the ankle by lowering the heel—an examination of the techniques used by outstanding performers (including Briesenick, Feuerbach, Gies, Hoffman, Komar, Long, Matson, Neider, and Woods) suggests that a landing on the ball of the foot is much more common among athletes of this caliber.

[3]Peter Tschiene, "Perfection of Shot Put Technique," *Track Technique*, No. 37, September 1969, pp. 1187–89.

[4]Geoffrey F. D. Pearson, "The Shot Put—I," in *Illustrated Guide to Olympic Track and Field Techniques*, ed. by Tom Ecker and Fred Wilt (West Nyack, N.Y.: Parker Publishing Co., Inc., 1966), pp. 123–24.

The position of the upper body during the glide has a considerable bearing on the final result. If the athlete allows his head and trunk to turn toward the front—and the backward swing of the left leg tends to encourage this—the magnitude of the forces he subsequently exerts on the shot and the distances over which these forces act are markedly reduced. To avoid these undesirable effects of turning to the front too soon, experienced athletes endeavor to maintain a back-facing (or "closed") position throughout the glide.

Delivery

As soon as his right foot lands near the center of the circle, the athlete begins his delivery, that final coordinated sequence of actions that culminates in the release of the shot.

The delivery begins with a strong lifting action produced by the contraction of the extensor muscles of the athlete's hip, trunk, and right knee. [The difference between the inclination of the athlete's trunk in Fig. 204(e) and (f) reflects this initial lifting action.]

The importance of the contribution of the legs during the initial phases of the delivery has been alluded to by Fischer and Merhaupt[5] following an electromyographic analysis of the actions of experienced and inexperienced shot-putters. They found that the leg muscles were active up to 72% of the time in the delivery of the experienced shot-putters and only 28% of the time in the case of the inexperienced ones. Futhermore, an analysis of the films taken in conjunction with the electromyographic recordings revealed that the inexperienced athletes threw "from a vertical position, bringing to bear only the body turn and the forward thrust of the shoulders." In other words, they made little or no use of the powerful lifting action that their legs were capable of producing.

This lifting movement is followed by a rotation of the athlete's body toward the front as the extension of the right leg continues and is supplemented by the contraction of the muscles producing trunk rotation [Fig. 204(g)].

The left arm, which has been allowed to hang loosely in a near-vertical position throughout the glide and the initial part of the delivery, is swung upward and backward, thereby contributing to the rotation of the trunk [Fig. 204(f) and (g)].

As the extension of the right leg and the rotation of the trunk to the front near completion, the shot moves away from its position against the athlete's

[5]A. Fischer and J. Merhaupt, results of a study cited in Toni Nett, "Foot Contact at the Instant of Release in Throwing," *Track Technique*, No. 9, September 1962, p. 272.

neck, and the right arm begins to make its contribution to the release speed of the shot [Fig. 204(h)]. This involves a coordinated forward rotation of the upper arm, a forceful extension of the right elbow, and a final flexion and pronation (or "snap") of the wrist.

The question of whether the athlete should retain contact with the ground during the release has been the subject of some debate. Dyson,[6] for example, has stated that "Theoretically, . . . the front foot should be firmly in contact with the ground, providing the necessary resistance for the hand to exert maximal force both vertically and horizontally." However, he concedes, that " . . . a majority—if not all—of the world's 60-ft shot putters *do* in fact break contact with this front foot fractionally before the missile leaves the hand." Nett[7] has produced photographic evidence that tends to confirm this latter point and has concluded that "the old idea of insisting that the feet be planted at the moment of release was simply an application of a 'brake' to the total effort." Herein, it seems, lies the crux of the matter—must the athlete reduce the magnitude of the vertical forces that he can exert in order to retain contact with the ground and, if he must, is the resulting loss in the release speed of the shot larger or smaller than would result from his being off the ground as the final "wrist snap" is executed? (*Note:* Only the vertical forces are of importance here, for these alone tend to cause the athlete to be projected into the air.) While to date there appears to be no objective basis for an answer to these questions, the empirically derived methods of the world's leading exponents of the shot put would seem to suggest that an athlete must reduce the vertical forces he exerts if he is to remain on the ground and that the resulting loss in the release speed of the shot is probably greater than that due to his being airborne in the concluding stages of the delivery.

Reverse

Once the shot has been released, the athlete's right foot comes forward to support his weight while his left leg swings back toward the center of the circle [Fig. 204(k)]. This reversing of the feet is employed to assist the athlete to remain in the circle once the shot has left his hand. The swinging back of the left leg (and, in extreme cases, the flexion of the hips and downward-backward-upward swing of the arms—see inset, Fig. 204) serves to produce a contrary angular reaction that tends to move the athlete's center of gravity back from the forward limit of his base.

[6]Geoffrey H. G. Dyson, *The Mechanics of Athletics* (London: University of London Press Ltd., 1967), p. 211.

[7]Toni Nett, "Foot Contact at the Instant of Release in Throwing," *Track Technique*, No. 9, September 1962, p. 274.

The rotational technique is essentially a combination of the techniques used in the first half of an orthodox discus throw with those used in the second half (or delivery phase) of an O'Brien-style shot put.

The argument most frequently advanced in favor of the rotational technique is that, because of the greater distance through which the shot travels, its velocity at the instant when the thrower lands in the front of the circle is greater than it would be if the thrower used the orthodox O'Brien style. This seemingly logical contention is not supported, however, by the available data. In a comparison of the techniques employed by Baryshnikov (who set a world record using the rotation techniques) and Al Feuerbach (a former world-record holder who used the O'Brien style), Kerssenbrock[8] reported that, although both achieved similar release speeds of the shot (44.2 fps and 45.3 fps, respectively) the speeds at the instant when the final putting stance was reached (i.e., when the left foot was grounded near the front of the circle) were vastly different (4.6 fps and 8.2 fps, respectively). Further, the difference was contrary to what the prevailing argument would suggest.

The reason for this discrepancy between theory and practice is not hard to find—the theory is at fault. When an athlete using the rotational technique makes his way across the circle, his body is simultaneously translated in the general direction of the throw and rotated about a vertical or near-vertical axis. In the first part of the turn [Fig. 205(d)–(f)] the shot moves forward relative to the athlete's center of gravity, which is itself moving forward across the circle. The effects of the rotation and translation thus complement each other and the shot moves forward at a relatively high speed. During the second part of the turn [Fig. 205(g)–(k)] the shot moves backward relative to the athlete's center of gravity while the latter continues to move forward across the circle. The respective contributions that the athlete's rotation and translation make to the speed of the shot thus tend to offset each other and the speed of the shot is relatively low. Indeed, if the shot is traveling backward relative to the athlete's center of gravity at a greater rate than the latter is traveling forward, the shot will actually be moving in a direction opposite to the ultimate direction of the throw! (These various effects are clearly evident in Fig. 206 which shows how the speed of the shot varies during the course of a throw using the rotational technique.)

On the basis of all this, it may reasonably be concluded that, *if* the rotational technique is as good or better than the O'Brien technique, it is *not* because the speed of the shot is greater when the athlete lands in the front of the circle. Instead, it seems likely that the success enjoyed by leading exponents of the rotational technique is due to the athlete being able to assume a more favorable position from which to begin the final putting action.

[8]Klement Kerssenbrock, "Potential of the Rotation Shot Put," *Track Technique*, No. 58, December 1974, p. 1848.

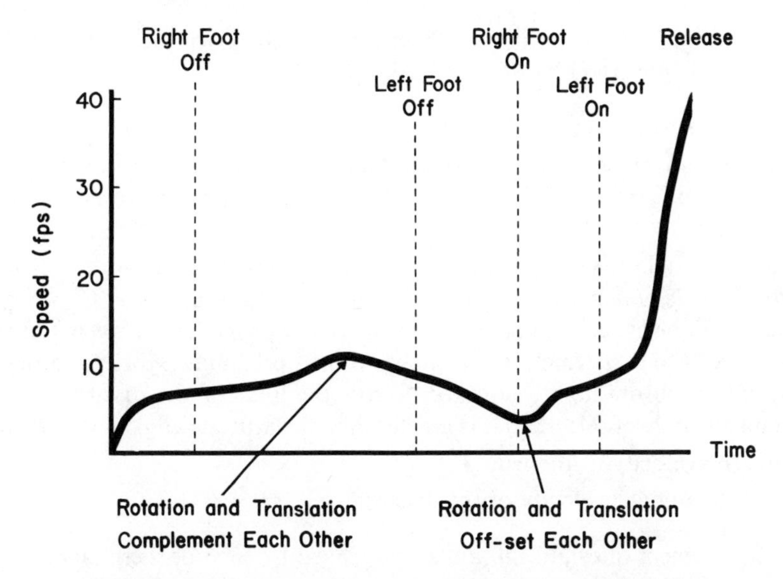

Fig. 206. Variations in the speed of the shot during the execution of a throw using the rotational technique (*Based on data in Rolf Geese, "Überlegungen zur Kugelstoss–Drehtechnik,"* Die Lehre der Leicht-athletik *xxv, 28 May 1974*)

DISCUS THROW

Basic Considerations

Assuming that his throw conforms with the rules governing the event, the distance with which a discus thrower is credited is determined by the speed, height, and angle at which he releases the implement and by the aerodynamic factors that influence its flight.

The speed and angle of release are determined by the magnitude and direction of the forces exerted on the discus and the time and distance over which these forces are applied.

The height of release is governed by the athlete's body position at that instant and by his physique. Although the height of release is a factor of relatively minor importance compared to the speed and angle of release, if all else is equal, a tall thrower who releases the discus from an erect position with his legs and trunk fully extended will have an advantage over other throwers who are shorter or who release the implement with their bodies in a less effective position.

Aerodynamic Factors

In addition to the ever-present gravitational force, a discus in flight is subjected to forces exerted upon it by the air through which it passes (pp.

171 *et seq.*). The magnitude of these forces, and hence the extent of their influence, is governed by:

1. The speed of release.

2. The angle of release, i.e., the angle between the direction in which the center of gravity of the discus moves immediately after release and the horizontal.

3. The angle that the discus is inclined to the horizontal—the so-called *attitude angle* or *angle of tilt*—at the instant it is released. [*Note:* Contrary to what might be expected, the angle at which a discus is released is generally not the same as that at which it is inclined to the horizontal. Most good throwers, in fact, throw slightly upward "across the line" of the discus and thus obtain an angle of release that is greater than the attitude angle—Fig. 207(a)].

4. The velocity of the wind.

5. The angular velocity of the discus at release.

The problem of establishing the role played by each of these factors (like most such problems in fluid mechanics) is exceedingly complex. Several attempts have been made to solve the problem by placing a discus in a wind tunnel and measuring the forces exerted upon it under varying conditions. Probably the first of these, at least in the western world, was a study conducted by Taylor[9] in order to determine the effect that head winds and tail winds had on the distance of a throw. He concluded that while head winds were increasingly helpful up to between 7 and 8 mph, at speeds greater than this the advantage decreased steadily until at 14.5 mph throwing into a head wind became a disadvantage. Tail winds of up to 14 mph were also found to detract from the distance that a thrower would otherwise achieve. Unfortunately, Taylor's reporting of the procedures followed in his experiments was very incomplete, and the validity of his findings has been called into question by others who have subsequently conducted research in this area.

Ganslen[10,11,12] conducted an extensive series of studies into the aerodynamic factors influencing the flight of the discus. Among the various conclusions he reached as a result of these studies were the following:

1. A relatively poor thrower will benefit more from a head wind of a given velocity than will a good thrower. This is because the percentage increase

[9]James A. Taylor, "Behavior of the Discus in Flight," *Athletic Journal*, XII, April 1932, pp. 9–10, 45–47.

[10]Richard V. Ganslen, *Aerodynamic Factors Which Influence Discus Flight*, Research Report, University of Arkansas, 1958.

[11]Richard V. Ganslen, "Aerodynamic Forces in Discus Flight," *Scholastic Coach*, XXVIII, April 1959, pp. 46, 77.

[12]Richard V. Ganslen, "Aerodynamic and Mechanical Forces in Discus Flight," *Athletic Journal*, LXIV, April 1964, pp. 50, 52, 68, 88–89.

in the relative wind will be greater for him than for a thrower who is attaining a high speed of release.

2. There is no such thing as an optimum wind velocity for maximum distance.

3. The discus stalls (i.e., experiences a marked reduction in lift) at angles of attack between 27° and 29°.

[*Note:* The *angle of attack* is the angle between the central plane of the discus and the relative wind. If the discus is thrown in still air, the relative wind is equal in magnitude and opposite in direction to the velocity of the implement. The angle of attack at the instant of release is therefore equal to the attitude angle minus the angle of release—Fig. 207(b). A positive angle of attack means that the underside of the discus is exposed to the oncoming airflow, while a negative angle of attack—by far the more common in good throwing—means that the upper side is thus exposed, Fig. 207(c). If the discus in thrown in other than still air, the velocity of the wind must also be taken into account to determine the magnitude and direction of the relative wind and the angle of attack—Fig. 207(d)].

Kentzer and Hromas[13] used a wind tunnel to examine the aerodynamic characteristics of a spinning discus—in none of the previous studies (those of Taylor and Ganslen) had the effect of the rotation of the implement been taken into account. They found that at an air velocity of 100 fps the maximum value of the lift/drag ratio was obtained with an angle of attack of 9°, a finding in close accord with those of Ganslen for speeds between 80 and 95.6 fps—see Table 13, p. 180.

Cooper, Dalzell, and Silverman[14] (introduced to the problem by Hromas) attempted to determine the attitude angle and angle of release at which a discus should be thrown in order to obtain the maximum possible distance for a given initial speed. Using lift and drag values obtained by Ganslen and a computer program designed for the purpose, they obtained the results depicted in Fig. 208. From these they concluded:

1. The speed of release is the most important factor in determining the length of a throw.

2. For any given speed of release, the angle of release is of prime importance. Good throwers (i.e., those who throw in the 150-200-ft range) should use an angle of release of between 35° and 40°, while throwers of lesser ability should increase this slightly but never beyond 45°.

3. The attitude angle should be between 25° and 35°.

[13]C. P. Kentzer and L. A. Hromas, Research Report, School of Aeronautical Engineering, Purdue University, July 2, 1958.

[14]Leonard Cooper, Donald Dalzell, and Edwin Silverman, *Flight of the Discus*, Division of Engineering Science, Purdue University, May 18, 1959.

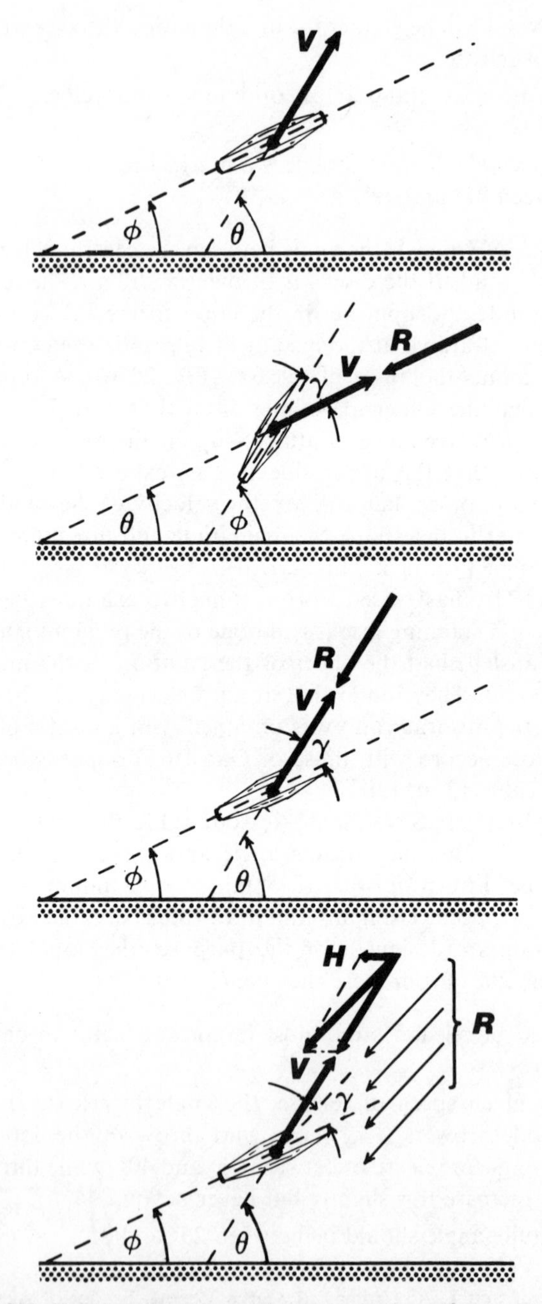

Fig. 207. Relationships between the angle of release (θ), the attitude angle (ϕ), and the angle of attack (γ) under differing conditions. (The release velocity of the discus is designated by V, the relative wind by R, and a head wind by H.)

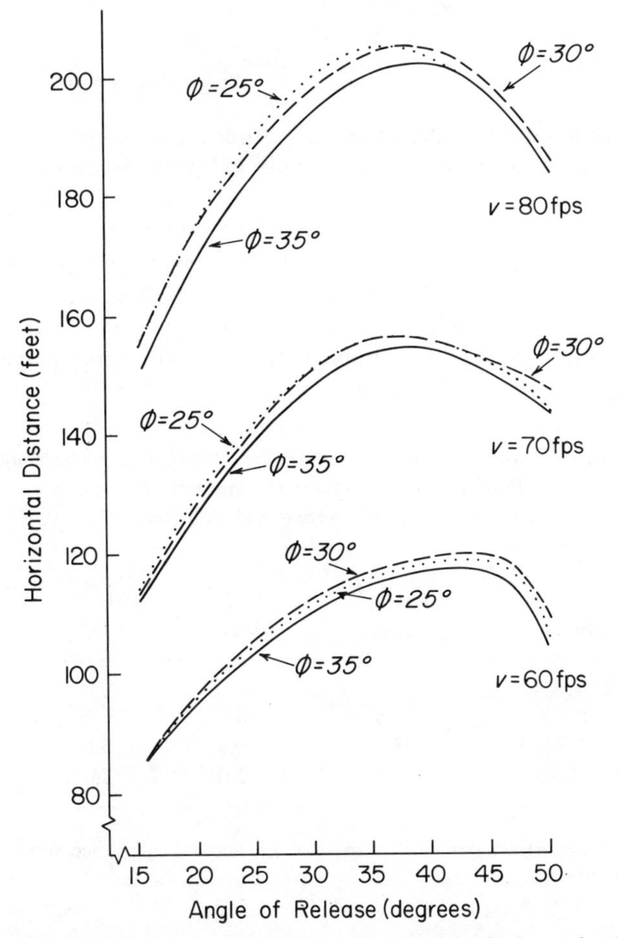

Fig. 208. The effect of variations in release speed (v), the angle of release, and the attitude angle (ϕ) ,on distances achieved in the discus throw (*Based on data in Cooper, Dalzell, and Silverman, "Flight of the Discus."*)

[*Note:* At first glance this conclusion—implying, as it does, an angle of attack of from 0 to − 15°—appears to conflict with the findings of Kentzer and Hromas and Ganslen, who have indicated that the angle of attack that yields the best lift/drag ratio is of the order of 9°–10°. The reason for this apparent conflict lies in the fact that while the attitude angle may remain essentially constant throughout the flight—a point attested to by Taylor[15] following an examination of slow-motion films of the discus in flight—the direction of the relative wind and hence the angle of attack is constantly changing as the discus rises to its peak height and then falls toward the ground. The angle of attack obtained at release should therefore be the one that will yield the best results overall rather than the one that merely happens to be the optimum for that instant.]

[15]Taylor, "Behavior of the Discus in Flight," p. 10.

Information on the angles of release, attitude angles, etc., used in practice is very limited indeed. What is available indicates, however, that the experimental results cited here are in excellent accord with those recorded in practice. Lockwood,[16] for example, studied slow-motion films of 10 experienced athletes throwing between 160 and 190 ft and found that all used angles of release between 30° and 45° and that six used angles between 34° and 37°. The angles of attack varied between 0° and −10°. Terauds[17] obtained similar results in a study of the four throwers competing in an international dual meet (Table 29).

Table 29. The Angle of Release, Angle of Attack, Release Velocity, and Distance Recorded in the Longest Throw of Each Competitor in an International Meet (after Terauds)

Athlete	*Angle of Release* (deg)	*Angle of Attack** (deg)	*Release Velocity* (fps)	*Distance of Throw* (ft)
Wilkins (U.S.A.)	35.0	−17.0	80.0	200.5
Drescher (U.S.A.)	37.5	−14.0	76.5	193.3
Zhurba (U.S.S.R.)	38.5	−19.5	79.6	182.4
Voikin (U.S.S.R.)	38.5	−2.0	72.0	161.0

*The wind ("5 mph from behind, 45° from right")was not taken into account in determining the angle of attack.

Summary

The relationships between the distance with which a discus thrower is credited and the factors that determine that distance are summarized in Fig. 209.

Techniques

Techniques in discus throwing have developed along much the same lines as those in shot-putting with the emphasis on increasing the distance through which force may be applied to the discus and, over the last two decades or so, on increasing the magnitude of the forces exerted by increasing the strength of the athlete through weight training.

[16] H. H. Lockwood, "Throwing the Discus," in *Athletics*, ed. by G.F.D. Pearson (Edinburgh: Thomas Nelson and Sons Ltd., 1963), p. 206.

[17] Juris Terauds, "Some Release Characteristics of International Discus Throwing," *Track and Field Quarterly Review*, LV, March 1975, pp. 54–57.

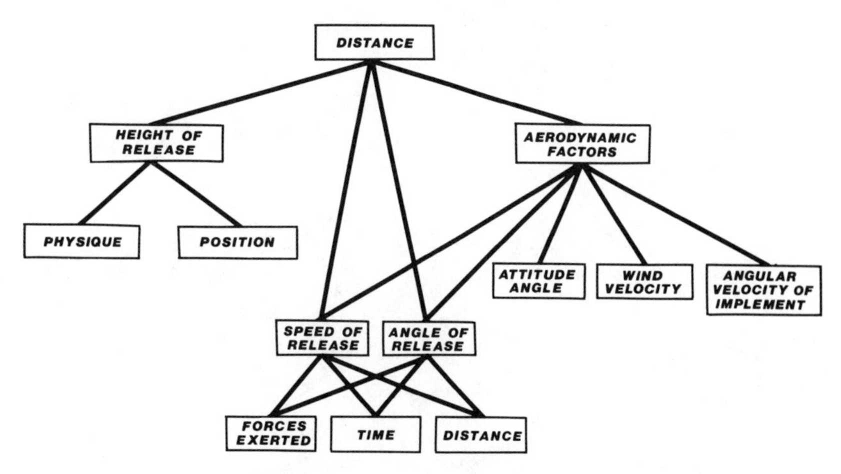

Fig. 209. Basic factors in discus throwing

Initial Stance

The first of these emphases has been most clearly apparent in the initial stance adopted by the thrower.

From the mid-1920s to the late 1930s the initial stance adopted by most leading exponents of the event was one in which the athlete stood at the back of the circle with his body sideways to the direction of the throw. From this position he pivoted on the ball of his left foot and executed a one-and-one-quarter turn to bring him to the point in the front of the circle at which he released the implement [Fig. 210 (a)].

Since that time the initial positioning of the feet relative to the direction of the throw has been progressively modified in an attempt to increase the distance through which force may be applied to the discus. Figure 210(b)–(d) show the initial foot placements and the path followed by the discus in those modifications that have gained some measure of support among top-class throwers. Many other possibilities have been explored—some being used occasionally in competition by top-class performers [Fig. 210(e) and (f)] and others seemingly far too fanciful to attract much support [Fig. 210(g)].

With regard to the initial placement of the feet it is well to recognize that, because the only reason a thrower adds a preliminary turn (or turns) before executing the final delivery is to increase the speed of release of the discus beyond what he can achieve with a standing throw, the efficacy of a given initial placement of the feet can be judged only in terms of the effect produced on the speed of release of the implement. [*Note:* Because the speed of the discus at release is a function of the forces exerted upon it during the turn(s) and delivery and the distance (and time) over which these forces act, it is often supposed that any initial stance that allows an increase in distance

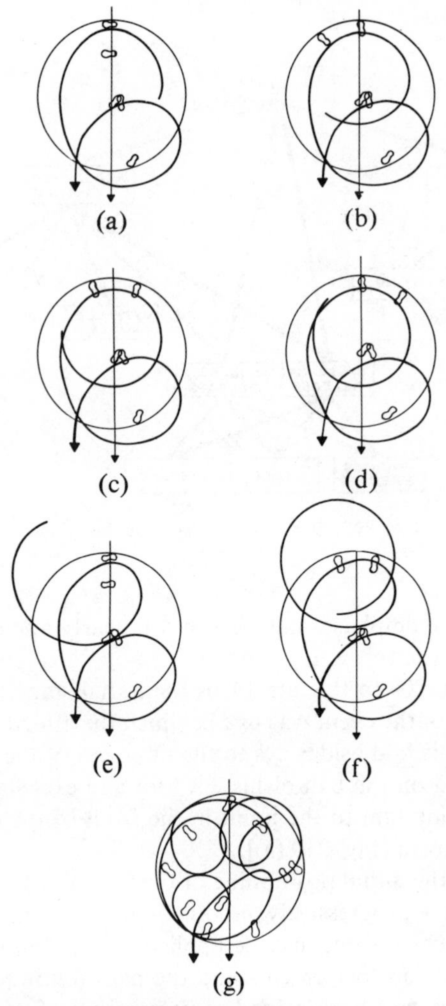

Fig. 210. Variations in the placement of the feet and the path followed by the discus. (a) Side-facing $1\frac{1}{4}$ turn used by leading throwers from mid-1920s to late 1930s. (b) Oblique back-facing, slightly less than $1\frac{1}{2}$ turn currently used by many leading throwers. (c) Back-facing $1\frac{1}{2}$ turn, also in widespread use at the present time. (d) Oblique, back-facing, slightly more than $1\frac{1}{2}$ turn used by some leading throwers. (e) Side-facing $1\frac{3}{4}$ turn advocated by Bosen[18] and used by Neu (East Germany) in 1968 Olympic Games final. (f) Back-facing $2\frac{3}{4}$ turn throw developed by Bob Humphreys (U.S.A.).[19] (g) Multiturn method suggested by Davenport.[20]

[18]K. O. Bosen, "Discus Throw Points to Ponder," *Track Technique*, No. 11, March 1963, p. 335.

[19]Bert Lockwood, "The Double-Turn Throw," *Track Technique*, No. 35, March 1969, pp. 1110–14.

[20]H. Davenport, "A New Discus Technique," *Modern Athletics*, V, January 1961, p. 11.

through which the discus travels before it is released is necessarily better than one for which the distance traveled is less. This is a fallacy. Only if the forces exerted in the case where the longer distance is involved do not decrease by an amount comparable, or more than comparable, to the gain in distance —in other words, only if the work done on the discus is greater—will there be an increase in the speed of release.]

Finanger[21] and Lockwood[22] have studied the manner in which the speed of the discus changes in the course of a throw and have arrived at very similar conclusions:

1. The speed of the discus at release is not the result of a steady acceleration from the end of the backswing to the point of release.
2. There is an initial increase in the speed of the discus until the athlete's right foot breaks contact with the ground. This is followed by a period during which the speed is relatively constant. Once the left foot leaves the ground (i.e., both feet are in the air), the speed of the discus decreases and continues to decrease until some time after the right foot lands near the center of the circle. Finally, once the left foot has been grounded, there is a marked acceleration of the discus that continues up until the instant the implement is released.

Thus, to be advantageous, any change in an athlete's initial position must accomplish one or more of the following without producing a comparable, or more than comparable, loss in speed at some other point in the throw:

1. increase the speed attained by the discus before the right foot is lifted,
2. increase the speed of the discus during the period after the right and before the left foot leaves the ground,
3. reduce the loss in speed that occurs between the time the left foot breaks contact with the ground at the back of the circle and the time it regains contact in the front of the circle,
4. increase the gain in the speed of the discus during the delivery.

Preliminary Swings

From an erect standing position with the discus held over his left shoulder or just in front of his chest, the athlete begins the throw with one or two preliminary swings, aimed at relaxing him and getting him mentally "set"

[21]Kenton E. Finanger, "An Electromyographic Study of the Function of Selected Muscles Involved in the Throwing of the Discus" (Ph.D. dissertation, The University of Iowa, 1969), p. 126.

[22]Lockwood, "Throwing the Discus," p. 189.

for the throw. On the completion of these preliminary swings (which, incidentally, are subject to considerable variation between throwers), the athlete moves into position to commence his turn. This involves a swinging of the discus downward and backward to a position behind the body and somewhere between hip and shoulder level, a twisting of the trunk to the right that carries the discus still farther back, a semiflexing of the knees in preparation for the movements to follow, and a shifting of the athlete's weight over his right foot [Fig. 211 (a)–(c)].

Transition

As the discus nears the limit of its backward swing (or just as it begins to come forward again), the athlete adjusts the position of his left foot by rotating it (heel raised and pivoting on the ball of the foot) toward the direction of the throw—Fig. 211 (d). This movement is accompanied by a shifting of his center of gravity to the left and over the left foot, a movement that is of considerable importance in determining the ultimate success of the throw. If the athlete fails to shift his weight sufficiently to the left, a moment (weight times distance from line of gravity to foot) is established that tends to rotate him sideways when his right foot is lifted from the ground. In an effort to offset or correct the unbalancing effects of this moment, an athlete usually drives rather more to the side than he should, as he pushes off from his left foot into the turn. This leads to a final placement of the left foot well to the left of the direction line and a correspondingly poor throw. [*Note:* The shifting of the athlete's center of gravity to the left during the transition and the resulting position with the center of gravity "balanced" over the left foot for the drive across the circle, are shown in Fig. 212(a)–(c) and Fig. 212(d), respectively.]

Turn

A discus turn is a combination of angular motion (the angular motion of the athlete rotating about his longitudinal, or a near-longitudinal, axis) and horizontal motion (the horizontal motion of the athlete's center of gravity as it moves forward across the circle). There are basically two types of turn in common use—the so-called "rotation" and "running-rotation" turns —distinguished primarily by a difference in the emphasis given to the angular and horizontal characteristics of the turn. In the rotation turn, the pivot on the ball of the left foot is prolonged until the foot has rotated through a near 360° angle. Only then does the athlete push off from the left foot toward the front of the circle—a push or drive that is relatively weak because of his near-back-facing position. In the running-rotation turn, the athlete pivots on the ball of his left foot until he is facing in the direction of the throw. He then drives vigorously downward and backward via his left foot—an action that momentarily gives the impression that he is sprinting across the circle.

Fig. 211. An example of good technique in the discus throw

Although the running-rotation turn is a relatively recent development in discus throwing, the majority of the world's leading male throwers currently use this technique in preference to the rotation turn that preceded it historically. (In the 1968 Olympic Games, 8 of the 12 finalists in the men's event, including the 3 medal winners, used a running-rotation turn. In the women's event—and techniques in women's track and field events invariably seem to lag behind those in the men's events—6 of the 13 finalists used a running-rotation turn.[23])

In a running-rotation turn, to which the following discussion will be confined, the athlete pivots on the ball of his left foot until he is facing in the

[23]P. Harper, "Discus (Men)" and "Discus (Women)," in *Mexico 68: A Technical Report by the National Coaches on the 1968 Olympic Games*, ed. by Tom McNab, mimeographed report, pp. 119, 125.

Fig. 212. The shifting of the athlete's center of gravity to the left during the transition from the last preliminary swing to the turn (*Photographs courtesy of Howard Payne*).

direction of the throw and then, lifting the knee of his right leg high, drives vigorously forward across the circle [Fig. 211 (g)]. The drive from the left leg, augmented by the quick lifting action of the right knee, projects him forward (and slightly upward) toward the center of the circle.

The instant at which the right foot is lifted from the ground and the action of the right leg prior to the initiation of the left leg drive have been the subject of some discussion in recent years. Some authorities[24,25,26] maintain that the right foot should be kept in contact with the ground for as long as possible—reasoning, presumably, that by so doing the athlete can exert small forces via the right foot to aid in controlling balance—and should then be brought close to the left foot as it is moved forward for the drive across the circle. Others,[27,28] influenced by former world-record holder Jay Silvester who first drew attention to the technique, prefer the right foot to be lifted much earlier and the near-straight leg to be brought around to the front in a wide-sweeping action [Fig. 211 (f)]. This action increases the athlete's moment of inertia relative to the axis about which he is rotating and, providing there is not a comparable or more than comparable loss in angular velocity, enables him to increase the angular momentum of his lower body. This gain in angular momentum can then be used to advantage later in the throw. The question as to which of these two techniques should be used by a given athlete is probably best viewed from the standpoint of experience—an inexperienced thrower is likely to fare better with the technique that emphasizes control and the experienced thrower, who has already acheived a high degree of control, is likely to benefit more from the method that affords the greater momentum.

During the short time that the athlete is in the air the hips move ahead of the shoulders, which in turn maintain their "lead" over the right arm and the discus. This movement of the hips to a position ahead of the shoulders is facilitated if the athlete:

1. Keeps the thighs fairly close together, thereby decreasing the moment of inertia and increasing the angular velocity of the legs.

2. Holds the right arm and the discus some distance away from the axis of rotation to increase the moment of inertia of the upper body and thus reduce its angular velocity.

3. Holds the left arm across the body in a deliberate attempt to keep the shoulders back—an angular action that yields a contrary reaction of the lower body. (*Note:* The athlete in Fig. 213 makes good use of all three of these techniques.)

[24] J. LeMasurier, *Discus Throwing* (London: The Amateur Athletic Association, 1957), p. 16.

[25] Frank Ryan, "Teaching the Discus Throw," *Scholastic Coach*, XXXI, February 1962, p. 25.

[26] David Pryor and H. H. Lockwood, "The Discus Throw," in *International Track & Field Coaching Encyclopedia*, ed. by Fred Wilt and Tom Ecker (West Nyack, N.Y.: Parker Publishing Co., Inc., 1970), p. 264.

[27] Tom Ecker, *Track and Field Dynamics* (Los Altos, Calif.: Tafnews Press, 1971), p. 55.

[28] Ralph B. Maughan, "Jay Silvester's Discus Form," *Track Technique*, No. 15, March 1964, pp. 477–78.

Fig. 213. The hips move ahead of the shoulders during the brief time the thrower is in the air (*Photograph courtesy of Howard Payne*).

Delivery

At the conclusion of the airborne phase of the turn, the athlete lands near the center of the circle on the ball of his right foot and with his center of gravity directly above or slightly behind the right foot. If he has correctly executed the preceding movements, his hips are well ahead of his shoulders and his shoulders are well ahead of his throwing arm at this instant [Fig. 211 (k)]. His next task is to get his left foot grounded in the appropriate position as quickly as possible. This is of critical importance because the discus is being decelerated at this time [presumably as a result of (1) a frictional couple applied via the right foot and (2) a decrease in the body's angular velocity as the left leg moves progressively farther from the axis of rotation], and the longer this deceleration is permitted to continue, the worse the final result is likely to be. In addition, the stretching of the musculature of the upper body to allow the hips to get ahead of the shoulders and the shoulders to get ahead of the throwing arm cannot readily be maintained unless powerful forces are exerted via the legs. Such forces can be exerted only if both feet are on the ground. In short, therefore, unless the thrower gets his left foot grounded quickly, he loses both the momentum that has been built up in the turn and the strong throwing position that he needs for the final, all-important acceleration of the discus.

Once the right foot has landed, and before the left foot is grounded in the front of the circle, the right leg drive initiating the delivery begins. This consists of a turning inward of the right knee, accompanied by a pivoting on the ball of the foot and an outward turning of the heel [Fig. 211 (k)–(m)]. This is followed, once the left leg is in position to provide resistance, by an extension of the hip, knee, and ankle joints that drives the athlete's hips forward and around toward the front [Fig. 211 (n)–(p)]. Well before the hips reach the front, the muscles that rotate the trunk, and which have been stretched by the preceding movements, contract forcefully to bring the shoulders around [Fig. 211 (n)–(q)]. Finally, the right arm—left well behind by these strong rotary movements of the hips and trunk—is swept forcefully out and forward; the near-straight left leg is extended in a short forceful movement that brings the athlete to his maximum height and adds to the vertical velocity of the implement; and the discus is brought around to the point where it is released [Fig. 211 (o)–(q)]. This point is some 5 ft 4 in. to 6 ft 3 in. above the ground on the average[29] and at the maximum distance possible from the axis of rotation—the latter to ensure the maximum linear velocity of the implement at release.

The question of whether or not the athlete should retain contact with the ground throughout the delivery (a question already discussed here with reference to shot-putting techniques—p. 467) is also the subject of some debate among those interested in discus throwing.

Basically the athlete has three methods of imparting the necessary vertical velocity to the implement:

1. by lifting the arm relative to the shoulder as the implement is swung forward,
2. by driving off the right leg and rotating forward and upward over the left,
3. by vigorously extending both legs (but particularly the left) immediately before the discus is released.

Which of the last two he gives the greater emphasis would appear to determine whether he completes the delivery while retaining contact with the ground or with both feet in the air. Throwers who stress driving up and over the firmly planted left leg generally release the implement with the left foot, and occasionally the toes of the right foot, in contact with the ground. Others who emphasize the vigorous driving action of both legs usually have both feet in the air at that instant. According to Lockwood,[30] throwers in this second group tend to have their feet closer together during the delivery than do those in the first group. He also states that "There does not seem to

[29] Robert E. Fitch, "Mechanical Analysis of the Discus Throw" (M.A. thesis, State University of Iowa, 1951), p. 76.

[30] Lockwood, "Throwing the Discus," p. 214.

be any correlation with the type of turn employed; there may be some correlation with build—the bigger and heavier men seem to favor the first style, probably because they have more bodyweight to lift into the air."

Reverse

To prevent fouling, most throwers use some form of reverse once the implement has been released. There are basically two types in common use: the orthodox reverse, which is essentially the same as that used in shot-putting [Fig. 211 (r)], and the spinning reverse, which involves one or more rotations on the ball of the right foot to dissipate angular momentum left over from the turn and throw. The former is unquestionably the easier to learn and execute; the latter is probably essential for a few highly skilled throwers.

JAVELIN THROW

Basic Considerations

The basic factors determining the distance with which an athlete is credited in the javelin throw—as always, assuming that his throw conforms with the various rules governing the event—are the same as those that apply in the case of the discus throw: the speed, height, and angle at which the implement is released and the aerodynamic factors that influence its flight. Although the techniques used in the two events are vastly different (at least since discus-style javelin throwing was banned by changes in the rules), the only one of these basic factors that warrants further consideration here is the last, the aerodynamic factors.

Apart from Ganslen and Hall's well-known study[31] and some more recent work by Terauds,[32,33] there have been very few reports of research conducted into the behavior of a javelin in flight. As a consequence considerable confusion exists as to the relative importance of the various factors that are known to influence the flight of the implement. In fact, such is the present state of knowledge in this area that even the most important fundamental questions—questions concerning the optimum angle of release, the optimum attitude angle, and the importance of the spin imparted at release—have yet to be answered satisfactorily.

[31]Richard V. Ganslen and Kenneth G. Hall, *Aerodynamics of Javelin Flight* (Fayetteville, Ark.: University of Arkansas, 1960).

[32]Juris Terauds, "Optimal Angle of Release for the Competition Javelin as Determined by Its Aerodynamic and Ballistic Characteristics," *Biomechanics IV* (Baltimore: University Park Press, 1974), p. 180.

[33]Juris Terauds, "Wind Tunnel Tests of Competition Javelins," *Track and Field Quarterly Review*, LXXIV, June 1974, p. 88.

The difficulty of arriving at appropriate conclusions concerning these matters has been further compounded by the development during the past two decades of so-called aerodynamic javelins. By making adjustments in the surface area, the location of the center of gravity, and other such parameters—all within the ever-narrowing limits imposed by the rules—javelin designers and manufacturers have been able to improve the aerodynamic characteristics of the implement to the point where the type of javelin used is now a very important factor in the event. Kuznetsov,[34] for example, has reported that a thrower who changes to the aerodynamic javelin may gain roughly a 10-20-ft improvement over the distance he could achieve with the "now obsolete javelin." Terauds,[35] referring to aerodynamic javelins thrown at 90–100 fps, has stated that the selection of the javelin "may result in a difference between two identical throws of 35.0 feet or more."

However, in spite of the widespread uncertainty concerning the aerodynamic behavior of javelins in flight, there appears to be some consensus on a number of points and at least initial indications concerning some others:

1. The speed of release is by far the most important single factor in determining the distance of a throw.

2. The optimum angle of release is a function of the type of javelin used. The optimum angle of release for the old nonaerodynamic javelin[36] is probably between 42° and 50° for good throwers. [Ganslen and Jarvinen[37] measured the angle of release on 15 throws by "world ranking throwers" and found that in none of these was the javelin released at less than 42°. Furthermore, in a total of 9 throws by Olympic champions Rautavaara (1948) and Jarvinen (1932) the angle of release ranged from 46° –50°].

The optimum angle of release for aerodynamic javelins has been variously estimated. Finnish throwers, it has been reported,[38] strive for an angle of 14° when using a javelin specifically designed for throws in excess of 70 m (approximately 230 ft). Kuznetsov[39] maintains that practice and research indicate the javelin must be released at 28° –30° if best results are to be

[34] V. L. Kuznetsov, "Mastery of Javelin Throwing Technique," *Track Technique*, No. 2, December 1960, p. 46.

[35] Terauds, "Wind Tunnel Tests of Competition Javelins."

[36] In the sense that all projectiles have specific flight (or aerodynamic) characteristics, the term nonaerodynamic is not strictly correct. It is used here, however, to describe the type of implement in use before the advent of the so-called *aerodynamic* javelin in the early 1950s.

[37] Richard V. Ganslen and Matti Jarvinen, "Finnish Javelin Throwing," *Scholastic Coach*, XIX, February 1950, p. 36.

[38] C. G. Smith, "Survey of Javelin Throwing in Finland," *The Athletics Coach*, II, December 1965, cited in Michael G. Wade, "A Javelin Refutation," *Track Technique*, No. 23, March 1966, p. 728.

[39] Kuznetsov, "Mastery of Javelin Throwing Technique," p. 46.

obtained. Paish analyzed[40] loop films of top Eurpoean throwers and found that the angle of release ranged from 30° –40°. The mean angle of release for the finalists in the 1966 European Championships, according to Paish, was 34°42′. These latter figures, incidentally, are in good agreement with those obtained by Nigg, Roethlin, and Wartenweiler[41] and by Terauds[42] in similar studies of top-class throwers.

3. While few have expressed any opinion on the optimum attitude angle, or angle of attack at the instant of release, most of those who have appear to favor an attitude angle equal to the angle of release (i.e., a zero angle of attack). Dyson,[43] however, states that the attitude angle of the aerodynamic javelin at release should possibly be a little less than the release angle.

4. The rotation imparted to the javelin about its long axis, of the order of 16–24 revolutions per second for top-class throwers,[44] is generally considered to have a beneficial stabilizing effect.

Techniques

Grip

Three grips are in common use:

1. the thumb and first finger (or American) grip—Fig. 214(a),
2. the thumb and second finger (or Finnish) grip—Fig. 214(b),
3. the first and second finger grip—Fig. 214(c).

Of these, the third appears to be most popular—70% of the leading throwers surveyed by Paish[45] used this grip; 25% used the thumb and second finger grip, and 5% used the thumb and first finger grip.

There appears, however, to be little evidence that one method is significantly better than another. Bankhead and Thorsen[46] measured the force and velocity produced in the final stage of the release ("that portion of the

[40]Wilf Paish, "The Javelin Throw," in *International Track and Field Coaching Encyclopedia*, ed. by Fred Wilt and Tom Ecker (West Nyack, New York: Parker Publishing Co., Inc., 1970), p. 282.

[41]B. Nigg, K. Roethlin, and J. Wartenweiler, "Biomechanische Messungen beim Speerwerfen," *Jugend und Sport*, XXXL, 1974, p. 172.

[42]Juris Terauds, "Javelin Release Characteristics," *Track Technique*, No. 61, September 1975, p. 1945.

[43]Dyson, *The Mechanics of Athletics*, p. 199.

[44]Terauds, "Javelin Release Characteristics," p. 1945.

[45]Wilf Paish, *Javelin Throwing* (London: Amateur Athletic Association, 1967), p. 43.

[46]William H. Bankhead and Margaret A. Thorsen, "A Comparison of Four Grips Used in Throwing the Javelin," *Research Quarterly*, XXXV, October 1964 Supplement, pp. 438–42.

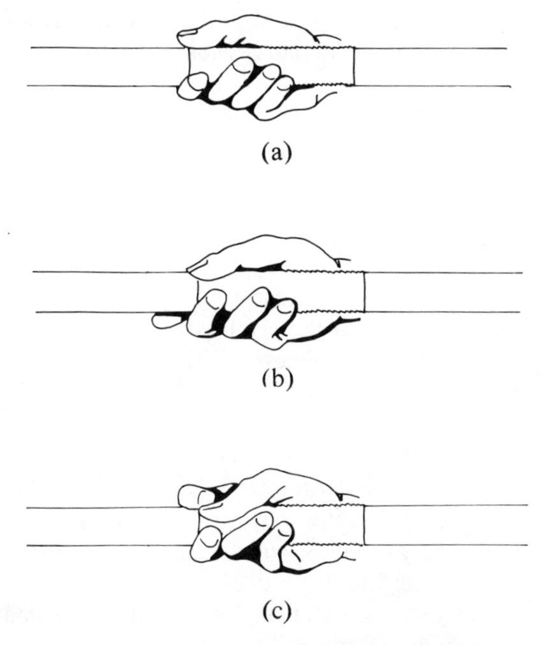

Fig. 214. Grips used in javelin throwing: (a) the thumb and first finger (or (American) grip, (b) the thumb and second finger (or Finnish) grip, and (c) the first and second finger grip

throw in which the muscles of the forearm and wrist are most fully used") using four different grips, including the three referred to above. Although their experimental procedure may be open to question, it is of some interest to note that they found no statistically significant differences in the power developed using the four grips.

Carry

During the approach run the javelin is usually carried in an over-the-shoulder position with the hand moving back and forth close to the ear and in unison with the athlete's leg action.

Run

In order to reach the maximum speed that they are capable of controlling during the transition from run to throw and during the throw itself, most good throwers use a run-up of some 10 to 12 steps (approximately 50-60 ft). This, together with (1) the 3 to 7 steps used in the transition from run to throw and (2) the recovery step used to bring the thrower to a halt after he has released the implement, generally results in a total run-up length of some 14 to 20 strides (approximately 70-100 ft).

The number of complete strides used in the transition from run to throw (i.e., from the moment the javelin starts back in preparation for the throw until the moment it is released) varies considerably among top-class throwers —in the 1968 Olympics this number ranged from 3 to 7 in both men's and women's events. The majority (56% in the men's event and 69% in the women's) used a 5-stride transition.[47]

Transition, Throw, and Recovery

An example of the sequence of movements performed by a top-class thrower, from the withdrawal of the javelin at the start of the transition to the recovery stride following the release, is shown in Fig. 215. The following points should be noted:

(a)-(e)[48] The withdrawal. As the thrower's right foot is grounded 5 strides before the final throwing position is reached, the withdrawal of the javelin begins. Although there are several ways in which this movement may be executed, the simplest, and the most widely used by top throwers, is that illustrated here—a direct pushing back of the throwing hand to a position in which the arm is fully extended and the hand is at approximately shoulder height. This movement, accompanied by a turning of the shoulders to maintain the javelin in line with the intended direction of the throw, occupies the first $1\frac{1}{2}$ to 2 strides of the transition. Throughout these strides and, indeed those that follow, the athlete's hips are kept essentially "square" to the front and his feet are grounded pointing directly forward—both measures aimed at maintaining the momentum developed in the approach run. (*Note:* The prime reason that the techniques used by leading throwers up until about the late 1950s—the so-called Finnish cross step and American hop—were superseded by the technique illustrated here is that these earlier techniques involved the placement of the feet across the line of the run-up and this almost inevitably resulted in a loss of forward momentum.)

(g) and (h) The cross step. The step preceding that in which the athlete adopts his final throwing position has been given this name because in the early techniques the legs actually crossed, one in front of the other, at this time. Although the term is no longer appropriate, at least for the majority of good throwers, it appears to have been retained by most writers on the subject.

The objective of the cross step is to get the athlete's feet forward of his upper body so that, as he lands on his right foot at the end of this step, he is able to move into the optimum position from which to execute the throw.

[47]W. Paish, "Javelin (Men)" and "Javelin (Women)," in *Mexico 68: A Technical Report by the National Coaches on the 1968 Olympic Games*, pp. 130–31, 137.

[48]The letters in parentheses refer to the corresponding positions in Fig. 212.

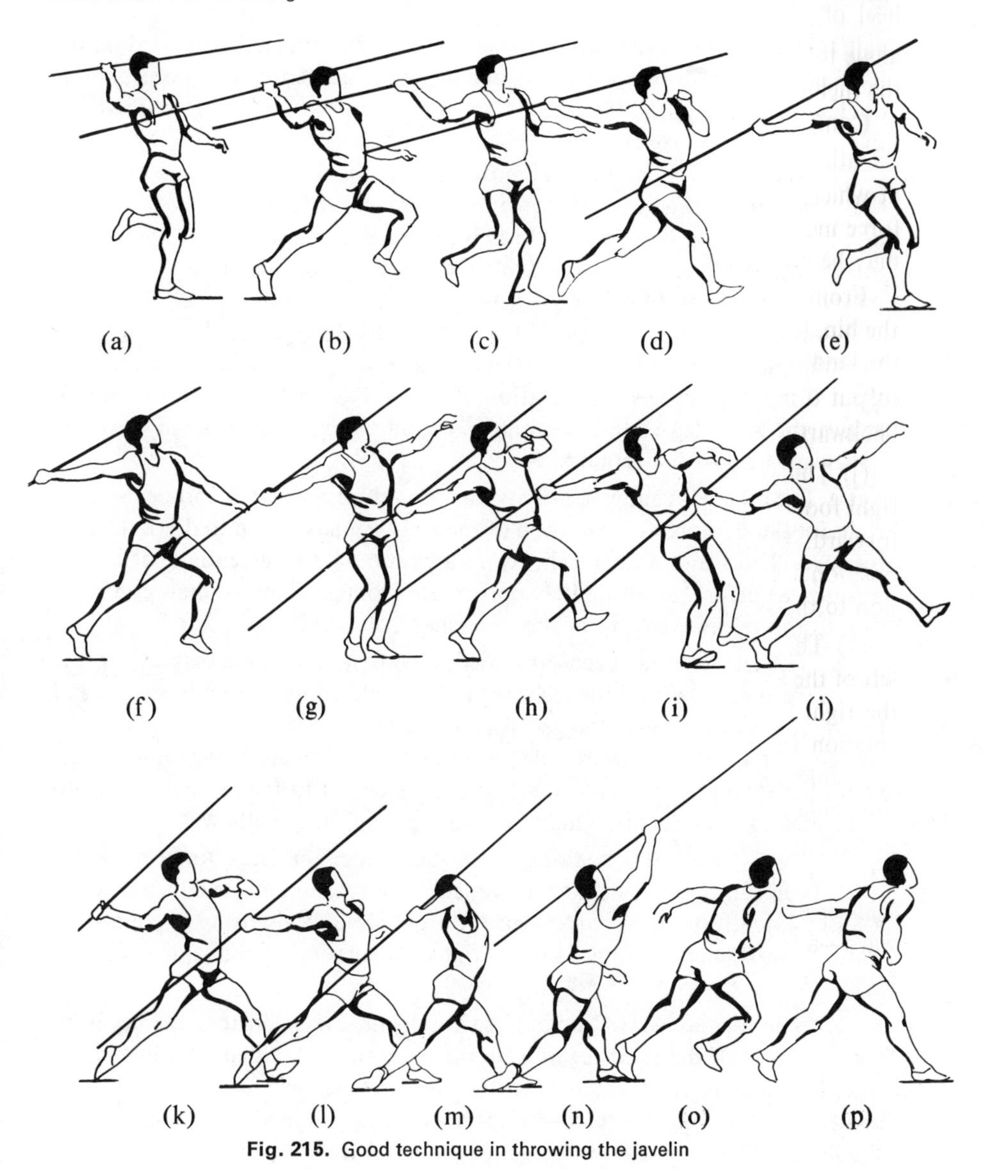

Fig. 215. Good technique in throwing the javelin

This he accomplishes with a forceful drive from the left foot and a fast action of both legs during the flight phase of the stride. This fast leg action brings the left foot forward of the right foot at the instant the latter is grounded and thus facilitates a rapid transition to the throwing position. [*Note:* In a normal running stride the foot of the recovery leg is well behind the grounded foot at the instant the latter touches down—compare positions (c) and (e) with position (i) in Fig. 215.]

(i) The landing at the end of the cross step. The athlete has landed on the heel of his right foot with his trunk inclined backward at a considerable angle to the horizontal. (Markov[49] has reported that "In the better throwers the inclination approaches 111-113 degrees.")

Compared with a more erect position, this backward inclination allows the athlete more time to exert force on the javelin before it reaches the point at which it should be released. It also increases the distance through which force may be exerted on the javelin—a distance which, according to Pugh,[50] may be as much as 14 ft in the case of a great thrower.

From this position the sole of the right foot is lowered to the ground and the hip, knee, and ankle joints of the right leg flex (1) to cushion the shock of the landing, (2) to speed the rotation of the body over the right foot, and (3) to put the right leg in the optimum position to exert force downward and backward against the ground.

(j) The athlete's center of gravity has passed forward over and beyond the right foot and a forceful extension of the right leg has begun to drive the hips forward. The trunk and the throwing arm are held in essentially the same position as before, awaiting the appropriate moment to make their contribution to the release velocity of the implement.

(k) The left foot has been grounded, heel first, and some distance to the left of the direction line; the right leg has completed its drive and the toes of the right foot are being dragged forward across the ground; and, with the rotation of the hips toward the front almost completed, the muscles responsible for the rotation of the trunk have begun to make their contribution, aided by an outward and backward swing of the left elbow.

(l) The left leg, after bending somewhat under the large force to which it has been subjected, is now firmly braced; the rotation of the trunk (and, to a lesser extent, the hips) continues; the right shoulder is being brought forward—a movement facilitated by an outward rotation of the athlete's upper arm.

(m) The hips are now "square" to the front, driven in that direction by the extension of the right leg and by the resistance provided later by the left leg; the trunk is also "square" to the front, chest well forward, and back arched slightly; and the throwing arm is being whipped forward with the shoulder leading, the elbow high, and the hand trailing.

(n) The release. The left leg has been extended to increase the height of release—5.51–6.59 ft for top-class throwers[51]—and, by contributing primarily to the vertical component of the velocity, to increase both the speed

[49] D. Markov, "Javelin Technique," *Track Technique*, No. 19, March 1965, p. 605.

[50] D. L. Pugh, *Javelin Throwing* (London: Amateur Athletic Association, 1960), p. 11.

[51] Terauds, "Javelin Release Characteristics," p. 1945.

and the angle of release. The extension of the elbow is all but complete and the javelin is about to be released, spinning about its long axis, from the fingers.

(o) and (p) The athlete takes a 5-6-ft long recovery step to dissipate the momentum left over from the preceding movements.

HAMMER THROW

Basic Considerations

As in each of the other throwing events, the horizontal distance with which a hammer thrower is credited (assuming, as always, that his throw is executed in strict accord with the rules) is governed by (1) the speed, (2) the height, and (3) the angle at which the implement is released, and (4) the air resistance encountered in flight.

Speed of Release

Since the hammer is initially at rest, the velocity that it possesses some time later clearly depends on the nature of the forces exerted on it to produce its subsequent motion (Newton's second law). The forces that the thrower exerts on the hammer are applied at the grip or handle and transmitted via the wire to the head of the hammer. Since the wire is light and flexible, only forces exerted along its length and toward the handle are effective in influencing the motion of the hammerhead.

If the wire is aligned so that it, or an imaginary extension of it, passes through the axis about which the hammer is rotating, any force exerted by the thrower acts inward toward the axis and thus may correctly be described as a centripetal force. Such a force (like all centripetal forces) serves only to change the direction in which the body is moving [Fig. 216 (a)]. If the wire of the hammer is aligned at an angle to a line joining the head of the hammer and the axis, the forces exerted along its length have both a centripetal and a tangential component [Fig. 216 (b)]. Such forces change both the direction in which the hammerhead is moving and the speed with which it is moving. The speed of release is determined in large measure by the thrower's ability to repeatedly get "ahead" of the hammer and into positions from which he can exert forces of this latter kind—forces that serve to increase the speed at which the hammer is moving.[52]

[52]The forces exerted along the length of the hammer wire—and the effects that they produce—are occasionally analyzed in terms of their horizontal and vertical components. Such analyses, although proceeding along slightly different lines from that presented here, invariably give rise to the same conclusions.

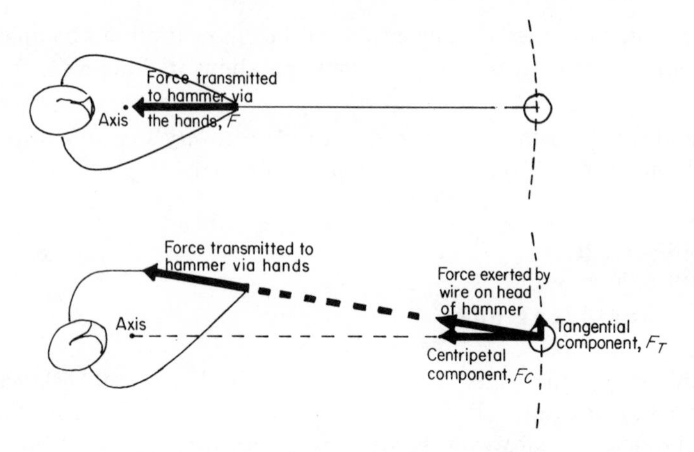

Fig. 216. Forces exerted and velocities attained in the hammer throw

The factors influencing the speed of release can also be considered from a purely kinematic standpoint by means of Eq. 26:

$$v = \omega r$$

From this equation the following conclusions can be drawn:

1. If the angular velocity is constant, the greater the radius, the greater the linear velocity of the hammer.

2. If the radius is constant, the greater the angular velocity at which the hammer is rotated, the greater its linear velocity.

3. In any given case, the greatest linear velocity is obtained when both the angular velocity and the radius are at a maximum.

As the linear velocity of the hammer increases, so too does the magnitude of the centripetal force that the athlete must exert in order to keep the hammer moving on a circular path relative to the axis of rotation (see Eq. 59). This increase in centripetal force is accompanied by an identical increase in the centrifugal force that the hammer exerts, in reaction, on the athlete. This buildup in centrifugal force necessitates that the athlete continually modify his body position if he is to retain his balance in a forward and backward sense. The athlete in Fig. 217 is acted on by three forces—W, the gravitational force, acting through his center of gravity; F_c, the centrifugal force, applied to his hands by the hammer; and R, the ground-reaction force, applied to his feet. The first two of these forces tend to rotate the athlete about a transverse axis through his feet—W with a moment Wx in a counterclockwise direction and F_c with a moment $F_c y$ in the opposite direction. As the speed with which the athlete rotates increases, F_c and the moment $F_c y$ increase accordingly, while the contrary moment Wx remains the same—unless, of course, some adjustment in the athlete's position alters the magnitude of the moment arm, x. Now, if the athlete makes no adjustment in x as his rate of rotation increases, he is subjected to a resultant clockwise moment *over and above that necessary to retain the body position he had initially* (i.e., at the slower rate of rotation). To overcome the serious problem that this excess moment represents, a good hammer thrower progressively modifies his body position by "sitting back against the hammer." This action, incorporating a lowering and a backward movement of the center of gravity as the hips and knees are flexed, decreases the moment arm of the centrifugal force and increases the moment arm of the body weight to ensure the appropriate relationship between the two influences (Fig. 217).

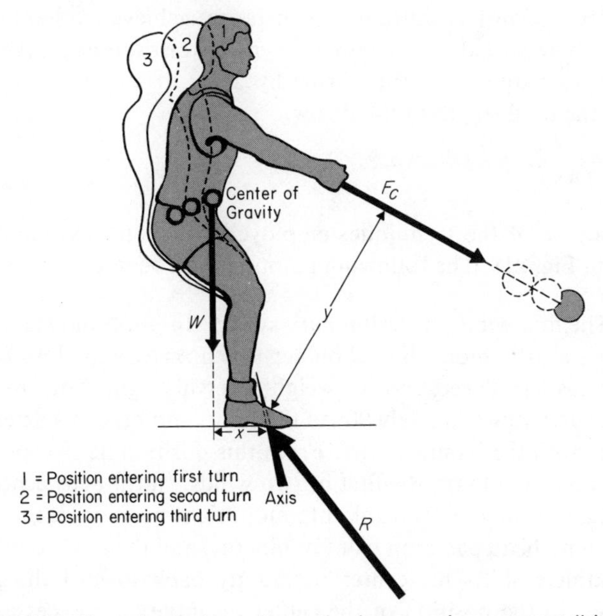

Fig. 217. To maintain his balance the hammer thrower must "sit back against the hammer" as the centrifugal force it exerts on him gets progressively larger.

Height of Release

The height of release depends primarily on the body position at that instant and on the physique of the thrower. The differences that exist, however, between one throw and another or between one thrower and another are generally so small as to be of little practical significance.

With regard to the effect of differences in physique, the observations of Payne[53] concerning the relative unimportance of body height are of some interest: "Provided that throwers reach a full leg and body extension at delivery, the few inches difference in height of different throwers is likely to make even fewer inches difference to distances thrown."

Angle of Release

Because the hammer is generally released at about shoulder level, a height of approximately 5 ft above the level at which it will land, the optimum angle of release is slightly less than 45°—approximately 43°-44° for throws in excess of 150 ft.

Air Resistance

The effects of air resistance on the distances achieved in hammer throwing are generally regarded as negligible. Payne,[54] for example, has reported that air resistance reduces a 200-ft throw by 2-3 ft at sea level—a reduction of 1-1½% in the total length of the throw.

Techniques

An example of the techniques employed by a top-class hammer thrower is shown in Fig. 218. The following points should be noted:

(a)[55] The first wind (or preliminary swing). In his initial stance the athlete has his feet a little more than shoulder width apart and close to the rim of the circle, his legs flexed and his weight over his right foot, his upper body twisted around approximately 90° to the right, and his arms extended behind him in line with the hammer wire. From this position he sweeps the hammer forward and around into the first of two winds, shifting his center of gravity to the left and bringing his trunk into an upright position in the process (a_1). As the hammerhead passes in front of him (a_2) and then out to his left (a_3 and a_4), the athlete shifts his center of gravity back toward the right. These adjustments in the position of the center of gravity are necessary to ensure that the moment of the athlete's weight is sufficient to offset the "overbalancing" moment of the centrifugal force. At the same time, however, the athlete endeavors to ensure that the hammerhead is moved through the largest arc possible, i.e., that its radius of rotation is as large as he can make it. To do this, he leans his upper body to the side (a_3) and slightly backward (a_4)—a movement known in hammer throwing as *countering*.

[53] Howard Payne, *Hammer Throwing* (London: Amateur Athletic Association, 1969), p. 53.

[54] *Ibid.*, p. 53.

[55] The letters in parentheses refer to the corresponding parts of the throw shown in Fig. 218.

As the hammer swings out to the thrower's left and before it begins to move around behind him, he quickly pulls his hands across the top of his head, turning his hips and shoulders well to the right in the process (a_4 and a_5). This places the athlete in a "wound-up" position in which his hips, his shoulders, and the hammer wire are each progressively farther around to the right. From this position, reached shortly after the hammerhead passes through the high point of its swing, the thrower exerts force along the hammer wire by contracting those muscles that rotate the hips and shoulders around to the front. This force, acting downward and inward along the length of the hammer wire, is supplemented by the component of the weight of the hammer acting in that direction (a_5 and a_6).

(b) The second wind. The second wind follows essentially the same pattern as the first, with the increased centrifugal force acting on the thrower (due to the increased speed of the hammer) necessitating slight adjustments in the athlete's body position.

(c) The first turn. Immediately after the hammerhead passes through the low point of its swing—a point that must be kept well to the thrower's right throughout the winds and turns, if the final delivery is to be fully effective—the thrower starts to move into the first turn by pivoting on the heel of his left foot and the ball of his right foot (c_1 and c_2). This action continues (c_3) as increasingly larger forces are exerted via the right foot to drive the thrower around to the left. Throughout this phase of the action the thrower's arms serve as passive extensions of the hammer wire (as indeed they do throughout all three turns), his shoulders are relaxed and pulled well forward, his trunk is inclined slightly forward in the direction of the hammer, and his center of gravity is low and well behind the heel of his left foot. Such a body position ensures that the radius of the arc through which the hammerhead swings is as large as it can be without upsetting the thrower's balance.

As the hammer continues upward toward the high point, the thrower's right foot is lifted quickly from the ground (c_4). Then, as the hammer moves through the high point (c_5) and begins to descend (c_6), the right foot is whipped quickly around the left leg to land about shoulder width away from the left foot and parallel with it (c_7). This movement, aimed at getting the thrower's feet ahead of his hips and his hips ahead of his shoulders, is facilitated if the right foot is kept close to the ground as it is brought around (thereby minimizing the distance it must travel) and if the right leg is kept close to the left leg (thus minimizing the moment of inertia of the thrower's lower body). Meantime, the pivot on the heel of the left foot (c_2–c_4) blends smoothly into a rolling movement on the outside border of the foot (c_5) and concludes with a pivot on the ball of the foot (c_6 and c_7).

With the right foot grounded, the thrower is once again in the wound-up position referred to previously and, as before, he powerfully accelerates the hammer as the muscles that rotate his hips and shoulders to the front are contracted forcefully (c_6 and c_7).

Fig. 218. Good technique in the hammer throw

(d) The second turn. The completion of the first turn and the start of the second blend smoothly into one another as the right foot starts to drive the athlete's hips and trunk once more around to the left (c_7 and d_1). This action is accompanied by a strong lifting action of the legs and trunk (cf. leg and trunk positions in c_7 and d_2) that serves to further accelerate the hammer as it passes through the low point and begins to ascend. The remainder of the second turn (d_3–d_6) follows much the same pattern as in the first turn.

(e) The third turn. Aside from the modifications in body position to

ensure that balance is maintained as the centrifugal force to which he is subjected increases, the thrower's actions during the third turn (e_1–e_5) are essentially the same as those in the preceding two turns.

(f) The delivery. The marked acceleration of the hammer during the delivery begins as the hammer passes through the high point of the third turn (e_4) and continues with a strong rotation of the hips and trunk—facilitated by a placement of the right foot farther around to the left than in the previous two turns—and a strong upward and slightly backward drive from the legs and back (f_1–f_4).

The extent to which the hammer may be accelerated during the delivery depends very largely on the thrower's body position as his right foot is grounded at the completion of the third turn. If he is in a strongly wound-up position at this time [i.e., with his feet rotated to a position ahead of his hips, his hips leading the shoulders in similar fashion, and the hammer trailing well behind all three (e_5)], he can make optimum use of the various muscular forces at his disposal. On the other hand, if he has failed to get his feet, hips, and shoulders progressively farther ahead of the hammer in this manner, the extent to which the available muscular forces can contribute to the speed at which the hammer is ultimately released is severely limited by the relatively short time and distance over which they can act.

Once the hammer has been released (somewhere between the positions shown in f_4 and f_5), the thrower's weight moves over his right foot (f_6) and the left foot is lifted and moved toward the left as he strives to recover his balance and avoid fouling.

APPENDIXES

Appendix A

ELEMENTARY TRIGONOMETRY

A knowledge of elementary trigonometry—the branch of mathematics that deals with the relationships between the sides and angles of triangles—is essential to a complete understanding of many of the basic concepts in biomechanics.

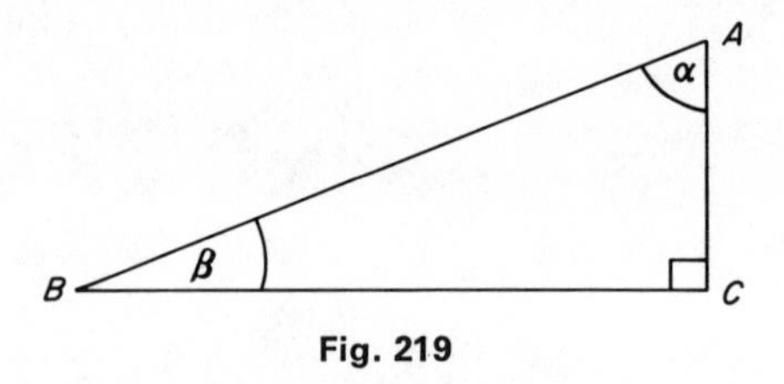

Fig. 219

Consider the right-angled triangle *ABC* in Fig. 219. If the lengths of the three sides of this triangle are measured, a total of six fractions (or ratios) can be obtained by putting the length of one side in the numerator and the length of another in the denominator:

$$\frac{AC}{AB}, \quad \frac{AC}{BC}, \quad \frac{BC}{AB}, \quad \frac{BC}{AC}, \quad \frac{AB}{BC}, \quad \frac{AB}{AC}$$

In trigonometry, these ratios are given special names according to how the sides are located relative to one of the acute (i.e., less than 90°) angles in the triangle. For example, if the angle *ABC* is designated by the Greek letter β (beta), the ratio formed by placing the side opposite β (i.e., *AC*) over the hypotenuse (i.e., *AB*, the side opposite the right angle) is called the sine of β. Thus:

$$\text{sine } \beta = \frac{\text{opposite}}{\text{hypotenuse}} = \frac{AC}{AB}$$

Although there are five other such ratios, only two of these need be considered here. These are

$$\text{cosine } \beta = \frac{\text{adjacent}}{\text{hypotenuse}} = \frac{BC}{AB}$$

and

$$\text{tangent } \beta = \frac{\text{opposite}}{\text{adjacent}} = \frac{AC}{BC}$$

The sine, cosine, and tangent ratios (normally abbreviated to sin, cos, and tan) for the angle BAC [here designated by the Greek letter α (alpha)] are similarly

$$\sin \alpha = \frac{\text{opposite}}{\text{hypotenuse}} = \frac{BC}{AB}$$

$$\cos \alpha = \frac{\text{adjacent}}{\text{hypotenuse}} = \frac{AC}{AB}$$

and

$$\tan \alpha = \frac{\text{opposite}}{\text{adjacent}} = \frac{BC}{AC}$$

Now, for an angle of a given size (say, 30°) the sine of that angle is a constant value regardless of the size of the right-angled triangle. In other words, the ratio formed by placing the length of the side opposite the 30° angle over the hypotenuse will always be the same—whether these sides be appropriately measured in millimeters or miles. The same holds true for each of the other trigonometrical ratios.

The values of the sine, cosine, and tangent of angles ranging from 0° to 90° are presented in Table 30. Thus, if the angle β (in Fig. 219) is 30°,

$$\sin \beta = \frac{AC}{AB} = 0.5000$$

$$\cos \beta = \frac{BC}{AB} = 0.8660$$

$$\tan \beta = \frac{AC}{BC} = 0.5774$$

A simple rearrangement of these equations shows clearly the relationships between the various sides of the triangle ABC:

$$AC = 0.5000 \times AB$$

$$BC = 0.8660 \times AB$$

$$AC = 0.5774 \times BC$$

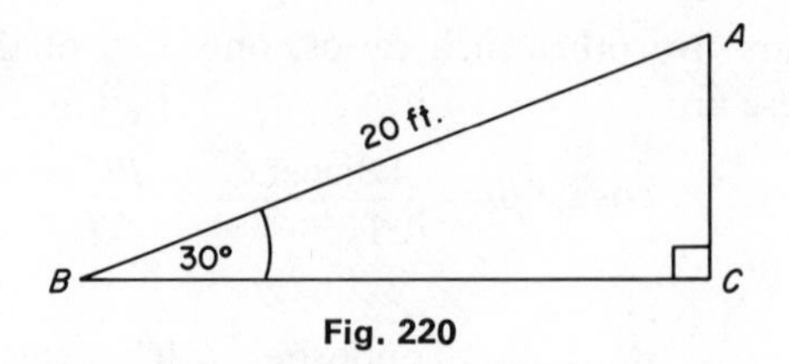

Fig. 220

Suppose now, that $\beta = 30^\circ$, $AB = 20$ ft, and nothing else is known about the right-angled triangle ABC (Fig. 220). The lengths of the remaining sides can readily be determined using the trigonometrical ratios defined here:

$$\frac{AC}{AB} = \sin 30^\circ$$

$$\frac{AC}{20 \text{ ft}} = 0.5000$$

$$AC = 0.5000 \times 20 \text{ ft}$$

$$= 10 \text{ ft}$$

Similarly,

$$\frac{BC}{AB} = \cos 30^\circ$$

$$\frac{BC}{20 \text{ ft}} = 0.8660$$

$$BC = 0.8660 \times 20 \text{ ft}$$

$$= 17.3200 \text{ ft}$$

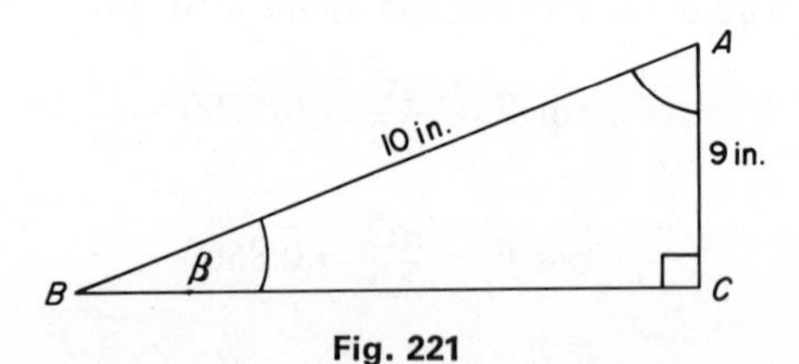

Fig. 221

Finally, suppose that the lengths of two sides of the triangle ABC are known (e.g., $AB = 10$ in., $AC = 9$ in., as in Fig. 221) and it is desired to find the size of the angles α and β. Now it is apparent that β is an angle, whose sine is given by the ratio AC/AB.[1] In the present example,

[1]This statement may be written in abbreviated form:

$$\beta = \arcsin \frac{AC}{AB}$$

in which the word *arcsin* is read "an angle whose sine is." Similar words (arccos and arctan) are used where the ratio involved is the cosine or tangent.

$$\sin \beta = \frac{9}{10} = 0.9000$$

An examination of the sine values listed in Table 30 reveals that

$$\sin 64° = 0.8988$$

and

$$\sin 65° = 0.9063$$

Thus, β is an angle just slightly larger than 64°. (*Note:* Tables that give values for fractions of a degree can be used if a more precise measure of β is required.)

Since the sum of the three angles in any triangle is 180°, α may be found simply by subtraction or, if preferred, by a similar use of trigonometry:

$$\cos \alpha = \frac{AC}{AB} = 0.9000$$

$$\alpha \approx 26°$$

Table 30. Trigonometric Functions

Degrees	*Sines*	*Cosines*	*Tangents*	*Degrees*	*Sines*	*Cosines*	*Tangents*
0	0.0000	1.0000	0.0000	46	0.7193	0.6947	1.0355
1	0.0175	0.9998	0.0175	47	0.7314	0.6820	1.0724
2	0.0349	0.9994	0.0349	48	0.7431	0.6691	1.1106
3	0.0523	0.9986	0.0524	49	0.7547	0.6561	1.1504
4	0.0698	0.9976	0.0699	50	0.7660	0.6428	1.1918
5	0.0872	0.9962	0.0875	51	0.7771	0.6293	1.2349
6	0.1045	0.9945	0.1051	52	0.7880	0.6157	1.2799
7	0.1219	0.9925	0.1228	53	0.7986	0.6018	1.3270
8	0.1392	0.9903	0.1405	54	0.8090	0.5878	1.3764
9	0.1564	0.9877	0.1584	55	0.8192	0.5736	1.4281
10	0.1736	0.9848	0.1763	56	0.8290	0.5592	1.4826
11	0.1908	0.9816	0.1944	57	0.8387	0.5446	1.5399
12	0.2079	0.9781	0.2126	58	0.8480	0.5299	1.6003
13	0.2250	0.9744	0.2309	59	0.8572	0.5150	1.6643
14	0.2419	0.9703	0.2493	60	0.8660	0.5000	1.7321
15	0.2588	0.9659	0.2679	61	0.8746	0.4848	1.8040
16	0.2756	0.9613	0.2867	62	0.8829	0.4695	1.8807
17	0.2924	0.9563	0.3057	63	0.8910	0.4540	1.9626
18	0.3090	0.9511	0.3249	64	0.8988	0.4384	2.0503
19	0.3256	0.9455	0.3443	65	0.9063	0.4226	2.1445
20	0.3420	0.9397	0.3640	66	0.9135	0.4067	2.2460
21	0.3584	0.9336	0.3839	67	0.9205	0.3907	2.3559
22	0.3746	0.9272	0.4040	68	0.9272	0.3746	2.4751
23	0.3907	0.9205	0.4245	69	0.9336	0.3584	2.6051
24	0.4067	0.9135	0.4452	70	0.9397	0.3420	2.7475
25	0.4226	0.9063	0.4663	71	0.9455	0.3256	2.9042
26	0.4384	0.8988	0.4877	72	0.9511	0.3090	3.0777
27	0.4540	0.8910	0.5095	73	0.9563	0.2924	3.2709
28	0.4695	0.8829	0.5317	74	0.9613	0.2756	3.4874
29	0.4848	0.8746	0.5543	75	0.9659	0.2588	3.7321
30	0.5000	0.8660	0.5774	76	0.9703	0.2419	4.0108
31	0.5150	0.8572	0.6009	77	0.9744	0.2250	4.3315
32	0.5299	0.8480	0.6249	78	0.9781	0.2079	4.7046
33	0.5446	0.8387	0.6494	79	0.9816	0.1908	5.1446
34	0.5592	0.8290	0.6745	80	0.9848	0.1736	5.6713
35	0.5736	0.8192	0.7002	81	0.9877	0.1564	6.3138
36	0.5878	0.8090	0.7265	82	0.9903	0.1392	7.1154
37	0.6018	0.7986	0.7536	83	0.9925	0.1219	8.1443
38	0.6157	0.7880	0.7813	84	0.9945	0.1045	9.5144
39	0.6293	0.7771	0.8098	85	0.9962	0.0872	11.43
40	0.6428	0.7660	0.8391	86	0.9976	0.0698	14.30
41	0.6561	0.7547	0.8693	87	0.9986	0.0523	19.08
42	0.6691	0.7431	0.9004	88	0.9994	0.0349	28.64
43	0.6820	0.7314	0.9325	89	0.9998	0.0175	57.29
44	0.6947	0.7193	0.9657	90	1.0000	0.0000	∞
45	0.7071	0.7071	1.0000				

Appendix B

NOMOGRAM FOR CONVERSION OF SPEED UNITS

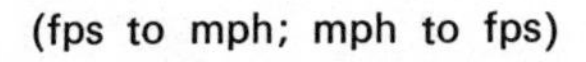
(fps to mph; mph to fps)

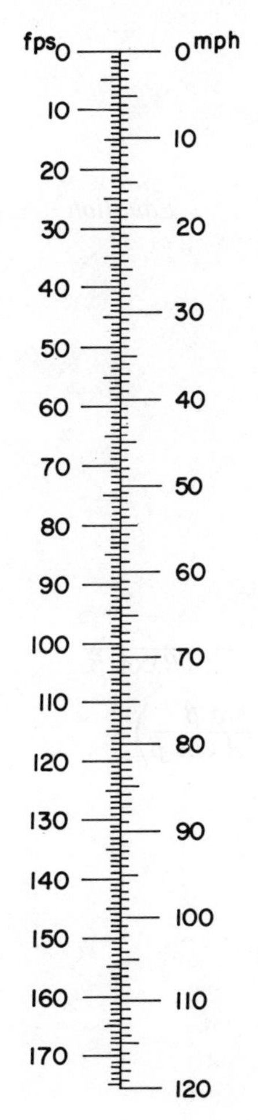

Appendix C

EQUATIONS

No.	*Equation*	*Page No.*
1.	$\bar{s} = \frac{l}{t}$	16
2.	$\bar{v} = \frac{d}{t}$	16
3.	$\bar{a} = \frac{v_f - v_i}{t}$	19
4.	$R = \sqrt{A^2 + B^2}$	25
5.	$\theta = \arctan\left(\frac{A}{B}\right)$	25
6.	$R = \sqrt{A^2 + B^2 + 2AB\cos\beta}$	26
7.	$\theta = \arctan\left(\frac{A\sin\beta}{B + A\cos\beta}\right)$	26
8.	$v_f = v_i + at$	29
9.	$d = v_i t + \frac{1}{2}at^2$	29
10.	$v_f^2 = v_i^2 + 2ad$	29
11.	$d_H = v\cos\theta \times t$	34
12.	$R = v\cos\theta \times T$	35
13.	$t_{\text{up}} = \frac{v\sin\theta}{g}$	35
14.	$t_{\text{down}} = \sqrt{\frac{2d_{\text{down}}}{g}}$	36

No.	*Equation*	*Page No.*
15.	$t_{\text{down}} = \frac{v \sin \theta}{g}$	36
16.	$T = \frac{2v \sin \theta}{g}$	37
17.	$d_{\text{down}} = \frac{(v \sin \theta)^2}{2g} + h$	37
18.	$t_{\text{down}} = \frac{\sqrt{(v \sin \theta)^2 + 2gh}}{g}$	38
19.	$T = \frac{v \sin \theta + \sqrt{(v \sin \theta)^2 + 2gh}}{g}$	38
20.	$R = \frac{v^2 \sin 2\theta}{g}$	38
21.	$R = \frac{v^2 \sin \theta \cos \theta + v \cos \theta \sqrt{(v \sin \theta)^2 + 2gh}}{g}$	39
22.	$\bar{\sigma} = \frac{\phi}{t}$	46
23.	$\bar{\omega} = \frac{\theta}{t}$	47
24.	$\bar{\alpha} = \frac{\omega_f - \omega_i}{t}$	47
25.	$s = \sigma r$	54
26.	$v_T = \omega r$	54
27.	$a_R = \frac{v_T^2}{r}$	54
28.	$\bar{a}_T = \frac{v_{Tf} - v_{Ti}}{t}$	55
29.	$F \propto \frac{m_1 m_2}{l^2}$	58
30.	$F = kma$	61
31.	$F = ma$	62
32.	$W = mg$	63
33.	$F = \mu R$	72
34.	$F_s = \mu_s R$	73
35.	$\text{Impulse} = F \times t$	75
36.	$Ft = mv_f - mv_i$	76

No.	*Equation*	*Page No.*
37.	$\dfrac{v_1 - v_2}{u_1 - u_2} = -e$	79
38.	$\dfrac{v_1}{u_1} = -e$	79
39.	$u_1 = \sqrt{2gh_d}$	79
40.	$v_1 = \sqrt{2gh_b}$	79
41.	$e = \sqrt{\dfrac{h_b}{h_d}}$	79
42.	$v_1 = \sqrt{\left[\dfrac{m_2u_2(1 + e) + u_1 \cos \alpha\,(em_2 - m_1)}{m_1 + m_2}\right]^2 + (u_1 \sin \alpha)^2}$	88
43.	$R = \arctan\left[\dfrac{(u_1 \sin \alpha)(m_1 + m_2)}{m_2u_2(1 + e) + u_1 \cos \alpha\,(em_2 - m_1)}\right]$	88
44.	$\text{Pressure} = \dfrac{\text{Force}}{\text{Area}}$	93
45.	$W = Fd$	95
46.	$P = \dfrac{W}{t}$	96
47.	$\text{K.E.} = \frac{1}{2}mv^2$	97
48.	$\text{P.E.} = Wh$	98
49.	$Fd = \frac{1}{2}mv_f^2$	101
50.	$Fd = \frac{1}{2}mv_f^2 + Wd$	103
51.	$M = F \times \text{M.A.}$	107
52.	$x = \dfrac{72(R_2 - R_1)}{W}\ \text{in.}$	128
53.	$x = \dfrac{(R_{B2} - R_{B1})h}{W}$	129
54.	$y = \dfrac{(R_{A2} - R_{A1})h}{W}$	129
55.	$I = \sum mr^2$	144
56.	$I_A = I_{CG} + md^2$	145
57.	$\text{Angular momentum} = I\omega$	148
58.	$T = I\alpha$	151
59.	$\text{Centripetal force} = \dfrac{mv_T^2}{r}$	163
60.	$\text{Centripetal force} = mr\omega^2$	163

INDEX

C

D

J

K

L

M

N

O

P

R

S

T

U

V

W

Y

Z